Seafood Processing

Seafood Processing

Technology, Quality and Safety

Edited by

Ioannis S. Boziaris

School of Agricultural Sciences, University of Thessaly, Volos, Greece

WILEY Blackwell

This edition first published 2014

Registered office: John Wiley & Sons, Ltd, The Atrium, Southern Gate, Chichester, West Sussex, PO19 8SQ, UK

Editorial offices: 9600 Garsington Road, Oxford, OX4 2DQ, UK
The Atrium, Southern Gate, Chichester, West Sussex, PO19 8SQ, UK
111 River Street, Hoboken, NJ 07030-5774, USA

For details of our global editorial offices, for customer services and for information about how to apply for permission to reuse the copyright material in this book please see our website at www.wiley.com/wiley-blackwell.

Library of Congress Cataloging-in-Publication Data

Seafood processing : technology, quality and safety / Ioannis S. Boziaris.
pages cm
Includes index.
ISBN 978-1-118-34621-1 (cloth)
1. Fishery processing. I. Boziaris, Ioannis S., editor of compilation.
SH335.S34 2014
664′.94 – dc23

2013024198

A catalogue record for this book is available from the British Library.

Wiley also publishes its books in a variety of electronic formats. Some content that appears in print may not be available in electronic books.

Cover image: Fish processing building in Sotra, Bergen © Simonas Vaikasas, courtesy of Shutterstock
Sea Crab © w-i-n-d, courtesy of iStock
Seafood © Cristian Baitg, courtesy of iStock
Cover design by www.hisandhersdesign.co.uk

1 2014

MIX
Paper from responsible sources
FSC® C013604

Contents

About the *IFST Advances in Food Science* Book Series

The Institute of Food Science and Technology (IFST) is the leading qualifying body for food professionals in Europe and the only professional organisation in the UK concerned with all aspects of food science and technology. Its qualifications are internationally recognised as a sign of proficiency and integrity in the industry. Competence, integrity, and serving the public benefit lie at the heart of the IFST philosophy. IFST values the many elements that contribute to the efficient and responsible supply, manufacture and distribution of safe, wholesome, nutritious and affordable foods, with due regard for the environment, animal welfare and the rights of consumers.

IFST Advances in Food Science is a series of books dedicated to the most important and popular topics in food science and technology, highlighting major developments across all sectors of the global food industry. Each volume is a detailed and in-depth edited work, featuring contributions by recognized international experts, and which focuses on new developments in the field. Taken together, the series forms a comprehensive library of the latest food science research and practice, and provides valuable insights into the food processing techniques that are essential to the understanding and development of this rapidly evolving industry.

The *IFST Advances* series is edited by Dr Brijesh K. Tiwari, Senior Research Officer in the Department of Food Biosciences at the Teagasc Food Research Centre, Dublin, Ireland.

Forthcoming titles in the IFST series

Emerging Dairy Processing Technologies, edited by Nivedita Datta and Peggy Tomasula
Emerging Technologies in Meat Processing, edited by Enda Cummins and James Lyng
Nutraceutical and Functional Food Processing Technology, edited by Joyce Irene Boye

List of Contributors

Mercedes Alonso
ANFACO-CECOPESCA, Vigo, Pontevedra, Spain

Sigurjon Arason
Faculty of Food Science and Nutrition, University of Iceland, Reykjavík, Iceland and Matís-Icelandic Food and Biotech R & D, Reykjavík, Iceland

Ioannis S. Arvanitoyannis
Laboratory of Food Technology, Department of Agriculture, Ichthyology and Aquatic Environment, University of Thessaly, Volos, Greece

Ioannis S. Boziaris
Department of Ichthyology and Aquatic Environment, School of Agricultural Sciences, University of Thessaly, Volos, Greece

Frank Devlieghere
Laboratory of Food Microbiology and Food Preservation, Member of Food2know, Department of Food Safety and Food Quality, Ghent University, Ghent, Belgium and Pack4Food, Department of Food Safety and Food Quality, Ghent University, Ghent, Belgium

Christina DeWitt
Oregon State University Seafood Research and Education Center, Astoria, OR, USA

Trygve Magne Eikevik
Norwegian University of Science and Technology – NTNU, Department of Energy and Process Engineering, Trondheim, Norway

Montserrat Espiñeira
ANFACO-CECOPESCA, Vigo, Pontevedra, Spain

George M. Hall
University of Central Lancashire, Centre for Sustainable Development, Preston, Lancashire, UK

Flemming Jessen
DTU Food, National Food Institute, Division of Industrial Food Research, Technical University of Denmark, Kgs. Lyngby, Denmark

Ikuo Kimura
Laboratory of Food Engineering, Faculty of Fisheries, Kagoshima University, Shimoarata, Kagoshima, Japan

Sevim Köse
Department of Fisheries Technology Engineering, Faculty of Marine Sciences, Karadeniz Technical University, Trabzon, Turkey

Fátima C. Lago
ANFACO-CECOPESCA, Vigo, Pontevedra, Spain

Erling Larsen
DTU Aqua, National Institute of Aquatic Resources, Technical University of Denmark, Charlottenlund, Denmark

Chengchu Liu
College of Food Science and Technology, Shanghai Ocean University, Shanghai, People's Republic of China

Catherine McLeod
South Australian Research and Development Institute, Adelaide, Australia and Seafood Safety Assessment, Tournissan, France

Michael Morrissey
Oregon State University Food Innovation Center, Portland, OR, USA

Minh Van Nguyen
Faculty of Food Technology, Nha Trang University, Nha Trang, Vietnam and Matís-Icelandic Food and Biotech R & D, Reykjavík, Iceland

Jette Nielsen
DTU Food, National Food Institute, Division of Industrial Food Research, Technical University of Denmark, Kgs. Lyngby, Denmark

Bert Noseda
Laboratory of Food Microbiology and Food Preservation, Member of Food2know, Department of Food Safety and Food Quality, Ghent University, Ghent, Belgium

Jörg Oehlenschläger
Seafood Consultant, Buchholz, Germany

Emiko Okazaki
Tokyo University of Marine Science and Technology, Department of Food Science and Technology, Tokyo, Japan

Foteini F. Parlapani
Department of Ichthyology and Aquatic Environment, School of Agricultural Sciences, University of Thessaly, Volos, Greece

Peter Ragaert
Laboratory of Food Microbiology and Food Preservation, Member of Food2know, Department of Food Safety and Food Quality, Ghent University, Ghent, Belgium and Pack4Food, Department of Food Safety and Food Quality, Ghent University, Ghent, Belgium

Tom Ross
University of Tasmania, Tasmanian Institute of Agriculture – School of Agricultural Science, Hobart, Australia

Dagbjørn Skipnes
Nofima AS, Stavanger, Norway

Yi-Cheng Su
Seafood Research and Education Center, Oregon State University, Astoria, Oregon, USA

John Sumner
University of Tasmania, Tasmanian Institute of Agriculture – School of Agricultural Science, Hobart, Australia

Somboon Tanasupawat
Department of Biochemistry and Microbiology, Faculty of Pharmaceutical Sciences, Chulalongkorn University, Bangkok, Thailand

Kristin A. Thorarinsdottir
Marel, Reykjavik, Iceland

Gudjon Thorkelsson
Faculty of Food Science and Nutrition, University of Iceland, Reykjavík, Iceland and Matís-Icelandic Food and Biotech R & D, Reykjavík, Iceland

Persefoni Tserkezou
Laboratory of Food Technology, Department of Agriculture, Ichthyology and Aquatic Environment, University of Thessaly, Volos, Greece

An Vermeulen
Laboratory of Food Microbiology and Food Preservation, Member of Food2know, Department of Food Safety and Food Quality, Ghent University, Ghent, Belgium and Pack4Food, Department of Food Safety and Food Quality, Ghent University, Ghent, Belgium

Wonnop Visessanguan
Food Biotechnology Research Unit, National Center for Genetic Engineering and Biotechnology (BIOTEC), Klong Luang, Pathum Thani, Thailand

Juan M. Vieites
ANFACO-CECOPESCA, Vigo, Pontevedra, Spain

Preface

Demand for fish and seafood has consistently increased during recent years and fish protein is the major animal protein consumed in many parts of the world. Seafood is a very perishable product and the risk of contamination of seafood products by biological hazards is very high. Processing is necessary to assure the prolonged shelf life and safety of seafood.

The seafood processing industry currently has to face new challenges. Production has increased and seafood products need to be transported over long distances. Increasing demands from legislation and from the consumer for better quality and safer products have to be taken into account. Seafood now has to be high quality, nutritious, safe and have the convenience of an extended shelf life. To meet these criteria, seafood processing has had to assimilate all the new advances in food science and technology and in quality and safety assurance. Current technologies have evolved rapidly (e.g. modified atmosphere packaging, minimal heat processing, rapid freezing, injection salting), while emerging technologies such as high-pressure processing are beginning to be used. Advanced quality and safety methods, such as modern and rapid techniques for assessing quality and safety, species identification techniques and risk assessment tools, all have significant applications in the seafood sector.

This book covers the whole range of technologies currently used for the main processing of seafood. Quality and safety aspects are also dealt with. The first part of the book covers primary processing, chilling and freezing, heat processing, irradiation, traditional preservation methods (salting, smoking, acidification, drying and fermentation) as well as packaging. Surimi production, fish waste treatment, sustainability and value-added seafood product development is also covered in this section. The second part of the book deals with the determination of seafood quality, microbiological examination, authenticity and risk assessment.

60

1

Introduction to Seafood Processing – Assuring Quality and Safety of Seafood

Ioannis S. Boziaris
Department of Ichthyology and Aquatic Environment, School of Agricultural Sciences, University of Thessaly, Volos, Greece

1.1 Introduction

Demand for seafood has consistently increased during recent years with fish protein being the major animal protein consumed in many parts of the world. According to the Food and Agriculture Organization (FAO, 2012), fresh seafood represents 40.5% of the world's seafood production, while processed products (frozen, cured, canned, etc.) represent 45.9%. To assure the quality of raw material used for processing, fish has to be treated carefully before and after harvest. Often fish and shellfish undergo some type of handling or primary processing (washing, gutting, filleting, shucking, etc.), before the main processing occurs, to assure their quality and safety, as well as to produce new, convenient and added-value products (e.g. packed fish fillets instead of unpacked, whole ungutted fish).

Processing of seafood mainly inhibits and/or inactivates bacteria and enzymes which results in shelf-life extension and also assures food safety. While the main role of processing is preservation, processing not only extends shelf life but also creates a new range of products.

Seafood processing uses almost all the processing methods available to the food industry. The most widely used methods to preserve fish involve the application of low temperatures (chilling, super-chilling, freezing). Improvements in packaging technology (modified atmosphere packaging, MAP) and the application of chilling maximise quality retention as well as extending shelf life. Heating inactivates bacterial pathogens and spoilage microorganisms, which contributes to the stability and safety of the products. Irradiation is a well-established, non-thermal method, while high-pressure processing of

Seafood Processing: Technology, Quality and Safety, First Edition. Edited by Ioannis S. Boziaris.

seafood is being continuously increased. Traditional methods of preservation (curing, fermentation, etc.) are also used in the production of a variety of products.

1.2 Seafood spoilage

Seafood deteriorates very quickly due to various spoilage mechanisms. Spoilage can be caused by the metabolic activity of microorganisms, endogenous enzymatic activity (such as autolysis and the enzymatic browning of crustaceans shells) and by the chemical oxidation of lipids (Ashie *et al.*, 1996; Gram and Huss, 1996; Huis in't Veld, 1996).

Seafood flesh has a high amount of non-protein nitrogenous (NPN) compounds and a low acidity (pH > 6), which support the fast growth of microorganisms that are the main cause of spoilage. The growth and metabolic activity of the spoilage microorganisms, especially specific spoilage organisms (SSOs), result in the production of metabolites that affect the organoleptic properties of the product (Ashie *et al.*, 1996; Gram and Huss, 1996). Briefly, SSOs may initially represent only a small proportion of the microbiota (indigenous and exogenous); however, they subsequently proliferate to become the part of the dominant microbiota that has spoilage potential (the qualitative ability to produce off-odours) and spoilage activity (the quantitative ability to produce metabolites) (Gram and Dalgaard, 2002). Inhibiting the growth of SSOs increases the shelf life of seafood. *Pseudomonas* and *Shewanella* species spoil marine fish and crustaceans stored aerobically at low temperatures, while *Photobacterium phosphoreum*, various lactic acid bacteria and *Brochothrix thermosphacta* usually predominate in spoilage associated with MAP (Gram and Huss, 1996; Dalgaard, 2000).

Immediately following death, autolysis resulting from the action of endogenous enzymes, initially causes loss of the characteristic fresh odour and taste of fish and then softens the flesh (Huss, 1995; Ashie *et al.*, 1996). The main changes that take place are initially the enzymatic degradation of adenosine triphosphate (ATP) and related products and subsequently the action of proteolytic enzymes. Enzymes are also responsible for colour changes. After microbial growth, enzymatic browning is the most important spoilage mechanism of crustaceans (Ashie *et al.*, 1996; Boziaris *et al.*, 2011). Browning of the crustacean shell is the result of the action of polyphenol oxidase on tyrosine and its derivatives such as tyramine (Martinez-Alvarez *et al.*, 2007). Inhibition or inactivation of polyphenol oxidase by various means (heating, additives, etc.) as well as oxygen reduction or exclusion can prevent the loss of the original colour of the crustacean shell.

Chemical oxidation of lipids (oxidative rancidity) is one of the most important spoilage mechanisms, especially in fatty fish. Oxygen is necessary for the development of oxidative rancidity; hence, oxygen reduction or exclusion limits the oxidation reaction (Ashie *et al.*, 1996).

All these mechanisms advance almost simultaneously contributing to the spoilage; however, fresh and lightly preserved seafood spoils mainly due to the action of microorganisms. For products in which microbial growth is retarded or inhibited, non-microbial mechanisms play a more determinative role.

1.3 Seafood hazards

Contamination of seafood by chemicals, marine toxins and microbiological hazards can be high. Various bacterial pathogens present in aquatic environments – either naturally (pathogenic *Vibrio, Clostridium botulinum, Aeromonas hydrophilla*), or as contaminants (*Salmonella* spp., pathogenic *Escherichia coli*) – can contaminate seafood, while contamination with other bacteria such as *Listeria monocytogenes, Staphylococcus aureus*, etc., can occur during processing (Feldhusen, 2000; Huss *et al.*, 2000). Seafood can also be contaminated by viruses (such as hepatitis A virus, Norwalk-like viruses, Astrovirus, etc.), marine biotoxins (which cause several diseases such as diarrhoeic shellfish poisoning (DSP), paralytic shellfish poisoning (PSP), neurotoxic shellfish poisoning (NSP), amnesic shellfish poisoning (ASP) and fish ciguatera poisoning) and chemical contaminants (such as heavy metals) (Huss, 1994). Generally, processing mainly controls microbiological hazards but leaves chemical hazards or biotoxins virtually unaffected. Effective control of chemical hazards and biotoxins has to be applied mostly during primary production and the pre-harvest stages.

From a safety point of view, seafood can be classified in to seven groups according to the risk of microbial contamination and the processing method (Huss *et al.*, 2000). Molluscs, especially those that are to be eaten without cooking, belong to the group with the highest risk. The second group contains the fish and crustaceans that will be consumed after cooking. The third and fourth groups contain lightly preserved (NaCl $< 6\%$ w/v in aqueous phase, pH > 5) and semi-preserved (NaCl $> 6\%$ w/v in aqueous phase, pH < 5) products, respectively. The fifth group contains the mild-heated products, such as pasteurized and hot-smoked seafood, while the sixth contains the heat processed products. Finally dried, dry-salted and smoke-dried seafood products have the lowest risk.

1.4 Getting the optimum quality of the raw material

Pre-harvest and post-harvest handling of fish affects its quality. A number of biochemical changes start immediately following the death of the fish. The most important change is the onset of rigor mortis, during which the initially relaxed and elastic muscles become hard and stiff. At the end of rigor mortis the muscles relax again but are no longer elastic. The mechanism of rigor mortis is described in Chapter 3. The significance of rigor mortis is important in post-mortem processing. Filleting fish in rigor may produce fillets with gaping and give lower yields, while whole fish and fillets frozen before the onset of rigor can give better products (Huss, 1995). The onset of rigor mortis and its duration depend on various factors such as the size of the fish, the temperature and the physical condition of the fish, including stress (Huss, 1995). For instance, in either starved or stressed fish the glycogen reserves are depleted and rigor mortis starts immediately. Rapid chilling of fish is important not only to inhibit bacterial growth but also for managing the onset and duration of rigor. Abe and Okuma (1991) suggested that the onset of rigor mortis depends on the difference between the sea temperature and the storage temperature. When this difference is high, the onset of rigor is fast and *vice versa*.

1.4.1 Pre-mortem handling

Handling of fish before death affects rigor mortis. It is important in wild fish to use methods of capture that do not stress and exhaust fish, while in farmed fish, pre-harvest starvation, harvesting and slaughtering practices that do not stress fish are essential to maximise seafood quality and shelf life (Bagni *et al.*, 2007; Borderias and Sanchez-Alonso, 2011). The digestive tract contains a high bacterial population that produces digestive enzymes that result in intense post-mortem autolysis giving strong off-odours in the abdominal area (Huss, 1995). Starvation reduces the amount of faeces in the intestines and delays spoilage. In general, the starvation period is 1–3 days. Harvesting, stunning and killing methods greatly affect post-mortem changes and subsequent fish quality. When fish are rapidly killed, stress can be reduced, improving quality (Ottera *et al.*, 2001; Bagni *et al.*, 2007). Many methods can be used for stunning and killing fish, such as asphyxiation, live chilling in ice slurry, electrical stunning and electrocution, carbon dioxide narcosis, knocking or spiking. Asphyxiated and electrically stunned fish are more stressed than spiked, knocked and live-chilled fish (Poli *et al.*, 2005). Knocking on the head is reported as the optimal killing method for obtaining the best quality flesh in turbot (Roth *et al.*, 2007).

For shellfish, suitable pre- and post-harvest handling is required to achieve a safe seafood product. Shellfish are filter-feeders and can concentrate contaminants from the aquatic environment. Preventive measures are required to deter the accumulation of pathogenic microorganisms, biological toxins and chemical contaminants. Water quality is one of the most important factors, while treatments such as depuration and subsequent suitable handling and processing are essential (see Chapter 2).

Regarding the handling of crustaceans, such as lobsters and crabs, considerably fewer studies have been published compared to finfish and molluscan shellfish. The quality and prolonged shelf life of lobsters and crabs can be maintained by keeping them alive as long as possible. Norway lobster and crab individuals stored at chilling temperature or in ice spoil rapidly mainly due to microbial growth, which occurs after their death (Robson *et al.*, 2007; Boziaris *et al.*, 2011). The effect of post-harvest handling on the quality of crustaceans has been recently reviewed (Neil, 2012).

1.4.2 Post-mortem handling

After killing and chilling fish, minimal processing such as washing, gutting or filleting can take place. The results of the effect of gutting on the quality and shelf life of fish are contradictory. Microbial counts in ungutted sea bream (*Sparus aurata*) and sea bass (*Dicentrarchus labrax*) were found to be slightly lower compared to gutted fish while the quality and shelf life assessed by sensory and chemical methods was found to be the same (Cakli *et al.*, 2006). Erkan (2007) reported that the shelf life of gutted and ungutted sea bream was similar. On the other hand, Papadopoulos *et al.* (2003) found that gutted sea bass have a shorter shelf life compared to their ungutted counterparts.

Fish are filleted to produce value-added products. In general the practice of filleting in rigor is avoided because the yield is low and may cause gaping. Filleting is usually performed before or after rigor with various advantages and disadvantages in each case. Pre-rigor fillets of Atlantic salmon had lower bacterial numbers, the odour-flavour scores were higher, the gaping was lower, but the water loss was higher compared to post-rigor fillets (Rosnes *et al.*, 2003).

1.5 Seafood processing

Processing imposes hurdles to the actions of microorganisms (Leistner and Gorris, 1995), hence inhibiting or inactivating them, which results in the prevention of spoilage and the extension of shelf life. Additionally, processing can also retard or inhibit non-microbial spoilage mechanisms (Table 1.1). From the safety point of view, processing can remove or eliminate pathogenic bacteria making seafood safer for consumption.

Table 1.1 Current food and seafood processing methods

Process	Hurdle	Objective
Chilling	Low temperature	Inhibition of microbial growth
Chilling and packaging under modified atmosphere	Low temperature, reduced O_2, elevated CO_2	Inhibition of microbial growth, slowing down chemical oxidations
Freezing	Low temperature	Inhibition of microbial growth, slowing down enzymatic activity and chemical oxidations
Freezing and glazing or vacuum packaging	Low temperature, reduced O_2	Inhibition of microbial growth, oxidative rancidity and enzymatic browning
Heating	High temperature	Inactivation of microorganisms and enzymes
Irradiation	Ionizing radiation	Inactivation of microorganisms and enzymes
Salting	Low a_w	Inhibition of microbial growth
Marination	Low pH, organic acids,	Inhibition and/or inactivation of microorganisms
Drying	Low a_w	Inhibition of microbial growth
Smoking	Low a_w, high temperature (in hot smoking), antimicrobial substances from smoke	Inhibition of microbial growth
Curing (combination of salting, smoking, acidification, drying)	Low a_w and pH, high temperature (in hot smoking), antimicrobial substances from the smoke	Inhibition of microbial growth
Fermentation	Low pH, organic acids, bacteriocins, bacterial antagonism	Inhibition of microbial growth

A range of processing methods can be used to preserve seafood. Processing methods can be applied either singly or in combination (Table 1.1). Cold storage is the simplest way to preserve seafood. Chilling of whole fish in ice takes place after harvesting and killing, while packed gutted or filleted fish are refrigerated. Fish shelf life can be extended by superchilling, where fish is stored a few degrees below 0 °C (see Chapter 3). Packaging under reduced O_2 and elevated CO_2 (modified atmosphere packaging, MAP) in combination with cold storage (0–2 °C) can extend the shelf life of various seafoods (see Chapter 10). Freezing is also one of the most widely used preservation methods for seafood (see Chapter 3).

Heat processing remains one of the major methods for extending the shelf life of seafood because as well as giving a long shelf life it also gives a high level of safety and convenience (see Chapter 4). Irradiation is a widely used non-thermal process for preserving fish and seafood. Irradiation guarantees the safety of the product and also increases its shelf life, despite the lack of trust for this method by consumers. It is a very effective method for inactivating microorganisms without considerably decreasing foodstuff quality (see Chapter 5).

Traditional methods of preservation such salting, smoking, marination, drying and fermentation (see Chapters 6, 7 and 8) are widely used throughout the world. Traditionally preserved fish is highly appreciated, mainly due to its excellent stability during storage, special organoleptic characteristics and nutritional value.

Quite considerable amounts of fish are used for surimi production. Surimi is a fish protein concentrate with gelling abilities that has become an important intermediate raw material for food production all over the world. Surimi is further processed into surimi-based products such as *kamaboko* and crab-meat analogues (see Chapter 10).

The seafood industry consumes a lot of energy and produces a considerable amount of waste. Methods to treat fish waste and to convert it into useful products such as feed, natural pigments and other products have been developed (see Chapter 11). The sustainable operation of fish processing plants, which involves not only waste treatment and disposal as well as the recovery of by-products, but also energy efficiency and water usage, are also of concern (see Chapter 12).

1.6 Quality, safety and authenticity assurance

Freshness and quality of seafood is assessed using sensory, microbiological and chemical methods (Olafsdottir *et al.*, 1997). Sensory assessment is subjective and requires highly trained personnel to be reliable, hence it is unattractive for routine examination, while microbiological results are retrospective, thus the determination of the chemical spoilage parameters related to microbial growth, is more practical for routine use (Dainty, 1996). Total volatile base-nitrogen (TVB-N) and trimethylamine-nitrogen (TMA-N) are the main chemical parameters related to the microbial growth of microorganisms such as *Pseudomonas* spp., *Shewanella putrefaciens* and *Photobacterium phosphoreum* (Gram and Huss, 1996; Gram and Dalgaard, 2002). However, the most used parameter, TVB-N, is considered as a poor indicator of teleostean fish freshness (Castro *et al.*, 2006). Current research is focusing on other metabolites produced during the storage of aquatic products, such as volatiles other than nitrogenous compounds (Duflos *et al.*, 2006; Soncin *et al.*, 2008). Additionally, a range of physical and automated instrumental methods that can give fast reliable measurements a without destruction of the sample are being developed, such as VIS/NIR spectroscopy, electronic nose, etc. (see Chapter 14).

A variety of microbiological parameters are also examined to assess the microbiological quality and safety of seafood. Despite the disadvantages of traditional culture techniques, they are still considered standard methods. However, advances in molecular microbiology and automated rapid methods offer alternative tools for quick and reliable analysis (see Chapter 15).

To protect consumers and prevent fraud in the marketing of fishery and aquaculture products, fast and reliable methods for species identification, even for processed products, are required. Recent developments in molecular biology and polymerase chain reaction (PCR)-based techniques, as well as the use of molecular markers and databases have greatly contributed to this field of study (see Chapter 16).

Finally, in the past two decades the concept of risk assessment has greatly improved the way that seafood hazards are evaluated and controlled, leading towards to an integrated approach of food/seafood safety. The presentation of the four elements of risk assessment (hazard identification, exposure assessment, hazard characterization and risk characterization) are analysed in Chapter 17.

1.7 Future trends

Current processing technologies are quickly evolving (modified atmosphere packaging, minimal heat processing, rapid freezing, etc.), while emerging technologies such as high pressure processing (HPP), radio-frequency heating, flexible retort packaging, and pH-shift processing, etc., will be applied extensively in seafood processing. HPP is currently applied mostly in oysters. Pressure severs the adductor muscle from the shell, which results essentially in a shucked oyster while inactivation of pathogenic *Vibrios* and other microorganisms occurs (see Chapter 2). HPP will soon extend its applications to various seafood products.

Value-added seafood products are becoming increasingly important in satisfying consumer demands for safe, high-quality, convenient, healthy and nutritious seafood throughout the world (see Chapter 13). Market requirements and the new technologies in seafood processing will soon be the driving force for many innovations in seafood processing.

References

Abe, H. and Okuma, E. (1991). *Rigor mortis* progress of carp acclimated to different water temperatures, *Nippon Suisan Gakkaishi*, 57, 2095–2100.

Ashie, I.N.A., Smith, J.P. and Simpson, B.K. (1996). Spoilage and shelf-life extension of fresh fish and shellfish. *Critical Reviews in Food Science and Nutrition*, 36, 87–121.

Bagni, M., Civitareale, C., Priori, A. *et al.* (2007). Pre-slaughter crowding stress and killing procedures affecting quality and welfare in sea bass (*Dicentrarchus labrax*) and sea bream (*Sparus aurata*). *Aquaculture*, 263, 52–60.

Borderias, A.J. and Sanchez-Alonso, I. (2011). First processing steps and the quality of wild and farmed fish. *Journal of Food Science*, 76, R1–R5.

Boziaris, I.S., Kordila, A. and Neofitou, C. (2011). Microbial spoilage analysis and its effect on chemical changes and shelf-life of Norway lobster (*Nephrops norvegicus*) stored in air at various temperatures. *International Journal of Food Science and Technology*, 46, 887–895.

Cakli, S., Kilinc, B., Cadun, A., Dincer, T. and Tolasa, S. (2006). Effects of gutting and ungutting on microbiological, chemical, and sensory properties of aquacultured sea bream (*Sparus aurata*) and sea bass (*Dicentrarchus labrax*) stored in ice. *Critical Reviews in Food Science and Nutrition*, 46, 519–527.

Castro, P., Padron, J.C.P., Cansino, M.J.C., Velazquez, E.S. and De Larriva, R.M. (2006). Total volatile base nitrogen and its use to assess freshness in European sea bass stored in ice. *Food Control*, 17, 245–248.

Dainty, R.H. (1996). Chemical/biochemical detection of spoilage. *International Journal of Food Microbiology*, 33, 19–33.

Dalgaard, P. (2000). Fresh and lightly preserved seafood, in *Shelf-Life Evaluation of Foods* (eds C.M.D. Man, and A.A. Jones), Aspen Publishers, London, pp. 110–139.

Duflos, G., Coin, V. M., Cornu, M., Antinelli, J.F. and Malle, P. (2006). Determination of volatile compounds to characterize fish spoilage using headspace/mass spectrometry and solid-phase microextraction/gas chromatography/mass spectrometry. *Journal of the Science of Food and Agriculture* 86, 600–611.

Erkan, N. (2007). Sensory, Chemical, and Microbiological Attributes of Sea Bream (*Sparus aurata*): Effect of Washing and Ice Storage. *International Journal of Food Properties*, 421–434.

FAO (2012). *The State of the World Fisheries and Aquaculture.* Food and Agriculture Organization, Rome, Italy. http://www.fao.org/docrep/016/i2727e/i2727e00.htm [accessed 30 May 2013].

Feldhusen, F. (2000). The role of seafood in bacterial foodborne diseases. *Microbes and Infections*, 2 (13), 1651–1660.

Gram, L. and Dalgaard, P. (2002). Fish spoilage bacteria – problems and solutions. *Current Opinion in Biotechnology*, 13, 262–266.

Gram, L. and Huss, H. H. (1996). Microbiological spoilage of fish and fish products. *International Journal of Food Microbiology*, 33, 121–137.

Huis in't Veld, J.H.J. (1996). Microbial and biochemical spoilage of foods: an overview. *International Journal of Food Microbiology*, 33, 1–18.

Huss, H.H. (1994). Assurance of seafood quality. FAO Fisheries Technical Paper – 334. FAO Press. Rome.

Huss, H.H. (1995). Quality and quality changes in fresh fish. FAO Fisheries Technical Paper – 348. FAO Press, Rome.

Huss, H.H., Reilly, A. and Karim Ben Embarek, P. (2000). Prevention and control of hazards in seafood. *Food Control*, 11, 149–156.

Leistner, L. and Gorris, L.G.M. (1995). Food preservation by hurdle technology. *Trends in Food Science and Technology*, 6, 41–46.

Martinez-Alvarez, O., Lopez-Caballero, M.E., Montero, P. and Gomez-Guillen, M.C. (2007). Spraying of 4-hexylresorcinol based formulations to prevent enzymatic browning in Norway lobsters (*Nephrops norvegicus*) during chilled storage. *Food Chemistry*, 100, 147–155.

Neil, D.M. (2012) Ensuring crustacean product quality in the post-harvest phase. *Journal of Invertebrate Pathology*, 110, 267–275.

Olafsdottir, G., Martinsdottir, E., Oehlenschläger, J. *et al.* (1997). Methods to evaluate freshness in research and industry. *Trends in Food Science and Technology*, 8, 258–265.

Ottera, H., Roth, B. and Torrissen, O.J. (2001). Do killing methods affect the quality of Atlantic salmon? in *Farmed Fish Quality* (eds S.C. Kestin and P.D. Warriss), Blackwell Science Ltd, Oxford, pp. 398–399.

Papadopoulos, V., Chouliara, I., Badeka, A., Savvaidis, I.N., Kontominas, M.G. (2003). Effect of gutting on microbiological, chemical, and sensory properties of aquacultured sea bass (*Dicentrarchus labrax*) stored in ice. *Food Microbiology*, 20, 411–420.

Poli, B.M., Parisi, G., Scappini, F. and Zampacavallo, G. (2005). Fish welfare quality as affected by pre-slaughter and slaughter management. *Aquaculture International*, 13, 29–49.

Robson, A.A., Kelly, M.S. and Latchford, J.W. (2007). Effect of temperature on the spoilage rate of whole, unprocessed crabs: *Carcinus maenas, Necora puber* and *Cancer pagurus*. *Food Microbiology*, 24, 419–424.

Rosnes, J.T., Vorre, A., Folkvord, L. *et al.* (2003). Effects of pre-, in-, and post-rigor filleted Atlantic salmon (*Salmo salar*) on microbial spoilage and quality characteristics during chilled storage. *Journal of Aquatic Food Product Technology*, 12, 17–31.

Roth, B., Imsland, A., Gunnarsson, S., Foss, A. and Schelvis-Smith R. (2007). Slaughter quality and rigor contraction in farmed turbot (*Scophthalmus maximus*): a comparison between different stunning methods. *Aquaculture*, 272, 754–61.

Soncin, S., Chiesa, M., L., Panseri S., Biondi, P. and Cantoni, C. (2008). Determination of volatile compounds of precooked prawn (*Penaeus vannamei*) and cultured gilthead sea bream (*Sparus aurata*) stored in ice as possible spoilage markers using solid phasemicroextraction and gas chromatography/mass spectrometry. *Journal of the Science of Food and Agriculture* 89, 436–442.

Part I

Processing Technologies

2

Shellfish Handling and Primary Processing

Yi-Cheng Su[1] and Chengchu Liu[2]

[1]*Seafood Research and Education Center, Oregon State University, Astoria, Oregon, USA*
[2]*College of Food Science and Technology, Shanghai Ocean University, Shanghai, People's Republic of China*

2.1 Introduction

The United States ranks as the third largest consumer of fish and shellfish, behind China and Japan. Americans consumed a total of 4.833 billion pounds of seafood with an average consumption of 7.2 kg (15.8 pounds) of fish and shellfish per person in 2009 (NOAA, 2010). In 2010, US commercial fishermen landed 8.2 billion pounds of seafood with a value of $4.5 billion (NOAA, 2011). It is estimated that about 86% of the seafood consumed in the United States is imported and nearly half of the imported seafood comes from aquaculture or farmed seafood. According to the United Nations Food and Agriculture Organization (FAO), aquaculture outside the United States has expanded dramatically in the past three decades and now supplies half of the world's seafood demand. The world production of molluscan shellfish increased from 1.7 million tonnes in 1950 to 9.1 million tonnes in 1990 and reached 20.8 million tonnes in 2010 (FAO, 2010, 2012). The rapid increase in aquaculture production of molluscs is attributable to an increase in consumption of molluscan shellfish over the past decade.

2.1.1 Health hazards associated with molluscan shellfish

Molluscan shellfish are filter-feeders and can concentrate toxic substances and microorganisms through filtering the water in which they grow for nutrients. Many studies have illustrated that contaminants, such as heavy metals (Eisenberg and Topping, 1984; Presley *et al.*, 1990; Jiann and Presley, 1997; Fang *et al.*, 2003), bacteria (Lipp *et al.*, 2001), viruses such as

Seafood Processing: Technology, Quality and Safety, First Edition. Edited by Ioannis S. Boziaris.

human noroviruses or hepatitis A virus (HAV) (Richards, 1987; Lewis and Metcalf, 1988; Beril *et al.*, 1996; Costantini *et al.*, 2006) and marine toxins produced by algae (USFDA, 2012b) can be accumulated in shellfish. Therefore, bivalves can serve as vehicles for human diseases.

Bacterial pathogens *Vibrio parahaemolyticus* and *Vibrio vulnificus* are halophilic bacteria that occur naturally in marine environments and have been frequently isolated from molluscan shellfish. Therefore, consumption of raw shellfish by humans is associated with a high risk of infections caused by *V. parahaemolyticus* or *V. vulnificus* infection. Consumption of raw or undercooked shellfish, particularly oysters, contaminated with *V. parahaemolyticus* or *V. vulnificus* may lead to development of acute gastroenteritis characterized by diarrhoea, headache, vomiting, nausea, abdominal cramps and low fever. Numerous outbreaks of *V. parahaemolyticus* infection linked to raw oyster consumption have been documented in the United States (CDC 1998, 1999, 2006; McLaughlin *et al.*, 2005). Recently, the US Center for Disease Control and Prevention (CDC) estimated that about 45 000 illnesses from *V. parahaemolyticus* occur each year in the United States (USFDA, 2012a).

The threat of *V. parahaemolyticus* infection following consumption of raw or undercooked oysters is a concern for public health. Outbreaks of *V. parahaemolyticus* infection typically cause substantial economic losses to the shellfish industry. In addition to *Vibrio* infections, gastroenteritis caused by human noroviruses has been implicated in outbreaks linked to raw oyster consumption (Kohn *et al.*, 1995). It is estimated that one in 2000 meals of raw molluscan shellfish serves as the vehicle for *Vibrio* infection in the United States and that consumption of raw oyster is responsible for about 95% of all deaths associated with seafood consumption (Oliver, 1989).

Viruses Viruses, such as noroviruses and hepatitis A virus, can be present in shellfish harvested from the growth environments contaminated with sewage (Fiore, 2004; CDC, 2012). Noroviruses are the most common cause of acute gastroenteritis in the United States with about 21 million illnesses, 70 000 hospitalizations and 800 deaths each year (CDC, 2012). Foods commonly involved in norovirus outbreaks include leafy greens, fresh fruits and shellfish (mainly oysters).

Infections caused by noroviruses typically trigger a sudden illness with symptoms of headache, nausea, vomiting, abdominal pain, diarrhoea and fever within 12–48 h. The symptoms usually last for a few days. Most people can recover from the sickness without medication. Infection with HAV can lead to the development of liver disease with jaundice as the typical symptom. The severity of the HAV infection varies among individuals and may be a mild sickness for a few weeks or a severe syndrome lasting several months. Humans are the main reservoir for HAV, which can be spread from infected persons through contact with foods during food preparation (Fiore, 2004).

Marine toxins Several marine toxins may accumulate in shellfish and cause toxicity (USFDA, 2012b). Saxitoxin, produced by planktonic algae (mainly dinoflagellates), can accumulate in shellfish and cause paralytic shellfish poisoning (PSP) with symptoms of tingling, burning, numbness and incoherent speech. In severe cases, death can occur due to respiratory paralysis. Brevetoxins, produced by a dinoflagellate (*Gymnodinium breve*), can cause neurotoxic shellfish poisoning (NSP) characterized by neurological and gastrointestinal symptoms, including tingling of the lips and tongue, dizziness, nausea, vomiting, diarrhoea and muscular aches. The symptoms typically occur within a few hours after consumption of contaminated shellfish and most illnesses do not require hospitalization.

Okadaic acid and related toxins, produced by certain dinoflagellates, can cause diarrhoeic shellfish poisoning (DSP), primarily observed as a mild gastrointestinal disorder with symptoms of nausea, vomiting, diarrhoea and abdominal pain accompanied by chills and headache. DSP is generally not a life-threatening illness and recovery is usually complete with no after-effects. In addition, domoic acid, a neurotoxin produced by a genus of diatoms (*Pseudo-nitzschia* spp.), can cause amnesic shellfish poisoning (ASP) with symptoms of gastrointestinal disorders (vomiting, diarrhoea and abdominal pain) followed by neurological problems (confusion, memory loss, disorientation, seizure and coma) (Jeffery *et al.*, 2004). ASP can be life threatening and most fatalities are elderly patients.

2.2 Shellfish harvesting

To minimize the health risks associated with consuming bivalve molluscan shellfish, it is critical for public health protection that the water quality in shellfish-growing environments is constantly monitored so that toxic substances and microorganisms can be identified.

2.2.1 Growing area

Pollution of shellfish-growing areas can come from a variety of sources and under different conditions. In general, sources of pollution can be divided into two categories: point and non-point (CSSP, 2011). Both point and non-point pollution sources can release chemical and/or microbiological contaminants of public health concern.

A point source of pollution is pollution that enters the receiving water at discrete and measurable locations, such as in discharges from wastewater treatment and collection systems, pulp mills and food processing establishments. A non-point source of pollution refers to contamination that occurs from sources related to human activity and natural processes in the watershed which are diffuse or dispersed. Such sources do not enter at discrete, identifiable locations and are difficult to measure or define. There are eight types of non-point source pollution identified by the US Food and Drug Administration (FDA) which may affect shellfish-growing environments (USFDA, 1995). These include urban run-off, agricultural run-off, animal faecal pollution, sewage discharges from boats, wildlife faecal matter, dredging operations, and mining and silviculture practices.

United Kingdom In the United Kingdom, areas for molluscan bivalve harvesting are classified according to the degree of contamination determined by monitoring levels of *Escherichia coli* in shellfish flesh (CEFAS, 2011). The Centre for Environment, Fisheries and Aquaculture Science (CEFAS) manages the classification programme for bivalve production areas in England and Wales and defines which areas are authorized for harvesting bivalves and if post-harvest treatments are required before human consumption. The classification categories are:

1. Class A ($\leq$230 *E. coli*/100 g) – molluscs can be harvested for direct human consumption.
2. Class B (90% of samples must be $\leq$4600 *E. coli*/100 g; all samples must be <46 000 *E. coli*/100 g) – molluscs can be sold for human consumption:
   - after purification in an approved plant, or
   - after relaying in an approved Class A relaying area, or
   - after an European Commission (EC) approved heat treatment process.

3. Class C (≤46 000 *E. coli*/100 g) – molluscs can be sold for human consumption only after relaying for at least two months in an approved relaying area followed, where necessary, by treatment in a purification centre, or after an EC-approved heat treatment process.

In all cases, the health standards in Annex III of EC Regulations 853/2004 and the microbiological criteria adopted under EC Regulation 2073/2005 must be met. Molluscs must not be subject to production or collected in prohibited areas. There are two classification systems in England and Wales:

1. the annual or 'temporary' classification system
2. the long-term classification (LTC) system.

New areas will initially be given an annual/temporary classifications until they meet the criteria for an LTC. Harvesting sites that do not meet LTC criteria are automatically classified under the annual/temporary classification system.

Canada In Canada, the Environment Canada's Marine Water Quality Monitoring Program is the first line of defence in the sanitary control of shellfish. The programme is designed to identify and evaluate all sources of pollution to shellfish growing and harvesting waters. The Canadian Shellfish Sanitation Program (CSSP) Manual of Operations (version 2) requires that the water quality in shellfish area needs to be surveyed and that actual and potential sources of pollution are identified (CSSP, 2011). In such surveys, the shellfish-growing areas are classified according to their suitability for harvesting shellfish using accepted water quality standards and the general sanitary conditions in the area. The CSSP recognizes five classification categories assigned to shellfish areas based on sanitary and water quality survey results:

1. approved
2. conditionally approved
3. restricted
4. conditionally restricted and
5. prohibited.

Approved classification Shellfish-growing areas may be classified as 'approved' if the area is not contaminated with faecal material, pathogenic microorganisms, poisonous or deleterious substances, to the extent that consumption of the shellfish might be hazardous. The following conditions must also be met:

- the median or geometric mean of faecal coliform determined as most probable number (MPN) of the water does not exceed 14/100 ml and not more than 10% of the samples exceed a faecal coliform level of 43 MPN/100 ml, for a five-tube decimal dilution test; or
- the chemical levels meet the standards/tolerances outlined in Appendix II of Shellfish Sanitation Program Manual of Operations and in the Fish Products Standards and Methods Manual, Appendix 3, Canadian Guidelines for Chemical Contaminants and Toxins in Fish and Fish Products.

Evidence of potential pollution sources such as overflows from sewage lift stations, direct sewage releases/discharges, septic tank seepage, etc., is sufficient to exclude the shellfish-growing waters from the approved classification.

Conditionally approved classification Conditionally approved classification refers to shellfish-growing areas which have been determined by the shellfish control authority to meet the 'approved' criteria for a predictable period. These areas are subject to intermittent pollution caused by releases/discharges from wastewater treatment and collection systems, seasonal populations, non-point source pollution or boating activity. The period meeting the approved criteria (for other than seasonal boating activity) is conditional upon established performance standards specified in a Conditional Management Plan. A conditionally approved area which does not meet the criteria for an approved shellfish-growing area is placed in closed status by the shellfish control authority.

Restricted classification Restricted classification is a shellfish-growing area where the harvesting of shellfish is not permitted, except by licence issued under the management of Contaminated Fisheries Regulations (DFO, 1990), due to contamination by faecal material, pathogenic microorganisms, or poisonous or deleterious substances, to the extent that consumption of the shellfish might be hazardous.

Conditionally restricted classification Conditionally restricted is the classification of a shellfish-growing area which has been determined by the shellfish control authority to meet, at a minimum, the 'restricted' classification criteria for a predictable period. These shellfish areas are subject to intermittent pollution caused by releases/discharges from wastewater treatment and collection systems, seasonal populations, non-point source pollution or boating activity. The period meeting the 'restricted' criteria (for other than seasonal boating activity) is conditional upon established performance standards specified in a Conditional Management Plan. Harvesting is prohibited when a 'conditionally restricted' shellfish-growing area is in the closed status.

Prohibited classification Shellfish shall not be harvested from prohibited areas for any purpose, with the exception of harvesting for seed, spat, bait and for scientific purposes, all of which require special licenses.

United States In the United States, the National Shellfish Sanitation Program (NSSP) Guide for the Control of Molluscan Shellfish allows a growing area to be classified using either a total or faecal coliform standard. Each growing area will be correctly classified as approved, conditionally approved, restricted, conditionally restricted, or prohibited (NSSP, 2009). All growing areas which are subjected to a sanitary survey must be correctly classified based on the 12-year sanitary survey and its most recent triennial or annual re-evaluation when available, as only one of the following:

1. approved
2. conditionally approved
3. restricted
4. conditionally restricted or
5. prohibited.

Growing areas not subjected to a sanitary survey every 12 years will be classified as prohibited.

Approved classification A growing area will be classified as approved when a sanitary survey finds that the following criteria are met for the area:

- safe for the direct marketing of shellfish;
- not subject to contamination from human or animal faecal matter at levels that, in the judgement of the Authority, presents an actual or potential public health hazard; and
- not contaminated with:
  - (i) pathogenic organisms
  - (ii) poisonous or deleterious substances
  - (iii) marine biotoxins or
  - (iv) bacteria concentrations exceeding the bacteriological standards for a growing area in this classification.

Conditionally approved classification A growing area in the conditionally approved classification will meet the requirements for:

- an approved area classification when the conditionally approved classification is in the open status; and
- a restricted or prohibited classification when the conditionally approved classification is in the closed status.

If the closed status meets the criteria for the restricted classification, designate in its management plan whether the shellstock may be harvested for relaying or depuration.

Restricted classification A growing area may be classified as restricted when:

- a sanitary survey indicates a limited degree of pollution; and
- levels of faecal pollution, human pathogens, or poisonous or deleterious substances are at such levels that shellstock can be made safe for human consumption by either relaying, depuration or low acid-canned food processing.

The Authority will have effective controls to assure that shellfish are harvested from restricted areas only:

- by special license; and
- under the supervision of the Authority.

The Authority will establish Shellstock Quality Criteria for use in placing an area in the restricted classification. Depending on the treatment process to be applied to the shellstock, the criteria will be established in accordance with guidelines for 'Shellstock Relaying' or 'Depuration'.

Conditionally restricted classification Any growing area in the conditionally restricted classification will:

- meet the requirements for:
  - (i) a restricted classification when the conditionally restricted classification is in the open status; and
  - (ii) a prohibited classification when the conditionally restricted classification is in the closed status.
- designate in its management plan whether the harvested shellstock are to be relayed or depurated.

Prohibited classification The prohibited classification is not required for harvest waters within or adjacent to marinas. The Authority, however, may use the prohibited classification for these waters.

A growing area will be classified as prohibited if:

- no current sanitary survey exists;
- a sanitary survey determines:
  - (i) the growing area is adjacent to a sewage treatment plant outfall or other point source outfall with public health significance;
  - (ii) pollution sources may unpredictably contaminate the growing area;
  - (iii) the growing area is contaminated with faecal waste so that the shellfish may be vectors for disease microorganisms;
  - (iv) the concentration of biotoxin is sufficient to cause a public health risk; or
  - (v) the area is contaminated with poisonous or deleterious substances causing the shellfish to be adulterated.
- a risk assessment performed in accordance with Risk Assessment and Risk Management indicates that the shellstock are not safe for human consumption.

The Authority will (a) not permit the harvest of shellstock from any area classified as prohibited, except for the harvest of shellstock for the gathering of seed for aquaculture or the depletion of the areas classified as prohibited; and (b) ensure that shellstock removed from any growing area classified as prohibited is effectively excluded from human consumption unless it is seed to be cultured as outlined in NSSP MO Chapter VI, Shellfish Aquaculture. 03 Seed Shellstock (NSSP, 2009).

2.2.2 Water quality

The quality, value and safety of molluscan shellfish can be compromised by pollution of marine environments. Bacteriological standards for water quality, based on faecal coliform levels, as determined by the MPN method, are commonly used for the classification of shellfish-growing waters. Faecal coliform in water is frequently monitored to ensure that the environments meet established sanitary standards for shellfish-growing water. In addition, the presence of faecal coliform in water can be used as an indicator for the potential accumulation of faecal pathogens in shellfish.

Approved classification of growing areas The US National Shellfish Sanitation Program requires that the water quality in the approved classification of shellstock growing areas meets the faecal coliform standards (NSSP, 2009). The median or geometric mean, determined by MPN or membrane filtration (MF) (mTEC), of faecal coliform of water samples will not exceed 14 per 100 ml and not more than 10% of the samples will exceed an MPN or MF (mTEC) of:

- 49 MPN/100 ml for a three-tube decimal dilution test;
- 43 MPN/100 ml for a five-tube decimal dilution test;
- 28 MPN/100 ml for a twelve-tube single dilution test; or
- 31 cfu/100 ml for a MF (mTEC) test.

Restricted classification of growing areas If the tidal stage increases the faecal coliform concentration, the Authority will use samples collected under that tidal stage to classify the area. For the restricted classification of growing areas affected by point sources and used as a shellstock source for shellstock depuration, the faecal coliform median or geometric mean MPN or MF (mTEC) of the water sample results will not exceed 88 per 100 ml and the estimated 90th percentile will not exceed an MPN or MF (mTEC) of:

- 300 MPN/100 ml for a three-tube decimal dilution test;
- 173 MPN/100 ml for a twelve-tube single dilution test; or
- 163 cfu/100 ml for a MF (mTEC) test.

For the restricted classification of growing areas affected by nonpoint sources and used as a shellstock source for shellstock depuration, the faecal coliform median or geometric mean MPN or MF (mTEC) of the water sample results will not exceed 88 per 100 ml and the estimated 90th percentile will not exceed a MPN or MF (mTEC) of:

- 300 MPN/100 ml for a three-tube decimal dilution test; or
- 260 MPN/100 ml for a five-tube decimal dilution test;
- 163 cfu/100 ml for a MF (mTEC) test.

2.3 Bivalve shellfish handling

2.3.1 Temperature control

Once bivalve shellfish is harvested, it should be placed immediately under appropriate temperature control. Exposure of fresh shellfish to extreme temperature changes during transportation and handling may result in a loss of quality or the death of the live shellstock. Temperature control is defined as the management of shellstock temperature with ice, mechanical refrigeration or other approved means that can lower the temperature to 10 °C (50 °F) or less (NSSP, 2009).

To preserve quality and limit the growth of human pathogens in shellfish post harvest, the NSSP Guide for the Control of Molluscan Shellfish has established time-to-temperature regulations that limit the maximum time for the exposure of oysters to elevated temperatures upon harvest (NSSP, 2009). Shellfish harvested for raw consumption need to be cooled to 10 °C (50 °F) within a certain time to control the growth of *Vibrio parahaemolyticus* (Table 2.1) or *Vibrio vulnificus* (Table 2.2).

Table 2.1 Time-to-temperature matrix for controlling *Vibrio parahaemolyticus* in shellstock shellfish (data adapted from National Shellfish Sanitation Program 2009 Revision)

Action level	Average monthly maximum air temperature	Maximum hours from exposure to temperature control (h)
Level 1	< 66 °F (≤ 18 °C)	36
Level 2	66–80 °F (19–27 °C)	12
Level 3	≥ 81 °F (≥ 27 °C)	10

Table 2.2 Time-to-temperature matrix for controlling *Vibrio vulnificus* in shellstock shellfish (data adapted from National Shellfish Sanitation Program 2009 Revision)

Action level	Water temperature	Maximum hours from exposure to temperature control (h)
Level 1	<65 °F (<18 °C)	36
Level 2	65–74 °F (18–23 °C)	14
Level 3	>74–84 °F (>23–28 °C)	12
Level 4	> 84 °F (>28 °C)	10

2.3.2 Transportation and storage

Refrigeration has been the most commonly used method for short-term storage of perishable products to preserve quality and extend the shelf life of shellstock shellfish. Maintaining the 'Cold Chain' from harvest through consumption is critical for preserving freshness and quality as well as for ensuring the safety of shellfish products. The International Institute of Refrigeration (IIR) estimates that 300 million tonnes of produce are wasted each year due to deficient temperature control and the food industry discards $35 billion worth of spoiled foods each year in the United States (IIR, 2008). In addition, a high percentage of foodborne diseases occur due to consumption of foods subjected to temperature abuse. Therefore, time and temperature are the two main factors that contribute to the quality and safety of shellfish products.

Shellfish shipment Shellfish should be shipped with proper identifying tags and/or labels as well as their shipping documents. According to the NSSP, live shellstock should be cooled to an internal shellstock body temperature of 10 °C (50 °F) or less, while shucked shellfish and in-shell products need to be cooled to 7.2 °C (45 °F) or less (NSSP, 2009). A time-temperature indicating device can be used to monitor that the shellstock internal body temperature remains at 10 °C (50 °F) or less during shipping. Shellfish can be rejected if:

- they are not properly identified with tags or shipping documents;
- the internal shellstock body temperature exceeds 15.6 °C (60 °F) unless the harvest initiation time can be documented and indicates that the time from harvest has not exceeded the time-to-temperature matrix for controlling *Vibrio parahaemolyticus* (Table 2.1) or *Vibrio vulnificus* (Table 2.2) in shellstock shellfish;
- shucked shellfish temperature or the internal body temperature of in-shell products exceed 10 °C (50 °F); or
- the Authority determines that the product is unwholesome or unsafe for human consumption.

The Authority should notify the shipping dealer, the receiving dealer and the Authority in the State where the shipment originated about the shipment's rejection.

Shellfish transportation Shellstock shellfish should be transported in clean storage bins with effective drainage and be shipped on pallets in a truck that is properly maintained to prevent contamination, deterioration and decomposition. When the shellstock have been previously refrigerated or when ambient air temperature and time of travel facilitate bacterial growth or shellfish deterioration, they should be transported in trucks with a refrigeration system equipped with an automatic temperature control capable of maintaining the air temperature at 7.2 °C (45 °F) or less (NSSP, 2009). Any ice used in the transportation of shellfish should be made on-site from potable water in a commercial ice machine. All containers used to transport shellstock shellfish should be constructed from safe materials and allow for easy cleaning. They should be cleaned with potable water, detergents and sanitizers after use.

Shellfish shipping time When the shipping time is four hours or less, all shellfish may be shipped in ice or an acceptable means of refrigeration (NSSP, 2009). Lack of ice or an acceptable means of refrigeration is considered an unsatisfactory shipping condition. When a mechanical refrigeration unit is used, the unit should be equipped with an automatic temperature control capable of maintaining the air temperature in the storage area at 7.2 °C (45 °F) or less. The dealers are not required to provide thermal recorders during shipment.

When the shipping time is more than four hours, the dealer should ship all shellfish in mechanically refrigerated conveyances which are equipped with automatic controls capable of maintaining the air temperature in the storage area as well as containers with an internal ambient air temperature at 7.2 °C (45 °F) or less.

The initial dealers need to assure that a suitable time-temperature recording device with date and time information accompanies each shipment of shellfish, unless the dealer has an approved HACCP plan with an alternative means of monitoring time-temperature (NSSP, 2009). Each receiving dealer should write the date and time on the temperature-indicating device when the shipment is received and the doors of the conveyance or the containers are opened. The final receiving dealer should keep the time-temperature recording chart or other record of time and temperature on file and make it available to the Authority upon request. An inoperative temperature-indicating device is considered as no recording device.

2.3.3 Retail handling

Good Retail Practices (GRPs) are key components for keeping shellfish fresh and safe for consumption. GRPs are similar to Good Manufacturing Practices (GMPs) and are the minimum sanitary and processing requirements for food companies (Anonymous, 2001). GRPs should be developed that are specific to each facility and operation according to local regulatory standards. Once fresh shellfish is received, it needs to be stored immediately under appropriate conditions. All cold rooms and freezers used for storing shellfish must have temperature-recording devices and alarm systems to monitor and ensure temperature control during storage. It is highly recommended that retailers uses a first product in, first product out (FIFO) inventory rotation system and moves older products to the front of storage shelves (Anonymous, 2001). Designated display cases for shellfish should be separate from ready-to-eat product displays to prevent cross contamination. When ice is used for temperature control in the display case, the shellfish should be placed in clean containers to avoid direct contact with the ice. All containers used for displaying shellfish need to be properly sanitized after use.

2.4 Shellfish primary processing

The levels of *V. parahaemolyticus* in oysters at the time of consumption are influenced by methods of harvesting and post-harvest handling. It has been reported that the intertidal harvest of oysters in the Pacific Northwest estuaries significantly influenced levels of *V. parahaemolyticus* in oysters at harvest. In the process of intertidal harvest, oysters are first placed in baskets at low tide and then harvested when the tide rises. This practice exposes oysters to ambient air for hours before being harvested, which allows naturally accumulated *V. parahaemolyticus* to proliferate rapidly in oysters, especially on a warm and sunny day.

Nordstrom *et al.* (2004) reported that total populations of *V. parahaemolyticus* in oysters increased from fourfold to eightfold while the *tdh*-positive *V. parahaemolyticus* counts also increased from $\leq$10 to as high as 160 cfu/g after being exposed to ambient air between tides. The study also demonstrated that an overnight submersion for a single tidal cycle reduced *V. parahaemolyticus* to levels similar to those determined prior to the intertidal exposure. Therefore, intertidal harvest of oysters should be conducted at high tide to avoid exposure of oysters to air temperatures before harvest.

In addition to harvest methods, post-harvest handling also affects the levels of *V. parahaemolyticus* in oysters. For example, holding time (unrefrigerated storage) for oysters after harvest and before refrigeration varies in different geographic areas and at different times of year. Although the populations of *V. parahaemolyticus* in oysters, at the time of harvest, are usually lower than 10^3 cfu/g (Kaysner and DePaola, 2000), *V. parahaemolyticus* can multiply rapidly in oysters upon exposure to elevated temperatures. Therefore, harvested shellfish should be stored at a low temperature as soon as possible to prevent rapid growth of *V. parahaemolyticus* in contaminated shellfish.

2.4.1 Shucking

Shellfish can be shucked by heat or high-pressure process (HPP). In the process of heat shuck, the process should be developed by the Authority or qualified persons with adequate facilities for conducting the appropriate studies and be approved by the Authority. The Authority will assure that the critical factors, which may affect the heat shock process, have been adequately studied and provided for in establishing the process (NSSP, 2009). The Authority should retain records covering all aspects of the establishment of the heat shock process. The critical factors include:

- type and size of shellfish
- time and temperature of exposure
- type of process
- size of tank, tunnel or retort
- water to shellfish ratios in tanks, and
- temperature and pressure monitoring devices.

In addition, the Authority should assure that heat shock process does not:

- change the physical and organoleptic properties of the species;
- kill the shellfish prior to shucking; and
- increase microbial deterioration of the shucked shellfish.

In Canada, all shellstock which have been subjected to the heat shock process are shucked and the meat cooled to at least 7 °C within 2 h after the heat shock process and stored at a temperature between −1 and 4 °C (CSSP, 2011). In addition, the heat shock water tank is completely drained and flushed at 3-h intervals or less so that all mud and detritus remaining in the dip tank from the previous dipping are eliminated.

In addition to heat shock, recent studies of HPP for inactivating *V. parahaemolyticus* in oysters revealed that the process also assists in oyster shucking by destroying the adduct muscle. He *et al.* (2002) reported that a HPP of 240–275 MPa for less than 1 min could be used for shucking Pacific oysters with minimum changes in appearance. A disadvantage of using HPP for shucking shellfish is that the process kills the animals and, therefore, oysters need to be banded before the HPP treatment to stop the shell opening during the process. In addition, initial investment costs of the high pressure system are high and so limit its application by the shellfish industry.

2.4.2 Packing

Shellfish should be cleaned by washing and rinsing followed by thorough draining to ensure that shellstock is reasonably free of sediment and is culled before packing. Shucked shellfish meats should be packed promptly after being delivered to the packing room.

When using heat shock to prepare shellstock for shucking, hot dipped shellstock need to be cooled immediately after the heat shock process by (i) dipping in a ice bath or (ii) useing flowing potable water (NSSP, 2009). If a heat shock tank is used and the water is maintained at or above 60 °C (140 °F), the tank should be completely drained and flushed at the end of each day's operation to eliminate all the accumulated mud and debris. If the water temperature is maintained at below 60 °C (140 °F), the tank should be completely drained and flushed at 3-h intervals (NSSP, 2009).

2.4.3 Post-harvest processes

Several processes, including low-temperature pasteurization, flash-freezing followed by frozen storage, HPP and low-dose irradiation, are capable of reducing *V. parahaemolyticus* or *V. vulnificus* in oysters. However, the oysters are often killed during processing, except by low-dose irradiation.

Andrews *et al.* (2000) reported that a low-temperature pasteurization process for banded in-shell oysters in water at 52 °C for 10 min to achieve an internal temperature of 48–50 °C for 5 min reduced levels of *V. vulnificus* and *V. parahaemolyticus* in oysters by 99.9% and to non-detectable levels, respectively. However, some strains of *V. parahaemolyticus* required keeping the oysters for nearly 9 min at 55 °C to achieve a 5-log reduction (Johnson and Brown, 2002). The pasteurized oysters had a raw-like quality as long as the pasteurization temperature did not exceed 52.5 °C and could be stored in ice for up to 3 weeks. Liu *et al.* (2009) reported that a process of flash-freezing followed by storage at -21 ± 2 °C for 5 months yielded greater than 3.52-log (MPN/g) reductions of *V. parahaemolyticus* in half-shell Pacific oysters.

Several studies have shown that HPP can be used to inactivate *Vibrio* in oysters. Most environmental strains of *V. parahaemolyticus* were totally inactivated at 300 MPa for 300 s (Chen, 2007). However, clinical strains are known to be more resistant than the environmental strains. A HPP treatment of 300 MPa for 180 s was required to achieve a >5-log reduction of clinical strains, including the O3:K6 strain, of *V. parahaemolyticus* in oysters (Cook, 2003). A similar study also reported that pressure treatments needed to be ≥350 MPa for 2 min at temperatures between 1 and 35 °C and ≥300 MPa for 2 min at 40 °C to achieve a 5-log reduction of *V. parahaemolyticus* in live oysters (Kural *et al.*, 2008). A recent validation study of HPP for inactivating *V. parahaemolyticus* demonstrated that a HPP of 293 MPa for 120 s at 8 ± 1 °C was capable of achieving greater than 3.52-log reductions of *V. parahaemolyticus* in Pacific oysters (Ma and Su, 2011). Oysters processed at 293 MPa for 120 s had a shelf life of 6–8 days when stored at 5 °C or 16–18 days when stored in ice.

Irradiation is a non-thermal process and can be utilized to destroy vibrios in shellfish. A study conducted by Andrews *et al.* (2003) reported that irradiation with cobalt-60 gamma ray at doses of 1.0–1.5 kGy reduced the *V. parahaemolyticus* O3:K6 strain in laboratory inoculated oysters from 4-log unit to non-detectable levels (<3.0 MPN/g). Sensory analysis, including difference tests, conducted with 146 volunteers concluded that low doses of irradiation did not kill oysters or affect their sensory qualities (Jakabi *et al.*, 2003). However, the reluctance among consumers to accept irradiated food limits its usage.

2.5 Bivalve shellfish depuration

Depuration is a process utilizing the natural water-filtering mechanism of shellfish to release contaminants from the digestive tract into clean and unpolluted water (Blogoslawski and Stewart, 1983). The process is intended to reduce the number of microorganisms, including bacterial pathogens, present in shellfish harvested from moderately polluted waters so that the shellfish are acceptable for human consumption without further processing. The practice is not intended for shellfish grown in heavily polluted waters nor to reduce the levels of poisonous or deleterious substances that may have accumulated in shellfish from a polluted environment (NSSP, 2009). The process has a long history as a post-harvest treatment to reduce microbial contaminants in shellfish (Canzonier, 1971).

In most instances, total aerobic plate counts in shellfish could be reduced by one log value by the depuration process. Nevertheless, reduction of bacteria to fewer than 10^4 cells per gram of shellfish is seldom seen for some species. Motes and DePaola (1996) reported that oysters naturally contaminated with *V. vulnificus* at levels of 3–4 log MPN/g were reduced to <10 MPN/g after being relayed to unpolluted offshore water for 7–17 days. The levels of *V. vulnificus* were further reduced to a mean level of 0.52 MPN/g by extending the process to 49 days. However, several studies have reported that depuration, normally conducted at ambient water temperatures, was not effective in reducing certain persistent bacteria, including *Vibrio* spp., in shellfish (Colwell and Liston, 1960; Vasconcelos and Lee, 1972; Eyles and Davey, 1984). Depuration of oysters at temperatures higher than 23 °C resulted in increased levels of *V. vulnificus* in the oysters (Tamplin and Capers, 1992). Other studies of depuration in combination with chlorine, ultraviolet light, ozone and iodophors, reported increased reductions of bacteria in oysters (Fleet, 1978). However, none of the treatments could effectively remove *V. parahaemolyticus* from shellfish. Depuration of experimentally contaminated blue mussels with ozonated water for 44 hours resulted in only about 1.0-log reduction of *V. cholerae* and *V. parahaemolyticus* while a 3-log reduction of *E. coli* in the mussels was observed using the same treatment (Croci *et al.*, 2002).

A few studies have been performed to determine the efficacy of depuration in eliminating viruses from shellfish. A small reduction of HAV (98.7%) in mussels was reported after 96 h of commercial depuration using a continuous flow of ozonated seawater

(Abad *et al.*, 1997). However, HAV remained detectable in the bivalves after the 96 h of process. Recently, a study reported reductions of noroviruses in oysters from >1000 to 492 and to <200 viral genome copies/g after 17 and 21 days, respectively, of relaying in clean seawater (Doré *et al.*, 2010). These data indicate that noroviruses are more difficult than HAV to eliminate from shellfish by depuration processes and that shellfish may contain a low level of noroviruses even after a prolonged process.

Attempts have been made to eliminate marine toxins in shellfish through depuration. A few studies indicated that shellfish could release toxins accumulated in tissues during exposure to clean seawater. However, the efficacy of the process varied substantially among shellfish species. An early study investigating depuration of clams with clean seawater for removing PSP toxins contaminated naturally at >1000 μg/100 g reported that a depuration for three months reduced the toxins in the mantle and gill of the clams by 35% and 52%, respectively (Blogoslawski and Stewart, 1978). Additional studies, which included a treatment with ozonized seawater in the last two weeks of the months of depuration, observed 68% and 76% removals of PSP toxins from the mantle and gill, respectively. However, the amounts of PSP toxins in depurated clams remained at levels higher than the US FDA action level of 80 μg/100 g. Depuration appears to be ineffective in eliminating marine toxins from shellfish.

2.5.1 Factors affecting depuration

Water temperature is a major factor that affects the survival and growth of *V. parahaemolyticus*. The water temperature also affects the water-pumping activity of oysters. It has been reported that the pumping activity of Gulf oysters (*Crassostrea virginica*) was at correspondingly high levels when water temperatures were between 8 and 36 °C (Loosanoff, 1958). Pumping activity gradually decreased when the water temperature dropped to 5 °C. Souness and Fleet (1979) studied the effects of water temperature on the depuration of Sydney rock oyster (*Crassostrea commercialis*) using a dye uptake procedure and found that the pumping rates reached a peak at 25 °C and gradually reduced when water temperatures decreased to 15 °C or increased to 30 °C. A recent study of the depuration of Pacific oysters (*Crassostrea gigas*) at various temperatures reported that depuration at temperatures between 7 and 15 °C reduced *V. parahaemolyticus* populations in oysters by >3.0 log MPN/g after 5 days with no loss of oysters (Phuvasate *et al.*, 2012).

The salinity of seawater may also affect the depuration processes for reducing *V. parahaemolyticus* in oysters. A study of the effects of sodium chloride concentration on the survival of *V. parahaemolyticus* showed that *V. parahaemolyticus* survived better in tryptic soy broth (TSB) containing >3% (particularly 6–9%) NaCl than in TSB without NaCl supplement (Covert and Woodburn, 1972). It has also been reported that oyster's feeding activity was greater in high-salinity seawater than in a low-salinity environment and that oysters appeared to stop their water-feeding activity at salinities below <7.4 parts per thousand (ppt) (Chestnut, 1946; Roderick and Schneider, 1994). Since oysters tend to get used to the salinity of the seawater in their growing environments, it is recommended that the salinity of the water used for depuration should not differ by more than 20% from the salinity of the water in which the oysters are grown and harvested (NSSP, 1990).

Oysters generally have a normal water-pumping rate when the pH is about 7.0, which is also the optimal pH for growth of *V. parahaemolyticus*. An earlier study reported that the pumping activity of oysters was influenced by pH changes and that activity gradually decreased when the pH value in the environment dropped to 6.5 and lower (Loosanoff and Tommers, 1947). The water-pumping rate of oysters was reduced to 10% of its normal level when the pH in the environment dropped to 4.1. Therefore, a decrease in the pH of water in the oyster-growing environments caused by an influx of acid swamp water or industrial pollution will result in a decrease of water-pumping activity by oysters.

2.5.2 Facilities

The facilities of a shellfish depuration plant, including the processing system, should be designed and constructed to prevent contamination of shellfish by animals and other pests during operation and to allow easy cleaning and sanitizing after processing (NSSP, 2009). The grounds of a depuration plant need to be free of conditions that may contribute to shellfish contamination during processing and storage. Properly graded floors constructed with durable and impervious material permit rapid disposal of liquid and solid wastes, which facilitates easy cleaning of the plant (NSSP, 2009). Sand and mud attached to shells may interfere with shellfish depuration and make cleaning the depuration tank difficult after processes. Shellfish depuration plants should have separate tanks for washing shellfish to remove sand and mud before the depuration processes.

2.5.3 Water disinfection

Water quality is a critical factor that influences the shellfish depuration process. Seawater for shellfish depuration should be taken from areas free of chemicals or biotoxins and with low levels of microorganisms. The seawater may have a salinity ranging from 19 to 35 ppt with a turbidity of less than 15 nephelometric turbidity units (NTU) depending on the species to be depurated (FAO, 2008). When seawater free of pollution is not readily available, artificial seawater can be used as an alternative. The water should be filtered or disinfected by ultraviolet light, chlorine, chlorine-containing compounds, ozone or other means to prevent it from becoming a source of bacterial contamination.

Filtration Filtration is a process of removing impurities, such as particles and bacteria, from water based on the pore size of membrane filters. Certain bacteria and viruses can persist in seawater making it a source of contamination during shellfish depuration, particularly in a re-circulating system, if the water is not filtered or sterilized during the process (Di Girolamo *et al.*, 1975). Both flow-through and re-circulating depuration systems can be equipped with a filtration device. The filters need to be cleaned and changed frequently so that they function properly and do not become a source of contamination. A study reported that >95% of poliovirus attached to suspended solids in water was retained on filters with porosities ranging from 3.0 to 0.25 μm (Rao *et al.*, 1984).

Ultraviolet light Ultraviolet (UV) light, generated by mercury-vapour pressure lamps, can inactivate microorganisms by breaking deoxyribonucleic acid (DNA) bonds and causing cell damage. UV light has been used for decontaminating seawater for shellfish purification. Levels of coliforms in seawater were reduced from 17 to <0.18 MPN/ml after being treated with UV light for 72 h (Vasconcelos and Lee, 1972). Another study reported that populations of *V. parahaemolyticus* in seawater decreased from 500 to 1 cfu/ml after seawater was exposed to UV light for 2 h at 25 °C (Greenberg *et al.*, 1982).

UV sterilizers usually have flow-through and re-circulating depuration systems to sterilize seawater for use in shellfish depuration. However, the efficacy of UV radiation in inactivating bacteria in seawater can be influenced by particles suspended in the water. Therefore, it is recommended that a filter is installed with depuration systems, particularly for re-circulating systems so that any sand, mud and faeces released by the shellfish can be removed from the seawater before it is pumped through a UV sterilizer.

Sanitizers Chlorine-containing sanitizers, such as liquid chlorine, hypochlorites and chlorine dioxide, can be used to inactivate microorganisms in water or on food contact surfaces. Chlorine can be added directly to seawater to eliminate or reduce populations of bacteria. Sodium hypochlorite solution was added to seawater to achieve a residual free chlorine concentration of 0.3–0.4 ppm for depuration of soft-shell clams (Wright, 1933). A study using electrolyzed water containing 30 ppm chlorine for oyster depuration reported reductions of *V. parahaemolyticus* and *V. vulnificus* in oysters by 1.13 and 1.05 log MPN/g, respectively, after 4 h of processes (Ren and Su, 2006). However, the presence of chlorine in seawater may be unfavourable for the growth and affect the water-filtering activity of shellfish. Dodgson (1928) reported that chlorine levels ranging from 0.25 to 5 ppm affected normal physiological activities of mussels. Oysters died after being kept in electrolyzed water containing 50 ppm of chlorine for 12 h (Ren and Su, 2006). Therefore, residual chlorine in seawater treated with chlorine-containing sanitizers should be eliminated before the water is used for depuration.

Ozone Ozone exhibits a broad spectrum antibacterial activity and has been used to decontaminate drinking water, recycled poultry chilling water and seawater for shellfish depuration (Broadwater *et al.*, 1973; Hon and Chavin, 1976; Rice *et al.*, 1981; Liltved *et al.*, 1995; Restaino *et al.*, 1995). Populations of *E. coli* and *V. parahaemolyticus* in mussels were reduced by approximately 3.0 and 1.0 log units, respectively, after being depurated in ozonated seawater (0.139 ppm) for 44 h (Croci *et al.*, 2002). A 2-log reduction of *V. vulnificus* was observed in clams depurated in seawater containing ozone (0.6–3.1 ppm) for 24 h (Schneider *et al.*, 1991). A recent study of depuration of naturally contaminated clams with filtered seawater and ozone treatment reported that *E. coli* and coliforms in the claims were reduced to <2.3 and 3.0 MPN/g, respectively, after 24 h of process (Maffei *et al.*, 2009).

Ozone has been used for many years in Europe for treating water for depuration processes (NSSP, 2009). Ozone reacts rapidly with organic components. Therefore, seawater intended to be used in combination with ozone for shellfish depuration should be filtered first to remove most organic materials before being mixed with ozone to allow ozonated water to exhibit strong bactericidal properties.

Sanitation Prompt and proper disposal of shell and wastes from shellfish processing plants is critical for keeping processing environments clean and for minimizing potential contamination of shellfish and contact surfaces. Shellfish depuration tanks and containers for handling processed shellfish need to be frequently cleaned and sanitized. Ice used for storing shellfish should be produced with potable water. All shellfish plants should establish sanitation standard operation procedures (SSOP) for proper sanitation practice and provide adequate toilet and hand washing facilities for employees (NSSP, 2009). All employees should practice good personal hygiene including frequent washing and sanitizing of hands. Any employees who appear to be ill or who have open lesions or infected wounds should be excluded from operations that might cause shellfish contamination.

2.6 Shellfish labelling

In an ideal traceability system, shellfish should be handled in containers with a label containing information about species, date and time of harvest, container number and a barcode. The data should be recorded on a computer and transmitted to the receivers before the shellfish are unloaded from the trucks. The receivers will then handle the shellfish according to the traceable units and attach a new label to each shipping box. Each new label may include the name of the receiver, the size and weight of the shellfish and a new box number. All the information should be entered into a database. The same procedures should be used when the boxes are distributed to a wholesaler and to a retailer.

According to the NSSP (2009), each shellfish harvester will affix a tag to each container of shellstock before transporting the shellstock to a dealer. If the shellstock was harvested at more than one location, each container will be tagged at its growing area. The harvester's tags must be durable, waterproof and sanctioned by the Authority prior to use and be at least 13.8 square inches. The harvester's tag will contain indelible and legible information in the order:

- the harvester's identification number as assigned by the Authority;
- the date of harvest;
- the most precise identification of the harvest location or aquaculture site as is practicable, including the initials of the state of harvest and the Authority's designation of the growing area by indexing, administrative or geographic designation;
- the type and quantity of shellstock; and
- the following statement in bold capitalized type on each tag 'THIS TAG IS REQUIRED TO BE ATTACHED UNTIL CONTAINER IS EMPTY OR IS RETAGGED AND THEREAFTER KEPT ON FILE FOR 90 DAYS.'

For shucked shellfish meats, each individual package of fresh or frozen shellfish meats should include (1) the common name of the shellfish, (2) net contents as net weight, (3) name and address of processor or distributor, (4) 'date shucked' and (5) 'best before' date (CSSP, 2011).

2.7 Conclusion

Molluscan shellfish are filter-feeders and can concentrate toxic substances and microorganisms through filtering water in the growing environments. Shellfish harvested from areas with good quality water generally contain lower levels of chemical and microbial contaminants than those harvested from polluted marine environments. Contaminated molluscan shellfish can serve as vehicles for human diseases, particularly *Vibrio parahaemolyticus* or *Vibrio vulnificus* infections, and need to be processed by feasible post-harvest treatments to ensure the product is safe to eat. Consumers with underlying medical conditions, such as liver disease, diabetes, or immune disorders, are more susceptible to *Vibrio* infection and should avoid eating raw or undercooked shellfish.

Acknowledgements

This work was supported by the Agriculture and Food Research Initiative of the USDA National Institute of Food and Agriculture (Grant Number: 2011-68003-30005), the National Natural Science Foundation of China (No. 31071557) and the National High Technology Research and Development Program of China (No. 2012AA101702).

References

Abad, F.X., Pintó, R.M., Gajardo, R. and Bosch, A. (1997). Viruses in mussels: public health implications and depuration. *Journal of Food Protection*, 60, 677–681.

Andrews, L., Jahncke, M. and Mallikarjunan, K. (2003). Low dose gamma irradiation to reduce pathogenic *Vibrio* in live oysters (*Crassostrea virginica*). *Journal of Aquatic Food Product Technology*, 12, 71–82.

Andrews, L.S., Park, D.L. and Chen, Y.P. (2000). Low temperature pasteurization to reduce the risk of *Vibrio* infections from raw shell-stock oysters. *Food Additives and Contaminants*, 19, 787–791.

Anonymous (2001). *A Total Food Safety Management System Guide for Molluscan Shellfish*. Food Marketing Institute.

Beril, C., Crance, J.M., Leguyader, F. *et al.* (1996). Study of viral and bacterial indicators in cockles and mussels. *Marine Pollution Bulletin*, 32, 404–409.

Blogoslawski, W.J. and Stewart, M.E. (1978). Paralytic shellfish poison in *Spisula solidissima*: anatomical location and ozone detoxification. *Marine Biology*, 45, 261–264.

Blogoslawski, W.J. and Stewart, M.E. (1983). Depuration and public health. *Journal of the World Mariculture Society*, 14, 535–545.

Broadwater, W.T., Hoehn, R.C. and King, P.H. (1973). Sensitivity of three selected bacterial species to ozone. *Applied Microbiology*, 26, 391–393.

Canzonier, W.J. (1971). Accumulation and elimination of coliphage S-13 by the hard clam, *Mercenaria mercenaria*. *Applied Microbiology*, 21, 1024–1031.

CDC (Centers for Disease Control and Prevention) (1998). Outbreak of *Vibrio parahaemolyticus* infections associated with eating raw oysters – Pacific Northwest, 1997. *Morbidity and Mortality Weekly Report*, 47, 457–462.

CDC (Centers for Disease Control and Prevention) (1999). Outbreak of *Vibrio parahaemolyticus* infection associated with eating raw oysters and clams harvested from Long Island Sound – Connecticut, New Jersey and New York, 1998. *Morbidity and Mortality Weekly Report*, 48, 48–51.

CDC (Centers for Disease Control and Prevention) (2006). *Vibrio parahaemolyticus* infections associated with consumption of raw shellfish – Three States, 2006. *Morbidity and Mortality Weekly Report*, 55, 854–856.

CDC (Centers for Disease Control and Prevention) (2012). About norovirus. http://www.cdc.gov/norovirus/about/index.html [accessed 7 June 2013].

CEFAS (Centre for Environment, Fisheries and Aquaculture Science) (2012). Cefas Protocol for the Classification of Shellfish Harvesting Areas – England and Wales. http://www.cefas.defra.gov.uk/media/560824/20120124%20cefas%20external%20classification%20protocol%20version%205.pdf [accessed 7 June 2013].

Chen, H. (2007). Use of linear, Weibull, and log-logistic functions to model pressure inactivation of seven foodborne pathogens in milk. *Food Microbiology*, 24, 197–204.

Chestnut, A.F. (1946). Some observations on the feeding of oysters with special reference to the tide. *Proceedings of the National Shellfisheries Association*, pp. 22–27.

Colwell, R.R. and Liston, J. (1960). Microbiology of shellfish: bacteriological study of the natural flora of Pacific oysters (*Crassostrea gigas*). *Applied Microbiology*, 8, 104–109.

Cook, D.W. (2003). Sensitivity of *Vibrio* species in phosphate-buffered saline and in oysters to high-pressure processing. *Journal of Food Protection*, 66, 2276–2282.

Costantini, V., Loisy, F., Joens, L., Le Guyader, F.S. and Saif, L.J. (2006). Human and animal enteric caliciviruses in oysters from different coastal regions of the United States. *Applied and Environmental Microbiology*, 72, 1800–1809.

Covert, D. and Woodburn, M. (1972). Relationships of temperature and sodium chloride concentration to the survival of *Vibrio parahaemolyticus* in broth and fish homogenate. *Applied Microbiology*, 23, 321–325.

Croci, L., Suffredini, E., Cozzi, L. and Toti, L. (2002). Effects of depuration of molluscs experimentally contaminated with *Escherichia coli, Vibrio cholerae* O1 and *Vibrio parahaemolyticus*. *Journal of Applied Microbiology*, 92, 460–465.

CSSP (Canadian Shellfish Sanitation Program) (2011). *Manual of Operations (version 2)*, Canadian Food Inspection Agency, Fisheries and Oceans Canada and Environment Canada. http://www.inspection.gc.ca/english/fssa/fispoi/man/cssppccsm/cssppccsme.shtml [accessed 7 June 2013].

DFO (Fisheries and Oceans Canada) (1990). *Management of Contaminated Fisheries Regulations*. http://www.dfo-mpo.gc.ca/acts-loi-eng.htm [accessed 30 May 2013].

Di Girolamo, R., Liston, J. and Matches, J. (1975). Uptake and elimination of poliovirus by West coast oysters. *Applied Microbiology*, 29, 260–264.

Dodgson, R.W. (1928). *Report on Mussels Purification. Fisheries Investigations. Series II*, His Majesty's Stationary Office, London.

Doré, B., Keaveney, S., Flannery, J. and Rajko-Nenow, P. (2010). Management of health risks associated with oysters harvested from a norovirus contaminated area, Ireland, February–March 2010. *Eurosurveillance*, 15, 1–4.

Eisenberg, M. and Topping, J.J. (1984). Trace metal residues in shellfish from Maryland waters, 1976–1980. *Journal of Environmental Science and Health, Part B*, 19, 649–671.

Eyles, M.J. and Davey, G.R. (1984). Microbiology of commercial depuration of Sydney rock oyster, *Crassostrea commercialis*. *Journal of Food Protection*, 47, 703–706.

Fang, Z.Q., Cheung, R.Y.H. and Wong, M.H. (2003). Heavy metals in oysters, mussels and clams collected from coastal sites along the Pearl River Delta, South China. *Journal of Environmental Sciences*, 15, 9–24.

FAO (2008). Bivalve depuration: fundamental and practical aspects. *FAO Fisheries Technical Paper No. 511*. Food and Agricultural Organization of the United Nations, Rome.

FAO (2010). *Fishery Statistical Collections: Consumption of Fish and Fishery Products*, Food and Agricultural Organization of the United Nations, Rome. http://www.fao.org/fishery/statistics/global-consumption/en [accessed on 30 May 2013].

FAO (2012). *Production statistics: Global Production Statistics 1950–2010*, Food and Agricultural Organization of the United Nations, Rome. http://www.gfcm.org/gfcm/topic/17105/en [accessed 30 May 2013].

Fiore, A.E. (2004). Hepatitis A transmitted by food. *Clinical Infectious Diseases*, 38, 705–715.

Fleet, G.H. (1978). Oyster depuration – a review. *Food Technology*, 30, 444–454.

Greenberg, E.P., Duboise, M. and Palhof, B. (1982). The survival of marine vibrios in *Mercenaria mercenaria*, the hardshell clam. *Journal of Food Safety*, 4, 113–123.

He, H., Adams, R.M., Farkas, D.F. and Morrissey, M.T. (2002). Use of high-pressure processing for oyster shucking and shelf-life extension. *Journal of Food Science*, 67, 640–644.

Hon, K.V. and Chavin, W. (1976). Utility of ozone treatment in the maintenance of water quality in a closed marine system. *Marine Biology*, 34, 201–209.

IIR (2008). RFID technologies for cold chain applications. 4th Informatory Note on Refrigeration and Food. International Institute of Refrigeration, Paris, France.

Jakabi, M., Gelli, D.S., Torre, J.C.M.D. *et al.* (2003). Inactivation by ionizing radiation of *Salmonella* enteritidis, *Salmonella* infantis, and *Vibrio parahaemolyticus* in oyster (*Crassostrea brasiliana*). *Journal of Food Protection*, 66, 1025–1029.

Jeffery, B., Balow, T., Moizer, K., Paul, S. and Boyle, C. (2004). Amnesic shellfish poisoning. *Food and Chemical Toxicology*, 42, 545–557.

Jiann, K.-T. and Presley, B.J. (1997). Variations in trace metal concentrations in American oysters (*Crassostrea virginica*) collected from Galveston Bay, Texas. *Estuaries*, 20, 710–724.

Johnson, M.D. and Brown, M.H. (2002). An investigation into the changed physiological state of *Vibrio* bacteria as a survival mechanism in response to cold temperatures and studies on their sensitivity to heating and freezing. *Journal of Applied Microbiology*, 92, 1066–1077.

Kaysner, C.A. and DePaola, A. (2000). Outbreaks of *Vibrio parahaemolyticus* gastroenteritis from raw oyster consumption: assessing the risk of consumption and genetic methods for detection of pathogenic strains. *Journal of Shellfish Research*, 19, 657.

Kohn, M.A., Farley, T.A., Ando, T. *et al.* (1995). An outbreak of Norwalk virus gastroenteritis associated with eating raw oysters. Implications for maintaining safe oyster beds. *Journal of the American Medical Association*, 273, 466–471.

Kural, A.G., Shearer, A.E.H., Kingsley, D.H. and Chen, H. (2008). Conditions for high pressure inactivation of *Vibrio parahaemolyticus* in oysters. *International Journal of Food Microbiology*, 127, 1–5.

Lewis, G.D. and Metcalf, T.G. (1988). Polyethylene glycol precipitation for recovery of pathogenic viruses, including hepatitis A virus and human rotavirus, from oyster, water, and sediment samples. *Applied and Environmental Microbiology*, 54, 1983–1988.

Liltved, H., Hektoen, H. and Efraimsen, H. (1995). Inactivation of bacterial and viral fish pathogens by ozonation or UV irradiation in water of different salinity. *Aquacultural Engineering*, 14, 107–122.

Lipp, E.K., Farrah, S.A. and Rose, J.B. (2001). Assessment and impact of microbial faecal pollution and human enteric pathogens in a coastal community. *Marine Pollution Bulletin*, 42, 286–293.

Liu, C., Lu, J. and Su, Y.-C. (2009). Effects of flash freezing, followed by frozen storage, on reducing *Vibrio parahaemolyticus* in Pacific raw oysters (*Crassostrea gigas*). *Journal of Food Protection*, 72, 174–177.

Loosanoff, V.L. (1958). Some aspects of behavior of oysters at different temperatures. *Biological Bulletin*, 114, 57–70.

Loosanoff, V.L. and Tommers, F.D. (1947). Effect of low pH upon the rate of water pumping of oysters, *Ostrea virginica. The Anatomical Record*, 99, 668.

Ma, L. and Su, Y.-C. (2011). Validation of high pressure processing for inactivating *Vibrio parahaemolyticus* in Pacific oysters (*Crassostrea gigas*). *International Journal of Food Microbiology*, 144, 469–474.

Maffei, M., Vernocchi, P., Lanciotti, R. *et al.* (2009). Depuration of striped Venus clam (*Chamelea gallina* L.): effects on microorganisms, sand content and mortality. *Journal of Food Sci* 74, M1–7.

McLaughlin, J.B., DePaola, A., Bopp, C.A., Martinek, K.A. and Napol, N.P. (2005). Outbreak of *Vibrio parahaemolyticus* gastroenteritis associated with Alaskan oysters. *New England Journal of Medicine*, 353, 1463–1470.

Motes, M.L. and DePaola, A. (1996). Offshore suspension relaying to reduce levels of *Vibrio vulnificus* in oysters (*Crassostrea virginica*). *Applied and Environmental Microbiology*, 62, 3875–3877.

NOAA (2010). *US Seafood Consumption Declines Slightly in 2009*, National Oceanic and Atmospheric Administration. http://www.noaanews.noaa.gov/stories2010/20100909_consumption.html [accessed 30 May 2013].

NOAA (2011). *US Domestic Seafood Landings and Values Increase in 2010*, National Oceanic and Atmospheric Administration. http://www.noaanews.noaa.gov/stories2011/20110907_usfisheriesreport.html [accessed 30 May 2013].

Nordstrom, J.L., Kaysner, C.A., Blackstone, G.M. *et al.* (2004). Effect of intertidal exposure on *Vibrio parahaemolyticus* in Pacific Northwest oysters. *Journal of Food Protection*, 67, 2178–2182.

NSSP (National Shellfish Sanitation Program) (1990). *Manual of Operations, Part I: Sanitation of Shellfish Growing Areas*, Public Health Service, US Food and Drug Administration, Washington DC.

NSSP (National Shellfish Sanitation Program) (2009). *Guide for the Control of Molluscan Shellfish (2009 Revision)*, US Food and Drug Administration. http://www.fda.gov/downloads/Food/GuidanceRegulation/FederalStateFoodPrograms/UCM350004.pdf [accessed 3 June 2013].

Oliver, J.D. (1989). *Vibrio vulnificus*, in *Foodborne Bacterial Pathogens* (ed. M.P. Doyle), Marcel Dekker, New York, pp. 569–599.

Phuvasate, S., Chen, M.H. and Su, Y.-C. (2012). Reductions of *Vibrio parahaemolyticus* in Pacific oysters (*Crassostrea gigas*) by depuration at various temperatures. *Food Microbiology*, 31, 51–56.

Presley, B.J., Taylor, R.J. and Boothe, P.N. (1990). Trace metals in gulf on Mexico oysters. *Science of the Total Environment*, 97, 551–593.

Rao, V.C., Seidel, K.M., Goyal, S.M., Metcalf, T.G. and Melnick, J.L. (1984). Isolation of enteroviruses from water, suspended solids, and sediments from Galveston Bay: survival of poliovirus and rotavirus adsorbed to sediments. *Applied and Environmental Microbiology*, 48, 404–409.

Ren, T. and Su, Y.-C. (2006). Effects of electrolyzed oxidixing water treatment on reducing *Vibrio parahaemolyticus* and *Vibrio vulnificus* in raw oysters. *Journal of Food Protection*, 69, 1829–1834.

Restaino, L., Frampton, E. W., Hemphill, J. B. and Palnikar, P. (1995). Efficacy of ozonated water against various food-related microorganisms. *Applied and Environmental Microbiology*, 61, 3471–3475.

Rice, R.G., Robson, C.M., Miller, G.W. and Hill, A.G. (1981). Use of ozone in drinking water treatment. *Journal American Water Works Association*, 73, 44–57.

Richards, G.P. (1987). Shellfish-associated enteric virus illness in the United States, 1934–1984. *Estuaries*, 10, 84–85.

Roderick, G.E. and Schneider, K.R. (1994). Depuration and relaying of molluscan shellfish, in *Environmental Indicators and Shellfish Safety* (eds C. R. Hackney, and M. D. Person), Chapman & Hall, New York, pp. 331–363.

Schneider, K.R., Steslow, F.S., Sierra, F.S., Rodrick, G.E. and Noss, C.I. (1991). Ozone depuration of *Vibrio vulnificus* from the southern quahog clam, *Mercenaria campechiensis*. *Journal of Invertebrate Pathology*, 57, 184–190.

Souness, R.A. and Fleet, G.H. (1979). Depuration of the Sydney rock oyster, *Crassostrea commercialis*. *Food Technology in Australia*, 31, 397–404.

Tamplin, M.L. and Capers, G.M. (1992). Presistence of *Vibrio vulnificus* in tissues of Gulf coast oysters, *Crassostrea virginica*, exposed to seawater disinfected with UV light. *Applied and Environmental Microbiology*, 58, 1506–1510.

USFDA (United States Food and Drug Administration) (1995). *Sanitary Surveys of Shellfish Growing Areas – Training Course Source Book*, US Department of Health and Human Services, Northeast Technical Service Unit, Shellfish Program Implementation Branch, CBC, North Kingstown, Rhode Island.

USFDA (United States Food and Drug Administration) (2012a). *Bad Bug Book – Foodborne Pathogenic Microorganisms and Natural Toxins*, 2nd edn. http://www.fda.gov/Food/FoodborneIllnessContaminants/CausesOfIllnessBadBugBook/ucm2006773.htm [accessed 3 June 2013].

USFDA (United States Food and Drug Administration) (2012b). *Various Shellfish-Associated Toxins. Foodborne Pathogenic Microorganisms and Natural Toxins Handbook* http://www.fda.gov/Food/FoodborneIllnessContaminants/CausesOfIllnessBadBugBook/ucm070795.htm [accessed 3 June 2013].

Vasconcelos, G.J. and Lee, J.S. (1972). Microbial flora of Pacific oysters (*Crassostrea gigas*) subjected to ultraviolet-irradiated seawater. *Applied Microbiology*, 23, 11–16.

Wright, E. (1933). Public health engineering: shellfish treatment plant at Newburyport, Mass. *American Journal of Public Health and the Nations Health*, 23, 266–270.

3
Chilling and Freezing of Fish

Flemming Jessen,[1] Jette Nielsen[1] and Erling Larsen[2]

[1]DTU Food, National Food Institute, Division of Industrial Food Research, Technical University of Denmark, Kgs. Lyngby, Denmark

[2]DTU Aqua, National Institute of Aquatic Resources, Technical University of Denmark, Charlottenlund, Denmark

3.1 Introduction

Fish is a highly perishable commodity and must be preserved by chilling, freezing or some other means before it is consumed or processed. Chilling and even freezing was used as early as 1000 BC by the Chinese in ice cellars beyond the winter months and both the Greeks and the Romans stored compressed snow in insulated cellars. It was as late as 1913 that the first industrial patent was taken out by Sabroe A/S and A.J.A. Ottesen for using circulated brine at −17 °C (Jul, 1984). In 1930, Clarence Birdseye and his American company made frozen foods a practical reality by the successful creation of the double belt freezer. But the road to consumer acceptance of frozen products has been long and even now fresh fish is considered of superior quality. It is, however, possible to produce frozen fish of high eating quality by freezing the fish quickly and subsequently storing it at low and stable temperatures.

The latest available statistics from the Food and Agriculture Organization (FAO, 2012) state that 85% (131 million tonnes) of world fish production was used for human consumption, while the rest was destined for non-food purposes such as bait, fishmeal and fish oil for feed and pharmaceutical uses as well as for direct feeding in aquaculture and for fur animals. In 2010, fish marketed live or fresh (chilled) for human consumption was the most important product, with a share of 40.5%, followed by frozen fish (29.3%) (Figure 3.1).

The actual percentage of fish that is frozen during the processing chain is higher than 25.4% because thawed raw material is often used for preservation (e.g. canned mackerel and tuna) and curing (salted, smoked and marinated fish). Even fish marketed as fresh might have been frozen; in some countries fish for sushi must be frozen in 24 h before consumption due to the risk of nematodes, or the fish is sold in the supermarkets as 'refreshed'. Fish used for non-food purposes is also chilled or frozen before processing.

Seafood Processing: Technology, Quality and Safety, First Edition. Edited by Ioannis S. Boziaris.

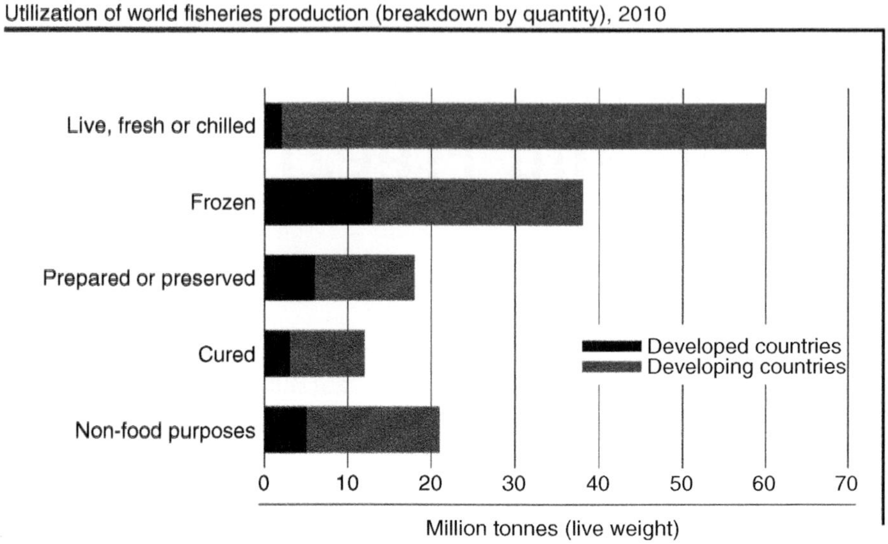

Figure 3.1 Utilization of world fisheries production (*Source*: FAO (2012). Reproduced with the permission of the Food and Agriculture Association of the United Nations.)

This chapter will attempt to give an overview of chilling and freezing of fish and the factors which influence product quality. Due to the diversity of the subject, detailed reviews will not be given, but key research papers will be cited in order to enable the reader to find more information.

Model supply chains for both chilling and freezing will be presented together with advice on suitable conditions for obtaining good product quality.

3.2 Post-mortem changes at chilled storage temperatures

A number of chemical reactions start immediately after the death of the fish. These include reactions that are responsible for how fish quality changes during chilled storage. Here we describe briefly the post-mortem processes predominantly influencing fish quality.

3.2.1 Rigor mortis

Death terminates the supply of oxygen to the muscle so the ongoing aerobic generation of ATP will close down as the oxygen present is quickly used up. The muscles attempt to keep 'alive' by trying to maintain the ATP level, initially by using the phosphocreatine stores, but soon after the ATP originates from the anaerobic glycolysis of glycogen present in the muscle. An outcome of this process is the formation of lactic acid and a concomitant drop in pH. Glycolysis will proceed as long as glycogen is present in the muscle or until the pH has reached such a low value that it decreases the activity of the glycolytic enzymes. As this happens, the ATP concentration in the muscle falls and when it is below certain values (1–2 μmol/g muscle), there is insufficient ATP to break the bindings between actin and myosin filaments (acto-myosin is formed), thus, the filaments will stay cross linked and the muscle becomes increasingly stiff and enters rigor mortis (Iwamoto *et al.*, 1988). After a while rigor will slowly resolve as the fish loses stiffness and the muscle becomes softer again. The exact mechanism of this muscle softening is still not known in fish. However, it is generally supposed that possible causes include proteolytic enzymes acting on the myofilaments either at the Z-disks (Seki and Tsuchiya, 1991) or at the myosin–actin junctions (Yamanoue and Takahashi, 1988), but a degradation of connective tissue (Ando *et al.*, 1993) is also considered to be involved.

The progress of rigor mortis varies between species and the rates of onset and resolution are affected by many other factors apart from temperature, such as handling (e.g. pumping, crowding), size, physical condition and the biological state of the fish (Huss, 1995). In general, high temperature results in a fast onset of rigor and vice versa. However, this seems only to be true for cold-water fish, whereas in fish living at higher temperatures an accelerated onset of rigor is also seen at 0 °C (Abe and Okuma, 1991). A more general rule could therefore be that the time from death to onset of rigor depends on the difference between the ambient temperature and the storage temperature, independent of which of the temperatures is the highest. The bigger this temperature difference, the stronger the muscle contraction and the shorter the time before the onset of rigor. Storage temperatures resulting in a fast and very strong rigor mortis should be avoided as strong rigor tensions can cause gaping – that is ruptures in the connective tissue of the fillet (Lavety *et al.*, 1988) – and it is also associated with an accelerated tenderization of the muscle (Figure 3.2).

This softening seems not to be a direct consequence of rigor resolution (Ando *et al.*, 1991) but is rather due to other proteolytic events. Furthermore, it is questionable whether rigor mortis development as such is responsible for the activation of the proteases involved in softening of the texture or if this is initiated by other mechanisms (Roth *et al.*, 2006).

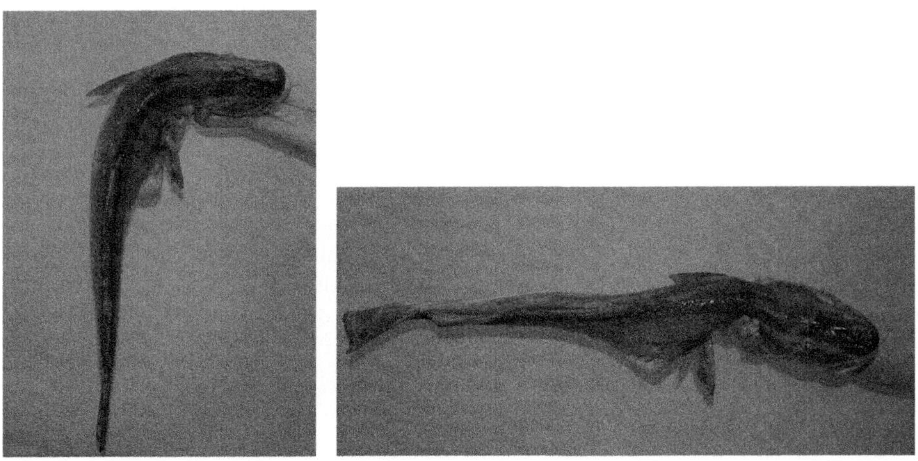

Figure 3.2 Rigor mortis. Post mortem, a relaxed pre-rigor cod (left panel) enters the hard and stiff rigor state (right panel) as the ATP concentration in the muscle falls to below 1–2 μmol/g muscle. The time from death to onset of rigor, and the strength of rigor depends strongly on the difference between the ambient temperature and the storage temperature. The bigger this temperature difference, the stronger the muscle contraction and the shorter the time before the onset of rigor

3.2.2 Protein changes

The post-mortem degradation of proteins is one of the most important processes influencing the textural quality of fish muscle (Delbarre-Ladrat *et al.*, 2006) and these autolytic events start shortly after death. Although there are great differences between fish species, some of the first proteins to be degraded within 1–2 days post mortem are dystrophin and other proteins linking the myofibrils to the cell membrane (Papa *et al.*, 1996b; Caballero *et al.*, 2009). Other structural proteins are also degraded by proteolytic enzymes in muscle from different fish species: titin (Seki and Watanabe, 1984), nebulin (Astier *et al.*, 1991), α-actinin (Papa *et al.*, 1996a), myosin (Wu *et al.*, 2010) and tropomyosin (Astier *et al.*, 1991). An unknown number of proteolytic enzymes, present in the muscle, are probably activated during the early post-mortem period when the normal physiological control of their activities gradually vanishes, that is proteases are released from cellular organelles, activated by Ca^{2+} leaking from stores or by pH changes. The mechanisms for the post-mortem protein degradation in fish muscle are far from understood, but it is generally accepted that different proteinases from the protease families of cathepsins and calpains are involved. A synergistic interaction is suggested by Delbarre-Ladrat *et al.* (2006) to the effect that cathepsins and calpains act in a complementary way at different levels in the breakdown of myofibrillar proteins. In meat there are indications that the proteolytic proteasome complex is involved in the ongoing degradation of peptides formed by other proteases (Houbak *et al.*, 2008), but this has not been shown in fish. As mentioned earlier (see 3.2.1) the connective tissue of fish muscle is also degraded post mortem and again the activity of cathepsins seems to be important (Sato *et al.*, 1997) but the involvement of other collagenases as matrix metallopeptidases has also been described (Kubota *et al.*, 2001).

Recently, investigations on how muscle is converted to meat has increasingly focused on the importance of the early physiological processes of programmed cell death (apoptosis) in contributing to the tenderization of meat through the activity of the protease family of caspases (for reviews see Herrera-Mendez *et al.*, 2006; Kemp *et al.*, 2010).

We think that evidence for the involvement of caspases in the early stages of the post-mortem protein changes that cause the softening of fish muscle has only appeared once in the literature (Ishida *et al.*, 2003). When no oxygen is present in the muscle cells post mortem, the cells will sense an ischaemic-like state that in many cell types is known to induce apoptosis and therefore it is possible that apoptosis has an important role in fish muscle softening. However, it has to be proved and this is an important challenge for future research in post-mortem alteration of fish muscle. Altogether this could suggest that post-mortem fish muscle softening is due to a sequential degradation of muscle proteins as illustrated in Figure 3.3.

3.2.3 Lipid changes

In live fish, the lipids are quite stable due to a balance between antioxidants and pro-oxidants. Several post-mortem events, including autolytic processes, favour the initiation of lipid hydrolysis and oxidation that produces a range of substances (Undeland, 1997). Some of these have a rancid taste and smell and some contribute to protein denaturation by binding to the proteins. These processes can include the increase or activation of pro-oxidants, the loss or inactivation of antioxidants, the activation of enzymes and

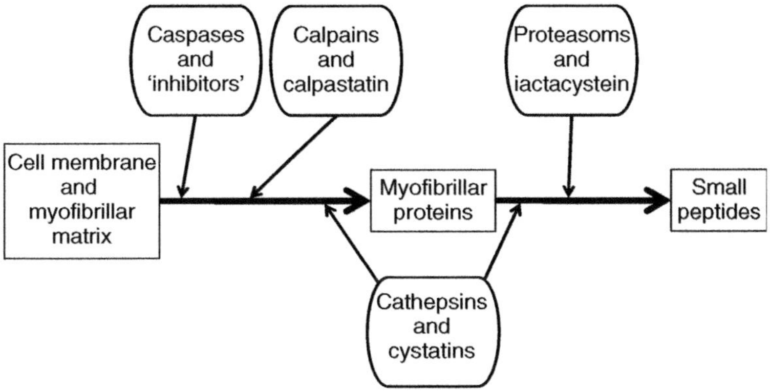

Figure 3.3 Sequential post-mortem degradation of fish muscle proteins (*Source*: Ishida *et al.*, 2003; Delbarre-Ladrat *et al.*, 2006; Houbak *et al.*, 2008)

the disintegration of membranes making them more susceptible to oxidation (Nielsen *et al.*, 2001).

3.2.4 Microbial changes

The bacterial influence on the chilled fish also starts just after death, but the effect on spoilage through the biochemical changes induced by the bacterial growth is not pronounced until the specific spoilage organisms have increased to a certain level (Gram and Huss, 1996). In addition, as microbial counts increase, enzymes secreted by microorganisms may also cause additional softening of the fish (Nielsen *et al.*, 2001).

3.3 Effect of freezing temperatures on quality-related processes

The rates of post-mortem processes that reduce quality are temperature dependent and these reaction rates decrease, in general according to the Arrhenius relation, as temperature is lowered. Thus, we need to know what mechanisms are involved in the diverse changes that occur during freezing and frozen storage so that the quality of frozen fish products can be maximised. As huge differences exist between fish species on the extent and rate of the various post-mortem processes, which also depend on fish handling, both before and after death. As these processes are also influenced differently by freezing temperatures, it is evident that fish species will behave differently as to how their quality changes during frozen storage.

3.3.1 The freezing process

From water to ice In muscle from other animals as well as in fish muscle, the water content is quite large and a water fraction of around 80% is not uncommon. During freezing, the core temperature of the fish quickly falls to just below 0 °C (Figure 3.4). At around −1 °C

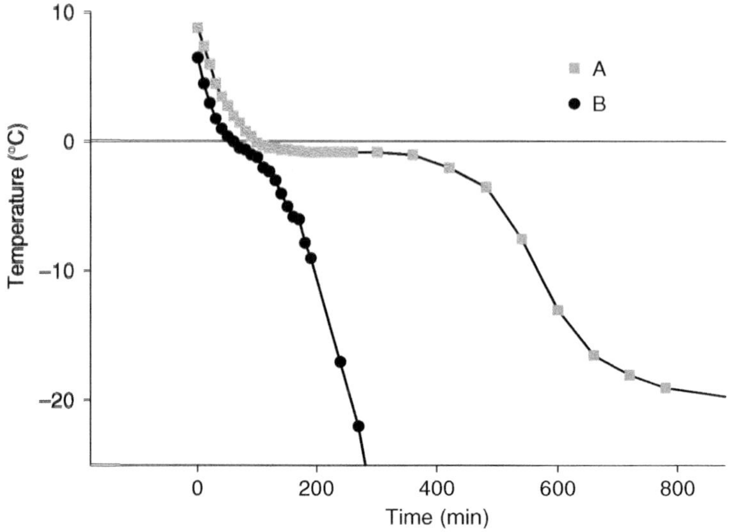

Figure 3.4 Temperature profiles during freezing of whole cod. A: Freezing at −20 °C in still air (home freezer). B: Freezing at −45 °C in air blast freezer (industrial freezer). The temperature was measured in the centre of the fish (core temperature). It is apparent from the figure that the fish can be brought fast through the critical temperature interval, 0 to −5 °C, when quick freezing takes place at a low temperature and when a fan at the same time makes sure that the heat from the fish is removed quickly from the surface (*Source*: Nielsen and Jessen, 2007. Reproduced with the permission of John Wiley and Sons.)

the water changes from liquid to solid as ice crystals are formed. As most of the water in the muscle turns to ice, the heat from this exothermic process has to be removed from the fish and therefore the temperature in the fish falls only slowly during this period of freezing. When the water capable of freezing is actually frozen, the temperature starts again to drop rapidly, until the preset temperature is reached. A part of the water will remain unfrozen due to a combination of the high freeze-concentrated amount of small solutes, the hydration of macromolecules as proteins, the small distances between structural elements and the very high viscosity. This unfrozen water fraction in the muscle constitutes the basis for the chemical quality deteriorating processes occurring in frozen fish.

Formation of ice crystals Whether the fish is frozen before, during, or after rigor mortis, as well as the speed of freezing has a great influence on where in the muscle tissue ice crystals are formed and on what their size will be. In pre-rigor fish muscle, the water is almost exclusively present inside the muscle cells and during freezing of such a fish the ice crystals will therefore be formed inside the cells. Independent of the freezing rate, the ice crystals will in general be small, but if the freezing process is very slow (e.g. in still air above −18 °C) large ice crystals will also be formed. In either in-rigor or post-rigor muscle, the water will also be present outside the cells and the speed of freezing will have a much greater influence on the formation of ice crystals. During quick freezing, small ice crystals will be formed both inside and outside the cells, whereas slow freezing will induce ice crystal formation outside the cells before they are formed inside the cells. The resulting higher extracellular salt concentration will create an osmotic drag of water from the cells and ice crystals already

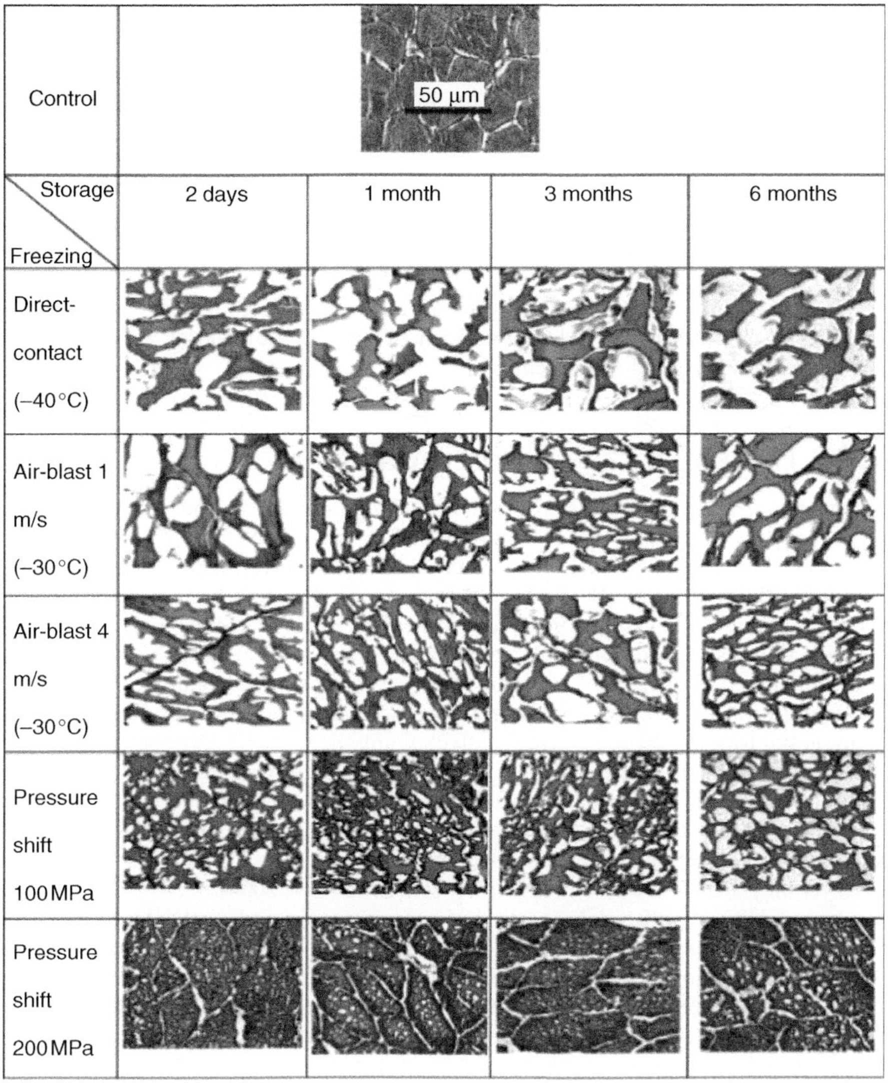

Figure 3.5 Ice crystals in muscle as influenced by freezing procedure and frozen storage time. It is evident that the size of ice crystals can be very much influenced by pressure (*Source*: Alizadeh *et al.*, 2007. Reproduced with permission of Elsevier.)

formed outside the cells will grow rather than allowing formation of new ice crystals. Thus, the speed at which in-rigor and post-rigor fish are frozen determines both the size and location in the muscle of the ice crystals, ranging from many small ice crystals both inside and outside the cells to fewer large ice crystals only located outside the cells (Figure 3.5).

At the frozen storage temperature, ice crystals still change in size and in distribution. Not many new ice crystals are formed, but those already present undergo recrystallization. In this process the surface of the primarily small ice crystals melts and the water formed subsequently freezes to make larger ice crystals. Fluctuating storage temperatures very much promote this recrystallization process, even when the fluctuations in question are only a few degrees.

The formation of ice within the muscle has of course an impact on the muscle and this can affect the quality of the thawed fish; the extent of which depends on where in the fish muscle the ice crystals develop and on their size. It is generally accepted that small ice crystals are the least detrimental to fish quality during frozen storage and this makes pre-rigor freezing very attractive due to the formation of small ice crystals independent of the freezing rate. Thus, when it comes to post-rigor fish, it is preferable to decrease the size of the ice crystals with a high freezing rate as this also minimizes the amount of ice formed outside the muscle cells. This extracellular ice will drain from the fish muscle during thawing (thaw drip or drip loss) and during cooking (boiling loss), meaning the fish will be drier when prepared and eaten.

3.3.2 Frozen storage temperatures

Freezing and frozen storage can never improve the quality of fish. Therefore the primary quality properties of fresh fish are essential to the final quality of frozen fish and the initial quality can be maintained within limits if catch handling and freezing processes are carried out properly (Boknaes *et al.*, 2001).

As mentioned earlier, the conversion of water into ice means that the soluble substances, such as salt and enzymes, in the muscle are present in high concentrations in the remaining 'non-freezable' water, which can affect the quality of the frozen fish. Denaturation of proteins and several chemical reactions, such as lipid oxidation, proteolysis and formaldehyde formation, affect quality negatively. Although the chemical reaction rates in general are reduced at lower temperatures, the effect on enzyme catalysed reactions of higher concentrations of substrates and enzymes in the unfrozen water might counteract this and keep the reactions running even at very low temperatures. Quality reducing reactions have been documented to occur at temperatures as low as −30 °C, yet they are only running very slowly (Hiraoka *et al.*, 2004).

It is unavoidable that fish deteriorate during frozen storage; however, the nature and extent of this deterioration varies between species and is especially dependent on the content and distribution of fat and protein. In lean fish species such as cod, a major quality reducing process is denaturation and aggregation of proteins (Mackie, 1993), resulting in loss of functional properties such as water holding capacity and consequentially resulting in a tough and dry texture in the prepared fish. In fatty fish species like herring, the main factors responsible for quality loss are oxidative changes of the lipids and the associated development of rancid flavour (Hultin, 1992). Rancidity is not always a problem as some fatty fish species like salmon can actually maintain a commercially acceptable quality for about 25 months. However, this demands that only the highest quality of fish is used. Furthermore, the fish has to be frozen shortly after slaughter, handled and frozen by use of best practice and afterwards glazed and vacuum packed before storage at a stable temperature of −30 °C or lower (Sørensen *et al.*, 1996).

Several factors are of importance for the protein changes occurring during frozen storage (Figure 3.6). Apart from the direct physical effect on the proteins of ice crystal formation, this formation also leads to the partial dehydration of proteins. Also, proteins are denatured due to interaction with salts in high concentration, oxidised lipids, free fatty acids, products from lipid oxidation and oxidative reactants interacting with, for example, thiol groups (Haard, 1992; Mackie, 1993).

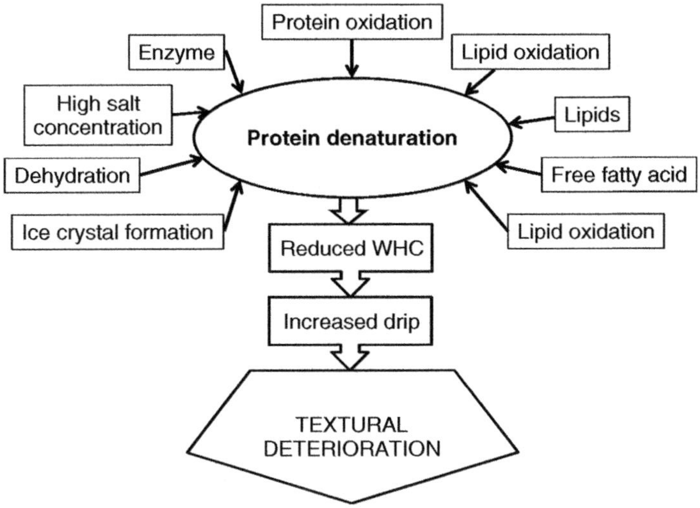

Figure 3.6 Factors influencing fish quality (texture) by a protein denaturizing effect (*Source:* Shenouda, 1980. Reproduced with permission of Elsevier.)

Gadoid species contain the enzyme trimethylamine-oxide aldolase (TMAOase; EC 4.1.2.32) that has higher activities at freezing temperatures than at temperatures above 0 °C. The activity of the TMAOase is maximal around −10 °C and decreases to very low values when the temperature approaches −30 °C (Howell *et al.*, 1996). The TMAOase catalyses a very quality reducing reaction, namely the degradation of trimethylamine oxide (TMAO) to formaldehyde and dimethylamine (DMA) leading to protein denaturation and extensive textural changes (Babbitt *et al.*, 1972; Sotelo *et al.*, 1995; Nielsen and Jorgensen, 2004). Formaldehyde is considered to be more detrimental to the proteins than DMA, but it is not yet fully understood how these components lead to tough, hard, fibrous and dry fish meat (LeBlanc and LeBlanc, 1988; Chapman *et al.*, 1993; Burgaard and Jorgensen, 2010).

When pre-rigor fish are frozen, the progress of rigor is stopped and there will still be ATP and glycogen present in the frozen muscle. This can generate a problem during thawing where the presence of ATP and the potential for synthesizing even more ATP from glycogen creates the energetic basis for a strong muscle contraction when the rigor process is resumed. The muscle contraction is especially strong because the concentration of Ca^{2+} (contraction activator) is very high as freezing induces a massive release of Ca^{2+} from the cellular stores (sarcoplasmatic reticulum). The phenomenon is called thaw-rigor and it can cause an increased thaw-drip (Jones, 1965) and lead to a hard texture. Although the exact mechanism of thaw-rigor is still unclear, much is already known about the processes involved (Cappeln and Jessen, 2001; Imamura *et al.*, 2012), which might make it possible to improve the process of thawing pre-rigor frozen fish in the future so they do not exhibit thaw-rigor (see 3.5.3 Thawing).

3.4 Fresh fish chain

Fish has been consumed fresh for millennia and there has always been a demand for methods to keep the surplus of fish fresh. Due to the relatively short shelf life of fish, different preservation methods have been developed. Traditionally only a few methods can be used for preserving the so-called fresh fish: either alive in a cage or container with water or by the use of ice. Production of ice is quite easy in the temperate climate zones – during wintertime Mother Nature takes over the production. But to keep the ice in the summertime is much more difficult. Special houses for storing ice were originally constructed and the ice was among other things used for keeping fish fresh. Before the era of electricity it was possible to produce ice, using salt brine, but only in small quantities. The general introduction of electricity made it possible to install ice making machinery, first as 'crystal ice' in blocks, that was crushed to use in fish crates or boxes – the larger the surface the more melting water. It is the melting water that cools the product, not the ice.

There are various advantages of using ice. Ice at a temperature of 0 °C needs 385 kJ/kg in order to melt into cold water, thus drawing that amount of heat from the surroundings. Ice has a very high cooling capacity, ice can maintain a very definite temperature, ice is harmless if the water source is harmless and water from melting ice can reach the whole batch of fish in a crate (FAO, 2003). And you can add the following economic benefits such as: ice is nowadays available in nearly all fishing ports, ice is relatively cheap, ice can be moved from place to place and is easily stored onboard fishing vessels and with today's techniques ice can be made by relatively small machinery that can be installed onboard the fishing vessels.

The development of transportation more than a century ago, with the building of railroad systems, made it possible to transport fresh iced fish over long distances, 1500 km in 48 h was not uncommon. So a fish that was caught on Monday in the northern North Sea, could be sold in Paris on Thursday/Friday. With an unbroken cooling chain, the fish is a very good fresh product. Today, transportation by lorry is not faster than by rail, but the transportation before the first train station and the last train station is much quicker. The logistics of transportation is now an independent scientific discipline.

Fish is a global commodity. Fresh fish as air cargo has been one of the fastest growing business in the past 15 years. Increasing amounts of fresh fish are flown into the European Union (EU) each year. The import to Japan of live fish and shellfish is steadily growing. So the fresh fish chain is not a single one, but is many chains, with many variables. Generally the chains can be broken down to the following main links/notes: (i) the fishing operation/primary production, (ii) landing and first sales, (iii) processing (mainly filleting), (iv) retailer, and (v) consumer. These links can also be regarded as levels and in between these there are the logistic activities, either as transportation or storage (Figure 3.7).

The increasing focus on traceability in the past 15 years has shown that the links in the fish chains often act independently of each other. Relevant information is lost in the chain, or the batch sizes are too big to handle in a situation where it is needed (Larsen and Villerreal, 2009). From 1 January 2012, the EU Commission has put into action a new legislation for fresh fish that tightens the demands on individual links in the chain, compared to the rest of the food sector in the EU (EU, 2011).

3.4.1 Handling and processing on board fish vessels

Fresh fish trawlers are still the backbone of the fishing fleets from major fishing countries. Their prime started in the 1950s and has continued until today. Fresh fish is landed from a variety of fishing operations. Trawl fishing belongs the group of activities that can be called active fishing alongside seining, from Danish seining over Scottish seining to purse seining. Different species are targeted, but all depend on cooling the catch down to around 0 °C. The passive fishing activities of gillnetting, long-lineing, pot fishing and stake netting

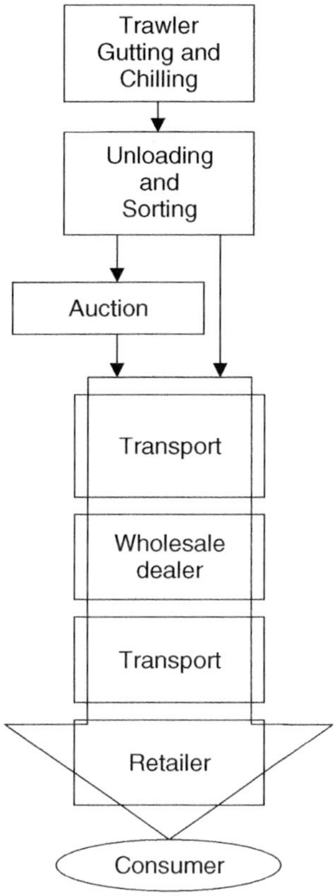

Figure 3.7 The common supply chain of fresh fish – simplified flow chart

are all activities that are considered to be traditional, but they have been developed just as much as active fishing activities. As a 'rule of thumb', fish from passive fishing methods is considered to be of a better quality, due to the less stressful fishing method. Studies have shown that rested fish has a much longer shelf life than stressed fish (Skjervold *et al.*, 2001). Commercially, these arguments are used by different trademarks to differentiate, for example, long-line caught fish from trawled caught fish.

The modern fishing vessels, especially in the old industrialized countries, have been heavily mechanized. This development has partly to do with the increasing size of the vessels and thereby the size of the fishing gear and the amount of fish caught. On the other hand, the increased wages to the crew members and the general demand for decent facilities and working hours have pushed the development in the direction of mechanization (Olsen, 1992). Fish pumps are widely used in fishing for pelagic fish, such as mackerel, herring and sardines. To avoid to large catches in the cod end of the midwater trawls, fish pumps are now connected to the trawl cod end and the fish can start to be pumped onto the vessels immediately the fishing operation is stopped. This can bring the time the catch has to be in the trawl down to a minimum compared to old-fashioned operation techniques. Deck facilities have improved from the old method of emptying the cod end on the deck to having specific receiving areas or tanks which can be filled with cooled seawater. Transportation can be carried out on conveyer belts and the slaughter and gutting stations are now constructed to reduce lifting and carrying the catch. The fish is washed and bled in specific washing stations and then conveyed to the fish hold and iced in boxes, with fresh ice made on the vessel's own ice machine (Frederiksen *et al.*, 1997). The hold is kept cooled to 2 °C (to ensure melting of ice) by mechanical refrigeration and is insulated. The boxes can be packed with fish that are sorted and weighed accordingly to the legislation and labels or other electronic devices can carry information for when the boxes are landed (Mission *et al.*, 2003).

The small pelagic fish species are normally stored in a tank system with refrigerated seawater (RSW-method). The normal method consists of seawater cooled in a mechanical refrigerator unit and circulated in the tank at specific points to avoid 'pockets' of warmer water. There are now EU rules for the cooling rate of the catch. Small pelagic fish are not gutted as the amount of fish makes it an impossible operation. In some tank systems, freshwater ice is added to the water in the tanks to make the cooling faster.

The quality of the pelagic fish stored in a properly designed tank system is of a good quality, especially during the first days. The mechanized catch handling procedure keeps the quality of the catch high. Because the small pelagic fishes are not gutted, samples of the catch are taken for evaluating the stomach contents. During the feeding time of the year, high levels of digestive enzymes in the stomach and intestine may cause autolysis. The result can be a strong off-flavour and ultimately belly-busting of the fish. In other cases the feed of the fish, which has a high enzyme activity, can result in belly-busting (Felberg and Martinez, 2006).

There are some very specialized fishing activities where processing takes place on board, such as fishing for sand shrimps *Crangon vulgaris*. After the shrimps are caught, they are cooked and then iced in fish boxes on board. Normally processing of shrimps on board is followed by freezing.

Different icing methods are used, depending on the fish species, the holding time on board, season and temperature. It is now very seldom to find wooden fish boxes in the modern fish sector, except for a few fisheries such as sardine/anchovy fishing, where the traditional 10 kg fish boxes are still in use. Modern fish boxes are made of plastic, have a lifetime duration of, for example, seven years, and differ in shape and size from region to region. Ice has it's cooling effect when it is melting, so the first generation of plastic fish boxes had drainage holes in the bottom, so the draining water could cool the box underneath. This principle has more or less been abandoned, because with sufficient refrigeration capacity in the fish hold, the temperature can be regulated precisely, so contamination from one fish box to the next is avoided. The ice that is added to the fish depends on the temperature of the fish and the temperature of the hold. During summer fishing, the water temperature is high and the fish have the same temperature – then it can be necessary to add up to 25% of the weight of the fish in ice.

The way the ice is placed in the fish box differs from region to region. Normally the bottom of the fish box is covered with ice, then the fish is placed and a cover of ice is put on top of the fish. But if the fish is going to be landed in a few days, then the fish, which is placed with the heads one way and the tail the other, is only iced on the heads and tails. This gives the fish a more appealing appearance when it is sold.

3.4.2 Landing, sorting and first sale

Fishing vessels that land their fresh chilled fish, follow different procedures due to the storage onboard, the fish species and the region where they land the fish. In the EU it is now mandatory that the fish is weighed at the time of landing and if this is not done there are specific procedures that have to be followed (EU, 2009). If the catch is iced in boxes with ice, then the boxes have to be emptied and the fish have to be deiced before the actual weight can be found. Then the fish will normally be sorted accordingly to EU regulations (EU, 1996) in different weight classes and quality class and then stored in a chilling room with an ambient temperature of 2 °C, until sale or further transportation. If the first sale of the fish takes place in the landing port, then the buyer transport the fish boxes to their premises. Here either processing takes place or preparation for further transport will be done. Fish that are to be exported can be repacked in polystyrene boxes, which require new icing. The boxes for export are then stored in a chilling storage or placed directly in a refrigerated lorry. The temperature in the lorry cargo room will be kept at 2–4 °C. Transport can take as long as three to four days, depending on reloading of the fish boxes at a cargo centre, normally at a border between two countries. Transportation of up to 3000 km is not that uncommon, for example, from northern Norway to France. Temperature measurements in the cargo room of the lorries shows that the temperature fluctuates, especially when part of the cargo is unloaded and the doors to the cargo room are left open for long (Frederiksen *et al.*, 2002).

If the chilled fish are processed (e.g. to fillets) then a typical procedure could be as follows. Arriving at the filleting facility, the fish boxes are stored in a cold storage (reception cold room). The storage time can be from one day to several days, before they are taken out to be processed. The fish is de-iced and then conveyed to the processing area. The temperature of the processing area is below 10 °C and in some modern factories the temperature is as low as 4 °C. The operating time for filleting and packing a fish could be up to 1 h, but then ice will be added or the product will be placed in a cold storage (Whittle, 1997). After filleting, the product is packed in polystyrene boxes and placed in another cold storage for further transportation to either a wholesaler or a fish retailer. If a wholesaler is repacking the fillets in smaller boxes, then they are either iced or kept in cold storage.

The description of the general chain from catch, landing, sale, transportation, processing, wholesaler and retailer, shows that one single fish can be iced and de-iced and iced again as many as six times before it finally is sold the end user – the consumer. The more links in the chain, the greater the chance that something goes wrong. A chilling chain such as this one can easily go wrong and then the shelf life of the product is reduced (Larsen *et al.*, 1992).

Pelagic fish such as herring, mackerel and anchovies are handled differently from round fish (e.g. codfishes and flatfishes). Pelagic fish catches normally consist of fish of the same size and species. The handling layout of the fishing vessels enables the fish to be pumped up from the tanks, the chilled water drained and the fish conveyed either directly to the processing facilities or to lorries equipped with tanks filled with one-third chilled water. The lorries go by road to the processing facilities. Fish that is not processed immediately is pumped to a tank storage facility, normally with the water from the lorry's tank. Handling as gently as possible and no rise in temperature are key to maintaining the good quality of the fish. The actual processing is normally very quick and the holding time in the processing area can be as short as 5–10 min.

3.4.3 Transport and wholesaler/central storage

Chilled fish and fish products are mainly transported by road in specialized lorries, as well as by sea and air also, or a combination of two or more of these methods. The different transportation systems are designed to keep or maintain a chosen temperature, not to cool down the cargo (IIR, 2003). The field of transportation seems to be developing very rapidly, mainly because there is room for energy reduction. There are some rules for road transport that can be recommended: (i) check product temperature before loading, (ii) maintain air temperature between −1 and +4 °C, (iii) permit free flow of cooling air around cargo, (iv) utilize forced air circulation, (v) monitor the air temperature (use temperature loggers), (vi) consider using insulated containers for short transportations (Margeirsson *et al.*, 2006).

In the distribution chain, wholesalers and or central storage facilities play a role in the cooling chain. Their main activities are repacking the different batches according to the orders of their clients.

3.4.4 Super-chilling

Extending the shelf life of fish by reducing the rates of spoilage can sometimes be achieved by super-chilling, where fish is stored below 0 °C. Fish can reached a super-chilled state, for example, in refrigerated seawater. The seawater allows fast heat removal from the fish and makes the chilling processes quicker, reducing the temperature of the fish to a temperature down to between −1 to −2 °C (Hedges and Nielsen, 2000; Magnussen *et al.*, 2008). Super-chilling can be used to protect the fish from temperature rises in poor cold chains for a short time (Lee and Toledo, 1984; Chang *et al.*, 1998). If the cold chain is out of control for a longer time, microstructural changes to the tissue are caused by crystallization and recrystallization causing extended drip loss, resulting in a dry and tough product. In practise it can be very difficult to keep the super-chilled state (Kaale *et al.*, 2011) due to temperature fluctuation in the cooling cabinets.

An extension to super-chilling, also called freeze-chilling (Fagan *et al.*, 2003), is to allow the fish to be partially frozen by using a temperature in the fish of −2 to −4 °C. This state can for example be obtained when the fish is tempered before thawing. Freeze-chilling for a short period can give some extension of shelf life compared with chilling when a deep frozen product is thawed during distribution in a cold chain (Jensen *et al.*, 2010; Adler-Nissen and Zammit, 2011).

3.5 Frozen fish chain

Freezing is the most widely used preservation method for fish. Nevertheless, the eating quality of fresh fish is considered superior to that of frozen fish. It is, however, possible to produce frozen fish of high eating quality by freezing the fish quickly and subsequently storing it at low and stable temperatures (Nielsen *et al.*, 2001).

Figure 3.8 shows a frozen fish supply chain, going from catch through to the retail cabinet. The main factors which determine frozen fish quality can be grouped into pre-freezing, the freezing process and frozen storage/product distribution. Good manufacturing practices for chilled fish (Section 4.4) must also be followed in the pre-freezing step. The least well-defined areas in any supply chain is the retail/consumer handling stages and it is during this phase that potentially severe thermal abuse may occur (Hedges and Nielsen, 2000), because consumers and shop assistants are unaware of the influence of fluctuating temperature (see 3.3.2).

Frozen fish products are usually processed in one of two ways: the fish is either frozen onboard a ship or on land. Onboard the fish is either frozen whole or as fillets. The frozen fillet is often used directly for sale, whereas the whole fish is thawed on land, processed

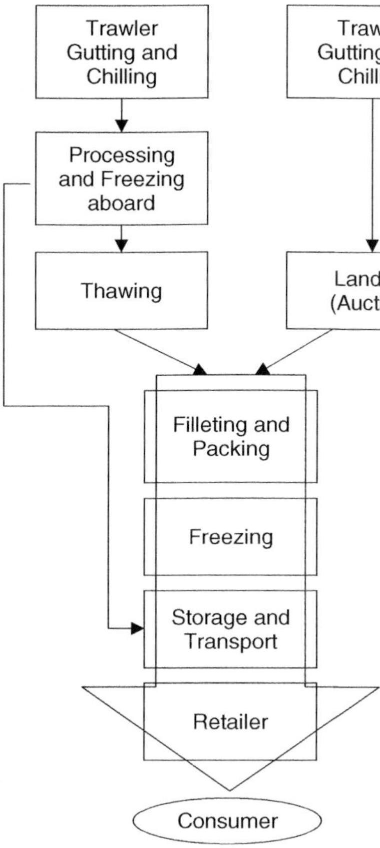

Figure 3.8 Frozen fish supply chain

and then frozen again (double freezing). Freezing and frozen storage of fish can give a storage life of more than one year, especially if the fish is frozen pre-rigor and the storage temperature is lower than −30 °C and non-fluctuating. Freezing fish onboard has enabled fishing vessels to remain at sea for a longer period in order to stockpile fish during periods of good fishing and has widened the market for fish products of high quality (Hedges and Nielsen, 2000; Nielsen *et al.*, 2001).

3.5.1 Freezing systems

Industrial freezing of fish products is carried out at various stages of production onboard the ship or on land. The great variety of species and processes gives many different frozen fish products that can be divided into two main groups: products intended for direct consumption and products intended for further processing (Johnston *et al.*, 1994).

Products for direct consumption can be individually quick frozen (IQF) as single units (e.g. fillets or shrimps and other shellfish) which need not be thawed for sub-division or perhaps even for cooking purposes. Other products such as small blocks of fish and other fish portions can also be packed for direct consumption without the need for reprocessing. The consumer can purchase this type of product from the retailer, still in the frozen state and either cook it in the frozen state or thaw it for immediate consumption (Johnston *et al.*, 1994).

The products for further processing are either frozen in bulk and thawed after storing, to be used as fresh fish (refreshed fish), or frozen in bulk and further processed without thawing, to be presented as a retail pack. Products frozen in bulk can be unprocessed, such as blocks of whole fish frozen in contact freezers. Blocks of frozen fish are usually glazed or wrapped after freezing and are then stored until required for further processing. Glazing is also used for single frozen whole fish and shrimps: once frozen, the fish is sprayed with vaporized water, which immediately freezes to a thin ice layer on the surface of the fish. In this way the entire surface of the fish is covered in an ice layer preventing air (oxygen) from getting in contact with the fish.

On board trawlers, whole fish is frozen in vertical plate freezers and fillets in horizontal plate-freezers as interleaved fillets or standardized blocks of 20 kg (Boknaes *et al.*, 2001) (Figure 3.9).

On shore, a modern fish industry will be equipped with quick freezing facilities such as air blast freezers, plate freezers and fluidized beds and cryogenic freezers used for individual quick freezing. Quick freezing means that the time elapsing between the change from 0 to −5 °C must not normally exceed 5–10 h and the warmest part of the product must be the temperature of the following frozen storage room (see Figure 3.4). Due to the flow in production, the total freezing period should be no more than 1–2 h. However, it is important for the sensory and physical quality of the product that the core temperature reaches the desired storage temperature otherwise slow freezing over several days will be a risk.

Figure 3.9 Freezer trawler *Ariadne*. *Ariadne* is 89.5 m long and has a capacity of 150 t/day. The frozen storage capacity is 1700 t fish blocks of 50 kg each packed in cartons (Photo: FiskerForum.dk/MHR)

Traditionally there are three methods used for freezing fish:

1. Blast freezers, where a stream of cold air is circulated at high speed over the fish usually in a small room or a tunnel. The air speed is usually 2.5–7.5 m/s. If the fish is unpacked and not glazed, the air velocity must be as low as possible to avoid drying of the surface (freezer-burning).
2. Contact or plate freezing, with direct contact between the fish and the refrigerated surface. Plate freezers do not have the same flexibility as air blast freezers and can only be used to freeze regularly shaped blocks and packages. There are two types of plate freezers – those with horizontal plates form a series of shelves and those with vertical plates arranged to form a series of bins. The vertical plate freezer is normally used for onboard freezing of bulk products (Figure 3.10). Fish is pumped into the bins and water is usually added to fill the void in the block. The horizontal plate freezer is used for freezing pre-packed cartons of fish and fish products for retail sale or for frozen fillet blocks in a regular-shaped block frozen within a treated cardboard carton and a metal retaining frame. The fish block (20 kg) is sliced into fish sticks and portions from the frozen state.
3. Cryogenic freezing, where refrigerated liquid is sprayed over the surface of the fish. The liquid can be nitrogen or carbon dioxide and the product is brought into direct contact with the refrigerant. The fish is placed on a stainless steel conveyor belt and comes into contact with a counter-current flow of gas at a temperature of about −50 °C. As the fish progress through the first stage of the freezer, the gas partially freezes the fish. The product then passes below a liquid spray where freezing is completed by the boiling liquid gas. The final stage in the freezer provides a few minutes for the fish temperature to reach equilibrium before the fish is discharged. The main advantage of the liquid nitrogen freezer is that freezing is very quick and the physical size of the freezer is correspondingly small (Johnston *et al.*, 1994).

Figure 3.10 Block of sandeel caught and frozen in a vertical plate freezer on board *Ariadne* wrapped in plastic

New development in freezers Freezing equipment continues to be developed but it will only be of interest to the established fishing industry if the cost is not prohibitive.

Super freezing Super frozen or ultra-frozen are terms normally associated with cryogenic preservation of biochemical, microbial and medical samples for long-term storage. In the fish sector, the term super frozen is used for very quick freezing and storage of fish at −60 °C or lower. This form of cryogenic preservation cannot be obtained with conventional freezing equipment. There is a need for special compressors, refrigerants and insulation. The aim is to stop all processes that lead to quality and texture issues either by preventing the formation of ice crystals or by the formation of very small crystals (see 3.3.1). This can be achieved by pressure shift (pressure shift freezing (PSF), and impingement freezing) or by a magnetic field (cell alive system (CAS)). Super freezing was originally introduced to a longer range of Japanese tuna boats mainly to preserve the dark red colour of the tuna meat for sushi and sashimi production.

Small or no ice crystals can also be obtained by freezing the fish pre-rigor using more conventional methods.

The fast freezing method is excellent for reducing ice-crystal formation due to the super cooling effect before freezing but it might also cause mechanical damage in larger objects such as a whole fish. It might also cause problems with the subsequent storage (low constant temperature) and, especially with thawing (thawing rigor), where the quality gained can be quickly lost (see 3.3.2).

PRESSURE SHIFT FREEZING – PSF In PSF the water–ice phase transition is depressed under pressure from 0 to −21 °C and the pressure is released causing instant freezing from the under-cooled state resulting in formation of very small and homogenous ice crystals (Kalichevsky *et al.*, 1995) (Figure 3.5). Several experiments on fish have shown that PSF gives good texture and water binding (Chevalier *et al.*, 2000; Schubring *et al.*, 2003; Alizadeh *et al.*, 2007) but commercial equipment is not yet available.

IMPINGEMENT FREEZING Impingement technology increases the surface heat transfer in air chilling. Velocity impingement air jets break up the static surface boundary layer of air around the fish product. The resulting medium around the product is more turbulent and the heat exchange becomes very effective (Figure 3.11). It can for example be used for single frozen fillets and shrimps and it can often replace cryogenic production lines that have a high production cost. The freezer is built around two straight product belts and equipped with a large number of small high-speed jets of air directed at the top and bottom surfaces of the product. The system reaches similar freezing rates as those achieved in a liquid nitrogen freezer.

Full-scale experiments for shrimps and fillets give a high-quality product as evaluated by sensory assessment and drip loss (authors' unpublished result). Impingement freezing technology is currently available in various freezer configurations from JBT (Frigoscandia) such as the Flat Product Freezers, Nautica and the Advantec.

CELL ALIVE SYSTEM – CAS FREEZING Normally a product freezes from the outside in, but in the CAS freezing system the process is reversed. CAS uses an oscillating magnetic field that causes water molecules to spin (as opposed to vibrate as in a microwave) and stops them from clustering and forming ice crystals that damage cell walls. This spinning motion also artificially lowers the freezing point of water to about −7 °C; when the product

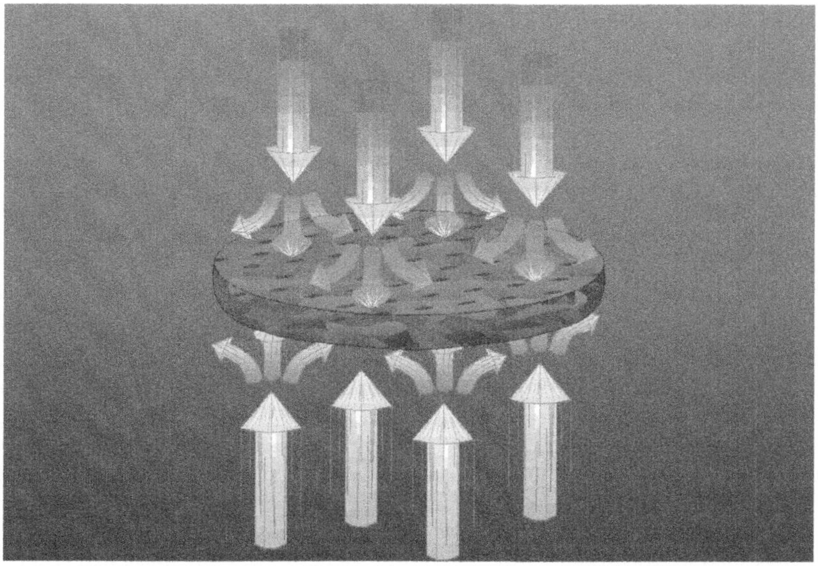

Figure 3.11 Air flow using impingement freezing of a flat product

reaches this temperature, the electrical field is turned off and freezing occurs almost instantly from the inside out. CAS freezing was developed in Japan (ABI Company) where it is commercially available and used in the tuna industry. It is used on an experimental basis in the Spanish, Irish and Alaskan seafood sector.

3.5.2 Frozen storage

Frozen fish is easily stored and distributed. Under ideal circumstances (good initial quality, good processing, low and stable storage temperatures during storage), some fish species may retain a fair eating quality for over a year. The spoilage of fish (see 3.3) due to protein denaturation, fat changes and dehydration can all be slowed down by reducing the storage temperature. Spoilage by bacterial action in any practical sense is normally not a problem as growth is arrested by freezing temperatures and the rate at which other undesirable changes proceed is greatly reduced (Johnston *et al.*, 1994). However, freezing is not sterilization and therefore some pathogenic bacteria and parasites could remain alive and general good practice must be kept for freezing (see 3.3), frozen storage and especially thawing time and temperature.

Most literature recommends a storage temperature for all fishery products of −30 °C or lower, but in practice all over the world higher storage temperatures are used, especially in the last part of the distribution chain. This might be because legislation in most countries and the International Institute of Refrigeration (IIR) recommends a storage temperature of −18 °C for lean fish such as cod and haddock and −24 °C for fatty species such as herring and mackerel (otherwise there is a risk of oil drainage from very fatty fish). Another factor is economic considerations, although it has been calculated by an authority on cold store design that under specific conditions, the total cost of operating a cold store at −30 °C is only 4% higher than when operating at −20 °C (Johnston *et al.*, 1994). It might therefore be an economical advantage to keep the storage temperature down when the raised quality and possible increased storage time of the fish are taken into consideration.

Container units that operate at −60 °C are available commercially and used in freezing, transporting and storing valuable products such as the tuna, sea urchins and swordfish demanded by the Japanese sushi and sashimi market.

3.5.3 Thawing

Frozen fish can be marketed as super-chilled or frozen thawed, freeze-chilled and refreshed with and without modified atmosphere packaging (Boknaes *et al.*, 2000; Olafsdottir *et al.*, 2006). The quality of freeze-chilled fish is dependent on the quality of the frozen fish and the thawing process. The process enables manufacturing of bulk products to be released into the chill chain on demand. Controlled thawing is therefore an important process.

Thawing takes a longer time than the freezing process because the rate of heat transfer decreases when the surface of the food thaws. The thermal conductivity of the unfrozen fish is a lower than that for the frozen fish. The recommended thawing rate depends on many things such as, for example, the type of product, thawing method, fish species and state of rigor (see Section 3.2.1). Thawing is complete when there is no ice remaining in the flesh of the product (Archer *et al.*, 2008).

There are several hazards to consider during and after thawing. As the surface ice (glaze) melts, the thawing rate slows down and there follows a long period when the temperature of the food is close to that of melting ice. At this time, the energy required to change the solid ice to liquid water needs to be overcome. It is also the period where any cellular damage caused by, for example, poor handling before freezing, excessively slow freezing or poor temperature controls, result in the release of cell constituents to form 'drip losses' (Archer *et al.*, 2008).

Most studies on thawed foods show very little difference in shelf life and microbial growth between never-frozen and pre-frozen/thawed foods (Archer *et al.*, 2008). In some cases, freezing at −20 °C may increase the storage life of the thawed product. Studies have shown that storage of cod from the North Atlantic Ocean and the Barents Sea at −20 °C extended its shelf life (microbiologically), when thawed, by several days compared with both fresh cod or cod stored at a lower temperature (Martinsdøttir and Magnússon, 2001; Boknaes *et al.*, 2002). This is because the two most important spoilage bacteria in cod (*Shewanella putrefaciens* and *Photobacterium phosphoreum*) die after eight weeks at −20 °C, but survive eight weeks at −30 °C or colder. Factors such as texture, taste and drip loss should also be considered and the thawing process must be carefully controlled.

There is continued development of equipment for thawing, but the established fishing industry will only be interested if the cost is not too high. Traditional thawing methods include:

- water (immersion or spray)
- forced air or air blast
- still or ambient air.

The newer developments in tempering and thawing equipment include different forms of electric thawing using a vacuum, microwaves or radio waves. The newest experimental method is pressure-assisted thawing. Schubring *et al.* (2003) saw a better quality of fish fillets thawed under 200 MPa pressure compared with conventional thawing at 15 °C.

For further information of different thawing methods, including recommendations for good practice, see Archer *et al.* (2008).

3.5.4 Storage life

The length of time that frozen food is stored has no affect on human health, so a 'use by' date is therefore not required for frozen fish products. The question is then how to define the storage life considering the sensory properties (odour, flavour, texture, appearance) and the physical properties (drip and water binding) (P. Howgate, personal communication, 2008, seafood@ucdavis.edu). The rate of changes during storage depends on several factors including the nature of the product and its packaging, but crucially upon temperature and the fluctuation of temperature during storage. The changes are progressive and there is no sudden step change at some point that could be used as a criterion of the product no longer being at its best. The IRR operates using practical storage life (PSL) and a high-quality storage life (HQL). PSL is defined as the time the fish can be stored without losing its characteristics and becoming unusable for consumption. HQL is determined by a trained taste panel as the time before a difference from the original fresh fish can be detected. HQL is normally two to three times shorter than PSL (Table 3.1). The literature describing changes in frozen stored fish shows many different information of acceptability and a wide range of criteria for deciding on the end of storage life (Reid, 1998). The definition of PSL and HQL is not used and the shelf life described is something in between. Most often texture is considered as the limiting property but also oxidation is important – even in a lean fish as cod an unpleasant oxidised freeze-house flavour is developed (Mcgill *et al.*, 1974). The storage life is also affected of the handling prior to freezing.

The EU legislation states that a manufacturer of pre-packaged frozen fish products has to put a best before date on the product. There is no official legislation for shelf life but as most of the frozen fishery products available at the retail level in Europe come from just a few large food companies, the general shelf life is determined by these companies as an agreement between trade partners. It is in their interests if they wish to maintain their reputations for supplying good-quality products and to ensure repeat sales that their products are at least acceptable to consumers and preferably liked by them (P. Howgate, personal communication, 2008). Note that the best-before date is printed on the package at the time of preparation of the product and is the storage time for that product from the time of preparation, not from the time the original material was frozen. It is therefore impossible to give general rules.

Table 3.1 High quality (HQL) and practical quality (PSL) shelf life in months for frozen fish stored at −30 °C constant temperature (own results)

Fish species	PSL	HQL
	Months	
Lean fish (cod, haddock)	12	3
Big fatty fish (salmon)	>12	3
Small fatty fish (herring, mackerel)	10	2

3.6 Legislation

For as long as fish has been traded there have been special rules guaranteeing that only fish of a high sensory quality are sold to the consumer. There has been legislation connected to specific fisheries, for example the herring fishery in the Middle ages, where salt content and weight was important. Almost a hundred years ago protection against poor-quality fish became part of the official general food laws. These laws are based on sensory assessment of the fish (Nielsen *et al.*, 2001).

There have always been good practice manuals for keeping fresh fish chilled. Standards are normally voluntary agreements between the supplier and the buyer. The Codex Alimentarius Standards of the Food and Agriculture Organization/World Health Organization (FAO/WHO) are, in effect, model laws for use in international trade. Various Codex Standards deal with definitions of different fish products to ensure safe and animal-friendly production. The General Food Law (European Union (EU) 2002/178; http://ec.europa.eu/food/food/foodlaw/traceability/index_en.htm) was the first example of a general law that, for example, put down the rules for traceability in the EU. Recently the EU has made the demands for traceability more rigorous in The Councils Regulation 1224/2009, articles 56, 57 and 58 (EU, 2009). One of the demands is that the fish or fish product must state if the fish has been frozen previous to sale. This is only valid for wild caught fish.

3.7 Recommendations

Temperature and temperature fluctuations are the most significant factors affecting the quality of both chilled and frozen fish. It is important to ice the fish as quickly as possible immediately after catch in order to minimize biochemical and microbiological reactions. Fishing methods and temperature are likewise significant in relation to rigor mortis. A general rule is that the time from death to the onset of rigor depends on the difference between the ambient temperature and the storage temperature, independent of which of the temperatures is the highest. The bigger this temperature difference, the stronger the muscle contraction and the shorter the time before the onset of rigor. In addition, it is essential that fish frozen for subsequent thawing and processing are frozen before rigor sets in, as this will result in the best eating quality and the greatest yield. The fish must preferably be gutted and bled before storing and before freezing, because compounds in the blood intensify the degradation processes and because the guts have a high enzyme activity. If gutting and bleeding is impossible for practical reasons, the fish should be chilled as quickly as possible.

A first-class frozen product can only be obtained using high-quality, preferably pre-rigor fish and freezing and storing them at low and stable temperatures. The product will be ruined if it is subjected to poor handling during freezing and chilled storage, for example if a freezing process lasts a whole day and cold storage temperatures fluctuate over a long period. Short-term storage, gentle catch and processing, quick freezing, short-term storage as a semi-processed product and short-term cold storage at temperatures lower than −30 °C, will result in the best product.

References

Abe, H. and Okuma, E. (1991). Rigor-mortis progress of carp acclimated to different water temperatures. *Nippon Suisan Gakkaishi*, 57, 2095–2100.

Adler-Nissen, J. and Zammit, G.Ø. (2011). Modeling and validation of robust partial thawing of frozen convenience foods during distribution in the cold chain. *Procedia Food Science*, 1, 1247–1255.

Alizadeh, E., Chapleau, N., De Lamballerie, M., and Le-Bail, A. (2007). Effect of different freezing processes on the microstructure of Atlantic salmon (*Salmo salar*) fillets. *Innovative Food Science and Emerging Technologies*, 8, 493–499.

Ando, M., Toyohara, H., Shimizu, Y., and Sakaguchi, M. (1991). Post-mortem tenderization of fish muscle proceeds independently of resolution of rigor mortis. *Nippon Suisan Gakkaishi*, 57, 1165–1169.

Ando, M., Toyohara, H., Shimizu, Y., and Sakaguchi, M. (1993). Post-mortem tenderization of fish muscle due to weakening of pericellular connective tissue. *Nippon Suisan Gakkaishi*, 59, 1073–1076.

Archer M., Edmons M., and Martins G. (2008). Seafood Thawing. Seafish Research and Development, Report SR598. http://www.seafoodacademy.org/Documents/SR598%20thawing.pdf [accessed 31 May 2013].

Astier, C., Labbe, J.P., Roustan, C., and Benyamin, Y. (1991). Sarcomeric Disorganization in postmortem fish muscles. *Comparative Biochemistry and Physiology B-Biochemistry and Molecular Biology*, 100, 459–465.

Babbitt, J.K., Law, D.K., and Crawford, D.L. (1972). Decomposition of trimethylamine oxide and changes in protein extractability during frozen storage of minced and intact hake (*Merluccius productus*). Muscle. *Journal of Agricultural and Food Chemistry*, 20, 1052.

Boknaes, N., Guldager, H.S., Østerberg, C., and Nielsen, J. (2001). Production of high quality frozen cod (*Gadus morhua*) fillets and portions on a freezer trawler. *Journal of Aquatic Food Product Technology*, 3, 33–37.

Boknaes, N., Jensen, K.N., Guldager, H.S., Osterberg, C., Nielsen, J., and Dalgaard, P. (2002). Thawed chilled Barents Sea cod fillets in modified atmosphere packaging-application of multivariate data analysis to select key parameters in good manufacturing practice. *Lebensmittel-Wissenschaft Und-Technologie – Food Science and Technology*, 35, 436–443.

Boknaes, N., Osterberg, C., Nielsen, J., and Dalgaard, P. (2000). Influence of freshness and frozen storage temperature on quality of thawed cod fillets stored in modified atmosphere packaging. *Lebensmittel-Wissenschaft Und-Technologie – Food Science and Technology*, 33, 244–248.

Burgaard, M.G. and Jorgensen, B.M. (2010). Effect of temperature on quality-related changes in cod (*Gadus morhua*) during short- and long-term frozen storage. *Journal of Aquatic Food Product Technology*, 19, 249–263.

Caballero, M.J., Betancor, M., Escrig, J.C., Montero, D., Monteros, A.E.D., Castro, P., Gines, R., and Izquierdo, M. (2009). Post mortem changes produced in the muscle of sea bream (*Sparus aurata*) during ice storage. *Aquaculture*, 291, 210–216.

Cappeln, G. and Jessen, F. (2001). Glycolysis and ATP degradation in cod (*Gadus morhua*) at subzero temperatures in relation to thaw rigor. *Lebensmittel-Wissenschaft Und-Technologie-Food Science and Technology*, 34, 81–88.

Chang, K.L.B., Chang, J.J., Shiau, C.Y., and Pan, B.S. (1998). Biochemical, microbiological, and sensory changes of sea bass (*Lateolabrax japonicus*) under partial freezing and refrigerated storage. *Journal of Agricultural and Food Chemistry*, 46, 682–686.

Chapman, K.W., Sagi, I., Hwang, K.T., and Regenstein, J.M. (1993). Extra-cold storage of hake and mackerel fillets and mince. *Journal of Food Science*, 58, 1208–1211.

Chevalier, D., Sentissi, M., Havet, M., and Le Bail, A. (2000). Comparison of air-blast and pressure shift freezing on Norway lobster quality. *Journal of Food Science*, 65, 329–333.

Delbarre-Ladrat, C., Cheret, R., Taylor, R., and Verrez-Bagnis, V. (2006). Trends in postmortem aging in fish: Understanding of proteolysis and disorganization of the myofibrillar structure. *Critical Reviews in Food Science and Nutrition*, 46, 409–421.

EU (1996). Council Directive 2406/96 of 26 November 1996 laying down common marketing standards for certain fishery products. *Official Journal of the European Union*, L 334, 1–15.

EU (2009). Council Regulation 1224/2009 of 20 November 2009, establishing a Community control system for ensuring compliance with the rules of the common fisheries policy. *Official Journal of the European Union*, L 343/1, 1–50.

EU (2011). Commission Implementing Regulation 404/2011 of 8 April 2011 laying down detailed rules for the implementation of Council Regulation (EC) No 1224/2009 establishing a Community control system for ensuring compliance with the rules of the Common Fisheries Policy. *Official Journal of the European Union*, L 112, 1–153.

Fagan, J.D., Gormley, T.R., and Mhuircheartaigh, M.U. (2003). Effect of freeze-chilling, in comparison with fresh, chilling and freezing, on some quality parameters of raw whiting, mackerel and salmon portions. *Lebensmittel-Wissenschaft Und-Technologie – Food Science and Technology* 36, 647–655.

FAO (2003). *The Use of Ice in Small Fishing Vessels*, Food and Agricultural Organization of the United Nations, Rome. ISBN 92-5-105010-4.

FAO (2012). *The State of World Fisheries and Aquaculture*, Food and Agricultural Organization of the United Nations, Rome. ISBN 978-92-5-107225-7.

Felberg H.S. and Martinez I. (2006). Protein degrading enzymes in herring (*Clupea harengus)* muscle and stomach, in *Seafood Research from Fish to Dish* (eds J.B. Luten, C. Jacobsen, K. Bekaert, A. Sæbø and J. Oehlenschläger), Wargeningen Academic Publishers, Wargeningen, The Netherlands.

Frederiksen, M., Popescu, V., and Olsen, K.B. (1997). Integrated quality assurance of chilled food fish at sea, in *Seafood from Producer to Consumer, Intergrated Approach to Quality* (eds J. Luten, T. Børresen and J. Oehlenschläger), Elsevier, Amsterdam.

Frederiksen, M., Østerberg, C., Silberg, S., and Larsen, E. (2002). Infofisk. Development and validation of an Internet based traceability system in a Danish domestic fresh fish chain. *Journal of Aquatic Food Product Technology*, 11(2), 13–34.

Gram, L. and Huss, H.H. (1996). Microbiological spoilage of fish and fish products. *International Journal of Food Microbiology*, 33, 121–137.

Haard, N.F. (1992). Biochemical reactions in fish muscle during frozen storage, in *Seafood Science and Technology* (ed. E.G. Bligh), Proceedings of the Conference Seafood 2000 celebrating the Tenth Anniversary of the Canadian Institute of Fisheries Technology of the Technical University of Nova Scotia, 13–16 May 1990, Halifax, Canada, Fishing News Books, Oxford.

Hedges N. and Nielsen J. (2000). The selection and pre-treatment of fish, in *Managing Frozen Foods* (ed. Christopher J. Kennedy), Woodhead Publishing, Cambridge.

Herrera-Mendez, C.H., Becila, S., Boudjellal, A. and Ouali, A. (2006). Meat ageing: reconsideration of the current concept. *Trends in Food Science and Technology* 17, 394–405.

Hiraoka, Y., Ohsaka, E., Narita, K., Yamabe, K., and Seki, N. (2004). Preventive method of color deterioration of yellowtail dark muscle during frozen storage and post thawing. *Fisheries Science*, 70, 1130–1136.

Houbak, M.B., Ertbjerg, P., and Therkildsen, M. (2008). In vitro study to evaluate the degradation of bovine muscle proteins post-mortem by proteasome and mu-calpain. *Meat Science*, 79, 77–85.

Howell, N., Shavila, Y., Grootveld, M., and Williams, S. (1996). High-resolution NMR and magnetic resonance imaging (MRI) studies on fresh and frozen cod (*Gadus morhua*) and haddock (*Melanogrammus aeglefinus*). *Journal of the Science of Food and Agriculture*, 72, 49–56.

Hultin H.O. (1992). Biochemical deterioration of fish muscle, in *Quality Assurance in the Fish Industry* (eds H.H. Huss, M. Jakobsen and J. Liston), Elsevier, Amsterdam.

Huss, H.H. (1995). *Quality and Quality Changes in Fresh Fish*, FAO Fisheries Technical Paper, 348, FAO, Rome.

IIR (2003). Refrigerated transport: progress achieved and challenges to be met. *16th Informatory Note on Refrigerating Technologies*, International Institute of Refrigeration.

Imamura, S., Suzuki, M., Okazaki, E. *et al.* (2012). Prevention of thaw-rigor during frozen storage of bigeye tuna (*Thunnus obesus*) and meat quality evaluation. *Fisheries Science*, 78, 177–185.

Ishida, N., Yamashita, M., Koizumi, N. *et al.* (2003). Inhibition of post-mortem muscle softening following in situ perfusion of protease inhibitors in tilapia. *Fisheries Science*, 69, 632–638.

Iwamoto, M., Yamanaka, H., Abe, H. *et al.* (1988). ATP and creatine phosphate breakdown in spiked plaice muscle during storage, and activities of some enzymes involved. *Journal of Food Science*, 53, 1662–1665.

Jensen, L.H.S., Nielsen, J., Jorgensen, B.M., and Frosch, S. (2010). Cod and rainbow trout as freeze-chilled meal elements. *Journal of the Science of Food and Agriculture*, 90, 376–384.

Johnston, W.A., Nicholson, F.J., Roger, A., and Stroud, G.D. (1994). *Freezing and Refrigerated Storage in Fisheries*, FAO Fisheries Technical Paper, 340, FAO Rome.

Jones, N.R. (1965). Problems associated with freezing very fresh fish, in *Fish Handling and Preservation*. OECD Paris, Scheveningen.

Jul, M. (1984). *The Quality of Frozen Food*. Academic Press, St Louis, Missouri.

Kaale, L.D., Eikevik, T.M., Rustad, T., and Kolsaker, K. (2011). Superchilling of food: a review. *Journal of Food Engineering*, 107, 141–146.

Kalichevsky, M.T., Knorr, D., and Lillford, P.J. (1995). Potential food applications of high-pressure effects on ice-water transitions. *Trends in Food Science and Technology*, 6, 253–259.

Kemp, C.M., Sensky, P.L., Bardsley, R.G., Buttery, P.J., and Parr, T. (2010). Tenderness – an enzymatic view. *Meat Science*, 84, 248–256.

Kubota, M., Kinoshita, M., Kubota, S. *et al.* (2001). Possible implication of metalloproteinases in post-mortem tenderization of fish muscle. *Fisheries Science*, 67, 965–968.

Larsen, E.P., Heldbo, J., Jespersen, C.M., and Nielsen, J. (1992). Development of a standard for quality assessment of fish for human consumption, in *Quality Assurance in the Fish industry* (ed. H.H. Huss), Elsevier, Amsterdam.

Larsen, E.P., and Villerreal, B.P. (2009). Traceability as a tool, in *Fishery Products: Quality, Safety and Authenticity* (eds H. Rehbein and J. Oehlenschläger), Wiley-Blackwell Publishing.

Lavety, J., Afolabi, O.A., and Love, R.M. (1988). The connective tissue of fish. Gaping in farmed species. *International Journal of Food Science and Technology* 23, 23–30.

LeBlanc, E.L. and LeBlanc, R.J. (1988). Effect of frozen storage-temperature on free and bound formaldehyde content of cod (*Gadus morhua*) fillets. *Journal of Food Processing and Preservation*, 12, 95–113.

Lee, C.M. and Toledo, R.T. (1984). Comparison of shelf-life and quality of mullet stored at zero and subzero temperature. *Journal of Food Science*, 49, 317–322.

Mackie, I.M. (1993). The effects of freezing on flesh proteins. *Food Reviews International*, 9(4), 575–610.

Magnussen, O.M., Haugland, A., Hemmingsen, A.K.T., Johansen, S., and Nordtvedt, T.S. (2008). Advances in superchilling of food – process characteristics and product quality. *Trends in Food Science and Technology*, 19, 418–424.

Margeirsson, S., Nielsen, A., Jonsson, G., and Arason, S. (2006). Effect of catch location, season and quality defects on value of Icelandic cod (*Gadus morhuea*) products, in *Seafood Research from Fish to Dish* (eds J.B. Luten, C. Jacobsen, K. Bekaert, A. Sæbø and J. Oehlenschläger), Wargeningen Academic Publishers, Wargeningen, The Netherlands.

Martinsdóttir, E. and Magnússon, H. (2001). Keeping quality of sea-frozen thawed cod fillets on ice. *Journal of Food Science*, 71, M69–M76.

Mcgill, A.S., Hardy, R., Burt, J.R., and Gunstone, F.D. (1974). Hept-cis-4-enal and its contribution to off-flavor in cold stored cod. *Journal of the Science of Food and Agriculture*, 25, 1477–1489.

Mission, T., Mitchell, P., and Steele, A. (2003). Weighing and labelling at sea, in *Quality of Fish from Catch to Consumer* (ed. J.B. Luten), Wageningen Academic Publishers, Wageningen.

Nielsen, J. and Jessen, F. (2007). Quality of frozen fish, in *Handbook of Meat, Poultry and seafood quality* (ed. L.N.L. Nollett), Blackwell Publishing, Oxford.

Nielsen, J., Larsen, E., and Jessen, F. (2001). Chilling and freezing of fish and fishery products, in *Advances in Food Refrigeration* (ed. Da-Wen Sun), Leatherhead Publishing.

Nielsen, M.K. and Jorgensen, B.M. (2004). Quantitative relationship between trimethylamine oxide aldolase activity and formaldehyde accumulation in white muscle from gadiform fish during frozen storage. *Journal of Agricultural and Food Chemistry*, 52, 3814–3822.

Olafsdottir, G., Lauzon, H.L., Martinsdottir, E., Oehlenschlager, J., and Kristbergsson, K. (2006). Evaluation of shelf life of superchilled cod (*Gadus morhua*) fillets and the influence of temperature fluctuations during storage on microbial and chemical quality indicators. *Journal of Food Science*, 71, S97–S109.

Olsen, K.B. (1992). Handling and holding of fish on fishing vessels in Denmark, in *Quality Assurance in the Fish Industry* (ed. H.H. Huss), Elsevier, Amsterdam.

Papa, I., Alvarez, C., Verrez-Bagnis, V., Fleurence, J., and Benyamin, Y. (1996a). Post mortem release of fish white muscle (alpha)-actinin as a marker of disorganisation. *Journal of Science and Food Agriculture*, 72, 63–70.

Papa, I., Ventre, F., Lebart, M.C., Poustan, C., and Benyamin, Y. (1996b). Post mortem degradation of dystrophin from bass (*Dicentrarchus labrax*) white muscle, in *Refrigeration and Aquaculture*. Proceedings of the Meeting of Commission C2, Bordeaux, 20–22 March 1996, AD: Refrigeration.

Reid, D. (1998). Freezing preservation of fresh foods. Quality aspects, in *Food storage stability* (eds I.A. Taub and R.P. Singh), CRC Press LLC, Cambridge, England.

Roth, B., Slinde, E., and Arildsen, J. (2006). Pre or post mortem muscle activity in Atlantic salmon (*Salmo salar*). The effect on rigor mortis and the physical properties of flesh. *Aquaculture*, 257, 504–510.

Sato, K., Ando, M., Kubota, S. *et al.* (1997). Involvement of type V collagen in softening of fish muscle during short-term chilled storage. *Journal of Agricultural and Food Chemistry*, 45, 343–348.

Schubring, R., Meyer, C., Schlüter, O., Boguslawski, S., and Knorr, D. (2003). Impact of high pressure assisted thawing on the quality of fillets from various fish species. *Innovative Food Science and Emerging Technologies*, 4, 257–267.

Seki, N. and Tsuchiya, H. (1991). Extensive changes during storage in carp myofibrillar proteins in relation to fragmentation. *Nippon Suisan Gakkaishi*, 57, 927–933.

Seki, N. and Watanabe, T. (1984). Connectin content and its post-mortem changes in fish muscle. *Journal of Biochemistry*, 95, 1161–1167.

Shenouda, S.Y.K. (1980). Theories of protein denaturation during frozen storage of fish flesh. *Advances in Food Research*, 26, 275–311.

Skjervold, P.O., Fjaera, S.O., Ostby, P.B., and Einen, O. (2001). Live-chilling and crowding stress before slaughter of Atlantic salmon (*Salmo salar*). *Aquaculture*, 192, 265–280.

Sørensen, N.K., Gundersen, B., Nyvold, T.E., and Elvevoll, E. (1996). Evaluation of sensory quality during freezing and frozen storage of whole, gutted Atlantic salmon (*Salmo salar*), in *Refrigeration and Aquaculture*. Proceedings of the Meeting of Commission C2, Bordeaux, March 20–22.

Sotelo, C.G., Piñeiro, C. and Pérez-Martín, R.I. (1995). Denaturation of fish proteins during frozen storage: role of formaldehyde. *Zeitschrift fur Lebensmittel Untersuchung und Forschung*, 200, 14–23.

Undeland, I. (1997). Lipid oxidation in fish: causes, changes and measurements, in *Methods to Determine the Freshness of Fish in Research and Industry*. Proceedings of the Final Meeting of the Concerted Action 'Evaluation of Fish Freshness', 12–14 November 1997, Nantes, France. EU-project FAIR programme (AIR3CT94 2283). International Institute of Refrigeration.

Whittle K.J. (1997). Opportunities for improving the quality of fisheries products, in *Seafood from Producer to Consumer, Integrated Approach to Quality* (eds J. Luten, T. Børresen and J. Oehlenschläger), Elsevier, Amsterdam.

Wu, G.P., Chen, S.H., Liu, G.M. *et al.* (2010). Purification and characterization of a collagenolytic serine proteinase from the skeletal muscle of red sea bream (*Pagrus major*). *Comparative Biochemistry and Physiology B-Biochemistry and Molecular Biology*, 155, 281–287.

Yamanoue, M. and Takahashi, K. (1988). Effect of paratropomyosin on the increase in sarcomere-length of rigor-shortened skeletal-muscles. *Journal of Biochemistry* 103, 843–847.

4

Heat Processing of Fish

Dagbjørn Skipnes
Nofima AS, Stavanger, Norway

4.1 Introduction

Preservation by heat remains one of the major methods for extending the shelf life of packaged fish because of the advantages of a high safety level, convenience and a healthy product. By carrying out heat preservation after the packaging is hermetically sealed, microorganisms are prevented from contaminating the processed food. In-pack, thermally processed fish products have been produced for almost two centuries. While Nicolas Appert started the first cannery for meats and vegetables in 1803 (Appert, 1810), the first production of canned sardines took place in France in 1830. Loss of market shares from the traditional canning business to the fresh fish business has been documented by the Food and Agriculture Organization (FAO, 2010), but the total volume for global canned fish production has increased every year since 1962 (FAO, 2012). The exported volume of canned seafood products increased from 11 805 000 tonnes in 1998 to 17 106 000 tonnes in 2008 (FAO, 2009). In addition, an increasing proportion of the fish sold as frozen or refrigerated (typically at 2–5 °C) has been heat processed before chilling to distribution temperature.

4.2 Basic principles

Thermal processing can be subdivided into several more or less overlapping groups based on temperature regime, method or equipment for thermal processing, fish species, packaging method or the microbial target of the process.

Sterilization is the classical method. The products undergo a process aiming to inactivate all pathogenic bacteria and their spores. The temperature regime during processing may vary from 110 to 135 °C. For low acid foods (pH > 4.5) the process aims to inactivate the spores of *Clostridium botulinum* type A. This is sometimes referred to as commercial sterility, as some spore forming non-pathogenic strains may survive this heat load. Other subgroups of sterilization are also in use but are not presented here.

Seafood Processing: Technology, Quality and Safety, First Edition. Edited by Ioannis S. Boziaris.

Pasteurization is intended to inactivate vegetative cells but is not intended to inactivate the spores of all pathogenic bacteria. The term is often related to the heat treatment of acid foods or refrigerated foods where growth of surviving spores is prevented by a pH below 4.5, a low temperature or by other means. A variant of pasteurization is sous vide processing, that is mild thermal processing of vacuum packaged products.

Minimally processed convenience foods are a growing segment in the European marketplace. Fish-based products are underrepresented among these foods due to a number of unsolved problems. The products from the traditional fish processing industry have an unpredictable quality and suitability for minimal processing due to seasonal variations, variations in freshness and handling between catch and filleting, as well as differences in functional properties depending on their raw material history, for example fresh versus frozen raw material.

4.3 Best available technology for thermal processing of fish

For pasteurization at temperatures from about 90 °C and up, a counter pressure may be desirable for flexible packaging materials and in some cases (e.g. easy peel top film) even necessary. This means an autoclave is required, but even at temperatures below 90 °C, an autoclave may be the preferred solution because it provides counter pressure and a temperature distribution that is normally much better than alternative methods. The pressure may also be important for heat transfer and for the safety of the product (Skipnes *et al.*, 2002). A low pressure may produce a dead space between the food and the packaging that will insulate the food. A sudden pressure change at the start of cooling may result in ebullition and an unexpected fast drop in temperature inside the product.

Alternative equipment used for mild heat treatment are water baths and steam cabinets. As for autoclaves, these solutions have their continuous processing variants with steam tunnels and water baths with conveyors. For a water bath with a sufficient circulation (at least 50% exchange of water per min) and spreading system, a temperature distribution comparable to a modern autoclave should be possible. For cabinets, the performance depends on the mixture of air and steam in the cabinet and the fan system. Large temperature deviations must be expected in cabinets (Sheard and Rodger, 1995), and the variations in steam/air ratio may also result in uneven heat distribution for a seemingly acceptable temperature distribution. Measurements reported by Nicolaii (1994) revealed oven temperature differences up to 15 °C between different locations for a pre-set temperature of 70 °C in a combi-steamer, using a hot air/steam mixture. There are usually also larger deviations from the set point temperature in cabinets than in autoclaves, but there are examples of improvements by advanced control strategies being used in spite of an inhomogeneous temperature distribution (Ryckaert *et al.*, 1999; Verboven *et al.*, 2000a, 2000b). The large production of cabinets for kitchens has resulted in moderate prices, but for industrial fish processing, high-capacity cabinets having a temperature uniformity within 1 °C are required.

Rapid heating methods are more suited to continuous processing. Microwave heating problems include uneven heating and limited penetration depth (a few millimetres) (Ryynänen, 2002).

Emerging thermal and non-thermal technologies have clear environmental benefits, by improving the overall energy efficiency of the process or by reducing the use of non-renewable resources (Pereira and Vicente, 2010). However, a fair comparison of the processes is not easy since the final product might be different or comparable data are lacking, as can be seen from the references mentioned later. Cold pasteurization of food using high-pressure processing (HPP) can potentially reduce energy use by 20% compared to traditional thermal processing.

The typical energy consumption of HPP is 20 kWh/t food (Heinz and Buckow, 2010), but proper process specifications have not been given. Pereira and Vicente (2010) claimed that the energy required for sterilization of cans could be reduced from 83 to 75 kWh/t when applying high pressure-assisted thermal processing instead of conventional thermal processing. The process could be further enhanced by energy recovery reducing the energy input to 67 kWh/t (Toepfl *et al.*, 2006). This was ten times higher than for roasting of beef in an oven at 220 °C for 40 min, found to be 7 MJ/kg (Goñi and Salvadori, 2012), which equals approximately 1900 kWh/t beef. Our own calculations of energy consumption during traditional canning of spaghetti and meatballs with tomato sauce in 850-ml cans amounts was 165 kWh/t, while for longitudinal agitation at 140 strokes per minute, the energy consumption was 154 kWh/t. The minimum thermal energy required for heating a food with a specific heat capacity of 3.8 kJ/(kg K) by 100 °C is 106 kWh/t, provided there are no losses to the ambient surroundings, which may be expected to be true for microwave heating (Thostenson and Chou, 1999).

The main energy consumption during production of canned fish (tuna) is for thermal processing and the production and transportation of tin cans (Hospido *et al.*, 2006). Similar findings were made when investigating canning of mussels (Iribarren *et al.*, 2010a, 2010b) when focusing on the post-harvest activities. Specific thermal processing studies have documented the potential for 15–25% energy reductions by improving autoclave insulation of autoclaves (Simpson *et al.*, 2006). Using longitudinal agitation during processing could reduce the processing time for semi-liquid products by up to 90% (Walden, 2008), giving a 13–22% energy saving for autoclaves with state-of-the-art insulation. Additional measures to be taken are the reduction of the thermal mass to be heated and cooled per amount of food produced, and the temperature difference from loading to peak temperature to unloading of the retorts, which is yet to be investigated. One measure that has been suggested (Walden, 2009) to reduce the temperature peaks of the thermal mass is to provide an insulating shield on the inside of the autoclaves. This could be effective provided the processing time is short enough to avoid the full temperature span from sterilization temperature to final temperature. Further reduction of this temperature span could be achieved by removing the product from the retort at a higher temperature and continued non-pressurized cooling. In this way combinations of existing technologies would result in a novel process. For some retort systems the energy savings could be especially high, such as the traditional steam retorts which require venting during come up and sterilization to avoid air-pockets, which lead to inadequate heat transfer.

Modelling has been extensively used in thermal processing, including for optimal energy control, but volumetric heating and HPP have not been studied or optimized for saving energy and water.

4.4 Quality changes during heat treatment of fish

Healthy eating trends have given rise to an increased demand for fish and seafood which has many health benefits. In addition, cardiac health has become a major concern among modern consumers as heart disease causes many deaths in the developed world. The benefits of fish oil over animal fats in the diet are widely recognized among consumers and this has also led to a shift away from red meat towards white meat and, more importantly, fish and fish products. The health aspects of eating seafood have primarily been linked to marine lipids (Larsen *et al.*, 2007) and marine ω-3 polyunsaturated fatty acids (PUFA) that are associated with a reduced risk of coronary heart disease (Schmidt *et al.*, 2006). In contrast to water soluble vitamins, the fat soluble vitamins are relatively heat stable and since heat-preserved fish is usually vacuum packed, the fatty acids are well preserved. The prevention of oxidation is also important for conservation of the many minerals for which fish is an important source. Finally, there is no need for preservatives in thermally processed fish. Even if there might be other issues related to the nutritional outcome, for example heat denaturation of proteins might increase digestibility, it is more challenging to balance the heat load between the desired sensorial quality and the microbial constraints for obtaining the required safety and shelf life. The major group of nutrients from fish muscle is proteins. Thermal processing below 100 °C will not alter the composition of amino acids, though their functional properties may be changed (Cheftel *et al.*, 1985) and may give enhanced digestibility.

Fresh fish quality is rapidly reduced by various microbial, biochemical and chemical breakdown processes, so thermal processing of the fish needs to occur before these breakdown processes result in a significant loss of quality. Any initial loss of quality is mainly due to post-mortem autolytic activity and chemical degradation processes, but from the mid stages of the product's shelf life, microbial activity becomes increasingly important for changes in quality (Huss, 1995). Several psychrotolerant gram-negative bacteria (e.g. *Pseudomonas* spp. and *Shewanella* spp.) grow on fresh chilled fish but may be inhibited by vacuum packaging in favour of *Photobacterium phosphoreum* and lactic acid bacteria, as reviewed by Gram and Dalgaard (2002). However, none of these spoilage bacteria are heat resistant and they could not survive mild thermal processing.

The quality of thermally processed products is highly dependent on the heat load, which is determined by the requirements for inactivation of microorganisms. Legislation on the inactivation of pathogenic microorganisms often includes a safety margin, which may lead to over-processing of some goods. In the United States, the Food and Drug Administration (FDA) was quick to announce concerns for the safety of pasteurized foods (Rhodehamel, 1992) and it has been common to distribute pasteurized fish products frozen. In the current recommendations from the FDA's 'Food Code', techniques like sous vide and cook-chill should only be used if the fish product is kept frozen from packaging/thermal processing until it is consumed (FDA, 2005). The safety of hermetically packaged and thermally processed foods has been an issue since the invention of the technology, and detailed legislations are in use. Sterilized foods have been used worldwide for several decades and one common requirement for most countries is that the lowest sterilizing value, F_0, should be 3.0 for a low-acid canned food. This has been a bearing principle for sterilized fish as well as other canned foods, but is a recommendation rather than a legislative limitation. There are several fish products that are given a higher heat load than an F_0 of 3 to obtain higher safety margins, special quality attributes (e.g. soft bones) or a long shelf life even under storage at temperatures above 40 °C. Internationally recognized guidelines are published by the United Nations Codex Alimentarius Comission (2001). In most legislation for hermetically packaged, heat-preserved foods, the following topics, in addition to more general issues such as hygiene, are of major concern:

- Determination of a safe heating procedure, i.e. requirements for sterilization or pasteurization values.

- How to achieve the required sterilization/pasteurization, i.e. heat penetration tests or other measuring techniques, and to determine a scheduled heating process (e.g. sterilizing time and temperature).
- How to reproduce a scheduled process, i.e. control of heat distribution and constant heat transfer conditions in the product.
- Validation of procedures and equipment (at least calibration of thermometers) and record keeping.
- End product control. For sterilized products this includes incubation and microbial sample testing.
- Integrity of packaging.

European Guidelines for canning have been published by the Campden BRI (May, 1997). Volunteer organisations, like the Grocery Manufacturers Association (formerly the National Food Processors Association) and the Institute for Thermal Processing Specialists (IFTPS) have produced several publications which are widely used (IFTPS, 1992, 1995, 2002). These guidelines describe how to perform heat penetration tests in canned products and heat distribution tests in autoclaves. These issues are also important to milder heat preservation techniques, and several elements can be transferred from the canning guidelines. The guidelines for heat penetration tests of canned foods could be successfully used for almost any heating regime for packaged foods.

In the sea, several pathogens are indigenous. Some are toxin-producing bacteria, such as psychrotrophic non-proteolytic *Clostridium botulinum* type B, E and F, and psychrotolerant histamine-producing bacteria (photobacteria). Other relevant microorganisms are *Listeria monocytogenes, Vibrio cholerae, Vibrio parahaemolyticus, Vibrio vulnificus, Aeromonas hydrophila* and *Plesiomonas shigelloides* (Nilsson and Gram, 2002). Fish may easily be contaminated by bacteria from the environment during processes like gutting and filleting. Psychrotropic *C. botulinum* type E and *L. monocytogenes* are examples of bacteria that may easily contaminate fish. Pathogenic toxin-producing *Bacillus cereus* is not associated with raw fish materials, but may be a risk factor from the ingredients in mixed or minced fish products or in marinades (Feldhusen, 2000). *C. botulinum, Vibrio* spp. and *L. monocytogenes* constituted the highest proportion of pathogen outbreaks originating from seafood as reported internationally from 1988 to 2007 (Greig and Ravel, 2009). In summary, a thermal process should target safe destruction of these pathogens if growth of these bacteria cannot be inhibited during the shelf life of the product, for example by refrigeration.

The heat resistance of bacterial spores has been the focus of several studies (Setlow and Johnson, 1997; Lindstrom *et al.*, 2006; Peleg *et al.*, 2008; Rajkovic *et al.*, 2010; Silva and Gibbs, 2010). There are large variations in the heat resistance reported depending on the strain, genotype, and growth medium paramters (pH, $a_{w,}$ fat content, etc.) among other things. Spores formed by the genera *Bacillus, Clostridium, Desulfotomaculum* and *Sporolactobacillus* are hot topics in food microbiology. *Clostridium* spp. receives special attention in relation to thermal processing and a desired 12 log reduction of *C. botulinum* is the basis for the already mentioned limitation for the sterilization value F_0 to be above 3 min. Several guidelines and codes of practice have been published for the safe production of ready-to-eat (RTE) packaged foods that have an extended shelf life under refrigeration (ACMSF, 1992, 2006; ECFF, 1996; Betts, 2009). Most of these are targeted at preventing growth and toxin production by non-proteolytic *C. botulinum*. A general recommendation from these guidelines is that the heat treatments or combination of processes used should reduce the number of viable spores of non-proteolytic *C. botulinum* by 6 log. Accordingly, a minimum heat treatment of 90 °C for 10 min or equivalent lethality in the slowest heating spot[1] of the product has been recommended by (ACMSF 1992, 1995). This is based on a $D_{90°C}$ of 1.6 min and a z-value of 7.5 °C when the temperature in the product is below 90 °C and a z-value of 10 °C at higher temperatures.

Listeria monocytogenes is a gram-positive bacterium, mobile by means of flagella, and has been considered as a leading cause of death from foodborne bacterial pathogens (Paoli *et al.*, 2005). Epidemiologic data indicate that foods involved in listeriosis outbreaks are those in which the organism has multiplied and in general the foods have contained levels significantly higher than 100 cfu/g (Buchanan *et al.*, 1997). The Codex Alimentarius also recommended that the maximum contamination level for *L. monocytogens* in food at consumption should be less than 100 cfu/g based on risk assessment for *L. monocytogenes* in RTE foods (Codex Alimentarius Comission, 2002).

The 6 D concept is also applicable for *L. monocytogenes*. Accordingly, a minimum heat treatment of 70 °C for 2 min or equivalent lethality in the slowest heating point of the product has been recommended (ACMSF, 1992, 2006). This is based on a $D_{70°C}$ of 0.33 min and a z-value of 7.5 °C. There is, however, a wide range of kinetic data reported for inactivation of *L. monocytogenes* depending on the strain and the model system used for determining heat resistance (Ben Embarek, 1993). For cod, Ben Embarek (1993) investigated the heat resistance of *L. monocytogenes* O62 and found a $D_{70°C}$ of 0.03 min and a z-value of 5.7 °C, while he found a $D_{70°C}$ of 0.05 min and a z-value of 6.1 °C for *L. monocytogenes* O57. This indicates that a 6 log inactivation would require a pasteurization value $P_{70°C}$ in the range 0.18–0.30 min, but further studies of heat resistance of *L monocytogenes* in fish are needed to draw a conclusion.

The heat resistance of *V. parahaemolyticus* has been reported by several authors (reviewed by Drake *et al.*, 2007) but kinetic data for thermal inactivation are scarce; only one detailed report on D-value stated a $D_{55°C}$ of 1.75 min (Johnston and Brown, 2002) and z-values are only found for other *Vibrio* spp. Based on this and the internationally recognized heat resistance for some of the target organisms mentioned, the required heat load for 6 log inactivation is shown in Figure 4.1 (based on kinetic data discussed earlier).

The data in Figure 4.1 are reported for log-linear inactivation kinetics, except for non-proteolytic *C. botulinum* type E which has a break point at 90 °C on the inactivation line and *V. parahaemotolyticus* where only one point is shown. In recent years the commonly used first order inactivation models described earlier have been challenged by more sophisticated non-log-linear modelling (Peleg and Cole, 1998; Peleg *et al.*, 2008). This may lead to more accurate optimization of thermal processes in the future if this can be adopted in legislations and standards, and more detailed knowledge is gained. Another aspect challenging the log-linear inactivation models is the detection of higher heat resistance of bacteria on the surface of foods compared to free-floating microorganisms (Lejeune, 2003). The surface of the fish is also expected to have a much higher microbial load than inside the fish meat, which together with a higher heat resistance could make a problem. On the other hand, the heat load is much higher on the surface of the product compared to the core when using conventional heating systems. For rapid heating technologies this is not always the case. Heat resistance of *L. monocytogenes* on a Teflon surface during steam pasteurization has also been shown to follow patterns other than all previously suggested inactivation models (Valdramidis *et al.*, 2007; Skåra *et al.*, 2011). However, for conventional pasteurization in an autoclave or steam cabinet, the well-established log-linear inactivation kinetics and calculation of pasteurization values in the core of the vacuum packaged fish product is still the method that is in practice.

[1] The slowest heating spot of a product is the place where the temperature is lowest during heating. This is not necessarily the geometric centre of the product and has to be determined for each product.

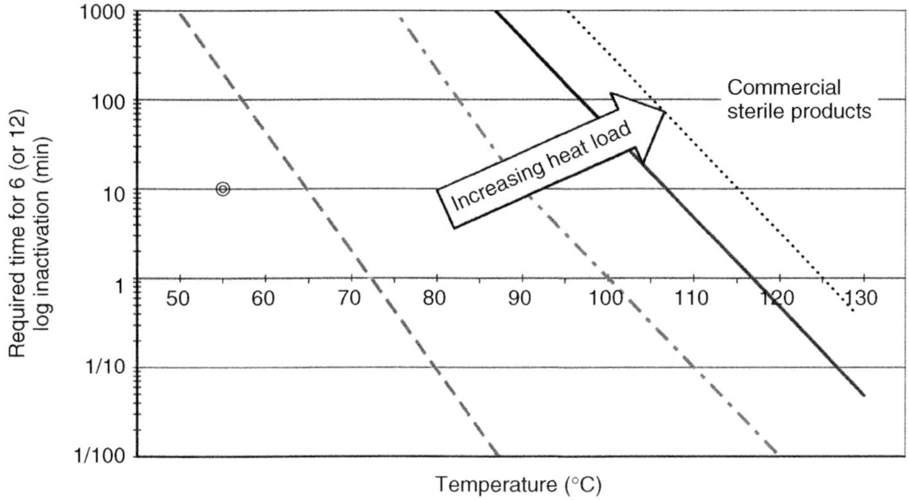

Figure 4.1 Required heat load for 6 log inactivation of some target organisms from left to right; *V. parahemolyticus* (double circle), *L. monocytogenes* (dashed line), psychrotropic non-proteolytic *C. botulinum* type E (dashed/dotted line), *B. cereus* (solid line) and *C. botulinum* type A (12 log inactivation, dotted line). The area on the right hand side of each line represents what is recognised as the 'safe side' for that organism

Even for a product intended for immediate consumption after cooking (e.g. fresh fish cooked and served in a restaurant or in a home), it is recommended by the FDA Food Code that the fish is heated to at least 63 °C for 15 s for food safety reasons (*Salmonella* spp.) (FDA, 2005). However, some chefs and cookbooks recommend using temperatures in the range 48–60 °C to achieve the desired quality. Temperatures below 55 °C may even be insufficient for killing nematodes (Huss, 1994). Nematodes may be avoided in farmed fish, for example in salmon (Lunestad, 2003), but have still been found in farmed cod (Mackenzie *et al.*, 2009).

Optimization of thermal processing for quality and safety is only possible if we are able to quantify the kinetics of microbial inactivation and quantify quality retention. While thermal inactivation kinetics of many pathogen microorganisms is well known, information on changes in quality during thermal processing is often lacking. Simple elementary chemical reactions can be well described, but interactions in the food matrix and individual biological variation are two complicating factors (van Boekel, 2008). The building of knowledge on the kinetics of quality factors is therefore one of the frontiers in thermal processing technology. A general overview on optimization including canning of fish can be found in Skipnes and Hendrickx (2008), but more specific studies are available for mild heat preservation as well (Skipnes *et al.*, 2011).

Sophisticated methods for optimization, including time-temperature integrators and computational fluid dynamics, have been introduced, as summarized by Richardson (2004). One of the easiest ways of performing optimization is to take advantage of the cook value (C-value) as described earlier to calculate the C-value for several time–temperature combinations that result in the same F-value. The time–temperature combination resulting in the lowest C-value correlates to the best quality retention (Tucker, 2003; Richardson, 2004). Quality changes are often too complex to be described in terms of a z-value, but there are exceptions. For instance, the denaturation enthalpy of fish proteins during heating has been studied by differential scanning calorimetry (Skipnes *et al.*, 2008).

Packaging geometry is another parameter determining the possible end qualities that may be achieved during thermal processing. The distance to the slowest heating spot of the product is crucial for the time necessary to achieve the desired sterilization/pasteurization through the whole product. The thermal conductivity of fish products depends on several parameters, most importantly the water content. The thermal conductivity of fresh fish is 0.5 W/m K. For a solid product, heat is transferred by conduction. At the boundaries of the product the temperature will be close to the water temperature during the heat treatment, resulting in a much higher heat load than at the cold spot.

Producers desiring a reproducible quality should note that continuous changes in the fish as raw material cannot be avoided, and therefore optimization should be regarded as a dynamic process to be regularly updated possibly for each product batch. Thus, the optimization tools need to be easy to use and give rapid results or even on-line process control.

4.4.1 Process design effects on product quality

Thermal processing of foods (e.g. packaged in cylindrical cans) with conventional methods results in a much higher heat load on the surface compared to the centre of the product and various methods may be used to reduce the difference between the surface and the cold spot. Achieving a similar temperature curve in the centre of the food as on the surface is often desirable as this would reduce the total heat load on the volume average of the product. In such cases the processing time is adjusted to achieve the required lethality of a target microorganism. This effect has been demonstrated using cans of different diameters to show that the optimum processing temperature decreases when the can diameter increases (Ohlsson, 1980).

Rapid heating can be achieved in many ways other than reducing the size of the food product. For liquid or semi-liquid products, rotation of the package during conventional heating is very effective (Eisner, 1988). Conventional heating can also be speeded up by vibration or by using ultrasound with the product. However, for some fish products shaking, rotating, and so on will make the fish fall apart limiting the use of these methods.

Rapid volumetric heating has been suggested as a method for reducing the heat load on the volume average of the product. Ohmic heating is a method that can generate temperature through an electrically homogen product, but the need for electrodes in contact with the food limits the method to aseptic packaging.

Heating with electric fields is more convenient for packaged products with suitable packaging, such as plastics. The use of metal containers is limited. Microwaves and radio frequency are more commonly used in the food industry for thawing or tempering of foods than for pasteurization or sterilization. Microwaves penetrate only a few millimetres into fish products, giving large temperature gradients through a cross-section of the product. Radio frequency penetrate several decimetres into the product, due to the longer wavelength, but may also result in uneven heating, especially if ions or minerals (like salt) are present in the product (Mckenna *et al.*, 2006). For radio frequency heating, water immersion can be used to stabilize the temperature at the product boundaries and may result in more widespread industrial use of this method in the future. Reduced cook loss and better control of texture attributes are proposed in the ongoing research (Lyng *et al.*, 2007). There might, however, be some limitations to the effect of rapid heating. Protein denaturation will typically be a process of low z-value, i.e. a small change in temperature will result in a large change in the time required for denaturation of proteins. Thus, changes in cook loss and water holding capacity (WHC) will be highly influenced by process temperature, while process time becomes less important. Most of the target microorganisms will have a higher z-value and therefore a low temperature and a relatively long process time will be favourable. For a process where this is the case, there is little to gain from rapid heating. An example of such a situation is shown by Kong *et al.* (2007b).

4.4.2 Biochemical changes during heating

Loss of liquid and texture changes are the major issues when optimizing the thermal processing of fish to maximize quality. Aitken and Connell (1979) reviewed the effects of heating on fish and reported that the cooking losses varied greatly with the fish species, the method of heating and the heating regime (i.e. sterilization, pasteurization, etc.). Few publications discuss the effects of heat on fish and the effects of filleting and heat processing pre-rigor and post-rigor. In the following text, this is discussed using farmed Atlantic cod as an example based on information from Skipnes (2011).

Heating converts the translucent, jelly-like cellular mass into an opaque, friable, slightly firm and springy form. The muscle shrinks during heating, resulting in a release of liquid. The proteins in this liquid may coagulate on the surface of the solid fish as a curd. The connective tissue holding the cells together is easily damaged and thus, cooked fish easily falls apart and becomes palatable on mild heating (Parry, 1970). Post-rigor processed fish does not fall apart as easily as pre-rigor processed fish, but it is more likely to break across the myotoma. Even at temperatures as low as 37 °C, the tensile strength is reduced to zero for codfish after 30 min (Aitken and Connell, 1979), and a visible softening of the connective tissue occurs after 15 min at 35 °C. Below this, no temperature effects were found by Aitken and Connell. However, Howgate and Ahmed (1972) showed that thermal effects on proteins were the most important parameter for texture in their study of drying of cod and hilsa at 30 °C. Mullet cooked for 20 min in boiling water showed an increase in denaturated proteins from 12.82 to 27.15% (Aman, 1983).

Approximately 95% of the water in muscle is mechanically immobilized water, often referred to as 'free water'. This water is free to migrate throughout the muscle structure and is of interest when measuring WHC in muscle. The fat content of cod muscle is about 0.3%. The average total liquid loss for non-fatty fish species is 18.6% (Aitken and Connell, 1979). Thus, the liquid loss must consist mainly of water and dissolved proteins. The nuance between water and liquid holding capacity is therefore less important for non-fatty fish species and WHC is in this text used to cover all liquid released from the fish even if it contains dissolved proteins and small amounts of fat.

Most of the fish muscle is segmented into disc-shaped myotoma each constituting a separate muscle. These discs are mainly fibres that are single cells with a membrane known as a sarcolemma (Bremner and Hallett, 1985). The fibres are composed of the myofibrils which contain the contractile sarcomere of actin and myosin in a cross-linked structure. Approximately 0.1% of the total tissue water content is chemically bound water, 5–15% is immobilized in the myofibrils, while the rest is extra-myofibrillar water which is the most critical for the WHC (Olsson, 2003). The extra-myofibrillar water is located between the myofibrils and between the myofibrils and the cell membrane (sarcolemma), as well as between muscle cells and between muscle bundles (groups of muscle cells) (Huff-Lonergan and Lonergan, 2005). The ratio between the immobilized myofibrillar water and the extra-myofibrillar water is important as the latter is easily lost.

It is generally accepted that the forces that immobilize 'free water' within the muscle are generated by surface tension (Hamm, 1986). More specifically, the water is trapped within the muscle by capillary action generated by small pores or 'capillaries' (Trout, 1988). The pores producing the capillary forces are located between myosin and actin, and are approximately 10 nm under normal conditions (Hermansson, 1983). Changes in the volume of the myofibrils will induce changes in water held by the muscle. In raw meat, absorption of water occurs by the entry of water into the myofibrils. Conversely, loss of water occurs by expulsion of water from the myofibrils as they shrink when filaments move towards each other. WHC depends on heat-induced structural changes, sarcomere length, pH, ionic strength, osmotic pressure and the state of rigor mortis (Ofstad *et al.*, 1996b). Changes in WHC of farmed cod are only broadly related to bacterial growth (Olsson *et al.*, 2007).

WHC of cod during heating has been studied by Ofstad *et al.* (1993), who showed that the main structural changes occur in the connective tissue at low temperatures (<40 °C) for both cod and salmon. They concluded that water loss at these temperatures is mainly due to denaturation and melting of collagen. These authors used a heating rate of 1 °C/min to the required temperature and a 10-min holding time. The solid material was placed on a net in a sample holder and centrifuged at 210 × g for 15 min. The maximum water loss was attained when the muscle cell had shrunk due to denaturation of myosin and the extracellular spaces were widened. The centrifugation loss decreased as a function of temperature when extracellular, granulated material became visible. Ofstad *et al.* (1993) found the centrifugation loss was at a maximum between 40 and 60 °C, while it was lower above 60 °C.

The properties of comminuted fish are affected both by the structure of the biological tissue and the new structures formed during processing (Ofstad *et al.*, 1996a). However, as the changes in microstructure showed a close relation to the WHC of the tissue during heating for coarsely chopped muscle, it was claimed that this would also be true for comminuted fish. As expected, it was also found that increasing the temperature from 30 to 60 °C increased the liquid loss. Ofstad *et al.* (1993) showed that liquid loss at 60 °C was much higher when the muscle was comminuted before heating.

When heating salmon at 121.1 °C, the cook loss increased rapidly to 14% after 5 min heating and then increased more slowly to 20% after 2 h (Kong *et al.*, 2007a).

For a solid portion of fish, for example a fillet cut of 140 g, a temperature gradient from the surface to the core of the fish will exist during a conventional heating process in a steam cabinet. This has been studied for transport of moisture in tuna muscle. Bell *et al.* (2001) found that moisture transport out of the tuna muscle primarily resulted from the denaturation of muscle proteins and the resulting pressure gradient. The changes in mass transfer properties in the near-surface region reduced the mass loss rates but enforced the pressure gradient. The ratio of surface area to total volume is not expected to be important for a vacuum packaged fish portion.

The most labile muscle proteins of fish are collagen and α-actin. Collagen denaturation starts at about 30 °C and α-actin becomes insoluble at 50 °C. Any temperature sufficient to cook fish is sufficient to disrupt collagen, so that it is not a factor in the toughness of the cooked material. Myosin becomes insoluble at about 55 °C and actin at 70–80 °C. Tropomyosin and troponin are the most heat resistant and become insoluble at about 80 °C (Damodaran, 1996).

Muscle proteins are generally classified as sarcoplasmic (water soluble), myofibrillar (salt soluble) or stromal (insoluble) (Skåra and Regenstein, 1990), of which myofibrillar constitute the largest fraction of the total protein in cod (76%), while the stromal (extracellular) fraction, which mainly consists of collagen, the smallest (3%) (Suzuki, 1981).

Cod muscle contains many proteins and 50% of the 446 unique proteins identified by Gebriel *et al.* (2010) were assigned general binding functions.

The sarcoplasmic proteins as such, do not contribute much to texture changes because they have a very low capacity for immobilization of water in their structure (Dunajski, 1979). The enzymes present in the sarcoplasmic fraction, however, may influence gelation of intact muscle.

Myosin and actin are the myofibrillar proteins directly involved in contraction–relaxation. Due to the low collagen content of fish, their role in gelation and texture is even more important than in meat (Brown, 1987). Much work has been done on thermal gelation of myofibrillar proteins, especially related to surimi (Arai, 2002).

The effects of proteolytic enzyme activity in fish products have been reviewed by Haard (1992) and Skåra and Olsen (1999). Quality changes in texture are the main problems related to residual enzyme activity in fish products after thermal processing. Deng (1981) studied textural changes in mullet during heating and found different profiles in shear force when changing the heating process. He suggested that the texture of cooked fish is tougher and is due to physical protein denaturation only, while a slower and/or stepwise heating process gave a tender/softer texture, which he explained by alkaline protease activity. Most enzymes are inactivated at temperatures above 50 °C (Svensson, 1977), while some enzymes are known to be heat stable (with remaining and stable activity over a certain range of time and temperature), and therefore have the potential to change the product quality. In fact some heat-stable alkaline proteases are more or less inactive at physiological temperatures and only activated at high temperatures (Toyohara *et al.*, 1987).

Cod meat of good quality has a mild taste which is not much altered by mild heat treatment. However, warmed over flavour (WOF), a phenomenon well known from mammal meats, has also been reported for fish. WOF sometimes develops 1–2 days after thermal processing and is characterized as the taste of carton or paint. WOF is caused by the oxidation of lipids and often associated with mammal meats but also fish (Pearson *et al.*, 1977). Thermal processing releases iron and other ions from the fish meat and increases the rate of the oxidation in the range 60–70 °C. Above 80 °C, and especially around 100–110 °C, Maillard reactions will slow down the reaction as reduction of sugar and proteins have an antioxidative effect (Yamauchi, 1972). Addition of antioxidant to the fish feed (e.g. E-vitamin, alfa-tokoferol) will reduce the rate of oxidation. The development of WOF is commonly measured by the content of thiobibuturic acid (TBA). Incidences of WOF in cod muscle have rarely been reported, possibly due to its relative low fat content or regarded as rancidity developed during storage. However, formation of TBA during cooking of comminuted cod has been found and may be controlled by the addition of chitosan (Shahidi *et al.*, 2002).

4.4.3 Cook loss

The primary mechanism leading to cook loss is thermal denaturation of muscle proteins (Bell *et al.*, 2001). Heating fish muscle causes denaturation of myosin and shrinkage of myofibrils, which leads to expulsion of water (Ofstad *et al.*, 1993). However, the cook loss continued to increase when the samples were heated to higher temperatures than required for the proteins to denaturate and the water in the sample could be expected to be available for expulsion by a weak mechanical force. Such a mechanical force could be due to gradients of temperature, pressure and water concentration. This will obviously lead to expulsion of water from any porous media that is heated, provided it contains free water.

Fish, like most foods, has small pores or capillary pores that may contain water. Most of the water is located within the myofibrils, in the narrow channels between thick and thin filaments (Offer and Trinick, 1983; Bertola *et al.*, 1994). There are primarily three transport mechanisms for fluids in food; molecular diffusion (for gases), capillary diffusion (for liquids) and convection (pressure driven or Darcy flow). A mechanistic model of this fluid transport has been described by Ni *et al.* (1999) and has been presented in the equations for total flux of vapour $\vec{n}_v$ and liquid $\vec{n}_w$ (kg/m^2 s) as shown in Equations 4.1 and 4.2. In addition, an equation was developed for total flux of air, $\vec{n}_a$, which can be neglected in our case since the amount of air in the vacuum pouch will be insignificant compared to the amount of water in the fish.

$$\vec{n}_v = -\rho_v \frac{k_g}{\mu_g} \nabla P - \frac{C_g^2}{\rho_g} M_a M_v D_{eff} \nabla \left(\frac{\rho_v}{P}\right) \tag{4.1}$$

and,

$$\vec{n}_w = -\rho_w \frac{k_w}{\mu_w} \nabla P - D_w \rho_w \phi \nabla S_w - D_T \nabla T \tag{4.2}$$

where C_g is molar gas concentration (mol/m^3), ϕ is relative humidity (% of saturated moist air), D_{eff} is effective capillary diffusivity (m^2/s), D_T is capillary diffusivity due to temperature gradient (m^2/s), M_a is moisture content (kg of water/m^3) with respect to air, M_v is moisture content (kg of water/m^3) with respect to vapour, P is pressure (Pa), S_w is saturation of water, T is temperature (K), k_w and k_g are permeability (m^2) of water and gas, respectively, μ_g is gas viscosity (kg/m s) and $\rho_a, \rho_g, \rho_v, \rho_w$ are density (kg/m^3) of air, gas, vapour and water, respectively.

Since air can be regarded as absent in the vacuum pouch, Equation 4.1 may be simplified by neglecting the flux of vapour due to binary diffusion in air. Thus, the vapour flux may be calculated by Equation 4.3.

$$\vec{n}_v = -\rho_v \frac{k_g}{\mu_g} \nabla P \tag{4.3}$$

The water saturation S_w was defined as in Equation 4.4 by Ni *et al.* (1999)

$$S_w = \frac{c_w}{\rho_w \phi} \tag{4.4}$$

where c is mass concentration (kg/m^3) and thus, Equation 4.2 can be simplified to Equation 4.5.

$$\vec{n}_w = \underbrace{-\rho_w \frac{k_w}{\mu_w} \nabla P}_{\text{flux of water due to pressure}} - \underbrace{D_w \nabla c_w}_{\text{concentration driven water flux}} - \underbrace{D_T \nabla T}_{\text{temperature driven water flux}} \tag{4.5}$$

To obtain effective heat transfer and predictable heat transfer coefficients between the heating medium and the fish, several processors choose to process the vacuum packed portions in an autoclave under an overpressure high enough to prevent formation of vapour and to assure good contact between the pouch and the product. Under such conditions the capillary diffusivity of water, D_w is given by Equation 4.6.

$$D_w = -\rho_w^2 g \frac{k_w}{\mu_w} \frac{\partial h}{\partial c_w} \tag{4.6}$$

where g is gravity (9.81 m/s^2) and h is enthalpy (J/kg).

To measure or determine the constants used in these equations is outside the scope of this thesis. Still, the principals of the water flux out of a vacuum packed portion of fish from the three terms in Equation 4.5 can be discussed in light of the observed cook loss:

1. Provided the density, permeability and viscosity of water is constant (true under isothermal conditions), the water flux is increasing linearly with increasing pressure. However, the pressure gradient is not increasing linearly because it is a result of:
   (a) muscle shrinkage caused by several steps of thermal protein denaturation, i.e. a function of temperature and time;
   (b) thermal expansion of water and gases in the fish meat, i.e. a function of temperature.

   Indeed, the observed cook loss increases with temperature as proposed by Equation 4.5, but it increases much faster at the end of the major protein transitions.
2. Water expelled will surround the fish and a concentration gradient will build up and counteract the expulsion of water. Depending on the time/temperature history and the condition of the fish, proteins will be dissolved in the expelled water and decrease the concentration gradient. Water will also be released from proteins and increase the concentration gradient in the opposite direction. The concentration gradient term may explain the observed cook loss at low temperatures or during heating to a higher temperature.
3. The temperature gradient is given by the heat equation and will be log-linearly decreasing from a maximum when heating is started and approach zero as the temperature in the fish evens out. This is consistent with the observation of little or no increase in cook loss over time at a constant temperature after protein denaturation, which is completed within a few minutes at constant temperature.

The three terms in Equation 4.5 could in principle fit well with the observed data but should incorporate the protein denaturation effects. It has previously been shown that heat-induced structural changes (Ofstad *et al.*, 1993) and protein denaturation (Skipnes *et al.*, 2008) is irreversible. During cooling the new configuration of the fish meat established during heating will remain unchanged. Only a small amount of the liquid expelled during heating will be absorbed by the fish during cooling and it is not expected to cause any swelling but rather follow Equation 4.6. Further research on protein denaturation would indeed help in establishing this model, but it is possible this could be studied indirectly by measuring the shrinkage of fish meat during heating. Studies of salmon heated at temperatures in the range 100–130 °C have shown that the cook loss is relatively constant after 30 min heating time and that this plateau level was a function of the temperature and correlated well with area shrinkage of a disc-shaped sample (Kong *et al.*, 2007b). This rapid increase before reaching the plateau level seems to be the same as behaviour seen in cod at temperatures below 100 °C.

4.4.4 Water holding capacity

Results for WHC of farmed Atlantic cod screened by isothermal heating at temperatures in the range 20–95 °C averaged over heating times of 1–60 min are shown in Figure 4.2. As expected, increasing temperature and increasing processing time result in a loss of WHC. There was a major drop in WHC from 20 to 45 °C. However, within a limited range of processing time and temperatures, the WHC increased in the range 45–70 °C. Time and temperature combinations resulting in excessive loss of WHC may be associated with

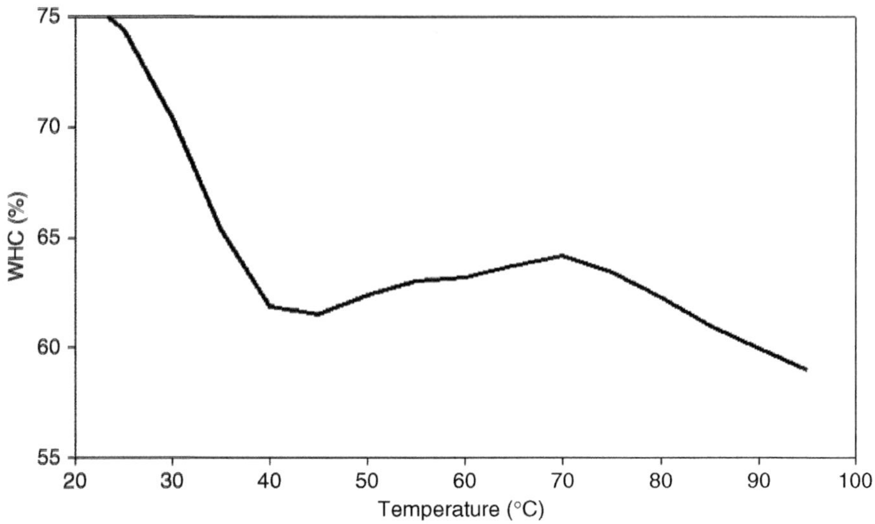

Figure 4.2 Water holding capacity (WHC) at heat treatment temperature from 20 to 95 °C averaged over cooking times from 1 min to 60 min

a high drip loss during distribution and storage, as well as loss of juiciness. Combinations with an intermediate heat load, for example 30 min at 60 °C, resulted in WHC in the range 66–70%. These time and temperature combinations coincide with the heat load required for inactivation of *L. monocytogenes*. In this range it is possible to keep WHC above 66% and the cook loss below 5.6%. Therefore pasteurization of cod should be carried out at approximately 68 ± 4 °C within a processing time of about 25–35 min.

Both WHC and cook loss originates from a combination of physical reasons (i.e. concentration and temperature gradients) and protein denaturation. Protein denaturation results in changes in the internal structure of the fish muscle and the capillaries therein. Shrinkage and mechanical expulsion of juice is another effect related to protein denaturation.

4.4.5 Texture and colour changes

Measurements of texture could be expected to correlate with water content as texture is closely related to protein denaturation (Dunajski, 1979), but this relation is not straight forward and is sometimes technically challenging. A texture profile analysis (TPA) on cooked material did not result in reproducible results because the angle of the myotoma is impossible to standardize after cooking (Skipnes, 2011). Texture measurements on single, intact myotoma has been used to include direct measurements of texture (Skipnes *et al.*, 2011) and a twice compression test on a small piece of food can be used to obtain a number of textural parameters that have been found to correlate well with sensory evaluation (Bourne, 1978, 1992). The primary parameters of hardness, cohesiveness, springiness (elasticity) and adhesiveness, and the secondary (or derived) parameters of fracturability (brittleness), chewiness and gumminess can be determined by the TPA (Szczesniac, 1966).

The hardness of samples heat treated at different temperatures is not necessarily significantly different from each other but all cooked samples were significantly harder than raw samples in a study by Skipnes *et al.* (2011). A tendency for hardness to increase as heat load increases can also be expected. Further measurements are needed to verify a possible correlation between texture and WHC.

When heated, the opaque cod muscle changes to white. This happens while heating from 30 to 40 °C. At temperatures from 70 to 90 °C there was no significant change in colour. The whiteness score of cod samples treated at temperatures above 60 °C was higher than for all samples treated at lower temperatures (Skipnes *et al.*, 2011). For salmon and trout, the colour changes from red to pink and then to pale pink as the heat load increases are more evident than the relatively small colour changes seen in white fish. The release of water-soluble proteins that later form a yellow to grey curd on the surface of the fish product is still a challenge to processors. It is well known that a brine containing a small percentage of salt and small amounts of citric acid will reduce this problem.

Acknowledgement

This work was part of the project 'Innovative and Safe Seafood – Processing, Hygiene, Spectroscopy' funded by the research council of Norway (NFR project no. 186905).

References

ACMSF (1992). *Report on Vacuum Packaging and Associated Processes*, Advisory Committee on the Microbiological Safety of Food, HMSO, London, UK.

ACMSF (1995). *Annual Report*, Her Majesty's Stationary Office, London, UK.

ACMSF (2006). *Report on Minimally Processed Infant Weaning Foods and Infant Botulism*, Advisory Committee on the Microbiological Safety of Food, Food Standads Agency, London, UK.

Aitken, A. and Connell, J.J. (1979). Fish, in *Effects of Heating of Foodstuffs* (ed. R.J. Priestley), Applied Science Publishers Ltd, Barking, Essex, UK, pp. 219–254.

Aman, M.B. (1983). Effect of cooking and preservation methods on the water holding capacity (WHC) of mullet fish in relation with changes occurred in muscle proteins. *Zeitschrift fur Lebensmittel-Untersuchung Und-Forschung*, 177, 345–347.

Appert, N. (1810). L'art de conserver pendant plusiers annes toutes les substances animales et vegetables (translated by B.G. Bitting, Chicago, IL, 1920), in *Introduction to Thermal Processing of Foods (1961)* (eds S.A.Goldblitch, M.A. Joslyn and J.T.R. Nickerson), AVI Publishing Inc., Westport, CT, USA.

Arai, K. (2002). Denaturation of muscular proteins from marine animals and its control. *Nippon Suisan Gakkaishi*, 68(2), 137–143.

Bell, J.W., Farkas, B.E., Hale, S.A. and Lanier, T.C. (2001). Effect of thermal treatment on moisture transport during steam cooking of skipjack tuna (*Katsuwonas pelamis*). *Journal of Food Science*, 66(2), 307–313.

Ben Embarek, P.K. (1993). Heat resistance of *Listeria monocytogenes* in vacuum packaged pasteurized fish fillets. *International Journal of Food Microbiology*, 20(2), 85–95.

Bertola, N.C., Bevilacqua, A.E. and Zaritzky, N.E. (1994). Heat-treatment effect on texture changes and thermal-denaturation of proteins in beef muscle. *Journal of Food Processing and Preservation*, 18(1), 31–46.

Betts, G. (2009). *Code of Practice for the Manufacture of Vacuum and Modified Atmosphere Packaged Chille Foods with Particular Regard to the Risk of Botulism (Guideline no. 11)*, 2nd edn, Campden BRI, Chipping Campden, UK.

Bourne, M.C. (1978). Texture profile analysis. *Food Technology*, 32(7), 62–66.

Bourne, M.C. (1992). Calibration of rheological techniques used for foods. *Journal of Food Engineering*, 16(1–2), 151–163.

Bremner, H.A. and Hallett, I.C. (1985). Muscle-fiber connective-tissue junctions in the fish blue grenadier (*Macruronus*-Novaezelandiae – a scanning electron-microscope study. *Journal of Food Science*, 50(4), 975–980.

Brown, W.D. (1987). Fish muscle as food, in *Muscle as Food* (ed. P.J. Bechtel), Academic Press Inc., Orlando, FL, pp. 405–451.

Buchanan, R.L., Damert, W.G., Whiting, R.C. and Van Schothorst, M. (1997). Use of epidemiologic and food survey data to estimate a purposefully conservative dose–response relationship for *Listeria monocytogenes* levels and incidence of listeriosis. *Journal of Food Protection*, 60(8), 918–922.

Cheftel, J.C., Cuq, J.L. and Lorient, D. (1985). Amino acids, peptides and proteins – VII Modifications of food proteins through processing and storage, in *Food Chemistry*, 2nd edn (ed. O.R. Fennema), Marcel Decker Inc., New York, USA, pp. 232–369.

Codex Alimentarius Comission (2001). *Fish and Fishery Products, Yearly Report*, Food and Agriculture Organization of the United Nations, Rome, Italy.

Codex Alimentarius Comission (2002). *Fish and Fishery Products, Yearly Report*, Food and Agriculture Organization of the United Nations, Rome, Italy.

Damodaran, S. (1996) Amino acids, peptides and proteins, in *Food Chemistry*, 3rd edn (ed. O.R. Fennema), Marcel Dekker, New York, USA, pp. 321–431.

Deng, J.C. (1981). Effect of temperatures on fish alkaline protease, protein interaction and texture quality. *Journal of Food Science*, 46, 62–65.

Drake, S.L., De Paola, A. and Jaykus, L.A. (2007). An overview of *Vibrio vulnificus* and *Vibrio parahaemolyticus*. *Comprehensive Reviews in Food Science and Food Safety*, 6(4), 120–144.

Dunajski, E. (1979). Texture of fish muscle. *Journal of Texture Studies*, 10, 301–318.

ECFF (1996). *Guideline for The Hygienic Manufacture of Chilled Foods*, European Chilled Food Federation, Brussels, Belgium.

Eisner, M. (1988). *Introduction into the Technique and Technology of Rotary Sterilization.* Private Author's Edition, Milwaukee, WI, USA.

FAO (2009). *Yearbooks of Fishery Statistics, Summary tables of Fishery Statistics, Capture – Aquaculture – Commodities*, Food and Agriculture Organization of the United Nations, Rome, Italy.

FAO (2010). *The State of World Fisheries and Aquaculture*, Food and Agriculture Organization of the United Nations, Rome, Italy.

FAO (2012). *The State of World Fisheries and Aquaculture*, Food and Agriculture Organization of the United Nations, Rome, Italy.

FDA (2005). *Food Code*, US Department of Health and Human Services, Food and Drug Administration.

Feldhusen, F. (2000). The role of seafood in bacterial foodborne diseases. *Microbes and Infection*, 2(13), 1651–1660.

Gebriel, M., Uleberg, K.E., Larssen, E. *et al.* (2010). Cod (*Gadus morhua*) muscle proteome cataloging using 1D-PAGE protein separation, nano-liquid chromatography peptide fractionation, and linear trap quadrupole (LTO) mass spectrometry. *Journal of Agricultural and Food Chemistry*, 58(32), 12307–12312.

Goñi, S.M. and Salvadori, V.O. (2012). Model-based multi-objective optimization of beef roasting. *Journal of Food Engineering*, 111(1), 92–101.

Gram, L. and Dalgaard, P. (2002). Fish spoilage bacteria – problems and solutions. *Current Opinion in Biotechnology*, 13(3), 262–266.

Greig, J.D. and Ravel, A. (2009). Analysis of foodborne outbreak data reported internationally for source attribution. *International Journal of Food Microbiology*, 130(2), 77–87.

Haard, N.F. (1992). Control of chemical composition and food quality attributes of cultured fish. *Food Research International*, 25(4), 289–307.

Hamm, R. (1986). *Muscle as Food*, Academic Press, New York, USA.

Heinz, V. and Buckow, R. (2010). Food preservation by high pressure. *Journal für Verbraucherschutz und Lebensmittelsicherheit*, 5(1), 73–81.

Hermansson, A.M. (1983). Quality of plant foods. *Human Nutrition*, 32, 369.

Hospido, A., Vazquez, M.E., Cuevas, A., Feijoo, G. and Moreira, M.T. (2006). Environmental assessment of canned tuna manufacture with a life-cycle perspective. *Resources, Conservation and Recycling*, 47(1), 56–72.

Howgate, P.F. and Ahmed, S.F. (1972). Chemical and bacteriological changes in fish muscle during heating and drying at 30 °C. *Journal of the Science of Food and Agriculture*, 23, 615–627.

Huff-Lonergan, E. and Lonergan, S.M. (2005). Mechanisms of water-holding capacity of meat: The role of postmortem biochemical and structural changes. *Meat Science*, 71(1), 194–204.

Huss, H.H. (1994). *Assurance of Seafood Quality*, Food and Agriculture Organization of the United Nations, Rome, Italy.

Huss, H.H. (1995). *Quality and quality changes in fresh fish*, Food and Agriculture Organization of the United Nations, Rome, Italy.

IFTPS (1992). *Temperature Distribution Protocol For Processing In Steam Still Retorts, Excluding Crateless Retorts*, Institute for Thermal Processing Specialists,Washington, DC.

IFTPS (1995). *Protocol for Carrying Out Heat Pentration Studies*, Institute for Thermal Processing Specialists, Washington, DC.

IFTPS (2002). *Temperature Distribution Protocol for Processing iIn Still, Water Immersion Retorts, Including Agitating Systems Operated in a Still Mode*, Institute for Thermal Processing Specialists, Washington, DC.

Iribarren, D., Moreira, M.T. and Feijoo, G. (2010a). Life cycle assessment of fresh and canned mussel processing and consumption in Galicia (NW Spain). *Resources, Conservation and Recycling*, 55(2), 106–117.

Iribarren, D., Moreira, M.T. and Feijoo, G. (2010b). Revisiting the life cycle assessment of mussels from a sectorial perspective. *Journal of Cleaner Production*, 18(2), 101–111.

Johnston, M.D. and Brown, M.H. (2002). An investigation into the changed physiological state of *Vibrio* bacteria as a survival mechanism in response to cold temperatures and studies on their sensitivity to heating and freezing. *Journal of Applied Microbiology*, 92(6), 1066–1077.

Kong, F., Tang, J., Rasco, B. and Crapo, C. (2007a). Kinetics of salmon quality changes during thermal processing. *Journal of Food Engineering*, 83, 510–520.

Kong, F., Tang, J., Rasco, B., Crapo, C. and Smiley, S. (2007b). Quality changes of salmon (*Oncorhynchus gorbuscha*) muscle during thermal processing. *Journal of Food Science*, 72(2), 103–111.

Larsen, R., Stormo, S.K., Dragnes, B.T. and Elvevoll, E.O. (2007). Losses of taurine, creatine, glycine and alanine from cod (*Gadus morhua* L.) fillet during processing. *Journal of Food Composition and Analysis*, 20(5), 396–402.

Lejeune, P. (2003). Contamination of abiotic surfaces: what a colonizing bacterium sees and how to blur it. *Trends in Microbiology*, 11(4), 179–184.

Lindstrom, M., Kiviniemi, K. and Korkeala, H. (2006). Hazard and control of group II (non-proteolytic) *Clostridium botulinum* in modem food processing. *International Journal of Food Microbiology*, 108(1), 92–104.

Lunestad, B.T. (2003). Absence of nematodes in farmed Atlantic salmon (*Salmo salar* L.) in Norway. *Journal of Food Protection*, 66(1), 122–124.

Lyng, J.G., Cronin, D.A., Brunton, N.P., Li, W.Q. and Gu, X.H. (2007). An examination of factors affecting radio frequency heating of an encased meat emulsion. *Meat Science*, 75(3), 470–479.

Mackenzie, K., Hemmingsen, W., Jansen, P.A., Sterud, E. and Haugen, P. (2009). Occurrence of the tuna nematode Hysterothylacium cornutum (Stossich, 1904) in farmed Atlantic cod (*Gadus morhua*) in North Norway. *Polar Biology*, 32(7), 1087–1089.

May, N. (1997). Guidelines for batch retort systems – full water immersion – raining/spray water – steam/air, in *Guideline No. 13* (ed. N. May), Campden and Chorleywood Food Research Association, Chipping Campden, Gloucestershire, UK, pp. 22–37.

Mckenna, B.M., Lyng, J., Brunton, N. and Shirsat, N. (2006). Advances in radio frequency and ohmic heating of meats. *Journal of Food Engineering*, 77(2), 215–229.

Ni, H., Datta, A. and Torrance, K. (1999). Moisture transport in intensive microwave heating of biomaterials: a multiphase porous media model. *International Journal of Heat and Mass Transfer*, 42(8), 1501–1512.

Nicolaii, B.M. (1994). Modeling and uncertainty propagation analysis of thermal food processes. PhD thesis. Katholieke Universiteit Leuven.

Nilsson, L. and Gram, L. (2002). Improving the control of pathogens in fish products, in *Safety and Quality Issues in Fish Processing* (ed. H.A. Bremmer), Woodhead Publishing Limited, Cambridge, England, pp. 54–84.

Offer, G. and Trinick, J. (1983). On the mechanism of water holding in meat – the swelling and shrinking of myofibrils. *Meat Science*, 8(4), 245–281.

Ofstad, R., Kidman, S. and Hermansson, A.M. (1996a). Ultramicroscopical structures and liquid loss in heated cod (*Gadus morhua*) and salmon (*Salmo salar*) muscle. *Journal of the Science of Food and Agriculture*, 72(3), 337–347.

Ofstad, R., Kidman, S., Myklebust, R. and Hermansson, A.M. (1993). Liquid holding capacity and structural-changes during heating of fish muscle – Cod (*Gadus morhua*) and Salmon (*Salmo salar*). *Food Structure*, 12(2), 163–174.

Ofstad, R., Kidman, S., Myklebust, R., Olsen, R.L. and Hermansson, A.M. (1996b). Factors influencing liquid-holding capacity and structural changes during heating of comminuted cod (*Gadus morhua*) muscle. *Food Science and Technology-Lebensmittel-Wissenschaft and Technologie*, 29(1–2), 173–183.

Ohlsson, T.H.O.M. (1980). Temperature dependency of sensory quality changes during thermal processing. *Journal of Food Science*, 45(4), 836–839.

Olsson, G.B. (2003). Water-holding capacity and quality – studies on Atlantic halibut (*Hippoglossus hippoglossus*) muscle. PhD thesis. University of Tromsø, Institute of Marine Biotechnology.

Olsson, G.B., Seppola, M.A. and Olsen, R.L. (2007). Water-holding capacity of wild and farmed cod (*Gadus morhua*) and haddock (*Melanogrammus aeglefinus*) muscle during ice storage. *LWT-Food Science and Technology*, 40(5), 793–799.

Paoli, G.C., Bhuni, A.K. and Byles, D.O. (2005). Listeria monocytogenes, in *Foodborne Pathogens: Microbiology and Molecular Biology* (eds P.M. Fratamico, A.K. Bhunia and J.L. Smith). Caister Academic Press, Norfolk, UK.

Parry, D.A. (1970). Fish as food, in *Proteins as Human Foods* (ed. R.A. Lawrie), Butterworths, London, UK, 365 p.

Pearson, A.M., Love, J.D. and Shorland, F.B. (1977). Warmed-over flavor in meat, poultry and fish. *Advances in Food Research*, 23, 1–74.

Peleg, M. and Cole, M.B. (1998). Reinterpretation of microbial survival curves. *Critical Reviews in Food Science and Nutrition*, 38(5), 353–380.

Peleg, M., Normand, M.D., Corradini, M.G. *et al.* (2008). Estimating the heat resistance parameters of bacterial spores from their survival ratios at the end of UHT and other heat treatments. *Critical Reviews in Food Science and Nutrition*, 48(7), 634–648.

Pereira, R.N. and Vicente, A.A. (2010). Environmental impact of novel thermal and non-thermal technologies in food processing. *Food Research International*, 43(7), 1936–1943.

Rajkovic, A., Smigic, N. and Devlieghere, F. (2010). Contemporary strategies in combating microbial contamination in food chain. *International Journal of Food Microbiology*, 141, 29–42.

Rhodehamel, E.J. (1992). FDA concerns with sous vide processing. *Food Technology*, 46(12), 73–76.

Richardson, P. (2004). *Improving the Thermal Processing of Foods*, Woodhead Publishing Ltd, Cambridge, England.

Ryckaert, V.G., Claes, J.E. and Van Impe, J.F. (1999). Model-based temperature control in ovens. *Journal of Food Engineering*, 39(1), 47–58.

Ryynänen, S. (2002). Microwave heating uniformity of multicomponent prepared foods. PhD EKT series 1260, University of Helsinki, Department of Food Technology (Dissertation).

Schmidt, E.B., Rasmussen, L.H., Rasmussen, J.G. *et al.* (2006). Fish, marine n-3 polyunsaturated fatty acids and coronary heart disease: a minireview with focus on clinical trial data. *Prostaglandins Leukotrienes and Essential Fatty Acids*, 75(3), 191–195.

Setlow, P. and Johnson, E.A. (1997). Spores and their significance, in *Food Microbiology. Fundamentals and Frontiers*, (eds M.P. Doyle, L.R. Beuchat and T.J. Montville), ASM Press, Washington, DC, pp. 30–65.

Shahidi, F., Kamil, J., Jeon, Y.J. and Kim, S.K. (2002). Antioxidant role of chitosan in a cooked cod (*Gadus morhua*) model system. *Journal of Food Lipids*, 9(1), 57–64.

Sheard, M.A. and Rodger, C. (1995). Optimum heat treatments for 'sous vide' cook-chill products. *Food Control*, 6(1), 53–56.

Silva, F.V.M. and Gibbs, P.A. (2010). Non-proteolytic Clostridium botulinum spores in low-acid cold-distributed foods and design of pasteurization processes. *Trends in Food Science and Technology*, 21(21), 95–105.

Simpson, R., Cortés, C. and Teixeira, A. (2006). Energy consumption in batch thermal processing: model development and validation. *Journal of Food Engineering*, 73(3), 217–224.

Skåra, T., Cappuyns, A.M., Johnsen, S.O., Van Derlinden, E., Rosnes, J.T., Olsen, Ø., Impe, J.F.M.V. and Valdramidis, V.P. (2011). A thermodynamic approach to assess a cellular mechanism of inactivation and the thermal resistance of *Listeria innocua*. *Procedia Food Science*, 1, 972–978.

Skåra, T. and Olsen, S.O. (1999) Proteolytic enzyme activity during thermal processing. Proceedings of the 29th WEFTA Meeting (ed. S.A. Georgakis), Greek Society of Food Hygienists and Technologists, Leptocarya, Pieria, Greece, pp 268–286,

Skåra, T. and Regenstein, J. (1990). The structure and properties of myofibrillar proteins in beef, poultry and fish. *Journal of Muscle Foods*, 1, 269–291.

Skipnes, D. (2011). Optimisation of thermal processing of fresh farmed cod. PhD dissertation. Norwegian University of Life Sciences.

Skipnes, D. and Hendrickx, M. (2008). Novel methods to optimise the nutritional and sensory quality of in-pack processed fish products, in *In-pack Processed Foods: Improving Quality* (ed. P. Richardson), Woodhead Publishing Inc., Cambridge, pp. 382–402.

Skipnes, D., Johnsen, S.O., Skåra, T., Sivertsvik, M. and Lekang, O. (2011). Optimization of heat processing of farmed Atlantic cod (*Gadus morhua*) muscle with respect to cook loss, water holding capacity, color and texture. *Journal of Aquatic Food Product Technology*, 20(3), 331–340.

Skipnes, D., Øines, S., Rosnes, J.T. and Skåra, T. (2002). Heat transfer in vacuum packed mussels (*Mytilus edulis*) during thermal processing. *Journal of Aquatic Food Product Technology*, 11(3/4), 5–19.

Skipnes, D., Van Der Plancken, I., Van Loey, A. and Hendrickx, M. (2008). Kinetics of heat denaturation of proteins from farmed Atlantic cod (*Gadus morhua*). *Journal of Food Engineering*, 85(1), 51–58.

Suzuki, T. (1981). *Fish and Krill Protein: Processing Technology*, Applied Science Publishers, London, UK.

Svensson, S. (1977). Inactivation of enzymes during thermal processing, in *Physical, Chemical, and Biological Changes in Food Caused by Thermal Processing* (ed. T.K.O. Høyem), Applied Science Publishers Ltd, London, UK, pp. 201–218.

Szczesniac, A.S. (1966). Texture measurements. *Food Technology*, 20(10), 1292–1293.

Thostenson, E.T. and Chou, T.W. (1999). Microwave processing: fundamentals and applications. *Composites Part A: Applied Science and Manufacturing*, 30(9), 1055–1071.

Toepfl, S., Mathys, A., Heinz, V. and Knorr, D. (2006). Review: potential of high hydrostatic pressure and pulsed electric fields for energy efficient and environmentally friendly food processing. *Food Reviews International*, 22(4), 405–423.

Toyohara, H., Nomata, H., Makinodan, Y. and Shimizu, Y. (1987). High molecular weight heat-stable alkaline proteinase from white croaker and chum salmon muscle: comparison of the activating effects by heating and urea. *Comparative Biochemistry and Physiology Part B: Comparative Biochemistry*, 86(1), 99–102.

Trout, G.R. (1988). Techniques for measuring water-binding capacity in muscle foods – a review of methodology. *Meat Science*, 23(4), 235–252.

Tucker, G. (2003). Cook value calculations and optimisation, in *Thermal Processing: A safe Approach* (ed. S. Emond), Campden and Chorleywood Food Research Association Group, Chipping Campden, UK, pp. 1–10.

van Boekel, M.A.J.S. (2008). Kinetic modeling of food quality: a critical review. *Comprehensive Reviews in Food Science and Food Safety*, 7(1), 144–158.

Valdramidis, V.P., Geeraerd, A.H., Poschet, F. *et al.* (2007). Model based process design of the combined high pressure and mild heat treatment ensuring safety and quality of a carrot simulant system. *Journal of Food Engineering*, 78(3), 1010–1021.

Verboven, P., Scheerlinck, N., De Baerdemaeker, J. and Nicolai, B.M. (2000a). Computational fluid dynamics modelling and validation of the isothermal airflow in a forced convection oven. *Journal of Food Engineering*, 43(1), 41–53.

Verboven, P., Scheerlinck, N., De Baerdemaeker, J. and Nicolai, B.M. (2000b). Computational fluid dynamics modelling and validation of the temperature distribution in a forced convection oven. *Journal of Food Engineering*, 43(2), 61–73.

Walden, R. (2008). The Zinetec Shaka^{TM} retort and product quality, in *In-pack Processed Foods: Improving Quality* (ed. P. Richardson), Woodhead Publishing Ltd, Cambridge, UK, pp. 86–101.

Walden, R. (2009). Energy saving strategies for batch retorts. Proceedings of the 2nd European IFTPS Symposium. Institute for Thermal Processing Specialists (IFTPS) Guelph, Canada, pp. 13–14.

Yamauchi, K. (1972). Effect of heat treatment on the development of oxidative rancidity in meats and its isolated tissue fraction. *Bulletine of the Faculty of Agriculture, Miyazaki University*, 19, 147.

5

Irradiation of Fish and Seafood

Ioannis S. Arvanitoyannis and Persefoni Tserkezou

Laboratory of Food Technology, Department of Agriculture, Ichthyology and Aquatic Environment, University of Thessaly, Volos, Greece

5.1 Introduction

Demand for fishery products is rising globally, therefore operative preservative techniques are needed (Rahman, 1999). The main disadvantage for the distribution of fishery products is their sensitivity to spoilage due to the growth of microorganisms (Colby *et al.*, 1993; Gram and Huss, 1996).

Food irradiation is considered a successful technique not only in guaranteeing the safety, but also in increasing the shelf life of foodstuffs such as fresh meats, despite the consumer lack of trust in this method (Arvanitoyannis *et al.*, 2009). Irradiation is very effective in inactivating pathogenic microorganisms without decreasing food quality (Josephson and Peterson, 2000; Mahapatra *et al.*, 2005). Foodstuffs are usually irradiated by gamma irradiation generated by a radioisotope source. However, X-rays, or electrons produced by an electron accelerator can also be used (Barbosa-Canovas *et al.*, 1998). Food irradiation under the recommended conditions does not include an atomic nucleus reaction, but a chemical reaction induced by the electron cloud surrounding the nucleus. The amount of energy absorbed by the foodstuff during irradiation is called the 'absorbed dose'. The international unit for the absorbed dose is the gray (Gy) (McLaughlin *et al.*, 1989).

In fact, the dose used to irradiate a foodstuff depends on the food type and the desired effect. The dose levels can be categorized in three main scales (Morehouse, 1998):

- 'Low' dose ($\leq$1 kGy) is applied to decrease physiological processes rhythm, such as ripening or sprouting of fresh fruits and vegetables, and to control pests in foodstuffs.
- 'Medium' dose (1–10 kGy) is used to decrease spoilage and pathogenic microorganisms on a variety of foodstuffs, to enhance foodstuff properties, such as decreased cooking times for dehydrated vegetables, and to extend the shelf life of many foodstuffs.
- 'High' dose ($>$10 kGy) is mainly applied for sterilization of meat, poultry, seafood and other ready-to-eat (RTE) foods in conjunction with mild heating to inactivate enzymes, and for the disinfection of foods or ingredients, such as spices.

Seafood Processing: Technology, Quality and Safety, First Edition. Edited by Ioannis S. Boziaris.

Table 5.1 Approval year and acceptable dose of a variety of food that can be irradiated in USA (http://www.cdc.gov/nczved/divisions/dfbmd/diseases/irradiation_food/ [accessed 1 June 2013])

Food	Dose (kGy)	Purpose	Approval year
Wheat flour	0.2–0.5	Mould control	1963
White potatoes	0.05–0.15	Reduce sprouting	1964
Pork	0.3–1.0	Inactivate Trichina parasites	1986
Fruit and vegetables	1.0	Pest control, increase shelf life	1986
Herbs and spices	30	Sterilization	1986
Poultry	3	Reduce bacteria	1990
Poultry	1.5–3.0	Reduce bacteria	1992
Meat	4.5	Reduce bacteria	1997
Meat	4.5	Reduce bacteria	1999

Source: US government (Center for Contagious Diseases).

Table 5.2 Authorized absorbed dose according to Directive 1999/2/EC (EPCEU, 1999) concerning fish and seafood ingredients treated with ionizing radiation

European Union Member State	Given maximum average absorbed dose (kGy) according to European Union legislation		
	Fish and shellfish (including eels, crustaceans and molluscs)	Frozen, peeld or decapitated shrimps	Shrimps
Belgium	3	5	–
France	–	5	–
Netherlands	–	–	3
United Kingdom	3	–	–

The approval year for a variety of foodstuffs to be irradiated and acceptable dose in the USA is given in Table 5.1. The maximum authorized radiation dose for fish and seafood products in the European Union member states is presented in Table 5.2.

This chapter aims to present a comprehensive review of irradiation research for fish and fishery products and also to examine the impact of irradiation on the survival of spoilage and pathogenic microorganisms.

5.2 Quality of irradiated fish and fishery products and shelf life extension

5.2.1 Fish

According to Badr (2012), irradiation of cold-smoked salmon by gamma irradiation had no significant impact on its moisture and salt content or on its pH value, total volatile base nitrogen (TVB-N), and trimethylamine nitrogen (TMA-N) content, but significantly reduced the amount of phenolic compounds and increased levels of thiobarbituric acid reactive substances (TBARS). Moreover, irradiation treatment at doses up to 3 kGy showed no significant effect on the sensory acceptability of samples.

Moini *et al.* (2009) studied the impact of gamma radiation (0, 1, 3, and 5 kGy) on the shelf life of farmed rainbow trout (*Oncorhynchus mykiss*) fillets treated with sodium acetate and vacuum packaged and then stored under refrigeration. Radiation considerably affected the populations of bacteria such as H_2S-producing bacteria and Enterobacteriaceae. The results showed that irradiation at a high dose (5 kGy) might encourage lipid and protein oxidation, although the growth of microorganisms was inhibited. Irradiation at a low dose (3 kGy) could control the microbial and safety biochemical indices of *O. mykiss* for up to 4 weeks under refrigeration without adverse effects on quality and acceptability.

According to Riebroy *et al.* (2007), irradiated samples of *somfug* (a Thai fermented fish mince) had higher TBARS values than control samples. In general, the increase in TBARS values was greater in samples irradiated at higher doses (6 kGy) during the first 25 days of the storage, than in control samples or samples irradiated at 2 kGy. The lightness (L*) value of all samples decreased, whereas the redness (a*) and yellowness (b*) values increased throughout storage. Samples irradiated at 2 kGy presented the smallest changes in a* and b* values. During storage, the pH values of all samples decreased gradually during the first 15 days of storage, compared with those stored at day 0.

The quality of non-irradiated and irradiated (2.5 and 5.0 kGy) sea bass (*Dicentrarchus labrax*) stored in ice was examined. Among the chemical indicators of spoilage, the TVB-N values increased to 36.44 mg/100 g for non-irradiated sea bass during ice storage, whereas for irradiated fish lower values of 25.26 mg/100 g and 23.61 mg/100 g were recorded at 2.5 and 5.0 kGy, respectively (day 17). TMA-N and TBARS values of irradiated samples were lower than those of non-irradiated samples (Ozden *et al.*, 2007).

Another study was conducted in fresh Atlantic horse mackerel (*Trachurus trachurus*) which were irradiated by gamma irradiation at 1 and 3 kGy, and stored in ice for 23 days (Mendes *et al.*, 2005). The non-irradiated samples showed an 8-day sensory shelf life, whereas the shelf life of irradiated samples was extended by 4 days. Total viable counts (TVC) and biogenic amine levels increased during storage time; however, their contents were significantly reduced in irradiated samples, even when the lower irradiation level (1 kGy) was used. Histamine in the irradiated lots was undetectable when the fish spoiled by the end of the day 23, while in the control lot, histamine concentration was higher but did not exceed the maximum level allowed (100 mg/kg).

The impact of gamma irradiation (1, 5 and 10 kGy) and post-irradiation ice storage ($2 \pm 0.5\,°C$) on horse mackerel (*Trachurus trachurus*) muscle proteins was studied by Silva *et al.* (2006). They concluded that irradiation doses from 1 to 10 kGy had no significant impact on the proteins. These results confirmed the potential of irradiation treatment up to 10 kGy for extending horse mackerel shelf-life and for ensuring health safety.

The effect of irradiation on colour changes for fish (*Oncorhynchus mykiss*) pieces was investigated by Dvorak *et al.* (2005). The parameters of colour, L*, a* and b*, were determined. The L* change was the same for both irradiated (3 kGy) and non-irradiated samples. This change was potentially caused by maturation of fish flesh. For a* their was no change, whereas b* decreased. The pH decrease was the same for both irradiated and non-irradiated samples. Therefore, it could be stated that pH was not affected by irradiation at 3 kGy.

Cozzo-Siqueira *et al.*, (2003) studied the effect of irradiation on *Tilapia nilotica* (*Oreochromis niloticus*). Various samples irradiated with 0, 1.0, 2.2 and 5.0 kGy. The samples were stored at temperatures ranging from 0.5 to 2 °C for 20–30 days. During storage, the moisture level in the non-irradiated samples reduced and the protein and lipid levels increased, whereas the irradiated samples remained stable. The TVB-N levels increased in the non-irradiated samples but tended to remain stable in the irradiated fish. The amino acids levels of fish muscle and fatty acids of fish oil remained stable in the irradiated samples, whereas they decreased in the non-irradiated ones. Lipid oxidation tended to increase when irradiation dose increased.

Bani *et al.* (2000) found that fish cutlets prepared at a laboratory scale according to a selected formulation and irradiated at a dose of 5 kGy could extend shelf life up to 5 weeks at room temperature. In commercially prepared fish cutlets, the maximum shelf life extension was 14 days for samples treated at 5 kGy and stored at ambient temperature on the basis of combined microbiological, chemical and organoleptic evaluation.

Mendes *et al.* (2000) studied the changes in quality of irradiated blue jack mackerel (*Trachurus picturatus*) stored at 3 °C for 23 days. In the irradiated samples (1, 2 and 3 kGy) TVB-N increased gradually with time and its level at the end of the storage was in general, three times lower compared to the non-irradiated lot. After 23 days storage, the trimethylamine (TMA) content in the 1, 2 and 3 kGy samples was 18.9, 30.9 and 10.0 mg/100 g, respectively. Analysis of results showed that irradiation substantially decreased the TMA content.

Lakshmanan *et al.* (1999) studied irradiated anchovy (*Stolephorus commersonii*) at a dose of 2 kGy. Non-packaged fish stored under melting ice had a 17-day shelf life compared to a non-irradiated counterpart which had a storage life of 13 days. Packaged irradiated fish had a longer shelf life of 20 days; however, packaging caused drip accumulation and poor appearance. The protein concentration was reduced after 10-day storage in both non-irradiated and irradiated samples, giving values of 0.97 ± 0.5 and 0.77 ± 0.01 mg, respectively. The protein values were further reduced to 0.21 ± 0.06 and 0.17 ± 0.06 mg, respectively, after 17 days of storage.

Al-Kahtani *et al.* (1998) studied amino acid and protein alterations in irradiated tilapia (*Tilapia nilotica* × *T. aurea*) and Spanish mackerel (*Scomberomorus commerson*), stored at 2 °C. The fish containers were irradiated with a ^{60}Co source at 1.5, 3.0, 4.5, 6.0 and 10.0 kGy. In general, amino acids of both fish either increased or decreased with increasing irradiation dose with no clear trend of change. For tilapia, an increase of lysine and methionine (g/100 g protein) was reported as the irradiation dose increased (2.0 ± 0.0, 2.4 ± 0.0, 2.3 ± 0.0, 2.0 ± 0.0, 2.4 ± 0.14 for methionine and 3.7 ± 0.0, 4.1 ± 0.3, 3.8 ± 0.0, 5.3 ± 0.0, 5.5 ± 0.1 for lysine at 1.5, 3.0, 4.5, 6.0 and 10 kGy, respectively, at the first day of storage). For Spanish mackerel stored up to 20 days, a high dose of 10.0 kGy led to a significant decrease of all essential and non-essential amino acids including lysine.

Al-Kahtani *et al.* (1996) determined the impact of gamma irradiation (1.5–10.0 kGy) and post-irradiation storage up to 20 days at 2 °C on some chemical criteria of tilapia and Spanish mackerel. The TVB-N formation was lower in irradiated fish than in the non-irradiated ones. Irradiation encouraged a larger increase in thiobarbituric acid (TBA) values. Some fatty acids reduced by irradiation treatments at all doses. Thiamin loss was more severe at higher doses (≥4.5 kGy), whereas riboflavin was not affected. Alpha and gamma tocopherols of tilapia and alpha, beta, gamma and delta tocopherols in Spanish mackerel, decreased with increased dose and continued reducing during 20-day post-irradiation storage.

Gamma-irradiation (1, 2 and 6 kGy) preservation of two species of Australian marine fish – black bream (*Acanthopagrus australis*) and redfish (*Centroberyx affinis*) –resulted in no significant changes in their fatty acid compositions. Vitamin E loss was observed in some fillets but could not be correlated with the treatment dosage. All irradiated fillets had a muscle content of vitamin E (mg/100 g) above levels considered appropriate for human consumption (0.936, 0.512, 0.510 for redfish and 0.246, 0.274, 0.253 for black bream irradiated with 1, 2 and 6 kGy, respectively) in relation to the amounts of accompanying polyunsaturated fatty acids (Armstrong *et al.*, 1994).

Kwon and Byun (1995) combined gamma irradiation with air-tight packaging to preserve and improve the quality of boiled-dried anchovies (*Engraulis encrasicholus*). After irradiation, polyunsaturated fatty acid content decreased about 5%, whereas saturated fatty acids slightly increased. However, no significant difference was found in the fatty acids profile of stored samples. The overall quantity of total and free amino acids slightly decreased with storage time. Regarding the sensory evaluation, gamma irradiation at or below 5 kGy did not induce any significant changes in appearance, rancid-flavour and palatability of the samples immediately after treatment. Prepackaging with air-tight laminated film (nylon 15 μm/PE 100 μm) and irradiation (5 kGy) proved to be effective for maintaining marketable quality over 6 months storage at ambient temperature and over 12 months at cooling temperature (5–10 °C).

Ionizing radiation was applied by Icekson *et al.* (1996) into two groups of refrigerated fish (*Cyprinus carpio*) to extend the shelf life. After irradiation, fish were kept at 0–2 °C. Non-irradiated fish became unacceptable after 16 days, whereas irradiated ones became unacceptable after 31 days, based on sensory evaluation.

Poole *et al.* (1994) irradiated sweetlip (*Lethrinus miniatus*), red emperor (*Lutjanus sebae*), mackerel (*Scomberomorus commerson*), whiting (*Sillago ciliate*), mullet (*Mugil cephalus*), barramundi (*Lates calcalifer*), sand crab (*Portunus pelagicus*), Moreton Bay prawns (*Metapenaeus* spp.) and king prawns (*Penaeus plubujus*) with 0, 1, 3 and 5 kGy. All samples were stored in ice. It was deduced that a 1 kGy dose resulted in a 1.5–4.0 log decrease of bacteria in comparison to 3.7–5.7 log decrease at 5 kGy. All species, except the Moreton Bay prawns and cooked king prawns, had satisfactory flavour, texture and odour after 5 kGy irradiation.

Indian mackerel (*Rastrellinger kanagurta*), white pomfret (*Scombermorus guttatus*) and seer (*Stromateus cinerius*) were treated with low dose gamma irradiation, and then they were stored on ice for 3–4 weeks. TBA values were raised in both irradiated and non-irradiated samples, mostly in the mackerel and the seer. Nevertheless, the TBA values in mackerel dropped with storage. Only the pomfret skin exhibited skin oxidation after treatment and this value continued to increase during storage (Doke *et al.*, 1992).

Fillets of tilapia (*Oreochromis mosambicus*) and silver carp (*Hypophthalmichythys molitrix*) were irradiated with 1 kGy at 2.4 °C. A significant decrease of thiamine content was determined in silver carp. Moreover, the nucleotide catabolite concentrations in the irradiated fish were not changed after irradiation. The counts of bacteria were retained for 5 days post-irradiation at 1 °C (Liu *et al.*, 1991).

Chuaqui-Offermanns *et al.* (1988) measured the radiation-induced lipid oxidation in whitefish (*Coregonus clupeaformis*) stored at 3 °C. The TBA for non-irradiated samples remained low and approximately stable throughout the study. In fish irradiated at 0.82 and 1.22 kGy, TBA values increased with storage time. However, on day 28, TBA still remained within an acceptable range, that is less than 20 g/kg. The average shelf life of samples based on chemical, sensory and microbiological analyses irradiated at 0.0, 0.82 and 1.22 kGy was 7.8 ± 1, 16.4 ± 3.6 and 20.9 ± 3.9 days, respectively.

European hake (*Merluccius merluccius*) can be irradiated optimally at 1.0–1.5 kGy, which gives a shelf life of 24–28 days at 0.5 °C (De la Sierra Serrano, 1970). The maximum acceptable dose for European hake is 2 kGy. Argentine whiting (*Merluccius hubsi*) has an optimum irradiation dose of 5 kGy with a shelf life achieving 48 days at 4 °C (Ritacco, 1976).

Emerson *et al.* (1964, 1965) found that the best irradiation dose for channel catfish (*Ictalurus punctatus*) was 1–2 kGy, which gave a 20-day shelf life (from 4 days) at 0 °C. The maximum irradiation dose was found to be 5 kGy. Aiyar (1976) reported that the optimum irradiation dose for threadifin (*Eleutheronesma tetradactylum*) ranged between 1.0 and 2.5 kGy leading to a threefold to fourfold increase in shelf life.

Ostovar and his colleagues (1967) found that non-irradiated whitefish (*Coregonus clupeaformis*) has a shelf life of 12–15 days. Otherwise irradiated whitefish with an optimum dose of 1.5–3 kGy, had a shelf life of 15–29 days, under refrigeration. According to Graikowski *et al.* (1968), the optimum irradiation dose for lake trout (*Salvelinus namaycush*) was 3 kGy, which achieved a 26-day shelf life at 0.6 °C in comparison to normal shelf life of 8 days. The pigment in irradiated samples does not vanish until doses of 7 kGy are used.

The optimum dose for ocean perch (*Sebastes marinus*) ranged between 1.5 and 2.5 kGy, thereby extending the shelf life to 30 days at 0.6 °C and 15 days at 7.8 °C. Moreover, doses higher than 1 kGy are not recommended due to sensory alternations because of the high concentration of fatty acids and the associated specific fine flavour (Reinacher and Ehlermann, 1978; Ronsivalli and Slavin, 1965).

The optimum dose for irradiating fillets of ocean perch (*Sebastodes alutus*) is 1–2 kGy, which yields a shelf life of 25–28 days at 0.6 °C. It should be said that the higher quality fish responded better to irradiation than the poorer quality fish (Miyauchi *et al.*, 1967; Teeny and Miyauchi, 1970).

Regarding air packed fillets of pollock (*Pollachius virens*), the optimum irradiation dose was 1.5 kGy accompanied with darkening of the fillet. The shelf life was 28–30 days at 0.6 °C and less than 20 days at 7.8 °C. The maximum acceptable dose was not clearly determined but scientific articles determined it at 2.3–2.5 kGy or 5–8 kGy, if blanched prior to irradiation (Slavin and Ronsivalli, 1964; Ampola *et al.*, 1969).

The optimal dose for white pomfret (*Stomateus cinereus*) and black pomfret (*Parastomatus niger*) was 1 kGy, with a shelf life of 4 weeks and 10–16 days, respectively, at 0–2 °C. The maximum acceptable dose was 3 kGy for both species (Kumta and Sreenivasin, 1970; Aiyar, 1976).

Sole (*Parophyrs vetulus*) fillets were found to be irradiated optimally at 2–3 kGy, achieving a 4–5 week shelf life (Teeny and Miyauchi, 1970). Moreover, grey sole (*Glyptocephalus cynoglossus*) fillets had an optimum dose of 1–2 kGy, which gave a 29-day shelf life at 0.6 °C or 10- to 11-day shelf life at 5.6 °C (Miyauchi *et al.*, 1968). Lemon sole (*Microstomus kitt*) packed under nitrogen had an optimum dose to 2.5 kGy, but, if frozen, this doubled to 5 kGy. The maximum dose was found to be 5–10 kGy (Coleby and Shewan, 1965).

According to Hussain (1980) and Ghadi *et al.* (1978), the optimal dose for mackerel (*Rastrellinger kanagurta*) was 1.5 kGy, which results in a shelf life of 21–24 days at 0 °C, 13–15 days at 5 °C, and 7–11 days at 7.8 °C. The optimal irradiation dose for mackerel (*Scomber scombrus*) was found to be 2.5 kGy giving shelf life of 30–35 days at 0.6 °C (Slavin *et al.*, 1966).

Sofyan (1978) found that Kembung fish (*Rastrelliger neglectus*) stored at 2–5 °C for 12 days imparts a significant increase in TVB-N and hypoxanthine but a decrease in the specific activity of SH-protease and acid phosphatase. Gamma irradiation at doses ranging from 1 to 2 kGy had no significant effect on the activity of SH-protease or acid phosphatase. There were significantly lower TVB-N values and hypoxanthine concentrations in samples irradiated at 2 kGy, compared to 2 kGy after 7 days storage. Herring (*Clupea herring*), which is an extremely fatty fish, has an optimum irradiation dose of 1–2 kGy, which gives a shelf life of 10–14 days at 2 °C. The maximal acceptable dose was less than 5 kGy, because of colour and natural flavour loss at $\geq$ 3 kGy (Snauwert *et al.*, 1977).

Furthermore, Carver and his co-workers (1969) found that herring smelt (*Argentina silus*) had an optimal dose ranging from 0.5 to 1.0 kGy which resulted in a 6-day shelf-life extension at 0.6 °C.

For haddock fillets (*Melanogrammus aeglefinus*) the optimal irradiation dose was found to be 1.5–2.5 kGy, achieving a shelf life of 22–25 days at 5.6 °C, and more than 30–35 days at 0.6 °C. The maximal acceptable dose was found to be 6–7 kGy (Rosnivalli *et al.*, 1968, 1970).

Skinned cod fillets were minced, hydrocolloids were added and then they were placed in a freezer in 1-kg portions, irradiated at 3 kGy and stored at −18 °C. No significant differences between irradiated and non-irradiated samples were observed over a 3-month storage period (de Ponte *et al.*, 1986).

Bombay duck (*Harpodon nehereus*) has a shelf life under refrigeration of about 5–7 days. Radiation doses of 1.0–2.5 kGy extended the shelf life to about 18–20 days. It was determined that 5 kGy was the maximum acceptable dose for Bombay duck even though an off-flavour increased at this dose, but it subsided after 4 days (Kumta *et al.*, 1973; Gore and Kumta, 1970). According to Cho and his coworkers (1992) irradiation of dried fish powders at 5–10 kGy did not change the amino acid concentration, TBA value, trymethylamine-nitrogen or colour of the samples. The organoleptic properties were more acceptable than the controls at 3-months post-irradiation.

The impact of irradiation dose on the quality and shelf-life extension of fish is given in Table 5.3.

5.2.2 Shellfish, crustaceans and molluscs

Chilled saucer scallops (*Amusium balloti*) were treated at 0.5, 1.5 and 3 kGy yielding a shelf life of 28 days (raw) and 43 days (cooked) at 0 °C, in comparison to 13–17 days for the non-irradiated samples. Doses of 1.5 kGy or above resulted in a soft, spongy and mushy texture (Poole *et al.*, 1990). Liuzzo *et al.* (1970) found that the optimal irradiation dose for shucked oyster meats was 2.5 kGy, which lead to a maximal shelf life of 7 days. The organoleptic quality was not reduced after 7 days storage on ice. Furthermore, they found that irradiation doses above 1 kGy changed the B-vitamin content, the moisture percentage, glycogen content and the soluble sugar content of oyster meats.

Novak *et al.* (1966) irradiated canned oysters in air at 2 kGy and stored them in ice for 23 days. The irradiated samples were acceptable throughout the 23-day storage period, whereas the non-irradiated controls were shown to be spoiled by day 7.

Nickerson (1963) found no detectable sensory differences between non-irradiated controls and clam meats irradiated at up to 8 kGy, after 40 days of storage at 6 °C. No significant difference was reported between non-irradiated controls and clam meats irradiated from 2.5 to 5.5 kGy (Connors and Steinberg, 1964). Moreover, for clams (*Venerupis semidecus sata*) the optimal irradiation dose was determined to be 1–4.5 kGy, thereby leading to a shelf life of 4 weeks at 0–2 °C.

Gardner and Watts (1957) treated oyster meats (*Crassostrea virginica, Crassostrea pacificus*) at 0.63, 0.83 and 3.5 kGy of ionizing irradiation and undesirable odours were observed. The raw irradiated oyster meat odour was described as 'grassy' and that of cooked raw irradiated oyster meats was described as 'oxidized'. Moreover, they reported that irradiation would not be beneficial to preservation because enzymic action was still active using 3.5 kGy irradiation at 5 °C.

The influence of gamma irradiation on the profile of volatile compounds in shrimp (*Solenocera choprii*) was investigated by Sharma *et al.* (2007). The whole shrimp, head or muscle portions were separately irradiated at a dose of 2 kGy. The volatile components from non-irradiated and irradiated portions were isolated. The quantitative analysis of the data revealed that the overall impact of irradiation on volatile flavour compounds of the whole shrimp or fractions was not significant, suggesting that the sensory value of shrimp muscle or shell waste was not affected by gamma irradiation.

Table 5.3 Irradiated fish quality and shelf life extension

Fish species	Irradiation type	Irradiation dose (kGy)	Temperature (°C)	Quality[a]	Shelf life (days)	References
Dicentrarchus labrax (sea bass)	–	–	−4	TVB-N values increased to 36.44 mg/100 g (day 17)	13	Ozden *et al.*, 2007
Dicentrarchus labrax (sea bass)	–	2.5	−4	TVB-N values increased to 25.26 mg/100 g (day 17)	15	
Dicentrarchus labrax (sea bass)	–	5	−4	TVB-N values increased to 23.61 mg/100 g (day 17)	17	
Bigeye snapper – Priacanthus tayenus (som-fug), a Thai fermented fish mince	–	–	4	L* values of all samples decreased, whereas a* and b* values increased throughout storage pH values of all samples decreased gradually	–	Riebroy *et al.*, 2007
Bigeye snapper – Priacanthus tayenus (som-fug), a Thai fermented fish mince	Gamma irradiation	2	4	L* values of all samples decreased, whereas a* and b* values increased throughout storage In these samples smallest changes in a* and b* values TBARS values were greater in samples with 6 kGy pH values of all samples decreased gradually	–	

Trachurus trachurus (horse mackerel)	–	1 5 10	2 ± 0.5	No significant effect on the proteins of this species		Silva *et al.*, 2006
Onchorynchus mykiss (rainbow trout pieces)	–	–	–	The reduction of the pH was identical for both irradiated and non-irradaited samples		Dvorak *et al.*, 2005
Onchorynchus mykiss (rainbow trout pieces)	Gamma irradiation	3	–		–	
Onchorynchus mykiss (rainbow trout pieces)	–	–	–	The change of L* was identical for both irradiated and non-irradiated samples This change may be caused by maturation of fish flesh a* was identical and b* reduced	–	
Onchorynchus mykiss (rainbow trout pieces)	Gamma irradiation	3	–		–	
Oreochromis niloticus (tilapia)	–	–	0.5–2 (for 20 and 30 days)	During storage, the level of moisture in the non-irradiated samples decreased and the levels of protein and lipid increased The levels of TVB-N increased	–	Cozzo-Siqueira *et al.*, 2003

(*continued overleaf*)

Table 5.3 *(continued)*

Fish species	Irradiation type	Irradiation dose (kGy)	Temperature (°C)	Quality[a]	Shelf life (days)	References
Oreochromis niloticus (tilapia)	–	1.0 2.2 5	0.5–2 (for 20 and 30 days)	During storage, the level of moisture and the levels of protein and lipid remain stable	–	
Oreochromis niloticus (tilapia)	–	1.0 2.2 5	0.5–2 (for 20 and 30 days)	The levels of TVB-N tended to remain stable.	–	
Oreochromis niloticus (tilapia)	Gamma irradiation	1.0 2.2 5	0.5–2 (for 20 and 30 days)	Lipid oxidation, showed a tendency to increase when the dose of irradiation increased	–	
Fish cutlets prepared at the laboratory scale	Gamma irradiation	5	Ambient temperature	–	5 weeks	Bani *et al.*, 2000
Fish cutlets commercially prepared	Gamma irradiation	5	Ambient temperature	–	14	
Trachurus picturatus (blue jack mackerel)	–	–	3 (for 23 days)	The TVB-N values of the 0 kGy meat exceeded the level regarded as acceptable (30–40 mg/100 g) after fourth day	Between 3 and 4	Mendes *et al.*, 2000

Trachurus picturatus (blue jack mackerel)	Gamma irradiation	1	3 (for 23 days)	The TVB-N contents increased more gradually with time, and levels at the end of the ice storage were, in general, three times lower After 23 days storage, the TMA content was 18.9 mg/100 g	8	
Trachurus picturatus (blue jack mackerel)	Gamma irradiation	2	3 (for 23 days)	The TVB-N contents increased more gradually with time, and levels at the end of the ice storage were, in general, three times lower After 23 days storage, the TMA content was 30.9 mg/100 g	8	
Trachurus picturatus (blue jack mackerel)	Gamma irradiation	3	3 (for 23 days)	The TVB-N contents increased more gradually with time, and levels at the end of the ice storage were, in general, three times lower After 23 days storage, the TMA content was 10.0 mg/100 g	8	
Stolephorus commersoni (anchovy)	–	–	13	The protein content decreased after 10 days of storage, giving values of 0.97 ± 0.5 mg The protein values dropped further to 0.21 ± 0.06 after 17 days of storage	13	Lakshmanan *et al.*, 1999

(*continued overleaf*)

Table 5.3 (*continued*)

Fish species	Irradiation type	Irradiation dose (kGy)	Temperature (°C)	Quality[a]	Shelf life (days)	References
Stolephorus commersoni (anchovy)	^{60}Co	2	13	The protein content decreased after 10 days of storage, giving values of 0.77 ± 0.01 mg The protein values dropped further to 0.17 ± 0.06 after 17 days of storage	17	
Scomberomorus commerson (Spanish mackerel)	Gamma irradiation	1.5	2 ± 2	Threonine 4.5 ± 0.5 g/100 g protein (at 0 days)	–	Al-Kahtani *et al.*, 1998
Scomberomorus commerson (Spanish mackerel)	Gamma irradiation	3	2 ± 2	Threonine 5 ± 0 g/100 g protein (at 0 days)	–	
Scomberomorus commerson (Spanish mackerel)	Gamma irradiation	4.5	2 ± 2	Threonine 5 ± 0.1 g/100 g protein (at 0 days)	–	
Scomberomorus commerson (Spanish mackerel)	Gamma irradiation	6	2 ± 2	Threonine 5.1 ± 0 g/100 g protein (at 0 days)	–	
Scomberomorus commerson (Spanish mackerel)	Gamma irradiation	10	2 ± 2	Threonine 4.4 ± 0 g/100 g protein (at 0 days)	–	
Tilapia nilotica × *Tilapia aurea* (tilapia)	–	1.5	2 ± 2	Threonine 4 ± 0.1 g/100 g protein (at 0 days)	–	
Tilapia nilotica × *Tilapia aurea* (tilapia)	–	3	2 ± 2	Threonine 4.2 ± 1 g/100 g protein (at 0 days)	–	

Tilapia nilotica × *Tilapia aurea* (tilapia)	–	4.5	2 ± 2	Threonine 3.7 ± 0 g/100 g protein (at 0 days)	–	
Tilapia nilotica × *Tilapia aurea* (tilapia)	–	6	2 ± 2	Threonine 3.9 ± 0 g/100 g protein (at 0 days)		
Tilapia nilotica × *Tilapia aurea* (tilapia)	–	10	2 ± 2	Threonine 4 ± 0 g/100 g protein (at 0 days)		
Cyprinus carpio (refrigerated fish)	–	–	0–2	From this study it is clear that chemical tests of freshness such as TVB-N and K value determination are not appropriate for the study od irradiated fish Organoleptic estimations and a new determination using an odour concentration meter seem to fit best the objectives determination of freshness	16	Icekson *et al.*, 1996
Tilapia nilotica × *Tilapia aurea* (tilapia) and	–	–	2 ± 2 (up to 20 days storage)	Total volatile basic nitrogen formation was lower in irradiated fish than in the non-irradiated Irradiation also caused a larger increase in thiobarbituric acid values – which continued gradually during storage Some fatty acids reduced by irradiation treatments at all doses	–	Al-Kahtani *et al.*, 1996
Scomberomorus commerson (Spanish mackerel)	–	1.5–10	2 ± 2 (up to 20 days storage)		–	

(*continued overleaf*)

Table 5.3 (*continued*)

Fish species	Irradiation type	Irradiation dose (kGy)	Temperature (°C)	Quality[a]	Shelf life (days)	References
Tilapia nilotica × *Tilapia aurea* (tilapia) and	–	–	2 ± 2 (up to 20 days storage)		–	
Scomberomorus commerson (Spanish mackerel)	–	1.5–10	2 ± 2 (up to 20 days storage)	Thiamin loss was more severe at higher doses (≥4.5 kGy), whereasriboflavin was not affected	–	
Acanthopagrus australis (black bream) and *Centroberyx affinis* (redfish)	Gamma irradiation	1	−22	0.246 mg/100 g	–	Armstrong *et al.*, 1994
Acanthopagrus australis (black bream) and *Centroberyx affinis* (redfish)	Gamma irradiation	2	−22	0.274 mg/100 g	–	
Acanthopagrus australis (black bream) and *Centroberyx affinis* (redfish)	Gamma irradiation	6	−22	0.253 mg/100 g	–	
Acanthopagrus australis (black bream) and *Centroberyx affinis* (redfish)	Gamma irradiation	1	−22	0.936 mg/100 g	–	
Acanthopagrus australis (black bream) and *Centroberyx affinis* (redfish)	Gamma irradiation	2	−22	0.512 mg/100 g	–	

Acanthopagrus australis (black bream) and *Centroberyx affinis* (redfish)	Gamma irradiation	6	−22	0.510 mg/100 g	–	
Coregonus clupeaformis (whitefish)	–	–	3	The TBA for unirradiated samples remained low and almost unchanged throughout the study	7.8 ± 1 average	Chuaqui-Offermanns *et al.*, 1988
Coregonus clupeaformis (whitefish)	Gamma irradiation	0.82	3	TBA values increased with the time of storage	16.4 ± 3.6 average	
Coregonus clupeaformis (whitefish)	Gamma irradiation	1.22	3		20.9 ± 3.6 average	
Bigeye snapper – *Priacanthus tayenus* (som-fug), a Thai fermented fish mince	Gamma irradiation	6	4	L* values of all samples decreased, whereas a* and b* values increased throughout storage TBARS values were lower in samples irradiated with 2 kGy pH values of all samples decreased gradually	–	
Kamaboko (fish paste) from Pollock and sardine	–	–	1–4 and 10–12	–	5	Oku, 1981; Oku, 1976; Oku and Kimura, 1975; Sasayama, 1973, 1972, 1977; Kume *et al.*, 1975
Kamaboko (fish paste) from Pollock and sardine	–	5	1–4 and 10–12	–	34	

(continued overleaf)

Table 5.3 *(continued)*

Fish species	Irradiation type	Irradiation dose (kGy)	Temperature (°C)	Quality[a]	Shelf life (days)	References
Rastrelliger kanagurta (mackerel)	–	1.5	0	–	21–24	Hussain, 1980; Ghadi *et al.*, 1978
Rastrelliger kanagurta (mackerel)	–	1.5	7.8	–	7–11	
Rastrelliger neglectus (kembung fish)	–	1–2	5	Significant increase in TVB-N and hypoxanthine, with a decrease in the specific activity of SH-protease and acid phosphatase There was significantly lower TVB-N values and hypoxanthine concentrations in samples irradiated at 2 kGy, compared to 2 kGy after 7 days storage	12	Sofyan, 1978
Sebastes marinus (ocean perch)	–	1.5–2.5	0.6	Due to the higher fat content and the associated specific fine flavour, higher doses than 1 kGy are not recommended because of sensory changes	30	Reinacher and Ehlermann, 1978; Teeny and Miyauchi, 1970; Miyauchi *et al.*, 1967; Ronsivalli and Slavin, 1965
Sebastes marinus (ocean perch)	–	1.5–2.5	7.8		15	
Sebastodes alutus (ocean perch fillets)	–	1–2	0.6		25–28	
Stomateus cinereus (white pomfret)	–	1	0–2		4 weeks	

Clupea herring (herring)	–	–	2	The maximum acceptable dose was found to be less than 5 kGy, due to loss of colour and natural flavour at 3 kGy and higher	10–14	Snauwert *et al.*, 1977
Merluccius hubsi (Argentine whiting)	–	5	4	–	48	Ritacco, 1976
Parastomatus niger (black pomfret)	–	1	0–2	–	10–16	Aiyar, 1976
Harpodon nehereus (Bombay duck)	–	–	Under refrig- eration	–	About 5–7	Kumta *et al.*, 1973; Gore and Kumta, 1970
Harpodon nehereus (Bombay duck)	–	1.0–2.5	Uner refriger- ation	5 kGy is the maximum acceptable dose for this fish in that an off-flavour develops at this dose, but will subside after 4 days	About 18–20	
Merluccius merluccius (European hake)	–	1–1.5	0.5	–	24–28	de la Sierra Serrano, 1970
Melanogrammus aeglefinus (haddock fillets)	–	1.5–2.5	5.6	–	22–25	Rosnivalli *et al.*, 1970;
Melanogrammus aeglefinus (haddock fillets)	–	1.5–2.5	0.6	–	30–35	Rosnivalli *et al.*, 1968
Glyptocephalus cynoglossus (grey sole fillets)	–	1–2	0.6	–	29	Miyauchi *et al.*, 1968

(continued overleaf)

Table 5.3 *(continued)*

Fish species	Irradiation type	Irradiation dose (kGy)	Temperature (°C)	Quality[a]	Shelf life (days)	References
Glyptocephalus cynoglossus (grey sole fillets)	–	1–2	5.6	–	10–11	
Salvelinus mamaycush (lake trout)	–	–	0.6	The pigment in irradiated samples does not diseappear until doses of 7 kGy are used	8	Graikowski *et al.*, 1968
Salvelinus mamaycush (lake trout)	–	3	0.6		26	
Coregonus clupeaformis (whitefish)	–	–	Under refrigeration	–	12–15	Ostovar *et al.*, 1967
Coregonus clupeaformis (whitefish)	–	1.5–3	Under refrigeration	–	15–29	
Scomber scombrus (mackerel)	–	2.5	0.6	–	30–35	Slavin *et al.*, 1966
Ictalurus punctatus (channel catfish)	–	1–2	0	–	20	Emerson *et al.*, 1964, 1965

[a]Abbreviations: TVB-N, total volatile base nitrogen; TMA, trimethylamine; TBARS, thiobarbituric acid reactive substances.

Sinanoglou *et al.* (2007) studied irradiated frozen molluscs (squid, octopuses and cuttlefish) and crustaceans (shrimp). The total lipids of non-irradiated mantle of cephalopod mollusc *Todorodes sagittatus* (squid), *Octopus vulgaris* (octopus) and *Sepia officinalis* (cuttlefish) represented 1.80, 2.32 and 1.55% of the wet tissue, respectively. The lipid contents of *Penaeus monodon* non-irradiated muscle and cephalothorax was 1.30 and 2.75%, respectively. At doses of 4.7 kGy, mollusc mantle displayed a loss of total lipid content of 4.5–6.9%, and in shrimp muscle and cephalothorax, the decrease was 6.2 and 5.8%, respectively, compared to non-irradiated samples. However, all changes reported were not statistically significant. The total fatty acid content and the ω-3 : ω-6 fatty acid ratio was not affected. A dose-dependent significant decrease in the ratio of polyunsaturated fatty acids to saturated fatty acids, was observed. Increasing the irradiation dose, the a* and b* values showed a variation, whereas the L* value significantly decreased in mollusc mantles and shrimp muscle and increased in shrimp cephalothorax. The total score of colour changes exhibited an increase with an increase in dose.

In a study by Lee *et al.* (2002), shrimps (*Acetes chinensis*) were sliced, washed and salted with 15 and 20% (w/w) sodium chloride. Salted shrimp was 0, 5 and 10 kGy-irradiated at two different stages: (i) irradiated directly after processing salted shrimp, (ii) irradiated at optimal fermentation period, and fermented at 15 °C for 10 weeks. Non-irradiated shrimps with 30% salt were also prepared as a control sample. Irradiated shrimps did not differ in proximate composition, salinity and water activity from non-irradiated shrimps with the same salt addition and irradiation time. During fermentation, volatile basic nitrogen (VBN) content increased as the salt concentration and irradiation dose decreased. According to analysis results of sensory characteristics, total bacterial count and pH, the combination of low salt concentration (15 or 20%) and gamma irradiation (5 or 10 kGy) was effective in processing low-salted and fermented shrimp. The results revealed no adverse sensory quality and improved microbial shelf stability compared to control samples (30% of salt addition).

The optimal dose for peeled European brown shrimp (*Crangon vulgaris* and *Crangon crangon*) was found to be 1.5 kGy, which yielded a shelf life of 23 days at 2 °C, in comparison to 9–16 days for non-irradiated shrimp (Vyncke *et al.*, 1976). The removal of oxygen helped toward maximizing the irradiation effects (Ehlermann, 1976; Ehlermann and Muenzer, 1976; Ehlermann and Diehl, 1977).

Scholz *et al.* (1962) found that Pacific shrimp (*Pandalus jordani*) irradiated with 5 kGy reached a shelf life of 3 weeks at 3 °C with no noticeable off flavours. The optimal dose for raw, beheaded, tropical shrimps (*Penaeus* spp.) has been established at 1.5–2 kGy, yielding a shelf life of about 42 days at 3 °C. When a 4-min blanching was employed prior to 1.5–2 kGy irradiation, the shelf life was extended to 130 days (Gueavara *et al.*, 1965; Kumta *et al.*, 1970). The impact of the irradiation dose on quality and shelf life for shellfish and crustaceans is given in Table 5.4.

5.3 Microflora of irradiated fish and fishery products

5.3.1 Fish

Irradiation at doses of 3 and 1 kGy inactivated *L. monocytogenes* and *V. parahaemolyticus* inoculated in cold-smoked salmon by 6.59 and 6.05 log cfu/g, respectively (Badr, 2012). Moreover, irradiation of the non-inoculated samples significantly reduced microbial

Table 5.4 Irradiated shellfish and crustaceans quality and shelf life extension

Seafood species	Irradiation type	Irradiation dose (kGy)	Temperature (°C)	Quality	Shelf life (days)	References
Solnocera choprii (shrimp head or muscle portions)	Gamma irradiation	2	–	–	–	Sharma *et al.*, 2007
Cragon vulgaris and *Cragon cragon* (peeled European brown shrimp)	–	–	2	–	9–16	Vyncke *et al.*, 1976
Cragon vulgaris and *Cragon cragon* (peeled European brown shrimp)	–	1.5	2	–	23	
Clam meat	–	–	6	No detectable sensory differences	40	Nickerson, 1963
Clam meat	–	Up to 8	6	No detectable sensory differences	40	
Pandalus jordani (Pacific shrimp)	–	5	3	No noticeable off flavours	21	Scholz *et al.*, 1962

populations of mesophilic aerobic, anaerobic, psychrophilic and lactic acid bacteria (LAB) as well as moulds and yeasts.

Medina *et al.* (2009) investigated the effectiveness of electron beam irradiation and high pressure treatment on cold-smoked salmon from two points of view: microbial safety and shelf life extension. From the response of *L. monocytogenes* to irradiation, a D value of 0.51 kGy was determined. For samples stored at 5 °C, 1.5 kGy achieves a Food Safety Objective (FSO) of 2 $\log_{10}$ cfu/g *L. monocytogenes* for a 35-day shelf life, while 3 kGy is needed for temperature mistreatment (5 + 8 °C).

Ozden *et al.* (2007) studied non-irradiated and irradiated (2.5 and 5.0 kGy) sea bass (*Dicentrarchus labrax*) and found that microbial counts (psychrotrophic bacteria, mesophilic aerobic bacteria, H_2S-producing bacteria, Enterobacteriaceae and pseudomonads) of non-irradiated sea bass samples were higher than irradiated ones. The results obtained showed that the shelf life of sea bass stored in ice was 13 days for non-irradiated samples, 15 days for 2.5 kGy irradiated samples and 17 days for 5 kGy irradiated ones.

The effects of irradiation at different doses (0, 2 and 6 kGy) of som-fug (Bigeye snapper (*Priacanthus tayenus*)), a Thai fermented fish mince, were studied by Riebroy *et al.* (2007). At day 0, the control sample had a TVC of 2.9 $\times 0^8$ cfu/g and LAB of 2.8 $\times$ 10^8 cfu/g. LAB, yeast and mould counts of samples irradiated at 6 kGy were not detectable throughout the storage of 30 days at 4 °C, whereas no growth was identified in the sample irradiated at 2 kGy within the first 10 days.

Jo *et al.* (2005) studied the irradiation impact to eliminate *Salmonella typhimurium, Escherichia coli, Staphylococcus aureus* and *Listeria ivanovii* on three prepared seafood products used to make a laver (dried seaweed) roll. The radiation sensitivity (D_{10} values or the dose required to inactivate 90% of a population) of these organisms ranged from 0.23 to 0.62 kGy in imitation crab leg, 0.31–0.44 kGy in surimi gel and 0.27–0.44 kGy in dried seaweed. The growth of these microorganisms inoculated (10^8 cfu/g) into these foodstuffs was inhibited by irradiation during 24 h of post-irradiation storage regardless of the storage temperature (10, 20 and 30 °C). *L. ivanovii* was not detected after the 3 kGy treatment but the other pathogens were not detected at 2 kGy.

According to Cozzo-Siqueira *et al.* (2003), irradiated (1.0, 2.2 and 5.0 kGy) samples of *Tilapia nilotica* (*Oreochromis niloticus*) stored at temperatures ranging from 0.5 to 2 °C for 20 and 30 days displayed microbiological loads below the levels established by the Brazilian seafood legislation, whereas the non-irradiated samples had a higher microbiological load and did not conform with the officially allowed levels.

Jaczynski and Park (2003) determined electron penetration and microbial inactivation with electron beam (e-beam) in surimi seafood. It was found that one- and two-sided e-beam could efficiently penetrate 33- and 82-mm thick surimi seafood, respectively. Modelling of microbial inactivation with e-beam showed that two-sided e-beam could control *Staphylococcus aureus* if the surimi seafood package was thinner than 82 mm. The $D_{\text{e-beam}}$ value for *Staphylococcus aureus* was 0.34 kGy. An e-beam dose of 4 kGy resulted in a minimum of a 7-log and most likely a 12-log reduction of *Staphylococcus aureus*. Microbial inactivation became slower when frozen samples were subjected to e-beam because of its reduced penetration.

According to Mendes *et al.* (2000), samples irradiated at different doses led to a relative reduction in the bacteria counts in blue jack mackerel (*Trachurus picturatus*) stored at 3 °C. Moreover, bacterial counts were lower in the 3 kGy lot and were increasingly higher in the 1-, 2- and 0-kGy irradiated samples. At spoilage, the non-irradiated 13-day-old fish samples had 5–10 times as many bacteria as their irradiated counterparts on the same day. The storage life of non-irradiated fish (0 kGy lot) was found to be between 3 and 4 days. Samples irradiation at the tested doses prolonged shelf life by 4 to 5 days, the fish being unacceptable after 8 days. Furthermore, the test panel data analysis showed that non-irradiated fish tended to deteriorate in quality quite rapidly, whereas the irradiated samples appeared to remain at borderline quality for a long time prior to becoming unacceptable.

According to Lakshmanan *et al.* (1999), the initial bacterial load (10^4 cfu/g) of anchovy (*Stolephorus commersonii*) increased with storage. During the entire storage period of over 20 days, the cfu levels of irradiated, non-packaged samples were 1 ± 2 log cycles less than the non-irradiated, non-packaged control samples. Packaged, irradiated samples had lower cfu scores as compared with packaged, non- irradiated samples. The increase in cfu levels in packaged samples was lower than the respective non-packaged samples. Packaged irradiated samples had a shelf life of 20 days, that is a week longer than the shelf life of 13 days for its non-irradiated counterpart.

Kamat and Thomas (1998) investigated the influence of low (0.39–1.1%), medium (4.25%) and high (7.1–32.5%) fatty acid concentration in fish on the irradiation inactivation of four food-borne pathogens. Cells of *Listeria monocytogenes* 036, *Yersinia enterocolitica* F5692, *Bacillus cereus* and *Salmonella typhimurium* at logarithimic phase were inoculated in 10% fish homogenates and subjected to gamma irradiation at ice temperature (0–1 °C) with doses ranging from 0.05 to 0.08 kGy. The D_{10} values in kGy ranged from 0.2–0.3, 0.15–0.25, 0.1–0.15 to 0.09–0.1 for *Listeria monocytogenes* 036, *Bacillus cereus, Salmonella typhimurium* and *Yersinia enterocolitica* F5692, respectively. The irradiation resistance per microorganism was not affected by the fat content of the fish.

Studies were carried out to evaluate the microbiological profile, shelf life and quality of gamma-irradiated Nagli fish (*Sillago sihama*). Non-irradiated samples had a shelf life of 7–8 days for storage at 1–2 °C while irradiated samples (2 and 3 kGy) were acceptable for up to 19 days. Dressing prior to the irradiation process did not increase shelf life compared to whole fish. *Salmonella* sp. was not detected in 3 kGy-irradiated samples, while 2 kGy destroyed *Vibrio parahaemolyticus* and *Staphylococcus aureus. Listeria monocytogenes* and *Yersinia enterocolitica* were not detected but non-pathogenic species such as *Listeria grayi, Listeria murrayi* and *Yersinia tuberculosis* were observed in fish prior to irradiation. Irradiation doses of 2 and 3 kGy destroyed *Yersinia* sp. and *Listeria* sp., respectively. None of these microorganisms were detected during storage of irradiated fish (Ahmed *et al.*, 1997).

Abu-Tarboush *et al.* (1996) studied tilapia (*Tilapia nilotica* × *Tilapia aurea*) and Spanish mackerel (*Scomberomorus commerson*) that were subjected to gamma irradiation doses of 1.5, 3.0, 4.5, 6.0 and 10.0 kGy. The irradiated and non-irradiated (control) fish were stored at 2 ± 2 °C. Doses of 3.0 and/or 4.5 kGy increased the organoleptic acceptability (appearance, odour, texture, flavour) and the microbial quality (total count and coliforms) by 8 days compared to the non-irradiated control samples. The numbers of H_2S producing bacteria were low in both fish and a dose of 1.5 kGy kept their population low throughout the storage period. Furthermore, this dose level was also sufficient to eliminate *Salmonella* spp. in both fish. *Yersinia* and *Campylobacter* species were effectively eliminated by doses of 1.5 and 3.0 kGy. Doses of 6.0 and 10.0 kGy caused decreased psychrotrophic counts but adversely affected fish quality.

The consumption of mullet (Mugilidae) as sashimi caused *Phagicola longa* infection in 10 people in Sao Paulo, Brazil. Antunes *et al.* (1993) investigated the radiolysis of *P. longa* in three mullet species, silver mullet (*Mugil nuema*), grey mullet (*Mugil plutunu*) and 'paratipema' (*Mugil* sp.) exposed to ionization ranging from 1.0 to 10.0 kGy in order to control such infections caused by commercial storage conditions and by consumption of raw fish. It was observed that 1.0 and 2.0 kGy caused a motility decrease in silver mullet parasites from 100% to 15 and 17%, and doses of 4.0 and 10.0 kGy caused metacercaria inviability. The parasite motility decreased in the grey mullet treated with doses of 2.0, 2.5, 3.0 and 3.5 kGy, from 56% to 31%, 9%, 18% and 5%, respectively, 4.0 kGy tending to be the control dose for *P. longa*. This dose also controlled other metacercaria found in the 'paratipema', without affecting the odour, colour or appearance of the treated mullet.

Hammad and El-Mongy (1992) studied smoked salmon fillets irradiated with 2 and 4 kGy. The main quality lost during the 4 kGy irradiation was the normal cherry red colour associated with smoked salmon. Colour loss was not observed in samples irradiated with 2 kGy. Both two and four doses were effective at reducing microorganism numbers in samples, while the 4 kGy dose eliminated all coliforms, fecal streptococci and *Staphylococcus aureus*. The non-irradiated samples reached an unacceptable plate count after 1 month of refrigerated storage, whereas the microbiological quality was retained for 3 and 4 months at 2 and 4 kGy, respectively.

Dried fish powders were inoculated with mesophilic aerobic bacteria, moulds and coliforms at 10^3–10^7, 10^2–10^3 and 10^2–10^6 cfu/g, respectively. It was determined that if the samples were irradiated with 5–10 kGy, all moulds and coliforms were eliminated but the mesophilic plate counts were reduced only to $<10^3$ cfu/g (Cho *et al.*, 1992).

According to Valdes and Szeinfeld (1989), hake fillets (*Merluccius merluccius hubsi*) irradiated at 2 kGy dose exhibited a 1 log cycle reduction in bacterial number versus the controls, whereas a 3 log cycle reduction was observed at 6–10 kGy. The optimized exposure was shown to be 6 kGy.

Indian mackerel (*Rastrellinger kanagurta*) was irradiated at 0–3 kGy and then stored at 1–3 °C by Hussain *et al.* (1985). The bacterial load was 1×10^5 at day 0 and increased to 2.5×10^7 in 28 days. *Pseudomonas* and *Proteus* spp. were the predominant genera in the controls, whereas *Acromobacter, Flavobacterium, Bacillus* and *Micrococcus* dominated in the irradiated fillets, but it was determined that a pretreatment with a dip in 10% sodium polyphosphate decreased associated drip loss. Optimum conditions were sustained for 3 weeks at 1.5 kGy with a predip in 10% sodium polyphosphate. Washed red hake fish mince (*Urophycis chuss*) was irradiated at 0.66 and 1.31 kGy and then stored aerobically at 3.3 °C. Total aerobic plate counts were monitored and found to be less than 10^6 cfu/g for 4, 10 and 17 days after irradiation at all doses. Organoleptic evaluation disclosed that irradiated samples were superior for 12–18 days longer than the non-irradiated control samples, as well as microbiologically for 6–13 days longer than the non-irradiated samples (Dymsza *et al.*, 1990).

Smoked dried fish were inoculated with spores of *Aspergillus flavus* (UI 81) and irradiated with doses of 0.625, 1.26, 2.60 and 6.00 kGy of gamma irradiation. The impact on aflatoxin Bl production was measured following incubation for 8 days as stationary cultures. The amount of aflatoxin B1 produced was found to decrease with increased gamma irradiation dose (4.73, 2.45, 1.90, 0.80, 0.51 mg/kg for 0, 0.625, 1.26, 2.60 and 6.00 kGy, respectively) while the non-irradiated control produced significantly greater amounts of aflatoxin B1 compared to the treated cultures (Ogbadu, 1988).

Kamaboko is a fish paste (from pollock and sardine) from Japan. Samples of kamaboko were irradiated with up to 5 kGy and stored at 1–4 °C and 10–12 °C. Shelf life was extended by about 5 days (non-irradiated samples) up to 34 days for samples irradiated with 5 kGy. The highest dose (5 kGy) used resulted in survival of the genus *Bacillus*. However, only a slight impact on flavour was observed and it was concluded that the optimal dose is 3 kGy, with shelf-life extension reported from 14 to 42 days at 10–12 °C (Oku, 1976, 1981; Sasayama, 1972, 1973, 1977; Kume *et al.*, 1975).

Laycock and Regier (1970) studied haddock (*Melanogrammus aeglefinus*) filleted after 2, 5 and 9 days in ice. The samples used were 2-, 5- and 9-day fish. Irradiation (1 kGy) reduced the total bacterial count by about 1 log order. *Achromobacter* reduced by about 1 log order with 1 kGy irradiation. Pseudomonads isolated after irradiation from fillets of 5- and 9-day fish, seem to be reduced by between 2 to 3 log orders. The shelf life (3 °C) of the non-irradiated fillets was determined with organoleptic evaluation to be 9 days for the 2-day fish, 7 days for the 5-day fish, and 5 days for the 9-day fish. The irradiated samples (3 °C) spoiled after about 18 days of storage for 2- and 5-day samples and after about 13 days of storage for 9-day samples.

'Screening' packs comprising 10 lots of codfish cake, each lot containing about 10^6 spores of different strains (five type A and five type B) of *Clostridium botulinum* per can, were irradiated at -30 ± 10 °C with a series of increasing doses (20 replicate cans/dose) of gamma irradiation. These packs were incubated for 6 months at 30 °C. The results concluded that the experimental sterilizing dose (ESD) for codfish cake was $27.5 < \text{ESD} < 30$ kGy (Anellis *et al.*, 1972).

Clostridium botulinum type E was inoculated into haddock fillets at 10^4 spores/g and irradiated at 1 and 2 kGy. Those irradiated at 1 kGy were acceptable because spoilage occurred before toxin production determination. However, at 2 kGy exposure, toxin was produced in haddock fillets prior to spoilage. Non-inoculated, naturally contaminated cod and haddock did not form toxins at doses as high as 2 kGy and stored at 10 °C (Nickerson and Goldblith, 1969/1971).

Irradiated (1 and 2 kGy) and non-irradiated yellow perch fillets (*Perca flavescens*) stored at 1 °C were studied by Kazanas and Emerson (1968). Microorganisms isolated from non-irradiated fillets consisted of *Flavobacterium, Microbacterium, Vibrio, Micrococcus-Sarcina, Pseudomonas, Aeromonas, Bacillus, Corynebacterium, Lactobacillus, Brevibacterium* and *Achromobacter-Alcaligenes-Mima*. Plate counts of fish prior to irradiation showed the presence of approximately 10^6 microorganisms/g of sample. Irradiation to 1 and 2 kGy produced 1.4 and 3 log initial count reduction, respectively. Irradiation to 1 and 2 kGy increased shelf life from 10 days for the control samples to 18 days, whereas irradiation of samples to 3 and 6 kGy extended shelf life to approximately 43 and 55 days, respectively.

Kazanas *et al.* (1966) studied the irradiation impact on yellow perch fillets (*Perca flavescens*). Samples were irradiated with 3 and 6 kGy and stored at 1 or 6 °C. Total plate counts prior to irradiation did not exceed 8.7×10^5 cfu/g of sample. This count was reduced nearly 100% with irradiation of either 3 or 6 kGy. Gradually, lower maximal bacterial populations and lengthened lag phases were attained as a higher radiation dose was used. The population growth rate did not decrease significantly. The shelf life of non-irradiated fillets at 10 °C was found to be about 9–13 days. The latter was extended 3.6- and 5-fold with doses of 3 and 6 kGy, respectively. Increasing storage temperature to 6 °C decreased the lag phase and considerably accelerated spoilage when compared with irradiated fillets stored at 1 °C. The shelf life of non-irradiated samples stored at 6 °C was about 6 days; this was extended 3.0 and 3.5 times with irradiation to 3 and 6 kGy, respectively.

The impact of irradiation dose on microflora and shelf life of fish is summarized in Table 5.5.

5.3.2 Shellfish, crustaceans and molluscs

Sommers and Rajkowski (2011) studied the ability of gamma irradiation to inactivate foodborne pathogenic microorganisms surface-inoculated to frozen seafood (scallops, lobster meat, blue crab, swordfish, octopus and squid). The radiation D_{10}-values for *Listeria monocytogenes, Staphylococcus aureus* and *Salmonella* inoculated to seafood samples that

Table 5.5 Effect of irradiation on fish microflora

Fish species	Irradiation type	Irradiation dose (kGy)	Temperature (°C)	Results	Shelf life (days)	References
Dicentrarchus labrax (sea bass)	-	-	−4	The higher the irradiation dose the lower the population of psychrotrophic bacteria, mesophilic aerobic bacteria, H_2S-producing bacteria, Enterobacteriaceae and pseudomonads	13	Ozden *et al.*, 2007
Dicentrarchus labrax (sea bass)	-	2.5	−4		15	
Dicentrarchus labrax (sea bass)	-	5	−4		17	
Bigeye snapper – *Priacanthus tayenus* (som-fug), a Thai fermented fish mince	-	-	4	TVC of 2.9×10^8 cfu/g and LAB of 2.8×10^8 cfu/g	-	Riebroy *et al.*, 2007
Bigeye snapper – *Priacanthus tayenus* (som-fug), a Thai fermented fish mince	Gamma irradiation	2	4	No Lactic acid bacteria (LAB), yeast and mould counts growth was found in the sample irradiated at 2 kGy within the first 10 days	-	
Bigeye snapper – *Priacanthus tayenus* (som-fug), a Thai fermented fish mince	Gamma irradiation	6	4	Lactic acid bacteria (LAB), yeast and mould counts in samples irradiated at 6 kGy were not detectable throughout the storage of 30 days	-	

(*continued overleaf*)

Table 5.5 (*continued*)

Fish species	Irradiation type	Irradiation dose (kGy)	Temperature (°C)	Results	Shelf life (days)	References
Oreochromis niloticus (tilapia)	–	1.0 2.2 5	0.5–2.0 (for 20 and 30 days)	Irradiated samples had a microbiological content below the levels established by the Brazilian seafood legislation, whereas the non-irradiated samples had a higher microbiological content and were not in conformity with the officially acceptable levels	–	Cozzo-Siqueira *et al.*, 2003
Blue jack mackerel	–	–	3	Irradiation of samples at different levels brought about a proportional reduction in the number of bacteria. At spoilage, the non-irradiated 13-day-old fish samples had 5–10 times as many bacteria as the irradiated counterparts on the same day	Between 3 and 4	Mendes *et al.*, 2000
Blue jack mackerel	Gamma irradiation	1 2 3	3		8	

Golden anchovy (0.39% fat)	Gamma irradiation	1	0–1	*Listeria monocytogenes* 036, *Yersinia enterocolitica* F5692, *Bacillus cereus* and *Salmonella typhimurium* had no difference in their survival	–	Kamat and Thomas, 1998
Indian sardine (7.1% fat)	Gamma irradiation	1	0–1		–	
Golden anchovy (0.39% fat)	Gamma irradiation	3	0–1		–	
Indian sardine (7.1% fat)	Gamma irradiation	3	0–1		–	
Sillago sihama (nagli fish)	–	–	1–2	–	7–8	Ahmed *et al.*, 1997
Sillago sihama (nagli fish)	Gamma irradiation	2	1–2	Elimination of *Vibrio parahaemolyticus* and *Staphylococcus aureus*. Elimination of *Yersinia* sp.	19	
Sillago sihama (nagli fish)	Gamma irradiation	3	1–2	*Salmonella* sp. was not detected. Elimination of *Listeria* sp.	19 days	
Tilapia nilotica × *Tilapia aurea* (tilapia) and *Scomberomorus commerson* (Spanish mackerel)	–	–	2 ± 2	–	–	Abu-Tarboush *et al.*, 1996

(*continued overleaf*)

Table 5.5 (*continued*)

Fish species	Irradiation type	Irradiation dose (kGy)	Temperature (°C)	Results	Shelf life (days)	References
Tilapia nilotica × *Tilapia aurea* (tilapia) and *Scomberomorus commerson* (Spanish mackerel)	-	1.5	2 ± 2	The population of H_2S producing bacteria remained low throughout the storage period. Elimination of *Yersinia* and *Salmonella* spp.	-	
Tilapia nilotica × *Tilapia aurea* (tilapia) and *Scomberomorus commerson* (Spanish mackerel)	-	3.0	2 ± 2	Elimination of *Campylobacter*	8 more than non-irradiated	
Tilapia nilotica × *Tilapia aurea* (tilapia) and *Scomberomorus commerson* (Spanish mackerel)	-	4.5	2 ± 2	-	8 more than non-irradiated	
Tilapia nilotica × *Tilapia aurea* (tilapia) and *Scomberomorus commerson* (Spanish mackerel)	-	6 10	2 ± 2	Reduction in psychrotrophic counts	-	
Mugil nuema (mullet)	No irradiation type	1	-	Motility decrease of *Phagicola longa* parasites from 100% to 15%	-	Antunes *et al.*, 1993
Mugil nuema (mullet)	No irradiation type	2	-	Motility decrease of *Phagicola longa* parasites from 100% to 17%	-	

Mugil nuema (mullet)	No irradiation type	4	–	*Phagicola longa* metacercaria inviability	–	
Mugil nuema (mullet)	No irradiation type	10	–	*Phagicola longa* metacercaria inviability	–	
Mugil plutunus (grey mullet)	No irradiation type	2 2.5 3 3.5	–	Motility decrease of *Phagicola longa* parasites from 56% to 31%, 9%, 18% and 5%	–	
Mugil plutunus (grey mullet)	No irradiation type	4	–	Tending to be control dose for *Phagicola longa*	–	
Dried fish powders inoculated with mesophilic aerobic bacteria, moulds and coliforms at 10^3-10^7 cfu/g, 10^2-10^3 cfu/g, 10^2-10^6 cfu/g, respectively	–	5–10	–	All moulds and coliforms were eliminated but the mesophilic plate counts were reduced only to $<10^3$ cfu/g	–	Cho *et al.*, 1992
Urophycis chuss (washed red hake fish mince)	–	0.66 1.31	3.3	Total aerobic plate counts were monitored and found to be less than 10^6 cfu/g for 4, 10, and 17 days after irradiation at all doses	6–13 more than non-irradiated	Dymsza *et al.*, 1990
Merluccius merluccius hubi (hake fillets)	Gamma irradiation	2	–	1 log cycle reduction in bacterial number versus the controls	–	Valdes and Szeinfeld, 1989

(*continued overleaf*)

Table 5.5 *(continued)*

Fish species	Irradiation type	Irradiation dose (kGy)	Temperature (°C)	Results	Shelf life (days)	References
Merluccius merluccius hubi (hake fillets)	Gamma irradiation	6–10	–	3 log cycle reduction in bacterial number versus the controls	–	
Smoked dried fish	Gamma irradiation	–	–	4.73 mg/kg aflatoxin B_1	–	Ogbadu, 1988
Smoked dried fish	Gamma irradiation	0.625	–	2.45 mg/kg aflatoxin B_1	–	
Smoked dried fish	Gamma irradiation	1.26	–	1.90 mg/kg aflatoxin B_1	–	
Smoked dried fish	Gamma irradiation	2.60	–	0.80 mg/kg aflatoxin B_1	–	
Smoked dried fish	Gamma irradiation	6	–	0.51 mg/kg aflatoxin B_1	–	
Coregonus clupeaformis (whitefish)	–	–	3	Initial count of 1.1×10^5 microorganisms/g	7	Chuaqui-Offermanns *et al.*, 1988
Coregonus clupeaformis (whitefish)	Gamma irradiation	0.82 1.22	3	Reduce 4.1×10^2 microorganisms/g Psychrotrophs were also reduced by two orders of magnitude Pseudomonads were not detected after irradiation	20	

Coregonus clupeaformis (whitefish)	Gamma irradiation	0.82 1.22	3	Reduce 1×10^2 microorganisms/g Psychrotrophs were also reduced by two orders of magnitude Pseudomonads were not detected after irradiation	20–28	
Melanogrammus aeglefinus (haddock) that filleted after 2 days in ice	–	–	3	Pseudomonas and to a lesser extent *Achromobacter* predominated throughout storage	9	Laycock and Regier, 1970
Melanogrammus aeglefinus (haddock) that filleted after 5 days in ice	–	–	3	–	7	
Melanogrammus aeglefinus (haddock) that filleted after 9 days in ice	Gamma irradiation	–	3	–	5	
Melanogrammus aeglefinus (haddock) that filleted after 2 days in ice	Gamma irradiation	1	3	Irradiation (1 kGy) reduced the total bacterial count by approximately 1 log order. *Achromobacter* was also reduced in numbers by about 1 log order by 1 kGy irradiation	18	

(*continued overleaf*)

Table 5.5 (*continued*)

Fish species	Irradiation type	Irradiation dose (kGy)	Temperature (°C)	Results	Shelf life (days)	References
Melanogrammus aeglefinus (haddock) that filleted after 5 days in ice	Gamma irradiation	1	3	Irradiation (1 kGy) reduced the total bacterial count by approximately 1 log order. *Achromobacter* was also reduced in numbers by about 1 log order by 1 kGy irradiation Pseudomonads, which were isolated immediately after irradiation from fillets of 5- and 9-day fish, seem to have been reduced by between 2 to 3 log orders	13	
Melanogrammus aeglefinus (haddock) that filleted after 9 days in ice	Gamma irradiation	1	3		9	
Codfish cake about 10^6 spores of a different strain (five type A and five type B) of *Clostridium botulinum* per can	Gamma irradiation	A series of increasing doses (20 replicate cans/dose)	30 (6 months)	Experimental sterilizing dose (ESD) was 27.5 $<$ESD$<$30 kGy	–	Anellis *et al.*, 1972
Perca flavescens (yellow perch fillets)	–	–	1	Approximately 10^6 microorganisms/g	10	Kazanas and Emerson, 1968
Perca flavescens (yellow perch fillets)	Gamma irradiation	1	1	1.4 log reduction of the initial count	18	
Perca flavescens (yellow perch fillets)	Gamma irradiation	2	1	3 log reduction of the initial count	18	

Perca flavescens (yellow perch fillets)	Gamma irradiation	3	1	–	43	
Perca flavescens (yellow perch fillets)	Gamma irradiation	6	1	–	55	
Perca flavescens (yellow perch fillets)	–	–	1	Initial total counts ranged from 0.93×10^5 to 8.62×10^5 per g	9–13	Kazanas *et al.*, 1966
Perca flavescens (yellow perch fillets)	–	–	6	–	6	
Perca flavescens (yellow perch fillets)	Gamma irradiation	3	1	Nearly 100% reduction of microbial count	3.6-fold shelf-life extension	
Perca flavescens (yellow perch fillets)	Gamma irradiation	6	1	Nearly 100% reduction of microbial count. Progressively lower maximal bacterial populations and lengthened lag phases were obtained as more radiation was used	5-fold shelf-life extension	
Perca flavescens (yellow perch fillets)	Gamma irradiation	3	6	Nearly 100% reduction of microbial count	3-fold shelf-life extension	
Perca flavescens (yellow perch fillets)	Gamma irradiation	6	6	Nearly 100% reduction of microbial count. Progressively lower maximal bacterial populations and lengthened lag phases were obtained as more radiation was used	3.5-fold shelf-life extension	

were frozen and irradiated in the frozen state (−20 °C) were 0.43–0.66, 0.48–0.71, and 0.47–0.70 kGy, respectively. The radiation dose needed to inactivate these food-borne pathogens on frozen seafood is significantly lower than that for frozen meat or frozen vegetables.

Collins *et al.* (2005) studied the e-beam irradiation efficacy on the viability of *Cryptosporidium parvum* oocysts in Eastern oysters (*Crassostrea virginica*), which were inoculated with the Beltsville strain of *C. parvum*. The impact of treatments was assessed by oral feeding of the processed oyster tissues to neonatal mice. A significant reduction in infectivity was observed for in-shell and shucked oysters treated with e-beam irradiation at doses of 1.0, 1.5 or 2 kGy in comparison to untreated control samples. A dose of 2 kGy completely reduced *C. parvum* infectivity and did not adversely affect the visual appearance of the oyster.

Live oysters (*Crassostrea virginica*) with naturally incurred and artificially inoculated pathogenic vibrios, were exposed to 0–3 kGy dose of gamma radiation. *Vibrio vulnificus* (MO-624) decreased from 10^6 cfu/g oyster meat to non-detectable levels (<3 MPN/g oyster meat) with 0.75–1.0 kGy irradiation exposure. *Vibrio parahaemolyticus*, 03:K6 (TX-2103), required 1.0–1.5 kGy for a decrease to non-detectable levels. Using triangle difference testing, sensory panelists were asked to identify differences between treated (1 kGy) and untreated oysters. They were unable to distinguish non-irradiated from irradiated oysters (Andrews *et al.*, 2003).

Jakabi *et al.* (2003) assessed the influence of a gamma radiation process on high counts of *Salmonella enteritidis, Salmonella infantis* and *Vibrio parahaemolyticus* incorporated by oysters (*Crassostrea brasiliana*). The oysters were gamma irradiated in doses ranging from 0.5 to 3.0 kGy. A dose of 3.0 kGy was adequate to reduce the level of *Salmonella* serotypes by 5 to 6 $\log_{10}$ units. A dose of 1.0 kGy was enough to produce a 6-$\log_{10}$ reduction in the level of *V. parahaemolyticus*. The highest irradiation dose did not kill the oysters or affect their sensory characteristics.

In oysters, according to Gelli *et al.* (1999), doses lower than 3 kGy may achieve practical safety levels even against *Salmonella enteritidis*, assuming that the number of potential viable *Salmonella enteritidis* cells is as low as it should be when good primary production practices and HACCP have been implemented.

Dixon (1992) investigated irradiated Florida shellstock oysters. A 2–3 log cycle reduction in bacterial numbers was reported at all doses immediately after irradiation. However, bacterial numbers were significantly reduced in irradiated oysters and *Vibrio vulnificus* demonstrated a high radiosensitivity. The shelf life of irradiated shellfish noticeably decreased at doses higher than 1 kGy.

Shiflett *et al.* (1966) studied the flora of Dungeness crabmeat (*Cancer magister*) and Pacific oysters (*Crassostrea gigas*). Their initial flora and the changes in flora during storage at 7 °C were determined. The microbial flora shifts in both shellfish were also determined after irradiation at 1 and 4 kGy and during consequent storage at 7 °C. The *Achromobacter* spp. prevailed in the initial crabmeat flora (77.0%). The predominant position of this group rose to 99.2% after 1 kGy and 100% after 4 kGy. A high percentage of *Lactobacillus* was detected in oysters (55.0%). *Lactobacillus* spp. were the main survivors after 1 kGy (92.4%) but the main survivors after 4 kGy were *Achromobacter* spp. (99.3%).

Sinanoglou *et al.* (2007) studied irradiated frozen molluscs (squid, octopus and cuttlefish) and crustaceans (shrimp) at different doses, to investigate the impact of gamma radiation on their microbial population. Irradiation of shrimp and squid with either 2.5 or 4.7 kGy reduced mesophilic bacteria contamination to low or non-detectable levels, respectively. Irradiation of octopus and cuttlefish with the same doses decreased the bacterial population.

Table 5.6 Microflora and shelf life of irradiated shellfish, crustaceans and molluscs

Seafood species	Irradiation type	Irradiation dose (kGy)	Temperature (°C)	Results	Shelf life (days)	References
Squid	Gamma irradiation	2.5 4.7	–	Reduced mesophilic bacteria contamination to low or nondetectable levels	–	Sinanoglou *et al.*, 2007
Octopus	Gamma irradiation	2.5 4.7	–	Reduction of the bacterial population	–	
Cuttlefish	Gamma irradiation	2.5 4.7	–	Reduction of the bacterial population	–	
Shrimp	Gamma irradiation	2.5 4.7	–	Reduced mesophilic bacteria contamination to low or non detectable levels	–	
Crassostrea virginica (live oysters)	Electron beam irradiation	2.0	–	Complete elimination of *C. parvum* infectivity	–	Collins *et al.*, 2005
Crassostrea brasiliana (live oysters)	Gamma irradiation	1.0	–	6-$\log_{10}$ reduction of *V. parahaemolyticus*	–	Jakabi *et al.*, 2003
Crassostrea brasiliana (live oysters)	Gamma irradiation	3.0	–	Reduction of *Salmonella* serotypes by 5 to 6 $\log_{10}$ units	–	

(*continued overleaf*)

Table 5.6 (*continued*)

Seafood species	Irradiation type	Irradiation dose (kGy)	Temperature (°C)	Results	Shelf life (days)	References
Crassostrea virginica (live oysters)	Gamma irradiation	0.75–1.0	–	*Vibrio vulnificus* (MO-624) was reduced from 10^6 cfu/g oyster meat to non detectable levels (<3 MPN/g oyster meat)	–	Andrews *et al.*, 2003
Crassostrea virginica (live oysters)	Gamma irradiation	1.0–1.5	–	*Vibrio parahaemolyticus*, 03:K6 (TX-2103), was reduced to non detectable levels.	–	
Crab products (white lump, claw and fingers)	–	–	Ice storage (14 days)		–	Chen *et al.*, 1996
Crab products (white lump, claw and fingers)	Gamma irradiation	2.0	Ice storage (14 days)	Irradiation effectively reduced spoilage	–	
Penaeus monodon (grass prawns)	–	Up to 7.5	–	A 2–3 log reduction in bacterial counts	–	Hau *et al.*, 1992; Haw and Liew, 1993
Frozen shrimp	–	3.0	–	Reduction *Vibrio spp*. and *Aeromonas hydrophila* by 4 log cycles	–	Rashid *et al.*, 1992

Frozen shrimp	–	3.5	–	Similar reduction of *Listeria monocytogenes* or *Salmonella* spp.	–	
Callinectes sapidus (crabmeat)	–	0.25	–	Reducing *Vibrio cholera* more than 3 log cycles	–	Grodner and Hinton, 1986
Callinectes sapidus (crabmeat)	–	0.5 1	–	Complete elimination of *Vibrio cholera* from the samples	–	
Crassostrea gigas (Pacific oysters)	–	–	7	A large percentage of *Lactobacillus* was detected in oysters (55.0%)	2	Shiflett *et al.*, 1966
Crassostrea gigas (Pacific oysters)	Gamma irradiation	1.0	7	*Lactobacillus* spp. was the main survivor (92.4%)	–	
Crassostrea gigas (Pacific oysters)	Gamma irradiation	4.0	7	The main survivor was *Achromobacter* spp. (99.3%)	–	
Cancer magister (dungeness crabmeat)	–	–	7	*Achromobacter* spp. predominated in the initial flora of crabmeat (77.0%)	2	
Cancer magister (dungeness crabmeat)	Gamma irradiation	1.0	7	The predominant position of *Achromobacter* increased to 99.2%	–	
Cancer magister (dungeness crabmeat)	Gamma irradiation	4.0	7	The predominant position of *Achromobacter* increased to 100%	–	

Chen *et al.* (1996) compared the microbiological and organoleptic quality of irradiated (2 kGy or less) crab products (white lump, claw and fingers) during 14 days of ice-storage. Irradiation effectively reduced spoilage bacteria, extending shelf life by more than 3 days in comparison to control samples. During storage, fresh crab odour and taste were similar for treated and control samples, while off-flavours and odours developed more rapidly in control samples. Overall acceptability scores for irradiated crab samples were higher than for control samples throughout the 14 days of ice storage.

Grass prawns (*Penaeus monodon*) were surface inoculated with *Vibrio cholera, Staphylococcus aureus, Escherichia coli* and *Salmonela enteritidis*, and irradiated at a wide variety of doses ranging from 0 to 10 kGy and stored for 48 days. A 2–3 log reduction in bacterial numbers was recorded up to 7.5 kGy and D_{10} values were found to be 0.11, 0.29, 0.39 and 0.48 kGy for *V. cholera, Staphylococcus aureus, E. coli* and *S. enteritidis*, respectively. Nutrient damage of samples was also evaluated and it was observed that C20:5, C22:6 and thiamine levels were decreased by 22, 25 and 32%, respectively (Hau *et al.*, 1992; Hau and Liew, 1993).

Rashid *et al.* (1992) reported that a dose of 3 kGy effectively decreased *Vibrio* spp. and *Aeromonas hydrophila* in frozen shrimp by 4 log cycles, whereas 3.5 kGy was able to achieve a similar reduction in counts of *Listeria monocytogenes* or *Salmonella* spp. *Vibrio cholera* was inoculated to crabmeat (*Callinectes sapidus*) 10^7 cfu/g homogenate. A dose of 0.25 kGy was effective in reducing *Vibrio cholera* by more than 3 log cycles, while 0.5 and 1 kGy completely eliminated *Vibrio cholera* in all samples (Grodner and Hinton, 1986).

Angel *et al.* (1986) irradiated freshwater prawns (*Macrobrachium rosenbergii*) at 1.45 kGy and 2.3 kGy. A 1 to 2 log cycle reduction in total numbers of bacteria was recorded. After 28 days of storage, total counts increased by about 2–3 log cycles.

The impact of irradiation dose on the microflora and shelf life of shellfish, crustaceans and molluscs is given in Table 5.6.

5.4 Conclusions

The advantages of irradiation as a preservation technique for extending the shelf life of fish and seafood are its effectiveness and its low cost compared to all other currently available methods. For fish products, shelf-life prolongation can be three to five times that of traditionally employed methods. Gamma irradiation of 2–7 kGy is considered a successful method of preservation since it can reduce the populations of food-borne bacterial pathogens (*Salmonella, Listeria,* and *Vibrio* spp.) as well as many fish-specific bacterial spoilers, and can extend the shelf life of fish.

References

Abu-Tarboush, H.M., Al-Kahtani, H.A., Atia, M. *et al.* (1996). Irradiation and postirradiation storage at 2 ± 2 °C of tilapia (*Tilapia nilotica* × *T. aurea*) and Spanish mackerel (*Scomberomorus commerson*): Sensory and microbial assessment. *Journal of Food Protection*, 9(10), 1041–1048.

Ahmed, I.O., Alur, M.D., Kamat, A.S., Bandekar, J.R. and Thomas, P. (1997). Influence of processing on the extension of shelf-life of Nagli-fish (*Sillago sihama*) by gamma irradiation. *International Journal of Food Science and Technology*, 32(4), 325–332.

Aiyar, A.S. (1976). Progress in food irradiation: India. *Food Irradiation Information*, 6, 35–38.

Al-Kahtani, H.A., Abu-Tarboush, H.M., Atia, M. *et al.* (1998). Amino acid and protein changes in tilapia and Spanish mackerel after irradiation and storage. *Radiation Physics and Chemistry*, 51(1), 107–114.

Al-Kahtani, H.A., Abu-Tarbous, H.M., Bajaber, H.S. *et al.* (1996). Chemical changes after irradiation and post-irradiation storage in tilapia and Spanish mackerel. *Journal of Food Science*, 61(4), 729–733.

Ampola, V.G., Connors, T.J. and Ronsivalli, L.J. (1969). Preservation of fresh unfrozen fishery products by low level irradiation. VI. Optimum radiopasteurization dose studies on ocean perch, pollock and cod fillets. *Food Technology*, 23, 357–359.

Andrews, L., Jahncke, M. and Mallikarjunan, K. (2003). Low dose gamma irradiation to reduce pathogenic vibrios in live oysters (*Crassostrea virginica*). *Journal of Aquatic Food Product Technology*, 12(3), 71–82.

Anellis, A., Berkowitz, D., Swantak, D.W. and Strojan, C. (1972). Radiation sterilization of prototype military foods: low-temperature irradiation of codfish cake, corned beef, and pork sausage. *Applied Microbiology*, 24(3), 453–462.

Angel, S., Juven, B.J., Weinberg, Z.G., Linder, P. and Eisenberg, E. (1986). Effects of radurization and refrigerated storage on quality and shelf life of freshwater prawns, *Macrobrachium rosenbergii*. *Journal of Food Protection*, 49, 142–145.

Antunes, S.A., Wiendl, P.M., Almeida, E.R., Arthur, V. and Daniotti, C. (1993). Gamma ionozation of *Phugicoiu longa* (Trematoda: Heterophyidae) in Mugilidae (Pisces) in Sao Paulo, Brazil. *Radiation Physics and Chemistry*, 30, 425–428.

Armstrong, S.G., Wyme, S.G. and Leach, D.N. (1994). Effects of preservation by gamma-irradiation on the nutritional quality of Australian fish. *Food Chemistry*, 50, 351–357.

Arvanitoyannis, S.I., Stratakos, A. and Mente, E. (2009). Impact of irradiation on fish and seafood shelf life: a comprehensive review of applications and irradiation detection. *Critical Reviews in Food Science and Nutrition*, 49, 68–112.

Badr, H.M. (2012). Control of the potential health hazards of smoked fish by gamma irradiation. *International Journal of Food Microbiology*, 154, 177–186.

Bani, M.L., Sabina, Y., Kusunoki, H. and Uemura, T. (2000). Preservation of fish cutlet (*Pangasius pangasius*) at ambient temperature by irradiation. *Journal of Food Protection*, 63(1), 56–62.

Barbosa-Canovas, G.V., Pothakamury, U.R. and Palou, E. (1998). *Nonthermal Preservation of Foods*, Marcel Dekker, Inc., New York, USA.

Carver, J.H., Connors, T.J. and Slavin, J.W. (1969). Irradiation of fish at sea. *Freezing and Irradiation of Fish*. Fishing News (Books) Ltd., London, UK, pp. 509–513.

Chen, Y.P., Andrews, L.S. and Grodner, R.M. (1996). Sensory and microbial quality of irradiated crab meat products. *Journal of Food Science*, 61(6), 1239–1242.

Cho, H.O., Byun, M.W. and Kwon, J.K. (1992). Preservation of dried fish powders and mix condiments by gamma irradiation, in Asian Regional Co-Operative Project on Food Irradiation: Technology Transfer. Proceedings of the Final Coordination Meeting in Bangkok, (1988), International Atomic Energy Agency, Vienna, pp. 51–63.

Chuaqui-Offermanns, N., McDougall, T.E., Sprung, W. and Sullvan, V. (1988). Raduriation of commercial freshwater fish species. *Radiation Physics and Chemistry*, 31(1–3), 243–252.

Colby, J.W., Enriquez-Ibarra, L. and Flick, G.J., Jr, (1993). Shelf life of fish and shellfish, in *Shelf Life Studies of Foods and Beverages – Chemical, Biological, Physical and Nutritional* (ed. G. Charalambous), Elsevier, Amsterdam, pp. 85–143.

Coleby, B. and Shewan, J.M. (1965). The radiation preservation of fish, in *Fish as Food*, Vol. 4 (ed G. Borgstrom), Academic Press Inc., New York, p. 419.

Collins, M.V., Flick, G.J., Smith, S.A. *et al.* (2005). The effects of e-beam irradiation and microwave energy on eastern oysters (*Crassostrea virginica*) experimentally infected with *Cryptosporidium parvum*. *Journal of Eukaryotic Microbiology*, 52(6), 484–488.

Connors, T.J. and Steinberg, M.A. (1964). Preservation of fresh unfrozen fishery products by low level irradiation. II. Organoleptic studies on radiation pasteurized soft-shell clam meats. *Food Technology*, 18, 1057–1060.

Cozzo-Siqueira, A., Oetterer, M. and Gallo, C.R. (2003). Effects of irradiation and refrigeration on the nutrients and shelf-life of tilapia (*Oreochromis niloticus*) *Journal of Aquatic Food Product Technology*, 12(1), 85–101.

De la Sierra Serrano, D. (1970). Prolonging the commercial life of fresh white (lean) fish: microbiological aspects, in *Preservation of Fish by Irradiation*, International Atomic Energy Agency, Vienna, pp. 27–48.

De Ponte, D.J.B., Roozen, J.P. and Pilnik, W. (1986). Effect of irradiation on the stability of minced cod, with and without hydrocolloids during frozen storage. *Lebensmittel-Wissenschaft und Technologie*, 19, 167–171.

EPCEU (European Parliament and the Council of the European Union) (1999) Directive (1999)/2/EC of the European Parliament and of the Council on the approximation of the laws of the Member States concerning foods and food ingredients treated with Ionizing radiation. *Official Journal of the European Communites*, L 66/16. http://eur-lex.europa.eu/LexUriServ/LexUriServ.do?uri=OJ:L:1999:066:0016:0022:EN:PDF [accessed 7 June 2013].

Dixon, D.W. (1992). The effects of gamma irradiation (^{60}Co) upon shellstock oysters in terms of shelf life and bacterial reduction, including *Vibrio vunificus* levels. *Master's thesis*. University of Florida, USA.

Doke, S.N., Sherekar, S.V. and Ghadi, S.V. (1992). Radiation preservation of white pomfret (*Scombermorus guttatus*) in tomato sauce. *Acta Alimentaria*, 21, 123–129.

Dvorak, P., Kratochv, B. and Grolichov, M. (2005). Changes of colour and pH in fish musculature after ionizing radiation exposure. *European Food Research and Technology*, 220, 309–311.

Dymsza, H.A., Lee, C.M., Saibu, L.O. *et al.* (1990). Gamma irradiation effects on shelf life and gel forming properties of washed red hake (*Urophycis chuss*) fish mince. *Journal of Food Science*, 55(6), 1745–1746.

Ehlermann, D. (1976). Preliminary studies on the radiation preservation of brown shrimp. *Chemie, Mikrobiologie und Technologie der Lebensmittel*, 4(5), 150–153.

Ehlermann, D. and Diehl, J.F. (1977). Economic aspects of the introduction of radiation preservation of brown shrimp in the Federal Republic of Germany. *Radiation Physics and Chemistry*, 9, 875–885.

Ehlermann, D. and Muenzer, R. (1976). Radurization of brown shrimp. *Archives Lebensmittelhygiene*, 27(2), 50–55.

Emerson, J.A, Greig, R.A., Kazanas, N., Kempe, L.L., Markakis, P., Nicholas R.C., Schweigert, B.S. and Seagran, H.L. (1964). Irradiation preservation of freshwater fish and inland fruits and vegetables, Michigan State University., Ann Arbor, COO-1283–12.

Emerson, J.A., Kazanas, N., Gnaedinger, R.H., Krzeczkowski, R.A. and Seagran, H.L. (1965). *Irradiation of freshwater fish*. Report to the US Atomic Energy Commission on Contract No. AT(11–1)-1283. Bur. Com. Fish. Tech. Lab., Ann Arbor, Michigan, USA.

Gardner, E.A. and Watts, B.M. (1957). Effect of ionizing radiation on southern oysters. *Food Technology*, 11, 329–331.

Gelli, D.S., Del Mastro., N., Rodrigues de Moraes, I. and Jakabi, M. (1999). Study on the radio-sensitivity of pathogenic Vibrionaceae and Enterobacteriaceae in vitro and after incorporation into oysters (*Crassostrea brasiliana*), in *Final Report, Co-ordinated Research Programme on Irradiation as a Public Health Intervetion Measure to Control Food-borne Desease (Cysticercosis/ Taeniasis and Vibrio Infections) in Latin America and the Carribean*, Pan American Health Organization, Washington, DC, pp. 5–6.

Ghadi, S.V., Alur, M.D., Ghose, S.K. *et al.* (1978). Quality assessment of radurized Indian mackerel (*Rastrellinger kanagurta*). Progress report of IAEA. Research contract No 1612/CF, presented at the Advisory Group Meeting on Radiation treatment of Fish and Fishery products, Manila, Philippines.

Graikowski, J.T., Kazanas, N., Watz, J. *et al.* (1968). Irradiation preservation of freshwater fish. Report to the US Atomic Energy Commission on Contract No. AT(49–11)-2954. Bur. Com. Fish. Tech. Lab., Ann Arbor, Michigan.

Gram, L. and Huss, H.H. (1996). Microbial spoilage of fish and fish products. *International Journal of Food Microbiology*, 33, 121–137.

Gore, M.S. and Kumta, U.S. (1970). Studies on textural stability of irradiated Bombay duck. *Food Technology*, 24, 286–289.

Grodner, R.M. and Hinton, A.H. (1986). Low dose gamma irradiation of *Vibrio cholera* in crabmeat (*Callinectes sapidus*). Proceedings of the Eleventh Annual Tropical and Subtropical Fisheries Conference of the Americas. Texas Aricultural Extension Service.

Gueavara, G., Borromeo, J. and Matias, V. (1965). Effect of gamma irradiation in the preservation of fish and fishery products. Report to IAEA on Contract number No. 1495/RB and 1495/R1/RB.

Hammad, A.A.I. and El-Mongy, T.M. (1992). Shelf life extension and improvement of the microbiological quality of smoked salmon by irradiation. *Journal of Food Processing*, 16, 361–370.

Hau, L.B. and Liew, M.S. (1993). Effect of gamma irradiation and cooking on vitamins B6 and B12 in grass prawns (*Penaeus monodon*). *Radiation Physics and Chemistry*, 42, 297–300.

Hau, L.B., Liew, M.S. and Yeh, L.T. (1992). Preservation of grass prawns by ionizing radiation. *Journal of Food Protection*, 55, 198–202.

Hussain, A.M. (1980). Radiation preservation of mackerel. *Food Irradiation News Letter*, 4, 14–15.

Hussain, A.M., Chaudry, M.A. and Haq, I. (1985). Effect of low doses of ionizing radiation on shelf life of mackerel (*Rastrellinger kanagurta*). *Lebensmittel, Wissenschaft und Technologie*, 18, 273–276.

Icekson, I., Pasteur, R., Drabkin, V. *et al.* (1996). Prolonging shelf-life of carp by combined ionizing radiation and refrigeration. *Journal of the Science of Food and Agriculture*, 72(3), 353–358.

Jaczynski, J. and Park, J.W. (2003). Microbial inactivation and electron penetration in surimi seafood during electron beam processing. *Journal of Food Science*, 68(5), 1788–1792.

Jakabi, M., Gelli, D.S., Torre, J.C.M.D. *et al.* (2003). Inactivation by ionizing radiation of *Salmonella* enteritidis, *Salmonella* infantis, and *Vibrio parahaemolyticus* in oysters (*Crassostrea brasiliana*). *Journal of Food Protection*, 66(6), 1025–1029.

Jo, C., Lee, N.Y., Kang, H.J. *et al.* (2005). Inactivation of pathogens inoculated into prepared seafood products for manufacturing kimbab, steamed rice rolled in dried seaweed, by gamma irradiation. *Journal of Food Protection*, 68(2), 396–402.

Josephson, E.S. and Peterson, M.S. (2000). *Preservation of Food by Ionizing radiation (II)*, CRC, Boca Raton, FL, pp. 102–103.

Kamat, A. and Thomas, P. (1998). Radiation inactivation of some food-borne pathogens in fish as influenced by fat levels. *Journal of Applied Microbiology*, 84, 478–484.

Kume, T., Tachibana, H., Aoki, S. and Sato, T. (1975). Dose distribution of 'Kamaboko' on package irradiation. *Shokuhin Shosha*, 1(1/2), 61–67.

Kumta, U.S., Mavinkurve, S.S., Gore, M.S. *et al.* (1970). Radiation pasteurization of fresh and blanced tropical shrimps. *Journal of Food Science*, 35, 360–363.

Kumta, U.S., Savagaon, K.A., Ghadi, S.V. *et al.* (1973). Radiation preservation seafoods: review and research in India. Radiation Preservation of Food, Proceedings of Symposium, Bombay, 1972, International Atomic Energy Agency, Vienna. pp. 403–424.

Kumta, U.S. and Sreenivasin, A. (1970). Preservation by gamma irradiation of Bombay duck, shrimps and white pomfet, in *Preservation of Fish by Gamma Irradiation*, International Atomic Energy Agency, Vienna, pp. 403–424.

Kazanas, N. and Emerson, J.A. (1968). Effect of γ irradiation on the microflora of freshwater fish III. Spoilage patterns and extension of refrigerated storage life of yellow perch fillets irradiated to 0.1 and 0.2 megarad. *Applied Microbiology*, 16(2), 242–247.

Kazanas, N., Emerson, J.A., Seagran, H.L. and Kempe, L.L. (1966). Effect of -γ-irradiation on TBW microflora of freshwater fish I. Microbial load, lag period, and rate of growth on yellow perch (*Perca flavescens*) fillets. *Applied Microbiology*, 14(2), 261–266.

Kwon, J.H. and Byun, M.W. (1995). Gamma irradiation combined with improved packaging for preserving and improving the quality of dried fish (*Engraulis encrasicholus*). *Radiation Physics Chemistry*, 46(4–6), 725–729.

Lakshmanan, R., Venugopal, V., Venketashvaran, K. and Bongirwar, D.R. (1999). Bulk preservation of small pelagic fish by gamma irradiation: studies on a model storage system using anchovies. *Food Research International*, 32, 707–713.

Laycock, R.A. and Regier, L.W. (1970). Pseudomonads and achromobacters in the spoilage of irradiated haddock of different pre-irradiation quality. *Applied Microbiology*, 20(3), 333–341.

Lee, K.H., Ahn, H.J., Jo, C., Yook, H.S. and Byun, M.W. (2002). Production of low salted and fermented shrimp by irradiation. *Journal of Food Science*, 67(5), 1772–1777.

Liu, M.S., Chen, R.Y., Tsai, M.J. and Yang, J.S. (1991). Effect of gamma irradiation on the keeping quality and nutrients of tilapia (*Orechromis mosambicus*) and silver carp (*Hypophthalmichthis molitrix*) stored at 1 °C. *Journal of the Science of Food and Agriculture*, 57, 555–563.

Liuzzo, J.A., Novak, A.F., Grodner, R.M. and Rao, M.R.R. (1970). Radiation Pasteurization of Gulf shellfish, Louisiana State University for the U. S. Atomic Energy Commission, Technical Information Division, Rep. No. ORO-676, Baton Rouge, LA, 42.

Mahapatra, A.K., Muthukumarappan, K. and Julson, J.L. (2005). Applications of ozone, bacteriocins and irradiation in food processing: a review. *Critical Reviews in Food Science and Nutrition*, 45, 447–461.

McLaughlin, W.L., Boyd, A.W, Chadwick, K.H., McDonald, J.C. and Miller, H. (1989). *Dosimetry for Radiation Processing*, Taylor & Francis, New York, p. 251.

Medina, M., Cabeza, M.C., Bravoa, D. *et al.* (2009). A comparison between E-beam irradiation and high pressure treatment for cold-smoked salmon sanitation: microbiological aspects. *Food Microbiology*, 26, 224–227.

Mendes, R., Silva, H.A., Nunes, M.L. and Empis, J.M.A. (2000). Deteriorative changes during ice storage of irradiated blue jack mackerel (*Trachurus picturatus*). *Journal of Food Biochemistry*, 24, 89–105.

Mendes, R., Silva, H.A., Nunes, M.L. and Empis, J.M.A.E. (2005). Effect of low-dose irradiation and refrigeration on the microflora, sensory characteristics and biogenic amines of Atlantic horse mackerel (*Trachurus trachurus*). *European Food Research and Technology*, 221, 329–335.

Miyauchi, D.T., Spinelli, J., Pelroy, G. and Steinberg, M.A. (1968). Radiation preservation of Pacific coast fishery products. *Isotope Radiation Technology*, 5, 136–141.

Miyauchi, D.T., Spinelli, J., Pelroy, G., Teeny, F. and Seman, J. (1967). Application of radiation pasteurization process to Pacific crab and flounder. Report to the U.S. Atomic Energy Commission on Contract No. AT(49–11)-2058. Bur. Com. Fish. Tech. Lab., Seattle, Washington. TID-24317.

Moini, S., Tahergorabi, R., Hosseini, S.V. *et al.* (2009). Effect of gamma radiation on the quality and shelf life of refrigerated rainbow trout (*Oncorhynchus mykiss*) Fillets. *Journal of Food Protection*, 72(7), 1419–1426.

Morehouse, K.M. (1998). Fish irradiation: results and perspectives. *Food Testing and Analysis*, 4(3), 9, 32, 35.

Nickerson, J.T.R. (1963). *The Storage Life Extension of Refrigerated Marine Products by Low Dose Radiation Treatment: Exploration of Future Food Processing Techniques*. MIT Press, Cambridge, MA.

Nickerson, J.T.R. and Goldblith, S.S. (1969/1971). Toxin production of naturally occuring levels of *Clostridium botulinum* in irradiated and nonirradiated seafood, U.S. Atomic Energy Commission, Division of Technical Information, Springfield, reports: MIT-4049–1. Progress Report, MIT-4049–2, Final Report (1969). Final Report MIT-4049–2 (1971), Final Report, 1970 (Revision).

Novak, A.F., Liuzzo, J.A., Grodner, R.M. and Lovell, R.T. (1966). Radiation pasteurization of Gulf Coast oysters. *Food Technology*, 20, 201–202.

Ogbadu, G.H. (1988). Use of gamma irradiation to prevent aflatoxin B1 production in smoked dried fish. *Radiation Physics and Chemistry*, 31(1–3), 207–208.

Oku, T. (1976). Changes in amino acid content of irradiated Kamaboko during storage *Shokuhin Shosha*, 11(1/2), 27–32.

Oku, T. (1981). Changes in amino acid content of irradiated Kamaboko during storage *Shokuhin Shosha*, 16(1/2), 13–18.

Oku, T. and Kimura, S. (1975). Effect of irradiation on amino acid content of tuna meat and steamed fish paste Kamaboko. *Shokuhin Shosha*, 10(12), 14–43.

Ostovar, K., Slusar, M. and Vaisey, M. (1967). Preservation of fresh whitefish with gamma irradiation. *Journal of the Fisheries Research Board Canada*, 24, 9–19.

Ozden, O., Inugur, M. and Erkan, N. (2007). Effect of different dose gamma radiation and refrigeration on the chemical and sensory properties and microbiological status of aquacultured sea bass (*Dicentrarchus labrax*). *Radiation Physics and Chemistry*, 76, 1169–1178.

Poole, S.E., Mitchell, G.E. and Mayze, J.L. (1994). Low dose irradiation affects microbiological and sensory quality of sub-tropical seafood. *Journal of Food Science*, 59(1), 85–87.

Poole, S.E., Wilson, P., Mitchell, G.E. and Willis, P.A. (1990). Storage life of chilled scallops treated with low dose irradiation. *Journal of Food Protection*, 53(9), 763–766.

Rahman, M.S. (1999). Irradiation preservation of foods, in *Handbook of Food Preservation* (ed. M.S. Rahman), Marcel Dekker, New York, pp. 397–419.

Rashid, H.O., Ito, H. and Ishigaki, I. (1992). Distribution of pathogenic vibrios and other bacteria in imported frozen shrimps and their decontamination by gamma irradiation. *World Journal of Microbiology and Biotechnology*, 8, 494–499.

Reinacher, E. and Ehlermann, D. (1978). Effect of irradiation on board on the shelf life of red fish. 1. Results of the sensory examination. *Archives von Lebensmittelhygiene*, 29(1), 24–28.

Riebroy, S., Benjakul, S.,Visessanguan, W. *et al.* (2007). Effect of irradiation on properties and storage stability of Som-fug produced from bigeye snapper. *Food Chemistry*, 103, 274–286.

Ritacco, M. (1976). Effect of vacuum on the radurization of hake (*Merluccius merluccius husbi*) fillets. *Comision Nacional de Energia Atomica*, 409, 1–16.

Rosnivalli, L.J., Ampola, V.G., King, F.G. and Holston, J.A. (1970). Study of irradiated pasteurized fishery products. Maximum shelf life study. Report to U. S. Atomic Energy Commission on Contract no. AT(49–11)-1889. Bur. Com. Fish. Tech. Lab., Glousester, Mass.

Rosnivalli, L.J., King, F.G., Ampola, V.G. and Holston, J.A. (1968). A study of irradiated pasteurized fishery products. Report to U.S. Atomic Energy Commission on Contract no. AT(49–11)-1889. Bur. Com. Fish. Tech. Lab., Gloucester, Mass, TID-24256.

Rosnivally, L.J. and Slavin, J.W. (1965). Pasteurization of fishery products with gamma rays from a Cobalt-60 source. *Commercial Fisheries Reviews*, 27, 1–9.

Sasayama, S. (1972). Irradiation preservation of fishery products -1. Effects of low dose irradiation and storage acceptability of Kamaboko. *Shokuhin Shosha*, 7(1), 64–72.

Sasayama, S. (1973). Irradiation preservation of fish meat jelly products -2. Classification of bacteria isolated from Kamaboko, *Shokuhin Shosha*, 8(1), 107–111.

Sasayama, S. (1977). Irradiation preservation of fish meat jelly products -3. Irradiation effect on freid Kamaboko (Age-Kamaboko). *Shokuhin Shosha*, 12(1), 55–56.

Scholz, D.J., Sinnhuber, R.O., East, D.M. and Anderson, A.W. (1962). Radiation pasteurized shrimp and crab meat. *Food Technology*, 16, 118–120.

Sharma, S.K., Basu, S. and Gholap, A.S. (2007). Effect of irradiation on the volatile compounds of shrimp (*Solenocera choprii*). *Journal of Food Science and Technology*, 44(3), 267–271.

Shiflett, M.A., Lee, J.S. and Sinnhuber, R.O. (1966). Microbial flora of irradiated dungeness crabmeat and pacific oysters. *Applied Microbiology*, 14(3), 411–415.

Silva, H.A., Mendes, R., Nunes, M.L. and Empis, J. (2006). Protein changes after irradiation and ice storage of horse mackerel (*Trachurus trachurus*). *European Food Research and Technology*, 224, 83–90.

Sinanoglou, V.J., Batrinou, A., Konteles, S. and Sflomos, K. (2007). Microbial population, physicochemical quality, and allergenicity of molluscs and shrimp treated with cobalt-60 gamma radiation. *Journal of Food Protection*, 70(4), 958–966.

Slavin, J.W., Carner, J.H., Connors, T.L. and Rosnivalli, L.J. (1966). Shipboard Irradiator Studies. Annual Report, TID-23398. Bureau of Commercial Fisheries, Gloucester, MA.

Slavin, J.W. and Ronsivalli, L.J. (1964). Study of irradiated pasteurized fishery products. Report to U.S. Atomic Energy Commission on Contract No.AT.(49–11)-1889. Bureau of Commercial Fisheries Laboratory, Gloucester, TID-21600.

Snauwert, F., Tobback, P., Maes, E. and Thyssen, J. (1977). Radiation induced lipid oxidation in fish. *Zeitscrift fur Lebensmittel-Unterforschung und Forschung*, 164, 28–30.

Sofyan, I.R. (1978). Biochemical effects of gamma irradiation on Kembung fish (*Rastrelliger neglectus*), in *Food Preservation by Irradiation*, International Atomic Energy Agency, Vienna, p. 407–420.

Sommers, C.H. and Rajkowski, K.T. (2011). Radiation inactivation of food-borne pathogens on frozen seafood products. *Journal of Food Protection*, 74(4), 641–644.

Teeny, F.M. and Miyauchi, D.T. (1970). Irradiation of Pacific coast fish at sea. *Journal of Milk and Food Techology*, 33, 330–340.

Valdes, E. and Szeinfeld, D. (1989). Microbiological analysis of *Merluccius merluccius hubsi* fillets treated with Co-60 ionizing radiation. *Bulletin of the PanAmerican Health Organization*, 23, 316–322.

Vyncke, W., Declerk, D. and Scietecatte, W. (1976). Influence of gas permeability of the packaging material on the shelf life of irradiated and nonirradiated brown shrimps (*Crangon vulgaris*). *Lebensmittel-Wissenschaft und Technologie*, 9, 14–17.

6
Preservation of Fish by Curing

Sigurjon Arason,[1,3] Minh Van Nguyen,[2,3] Kristin A. Thorarinsdottir[4] and Gudjon Thorkelsson[1,3]

[1]*Faculty of Food Science and Nutrition, University of Iceland, Reykjavik, Iceland*
[2]*Faculty of Food Technology, Nha Trang University, Nha Trang, Vietnam*
[3]*Matís-Icelandic Food and Biotech R & D, Reykjavik, Iceland*
[4]*Marel, Reykjavik, Iceland*

6.1 Introduction

Prehistorically, fish has been caught and eaten first by hominids (*Australopithicus* and *Homo erectus*) and then by men (*Homo sapiens*) (Stewart, 1994). There is enough archaeological evidence to show that fish has been used as food for humans for 380 000 years (Toussaint-Samat, 2009). Due to the absence of refrigeration in the past, curing was the best means of preserving fish until it could be taken ashore, consumed or shipped to market.

Fish curing generally comprises all methods of preservation except refrigeration and canning. It includes (i) drying, salting, smoking, pickling and marinating of fish, (ii) various combinations of these methods and (iii) preservation of fish by fermentation (Food and Agriculture Organization (FAO), 1983; Jarvis, 1988). The origins of preserving food using salt can be traced back at least to ancient Egypt, where salt was used as part of the embalming process, as well as in food preservation. Salted fish found in Egyptian tombs pre-dates the first Chinese record of preserving fish in salt from approximately 2000 BC. Fish salting was already a flourishing integrated industry in almost contemporary terms (capture, farming, processing, packaging, transport and distribution) in the Roman Empire from about 100 BC (McCann, 1988). Curing techniques have been developed and modified many times during the history of humans and are still widely used.

The fish curing industry has flourished through the ages and has not been affected to any great extent by modern fish preservation and processing techniques. The is because cured fish is a highly appreciated and traditional product in many countries, mainly due to its excellent storage stability, special organoleptic characteristics and nutritional value (Lauritzsen *et al.*, 2004a). Nowadays, cured fish plays an important role in the fish destined

Seafood Processing: Technology, Quality and Safety, First Edition. Edited by Ioannis S. Boziaris.

for direct human consumption, with a share of 9.8% compared to a share of 46.9% for live, fresh or chilled fish and 29.3% for frozen fish (FAO, 2012). The world production of cured fish products remained unchanged and was on average 12 million tonnes per annum in 2001–2010 (FAO, 2010). Europe produces about two-thirds of the total, the largest producers and exporters in order of importance being Norway, Iceland and the Faroe Islands, whereas the biggest importers are Portugal and Spain, followed by Brazil. The most important producing countries outside Europe are Canada and the USA (Bjarnason, 1987; Thorarinsdottir *et al.*, 2010b).

Traditionally, groundfish species (cod, saithe, haddock, ling, blue ling and tusk) are used for salting processes, including light salting and heavy salting, mainly because fish muscle has a low lipid content. Pelagic species (herring, sardine, capelin, blue whiting and mackerel) and salmonids (salmon, trout and arctic char) with a higher lipid content are more suited to other curing processes (smoking, marinating).

6.2 Salting

Salting is one of the oldest methods of fish preservation. It is based on the penetration of salt into the fish muscle and extraction of water from the fish muscle, thereby lowering water activity and pH. The water extraction is due not only to concentration differences in salt and water between the fish muscle and the surrounding media, but also to structural changes within the muscle (the effects of salt on the proteins). Traditionally, salting is performed either by dry salting, pickling or brining. During the past decades, the use of brine injection has increased to accelerate salting, increase automation, obtain a more equal salt distribution within thick muscles and improve processing yields (Thorarinsdottir *et al.*, 2010b).

The salting process has been developed from a single-step to a multi-step process, involving a pre-salting step to meet new demands from buyers and consumers, as well as to increase productivity and automation. The new salting method consists of the injection of brine into the fish muscle followed by dry salting or by brining and again dry salting. In this chapter, the main focus is on salting species of fish from the Gadidae family.

6.2.1 Salting methods

Dry salting (kench salting) Dry salting is the original salt-curing technique used during the processing of salted fish in many countries (Lauritzsen *et al.*, 1999; Gallart-Jornet *et al.*, 2007a). In traditonal dry salting, butterfly (split) fish or fillets were stacked with alternating layers of dry, coarse salt and kept for several weeks. Nowadays, the fish is piled with alternating layers of salt into a plastic tub with a hole in the bottom for draining away the liquid extracted from the fish. The salt uptake is initiated by extraction of liquid from the fish muscle and solubilization of the salt. During the salting process, salt diffuses into the fish muscle, whereas liquid diffuses out to the surrounding media (van Klaveren and Legrende, 1965). The driving forces for water and salt diffusion are mainly concentration gradient between the fish muscle and the surrounding media and within the fish muscle, as well as pressure gradient and water activity gradient. Because of increased salt concentration in the fish muscle, protein denaturation and aggregation occurs, affecting further penetration of the salt into the fish muscle.

Pickle salting In pickle salting, the procedure is the same as in the dry salting method. However, the liquid extracted while the salt is penetrating the fish muscle is not drained away. The brine drains to the bottom of tub, thus the fillets are gradually immersed in saturated brine. However, the ratio of brine to fish is much lower than that usually used in brine salting. The salting rate and the salt concentration cannot be controlled in the same way as during brining. Due to the high salt concentration, muscle proteins aggregate rapidly, resulting in lower weight yields compared to brine salting (Andres *et al.*, 2005a).

Brine salting Recently, brine salting has become popular for processing salted fish as a pre-salting step, followed by dry salting (Thorarinsdottir *et al.*, 2004). Brine salting is performed by immersing butterfly (split) fish or fillets into ready-made brine prepared from coarse salt and tap water. The brine concentration used and other conditions can vary between producers. The brining process is usually carried out for 1–3 days, thereafter dry salting is applied for 14–21 days. The salting rate depends on the initial brine concentration used and the ratio of brine to fish. Common initial ratios are 1.6 : 1 or 1 : 1, meaning that significant changes are observed in the salt concentration of the brine due to the dilution effects from liquid diffusing from the muscle during brining. The changes in brine concentration and salting rate during the brining process can be controlled either by increasing the brine to fish ratio or by adding salt to the brine to compensate for the salt taken up by the fish muscle (Barat *et al.*, 2002, 2003; Andres *et al.*, 2002, 2005b).

The use of brine salting as a pre-salting step before dry salting is believed to have several advantages over dry salting and pickling including a shorter processing time and higher weight yields as there is better control over the rate of salt uptake and water loss in the muscle (Beraquet *et al.*, 1983; Thorarinsdottir *et al.*, 2004; Andres *et al.*, 2005a). The rate of salt uptake and water loss can also be influenced by the conformational changes in proteins induced by the increased salt concentration in the fish muscle. Brine salting can also improve the colour and appearance of salted fish products when applied at levels of less than 20% NaCl (w/v) (Thorarinsdottir *et al.*, 2004).

Injection salting Salting by automatic injection has been in common use during past decades. Brine is injected into the fish flesh by multi-needle injector machines. The brine is believed to be located initially in brine pockets around the injection site. The salt concentration is quickly diluted by the liquid in the muscle as the salt migrates from the injection site to other parts of the muscle (Offer and Knight, 1988). The effects of injection can partly be described as brining, where the salt migrates from the injection site due to the concentration gradient. What makes the process different from brining is mainly the shorter diffusion distance of the brine. The volume of brine injected, distribution and retention of brine in the fish muscle depend on the characteristics of the raw materials, the brine composition and the equipment settings used, such as needle types, needle density, injection speed and the pressure applied. Needles with large diameters, high pressures and continuous jets during injection are factors that can lead to breakdown of the muscle structure, as well as the formation of channels or brine pockets in the muscle. Brine that is located in the holes made by the needles is not fixed within the structure and can easily drip off. High needle density can be more effective than increased pressure for improving the weight yields of fish muscle without increasing the risk of structural defects (Birkeland *et al.*, 2003). Pressure that is too high can easily cause damage to the muscle structure, such as increased gaping of the muscle segments and breakage of muscle fibres. As the structure of fish muscle is more delicate than that of meat muscle, milder treatments must be used.

Advantages of injection salting are (i) uniform salt concentration throughout the flesh, (ii) increased speed of salt penetration into the fish muscle and (iii) higher weight yields than for traditional dry salting or pickling methods, due to a lower moisture loss from the fish during the following salt-curing stages (Akse *et al.*, 1993). However, disadvantages of this method are (i) increased risk of microbial and metal contamination (e.g. iron and copper) from the needles and equipment used and (ii) possible damage to the fish muscle from the high pressure applied and needle holes (Bakowski *et al.*, 1970; Boles and Swan, 1997; Birkeland *et al.*, 2003, 2007).

6.2.2 Processes for salted fish products

Process for heavily salted fish products Heavily salted fish products are traditional products from the North-Atlantic fisheries and are highly regarded as ripened fish products in many countries (Lauritzsen *et al.*, 2004b). Heavy salting and curing of fish is a long process that can be divided into several steps as depicted in Figure 6.1. Originally, the fish was only pile salted by stacking split fish with alternating layers of coarse salt (kench salting, see 6.2.1). The fish was restacked several times to obtain an even pressure and curing. Finally, the salted fish was sun dried to remove more water. However, salting procedures have changed significantly, with new controlled processes designed not only to preserve the fish, but also to improve its sensory characteristics such as aroma, flavour, texture, juiciness and appearance (Martinez-Alvarez and Gomez-Guillen, 2005).

In the late twentieth century in Iceland, producers started to pre-salt fish before the pile salting step, using pickling (from ~1980) and brining (from ~1990). In the 1990s, exporters started to control the ambient temperature in storage, which made it possible to shorten the salting process. The fish is pile-salted for 10–14 days after pre-salting and then packed as lightly/wet cured. This curing stage of the products is also known 'wet salted'. Today, different combinations of injection, brining, pickling and dry salting/pile salting/stacking are used, varying with producers and production countries.

The changes in salting procedures and curing conditions have altered the characteristics of the products, increased weight yields and improved commercial quality (e.g. colour/appearance, juiciness) (Lindkvist *et al.*, 2008). The shorter curing time and lower temperature during curing and storage conditions have resulted in milder curing flavours and a whiter appearance (Barat *et al.*, 2003; Lindkvist *et al.*, 2008). Another important factor is the improved quality of the raw material as a result of improvements to the catching, handling and storage techniques for fish.

Storage and transport of heavily salted fish After salting, the fish is normally packed with excess salt into 25 kg waxed carton boxes and stored at 0–4 °C and a relative humidity (RH) of 78 ± 4%. To minimize weight changes during storage, equilibrium must be maintained between the water activity (a_w) in the salted fish and the RH in air where the fish is stored. When the RH is higher than a_w in fully salted muscle (0.77), the salted fillets absorb water and gain weight, whereas the opposite is observed when RH < a_w (Doe *et al.*, 1982). Precise control of temperature and humidity is essential for the growth of halophilic bacteria and the stability of quality parameters. Some microorganisms and enzymes are still active despite the high salt content and chilled storage conditions (Pedro *et al.*, 2004; Rodrigues *et al.*, 2003). Curing of the fish continues after packing, as well as during storage and transport of the products. The curing time, and whether or not further processing such as drying is applied, depends on the consumer demand.

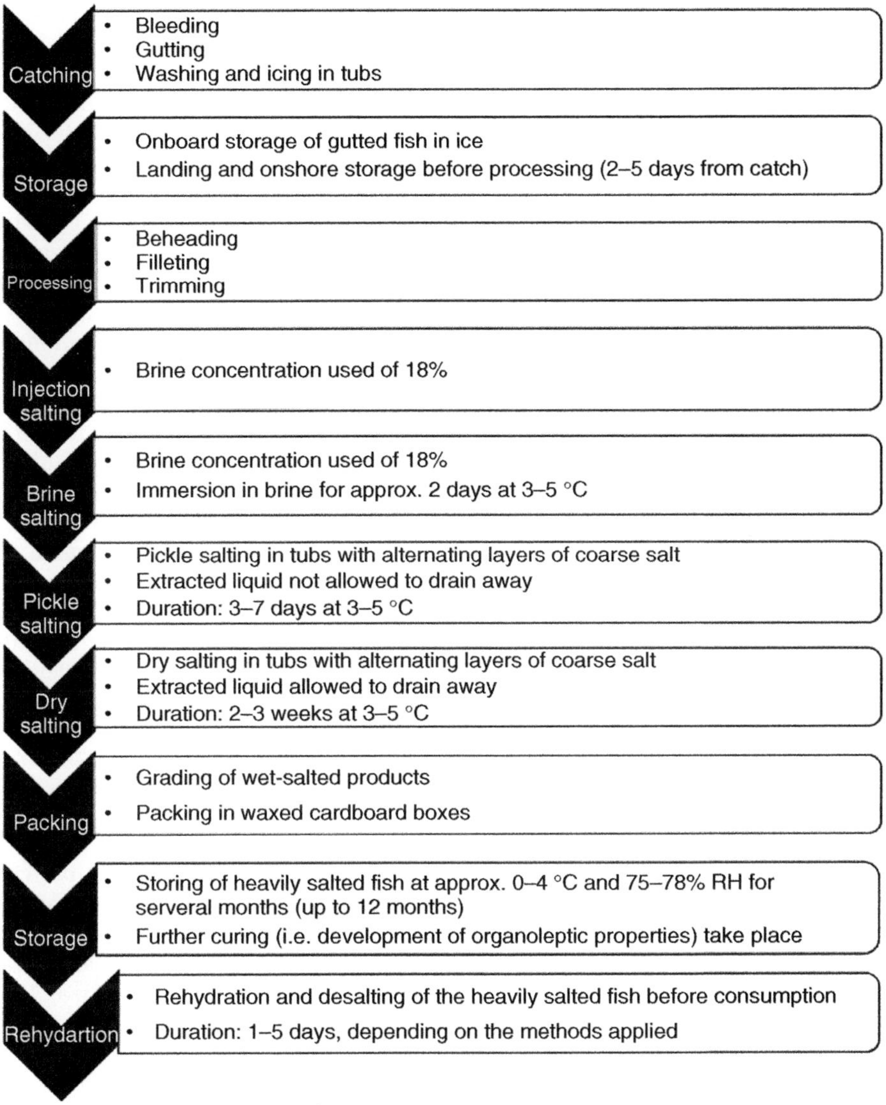

Figure 6.1 Heavy salting process of fish, from catch to a ready to cook product

Rehydration method Rehydration is needed before heavily salted fish is cooked. This is achieved by soaking the heavily salted fish in fresh water to reduce the salt concentration to an acceptable level for human consumption. Normally, the final salt concentration of the rehydrated fish can vary from 1 to 3% NaCl (w/w). The main components transferred during rehydration are water, salt and some nitrogenous compounds, as during the salting process. From an engineering point of view, the rehydration process is considered as a solid–liquid extraction, in which Na^+ and Cl^- are the solutes and water is the solvent (Barat *et al.*, 2004c). The fluxes of salt and water are reversed during rehydration compared to that during salting. Mass transfer during rehydration is accomplished not only by diffusion, but also by a hydrodynamic mechanism accelerated mainly by the pressure gradient, which promotes the bulk transport of the solution including water and salt (Barat *et al.*, 2004c; Thorarinsdottir *et al.*, 2010a).

The rates of chemical fluxes depend on the raw material used, the salting methods, the size of fish and the rehydration methods applied (Andres *et al.*, 2005b; Barat *et al.*, 2004a, 2004b, 2004c, 2006). The quality of water used and water management during rehydration are believed to affect the rehydration process in cod (Andres *et al.*, 2005b; Barat *et al.*, 2004b). Other process conditions that could affect rehydration are the ratio of water to fish and the temperature (Andres *et al.*, 2005b). In order to accelerate the rate of water uptake and salt loss during rehydration, other techniques have been applied such as injection and tumbling technologies (Bjørkevoll *et al.*, 2004), and rehydration using vacuum pulses (Andres *et al.*, 2005b). Skjerdal *et al.* (2002) have reported that the rehydration time of salted cod is reduced by 70–90% when tumbling technology is applied.

Process for lightly salted fish products Lightly salted fish with a salt concentration of approximately 2% is a new product that has become a popular recently in Southern Europe, especially in Spain, Italy and Greece. The fish fillets are lightly salted by brine injection and/or immersion into brine. Light salting is commonly used prior to freezing in order to prevent the negative effects of freezing. The purpose is to increase the water holding capacity and the weight of frozen products and to decrease water drip during storage (Mahon and Schneider, 1964; Boyd and Southcott, 1965). The flavour and texture characteristics, which are formed as proteins and lipids degrade during the traditional curing of heavily salted products, are missing. After salting, the fillets are frozen and exported. The light salting process can be divided into steps as depicted in Figure 6.2.

6.2.3 Changes in fish muscle during salting

Chemical composition of fish The main components of fish muscle are water, proteins, lipids and connective tissue. The chemical composition of fish varies widely, not only from fish to fish of the same species, but also within an individual fish (FAO, 2001). It is well known that fish composition is subject to seasonal variations, sex, age, water temperature, as well as the type and amount of dietary ingredients (Ross and Love, 1979; Shearer, 1994). The protein content and the water content in fish muscle have an inverse relationship, particularly when spawning time approaches. The lipid content in fish muscle varies much more widely compared to its protein and water content. Depending on the lipid content, fish is classified into four different groups (Ackman, 1990): (i) lean fish (e.g. cod, haddock) – lipid content is less than 2%, (ii) low fat fish (e.g. flounder, halibut) – lipid content ranges from 2 to 4%, (iii) semi-fatty fish (salmon) – lipid content is in the range of 4 to 10%, and (iv) fatty fish (mackerel) – lipid content is more than 10%. The chemical composition of some common species used in curing is showed in Table 6.1.

Changes in physical properties of fish muscle

Water activity and pH of fish muscle The water activity (a_w) and pH of fish muscle decrease during the salting process, due to the changes in the water and salt content in the fish muscle. Generally, the water activity of the heavily salted fish is in the range 0.70–0.75 (Troller and Christian, 1978; Thorarinsdottir, 2010). The water activity slightly decreases during pre-salting (injection salting and/or brining). However, after dry salting, the water

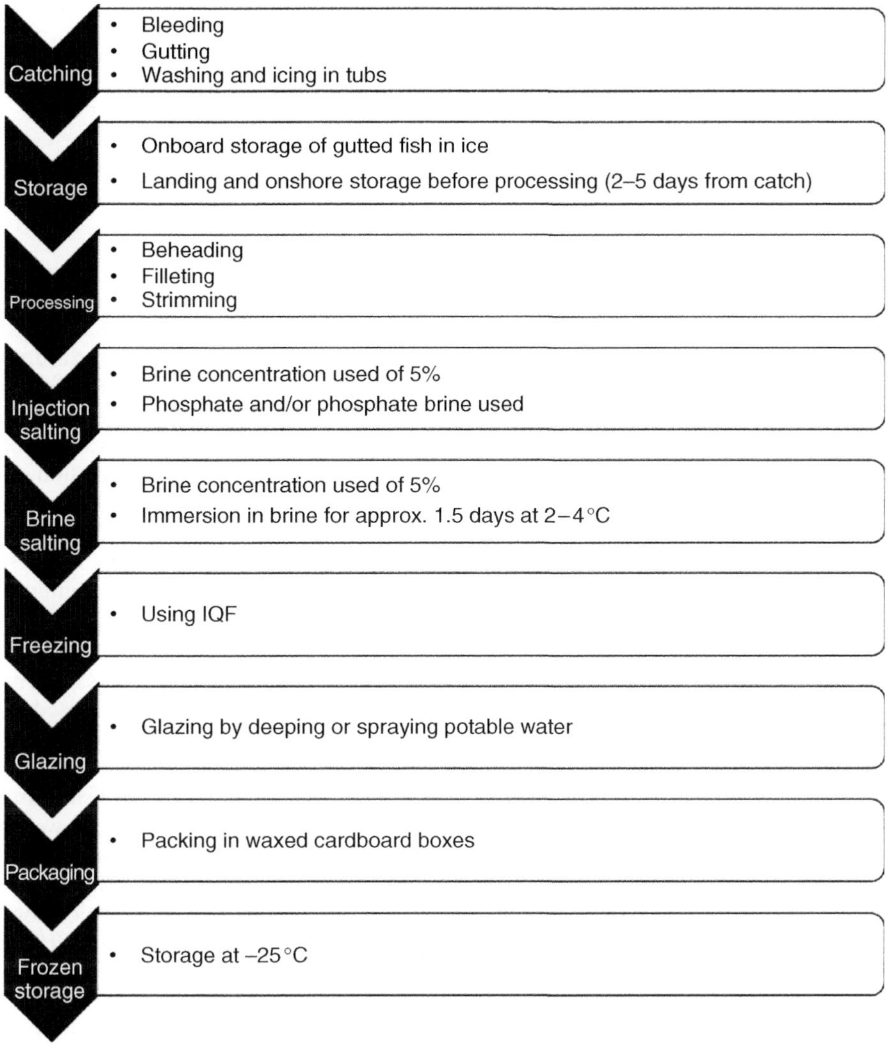

Figure 6.2 Light salting process of fish, from catch to final product

activity significantly decreases. The changes in pH of the fish muscle depend on many factors such as the state of raw materials (i.e. pre-/post-rigor, chilled/frozen fish, fresh fish), and the pH of the brine/salt used during the salting process. The pH of salted fish with added phosphates is lower than that of salted fish without added phosphates throughout the salting process (Thorarinsdottir *et al.*, 2001). The muscle pH of cured fish is positively correlated with the pH of the salt used and negatively correlated with the relative protein content (Lauritzsen *et al.*, 2004b).

Texture of the fish muscle Due to the changes in salt and water content during the salting process, fish muscle proteins and lipids are degraded, resulting in changes in the texture and microstructure of the fish muscle. Generally, the muscle shear force increases and

Table 6.1 Chemical composition of fish flesh[a]

Species	Scientific name	Water (%)	Fat (%)	Protein (%)
Cod	*Gadus morhua*	80.0–83.0	0.2–0.8	17.0–19.0
Haddock	*Gadus aeglefinus*	79.0–82.0	0.2–0.6	18.0–20.0
Saithe	*Pollachius virrens*	79.0–81.0	0.3–0.6	19.0–20.0
Ling	*Molva molva*	79.0–81.0	0.2–0.4	19.0–21.0
Tusk	*Brosme brosme*	80.0–81.0	0.2–0.5	18.0–20.0

[a]The data came from database of Matis.

the elasticity decreases with increasing salt concentration and brining time (Barat *et al.*, 2002, 2003; Jittinandana *et al.*, 2002; Gallart-Jornet *et al.*, 2007a, 2007b; Thorarinsdottir *et al.*, 2011a). The texture of salted fish depends on the status of the raw materials and the salting methods applied. Fish salted in saturated brine is harder than fish salted at lower brine concentrations (Barat *et al.*, 2002; Gallart-Jornet *et al.*, 2007b), mainly because more water is lost leading to a lower liquid phase in the protein matrix and the formation of a boundary layer on the surface of fish flesh. The presence of calcium and magnesium in the salt used also has an effect on the texture of the salted fish. Low concentrations of calcium and magnesium ($<1\%$) in salt causes a remarkable whitening and softening of the flesh. However, the firmness of cured fish muscle increases with increasing calcium content in the fish muscle (Lauritzsen *et al.*, 2004b).

Water holding capacity Changes in the water holding capacity (WHC) of salted fish are mainly due to changes in the properties of proteins resulting from an increased salt content and a decreased water content in the fish muscle (Thorarinsdottir *et al.*, 2002, 2004; Nguyen *et al.*, 2010). There is a correlation between changes in WHC and structural changes in the fish muscle during salting (Akse *et al.*, 1993). The WHC of fish muscle increases after pre-salting (injection and/or brine salting), mainly due to protein swelling (salting-in effect). However, the WHC of fish muscle decreases after dry salting, due to protein aggregation and precipitation (salting-out effect) (Fennema, 1990; Gallart-Jornet *et al.*, 2007b; Martinez-Alvarez and Gomez-Guillen, 2005, 2006; Offer and Trinick, 1983; Thorarinsdottir *et al.*, 2004; Chaijan, 2011; Nguyen *et al.*, 2010). Salt affects the WHC by binding cations to the proteins at low salt concentrations (<0.1 M $\sim0.58\%$ w/v) and by binding anions to the proteins at higher salt concentrations (0.1–1 M). The factors affecting the WHC of salted fish include raw materials, salting method, brine concentration and pH. Generally, salted fish produced from frozen fish has a lower WHC than that of salted fish produced from chilled fish (Lauritzsen *et al.*, 2004a). The WHC of heavily salted fish is reversely related to water yield in the fish muscle. Therefore, a lower WHC is obtained for injected fillets compared to other salting methods. This indicates that the additional water is relatively loosely bound in the fish muscle (Thorarinsdottir, 2010). Furthermore, divalent cations (Ca^{2+}, Mg^{2+}, Zn^{2+}) are also thought to affect the WHC of the muscle by cross-linking myofibrils, which decreases the space for water in the muscle cells (Asghar *et al.*, 1985).

Changes in chemical properties of fish muscle

Water and salt content During the salting process, water and salt are the main components transferred between the fish muscle and its surroundings, resulting in decreased water content and increased salt content in the fish muscle. Increase in salt content has a high correlation with decrease in water content. Diffusion is the most important mass transfer mechanism taking place during the salting process (Barat *et al.*, 2002; Wang *et al.*, 2000). Diffusion results from differences in concentration and osmotic pressures between the muscle and the salting agent (Raoult-Wack, 1994; Yao and Le Maguer, 1996; Barat *et al.*, 2003) and from a concentration gradient within the muscle (Erikson *et al.*, 2004). During brine salting, chemical fluxes may mainly be due to the concentration gradient between the muscle and brine, and within the fish muscle during the late stages of brining. For dry salting, the pressure gradient caused by denaturation/aggregation of the muscle proteins is thought to be the main driving force for water and salt diffusion (Jittinandana *et al.*, 2002; Sannaveerappa *et al.*, 2004; Thorarinsdottir *et al.*, 2004). The rate of water and salt fluxes depends on the salting methods applied, the quality and chemical composition of raw materials, as well as the status of proteins (Nguyen *et al.*, 2011) and muscle structure (Thorarinsdottir, 2010). The presence or absence of skin may affect mass transfer during salting (Wang *et al.*, 2000; Fuentes *et al.*, 2007). Divalent cations (Ca^{2+}, Mg^{2+}, Zn^{2+}) are thought to have a negative effect on the salting rate at high concentration. This may mainly be due to strong binding to proteins, resulting in formation of a barrier on the surface that reduces salt uptake and dehydration of the muscle. In addition, salt and water diffusion depend on temperature and muscle fibre orientation (Zhang *et al.*, 2011).

Proteins The increase in the salt content of fish muscle during salting results in conformational changes in proteins (i.e. protein denaturation/aggregation). The suggested mechanisms of salt-induced protein denaturation/aggregation is that at low salt concentrations (<1 M), chloride ions bind strongly to opposite charges of amino acid residues, raising the electrostatic repulsion between myofilaments, thus causing expansion of the filament lattice (swelling) (salting-in effect) (Offer and Trinick, 1983; Schmidt *et al.*, 2008). At high salt concentrations (>1 M), the solvation competition between salt ions and proteins becomes stronger. The higher solvation capacity of ions reduces the hydrodynamic radius of proteins, and protein–protein interactions become stronger than protein–water interactions. This leads to a decrease in the stable hydrophilic surface and an increase in polar and hydrophobic protein interactions, favouring their hydrophobicity, aggregation and precipitation (salting-out effect) (Offer and Trinick, 1983; Stefansson and Hultin, 1994; Caflisch and Karplus, 1994). Protein denaturation/aggregation is accompanied by changes in protein solubility, protein functionality, protein digestibility, texture and nutritional quality (Nguyen *et al.*, 2011; Sotelo *et al.*, 1995; Matsumoto, 1979; Love and Mackay, 1962). Protein solubility increases when the salt concentration increases from 0.15 to 1.0 M and reaches a maximum value in the range 0.8–1.0 M. At higher salt concentrations (>1.0 M), protein solubility decreases (Curtis and Lue, 2006; Stefansson and Hultin, 1994). The degree of protein denaturation/aggregation depends on the salting method and seems to increase in the following order for the different pre-salting methods: injecting + brining < brining < pickling (Thorarinsdottir *et al.*, 2011a). The functional characteristics of the myosin rod are better retained when injection and brining are used for pre-salting compared to pickle salting (Thorarinsdottir *et al.*, 2011a, 2011b). The degree of protein denaturation/aggregation of salted fish may also increase if fish is previously frozen and thawed, mainly due to the interaction effects from crosslinking of the muscle protein and the salting-out effect (Lauritzsen, 2004).

Lipids Lipids in fish muscle typically have a high percentage of polyunsaturated fatty acids that are known to be susceptible to oxidation and this is a major cause for quality loss (Masniyom *et al.*, 2005). The increase in salt content in the fish muscle results in muscle lipid degradation (i.e. lipid hydrolysis/oxidation). Lipid oxidation occurring during processing and storage of salted cod is a major cause of quality deterioration such as yellow/brownish discolouration of the flesh surface (Lauritzsen *et al.*, 1999; Nguyen *et al.*, 2012). A proposed mechanism of salt-induced lipid oxidation in muscle is the fusion of intracellular compounds and the destruction of the cell structure by salt (Shomer *et al.*, 1987). Furthermore, salt may displace iron ions from the binding macromolecules (Kanner *et al.*, 1991; Osinchak *et al.*, 1992). Another theory is that salt disrupts the structural integrity of the membranes, bringing the catalysts into closer contact with lipid reactants (Rhee and Ziprin, 2001). Salt may also initially promote the formation of metmyoglobin, which reacts with H_2O_2 to form ferrylmyoglobin to catalyse lipid oxidation. Hydroperoxides have more polarity than lipids from which they are generated, and ionic compounds such as salt may change the hydroperoxide conformation, affecting the oxidation kinetics (Calligaris and Nicoli, 2006). Lipid oxidation of fish during the salting process depends on different factors, such as status of raw materials, pH, salt composition and salting method applied. The highest lipid oxidation is obtained at the lowest ultimate muscle pH prior to heavy salting in cod (Lauritzsen *et al.*, 1999). However, the mechanisms on how pH controls the lipid oxidation of fish are not clear. Metal impurities, especially iron and copper, in salt are also believed to cause lipid oxidation. The mechanism of lipid oxidation induced by metal ions involves a one-electron redox reaction, which leads to hydroperoxide decomposition (Kanner *et al.*, 1988; Bondet *et al.*, 2000). Pro-oxidant effects of metals depend on ion type and valence of the ions and the relative activity of metals and hemin on lipid oxidation is as follows: Fe^{2+} > hemin > Cu^{2+} > Fe^{3+}. In addition, the potassium ferrocyanide used as an anti-caking agent in salt is thought to promote lipid oxidation (Andersen *et al.*, 1990; Nguyen *et al.*, 2012). The mechanism of lipid oxidation promoted by potassium ferrocyanide is that firstly, ferrocyanide ($[Fe(CN)6]^{-4}$) is oxidised to ferricyanide ($[Fe(CN)6]^{-3}$), which oxidises myoglobin to form perferryl-myoglobin and ferryl-myoglobin radicals. Thereafter, both perferryl-myoglobin and ferryl-myoglobin radicals catalyse lipid oxidation (Kanner *et al.*, 1987). Moreover, cyanide and iron ions in the potassium ferrocyanide molecule are thought to affect lipid oxidation development (Daya *et al.*, 2000; Nguyen *et al.*, 2012; Tokur and Korkmaz, 2007).

6.2.4 Heavily salted fish products

Classification of heavily salted fish Heavily salted fish is mainly classified by according to the size of the fish, processing method, whether it is dried or not and the quality of the products. The chemical content of salted fish products depends on the curing time, the frequency of restacking and the drying procedures (Table 6.2). Generally, the water and salt content of lightly cured/wet salted products are about 55–58% and about 18–21%, respectively. Medium-cured products are obtained by restacking the fillets and further salting for 1 to 3 weeks. Fully cured fish of this type has about 50–52% water and 20–22% salt content. Drying of the salted products can be carried out to obtain semi-dried or extra-dried salted fish products. However, no official standards are available for the chemical content of salted fish. The conditions of the products are usually agreed upon between exporters and buyers and in accordance to consumer preference.

Important quality criteria of heavily salted fish The quality of the final product depends on many factors, such as the condition of the raw material used, processing method and equipment and, finally, the methods used for salting, storage and rehydration

Table 6.2 Water and salt content of heavily salted cod products at different curing stages. These values are presented to give a basic reference, although no official standards are available for salted fish products (Thorarinsdottir, 2010)

Product	Water (%)	Salt (%)
Lightly cured/wet salted	55–58	18–21
Medium cured	53–54	19–21
Fully cured	51–53	20–22
Semi-dried	44–47	22–24
Dried	36–40	22–25
Extra dried	30–35	23–35

(Lindkvist *et al.*, 2008; Thorarinsdottir *et al.*, 2001; van Klaveren and Legrende, 1965). A commercial quality rating is carried out by trained quality assessors at the packing stage or before export. Rating is mainly by visual appearance of the product, such as colour (i.e. whiteness, yellowness), thickness and the presence of gaping/cracks, bruises and blood stains. The guidelines used for quality grading vary with export and import country and with different salted fish products. Furthermore, even for the same product, regional differences exist with each country (Gallart-Jornet and Lindkvist, 2007; Lindkvist *et al.*, 2008). The commercial quality ratings of heavily salted cod in Iceland and in Norway are presented here.

Commercial quality rating of heavily salted cod in Iceland At packing, the salted products are graded according to their size and commercial quality (PORT/SPIG). PORT refers to export of the products to Portugal and SPIG to Spain, Italy and Greece. Some basic rules for quality rating are described here, but the experience and training of the grader are essential factors for a consistent performance over time and between producers. The appearance or colour of the fish is the most important variable. Higher standards are used for SPIG fish; it is supposed to be whiter and thicker than the PORT fish.

Within the SPIG category are three quality grades, I, II and III:

- SPIG I: Fillets should be thick; with light appearance but no defects (PORT AB may have small defects and a slightly darker colour).
- SPIG II: Fillets are similar to SPIG I, except that minor defects are allowed.
- SPIG III: Fillets are allowed to have more gaping than fish in SPIG II and minor defects due to heading and filleting machines.

Three quality grades are usually used for assessing the PORT fish, A, B and C:

- PORT A: Fillets that are a light colour, thick, without blood stains; only minor gaping is allowed (gaping may appear as openings or ruptures between the myotomes because of the weakening of the connective tissue).
- PORT B: Fillets that are not of grade A quality because of small defects. The colour is darker and the fillets are thinner or with long gaps along the fillets.
- PORT C: Fillets that have quality defects, like gaping or other apparent mechanical defects in the fish flesh. The colour of the fillets is too dark to be graded as B quality. Fillets that have been washed due to slight red discoloration of the fillets (caused by the growth of halophilic bacteria).

- PORT AB: Fillets of A and B grades can be packaged together. The maximum ratio of B fillets is 50% in each unit.
- PORT CD and E: Grades D and E are sometimes also used. Then C and D fillets are packaged together. Fillets of grade D are of lower quality than C due to stronger gaping, discoloration or other visible defects. Fillets of grade E have greater defects but are still suitable for human consumption. The parts of the fillets with visible defects can be cut away.

Fillets that have yellow staining, which is caused by metal-catalysed oxidation of fat, must be sorted from fillets of other quality grades (Thorarinsdottir *et al.*, 2010b).

Commercial quality rating of heavily salted cod in Norway According to the 'Norwegian Industry Standard for salted and dried salted products', salted split fish, is classified in the following commercial quality categories:

- Imperial/Superior: Made from fish that is thoroughly bled, well washed and rinsed to remove remains of blood and entrails and with nape skin attached. The fish is to be properly split and evenly salted, well pressed and restacked during processing. The fish is to be light-coloured and firm, and without blemishes.

  This category may include fish with the following characteristics: poorly bled bellies, small tears or longitudinal cracks, not properly rinsed, some blood clots, somewhat unevenly salted. When assessing the fish, special consideration will be given to fish that has been thoroughly bled and properly restacked during production. In this case, somewhat larger defects will be tolerated if the overall impression justifies this, particularly if the fish is light-coloured and firm.
- Universal: Fish that do not meet the requirements to Imperial/Superior are to be classified as Universal.

  This category may include fish with the following characteristics: inadequately split, round tail, inadequately washed or rinsed, insufficient removal of backbone, moderate blood clot, major tears or longitudinal cracks, moderate cracking, minor blood, liver and/or bile stains. The fish must retain its natural shape. Disfiguring blemishes such as stains/lumps of dried blood or remains of entrails shall be removed.
- Popular: Fish that does not satisfy the requirements to Universal, which, nevertheless, is fit for human consumption, is to be categorized as Popular. However, this category must not contain fish that is sour, has been exposed to contamination, has ragged bellies, bile or gut content, or fish that is badly cracked/loose fleshed.
- Mix: All the products in this standard may be packaged in a common category under the description 'Mix' if the fish is distributed fairly equally on the different categories. All categories may include fish with black membrane. In the case of 'klipfish' (i.e. dried salted cod) products, this information is to be included on the package labels.

Some modifications have been added to the standard, for the rating of salted fillets:

- Imperial/Superior: In this category, the product comes from fish that has been emptied of blood, but includes fillets that are slightly darker and bear the traces of incomplete bleeding. This category may include fish with the following characteristics: small tears or longitudinal cracks, insignificant mechanical damage. The fish is to be light-coloured and firm, and without blemishes.
- Salted fillets, which do not satisfy the requirements to Imperial/Superior are to be classified as Universal. This category may include fish with the following characteristics:

signs of yellowing, larger tears or longitudinal cracks, moderate cracking, minor blood, liver and/or bile spots, darker colour (the degree of darkness acceptable in Universal is to be assessed based on the overall impression of the fish), minor marks from mechanical damage.

Factors affecting the quality

Raw materials The characteristics and condition of raw material are important for the processing yield and the quality of salted products. These factors vary with fishing grounds, fish species and fishing stocks as well as catching method and handling of the fish. The quality attributes affected are, for example, chemical content, WHC, gaping and texture of the salted products. Chemical composition, particularly the lipid content of the raw material affects the mass transfer and the quality of salted fish products. Lipid plays a role as a barrier to transport of water and salt during the salting process (Gallart-Jornet *et al.*, 2007a). The yellow/brownish discolouration of salted fish is closely associated with lipid oxidation during the salting process and storage time (Lauritzsen *et al.*, 1999; Nguyen *et al.*, 2012). The chemical composition of fish muscle not only depends on the season, but also on the size and age of the fish. The lipid and protein content of fish is the lowest in the spawning period, mainly due to the build up of reproductive organs (roe or milt), and so fish muscle is replaced with water. Therefore, after spawning, the fish is in bad condition for salting, the risk of quality defects like gaping is higher and the WHC of the muscle is lower than at other seasons, resulting in lower weight yields and reduced quality of the final products.

The quality of raw material is also affected by the catching methods and handling procedures on board. Quality defects caused by incorrect handling methods are, for example, bruises, blood clots and gaping (Botta *et al.*, 1987). These defects remain visible in the final products after salting and rehydration. The risk of such defects is higher when the fish is caught in a net trawl or seine in comparison to line. The risk of defects is also influenced by factors like time in the fishing gear, size of trawl and weather conditions during catching.

The state of rigor and post-mortem storage (chilled or frozen) of fish prior to salting affect the weight yields and the quality of the final products. Pre-rigor salting leads to a larger reduction in weight, a higher water loss and a lower salt uptake compared to fish salted post-rigor and fish salted after frozen storage (Lauritzsen *et al.*, 2004a). Freezing and thawing of fish prior to salting post-rigor increases the firmness of the salt-cured product compared to chilled fish salted post-rigor (Lauritzsen *et al.*, 2004a). This is mainly due to the protein denaturation/aggregation that occurs during freezing. Ice formation and melting during the freezing and thawing cycle may alter the muscle structure, resulting in higher salt diffusivity (Wang *et al.*, 2000).

Salt quality The quality of salt used to produce salted fish is one of the most important factors affecting the salt curing process and the quality of the final products. Metal impurities particularly iron and copper, within salt may cause lipid oxidation, increase salt-soluble protein levels and change the texture and colours of salted fish products. Metal impurities also affect the rate of water and salt diffusion during the salting process. Therefore, salt used in fish salting process must be clean, free from foreign matter and foreign crystals and show no visible signs of contamination with dirt, oil, bilge or other extraneous materials. The salt composition may vary among salt processing plants within each country and between processing batches. Producers of salted fish in Nordic countries mainly use coarse sea salt imported from Southern Europe (Spain) and Northern Africa (Tunis) (see Table 6.3, Table 6.4 and Table 6.5).

Table 6.3 Typical composition of different types of salt used in productions of heavily salted fish

Chemical component	Type of salt			
	Torrevieja	Tunis	Ibiza	Earth
Water (%)	2.50–1.80	2.30–1.30	2.60–3.10	0.40–0.50
NaCl (%)	97.3–97.7	97.1–97.8	96.2–94.5	98.0–99.0
$CaSO_4$ (%)	0.15–0.32	0.17–0.30	0.82–1.58	0.53–0.72
$MgSO_4$ (%)	0.00–0.10	0.10–0.30	0.16–0.38	0.04–0.17
$MgCl_2$ (%)	0.00–0.12	0.10–0.30	0.16–0.48	0.00–0.02
Na_2SO_4 (%)	0	0	0	0.04–0.49
Insoluble	0.1	0.1	0.1	0.10–0.40
Fe (mg/kg)	12–25	10–15	13–17	4–30
Cu (mg/kg)	<0.01	<0.01	<0.01	–

Source: Lauritzsen, 2004. Reproduced with the permission of Kristin Lauritzsen.

Table 6.4 Producer's criteria for salt used in production of heavily salted cod in Iceland

Chemical component	Faulted	Good	Faulted	Unusable
Water (%)		<3.5	≥3.5	
Salt (%)		≥98	<98	
Ca (%)	<0.05	0.05–0.20	0.20–0.35	>0.35
$CaSO_4$ (%)	<0.17	0.17–0.70	0.70–1.19	>1.19
Mg (%)		<0.1		≥0.5
$MgSO_4$ (%)		<0.5		≥0.5
Fe (mg/kg d.w.)		20		≥20
Cu (mg/kg d.w.)		<0.03	0.03–0.05	>0.05
Mn (mg/kg d.w.)		<2		≥2

Table 6.5 The CODEX standard for food grade salt

Chemical component	Limit
NaCl (%)	≥97
Arsenic (mg/kg expressed as As)	0.5
Copper (mg/kg expressed as Cu)	2.0
Lead (mg/kg expressed as Pb)	2.0
Cadmium (mg/kg expressed as Cd)	0.5
Mercury (mg/kg expressed as Hg)	0.1

Source: Food and Agriculture Organization of the United Nations, CODEX ALIMENTARIUS/CODEX STANDARD FOR FOOD GRADE SALT 150-1985 adapted into a table.

6.3 Marinating

6.3.1 Introduction

Marinating is a semi-preservative method for fish based on the treatment of muscles with marinade solutions (marination bath) containing salt, sugar, spices, oil and acids (from vinegar, fruit juice, wine), etc. Marinades are used to provide high sensory acceptability (e.g. tenderness, juiciness, flavour and aroma) and to extend the shelf life of the products (Cadun *et al.*, 2005; Duyar and Eke, 2009; Yashoda *et al.*, 2005). Marinade solutions are also used to prevent the growth of microorganisms and to inhibit the activity of enzymes, mainly due to the effect of salt, low water activity, acidic pH and the use of antimicrobial agents. The inhibitory effects of marinade solutions on bacteria and enzymes increase with concentration and marinating time. At acidic pH ($1.0 < pH \leq 4.5$), all food poisoning bacteria and most spoilage bacteria are prevented from growing. Therefore, the pH of the marinade solution should be kept lower than 4.5. Acetic acid is the most frequently used acid as well as the most efficient. It does not destroy the microflora in food; however, it is a strong growth inhibitor. Acetic acid also inhibits most enzymes but some still remain active.

Marinated fish are semi-preserved fish products that are highly appreciated in many countries. Generally, fatty fish like herring, sardine, mackerel and anchovies, as well as several kinds of crustaceans and bivalves are usually used in the marinating process. Traditionally, marinating involves immersing fish into a ready-made marinade solution. Nowadays, there are several methods to elaborate to fish marinating including cold, cooked, fried and pasteurised marinating (Yeannes, 1991; Capaccioni *et al.*, 2011). The quality of the final product depends on the composition of the marinade solution, the ratio of fish to liquid and how the fish is treated during marinating. The composition of the marinade solution used varies depending on the producers and the characteristics expected of marinated products. During the marinating process, salt and other substances into the fish muscle, whereas water diffuses out to the surrounding media. Concentration gradient is thought to be the main driving force for the diffusion of water, salt and other substances. Furthermore, a pressure gradient caused by protein denaturation/aggregation and a water activity gradient also contribute to the diffusion of water, salt and other substances. The diffusion of salt, vinegar and other ingredients depends on many factors such as fish species, muscle type, fish size, fillet thickness, weight, chemical composition (lipid content and contribution), and physiological state (Baygar *et al.*, 2010).

6.3.2 Marinating methods

Cold marinating Cold marinating is the most commonly used method for marinated fish processing. In Europe, cold marinating makes up about 92% of the market (Szymczak, 2011). It is carried out by first curing fish in a proper marinade solution of vinegar and salt for up to 12 weeks, depending on the products. Then the cured fish is packed in glass or plastic containers with spices and covered with a salty vinegar solution. The salt and vinegar concentrations used during the curing step depend on the type of fish and the expectation of the final products. The salt and vinegar concentrations of the marinade solution used for packaging are lower and similar to the salt and vinegar concentrations in the fish muscle after curing step. For example, for marinated herring, the initial composition of the marinade solution should be 4% (v/v) of vinegar and 10% (w/v) of salt if the process is carried out in open containers and 7% (v/v) of vinegar, 14% (w/v) of salt in closed containers. The composition of the marinade solution used for packaging marinated herring consists of 1–2% (v/v) of vinegar and 2–4% (w/v) of salt (FAO, 2001). The concentration of vinegar used determines the degree of preservation and sufficient salt has to be used to keep the fish flesh firm. The ratio of fish to marinade solution in open containers usually ranges from 1 : 1 to 1.5 : 1. However, in closed containers, the ratio of fish to marinade solution can be increased to 2.3 : 1. One of the most important conditions is that the fish are completely immersed in the marinade solution (marination bath) during the entire production process.

The marinated fish process can be divided into steps as depicted in Figure 6.3.

Cooked marinating Cooked marinating involves cooking or braising fish in a salty vinegar solution. Then fish are packed into boxes and covered with the cooking marinade solution including water, vinegar, salt and spices.

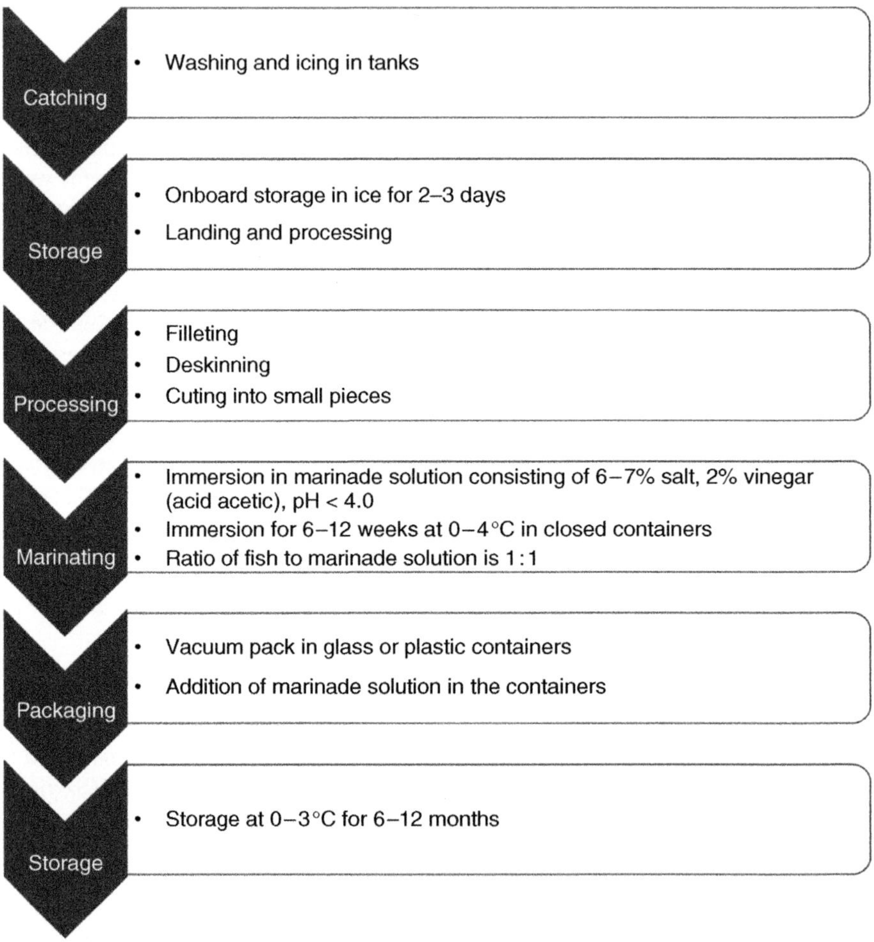

Figure 6.3 Marinating process of fish from catch to final product

Fried marinating In fried marinating, fish is firstly immersed in brine and flour and then brined fish is fried. Fried fish is cooled down and placed into the marinade solution containing salt, acetic acid and spices.

6.3.3 Ingredients used in marinating

Salt Salt is an important ingredient for marinating fish. Salt not only prevents the growth of microorganisms, but also changes the taste, texture and structural properties of the fish. The salt used in fish marinating process must be clean, free from foreign matter and foreign crystals and show no visible signs of contamination with dirt, oil, bilge or other extraneous materials. The salt composition may vary salt processing plants within each country and between processing batches. The salt concentration used in the marinating process varies depending on the products and the markets, salt levels.

Sugar Sugar is used in the marinating process of fish mainly due to its sweet taste. Sugar has many functions in food processing such as a sweetener, preservative, texture modifier, fermentation substrate, flavouring and colouring agent. The concentration of sugar used depends on the products and the markets. For instance, marinated fish products with sugar are favoured in the Scandinavian countries, whereas consumers in southern Europe (e.g. Portugal, Spain) prefer marinated fish products with a sour taste (vinegar of acetic acid marinades).

Vinegar Vinegar is a liquid consisting mainly of acetic acid and water. Vinegar is commonly used in food preservation such as pickling, and marinating as well as in salad dressings. When used for marinating fish, vinegar acts as a preservative, tenderises the fish muscle and gives a special taste to the final products. The concentration of vinegar used in the marinating process depends on the type of products and the market where marinated products are sold.

6.3.4 Factors affecting the quality of marinated products

Raw materials The quality of the raw materials affect the quality of the final products. The chemical composition of fish, especially the lipid content of fish muscle, varies according to the catching season. As the fish used in the marinating process is fatty fish, the catching method and the handling procedures on board will affect the quality of the raw material. Fish should be iced as soon as possible after catching to prevent the progress of lipid oxidation. The marinating process also needs to be carried out as soon as possible after landing.

Ingredients The quality of the ingredients used in the marinating process also affects the quality of the marinated fish products. For example, metal impurities, especially iron and copper, present in salt could induce lipid oxidation during the marinating process, giving a rancid flavour. Therefore, all ingredients used in the marinating process should be food grade.

6.3.5 Changes in fish muscle during marinating

During the marinating process, salt, vinegar and other ingredients diffuse into the fish muscle and water diffuses out. The changes in the amounts of water, salt, vinegar and other substances in fish muscle lead to changes in the texture, flavour, taste and pH, etc. The WHC of the fish muscle increases mainly due to protein swelling, caused by an increase in the salt content in the fish muscle. In addition, the decrease in muscle pH also increases the WHC. The hardness of the fish muscle decreases with increasing marinating time and is associated with improved sensory characteristics (Szymczak, 2011).

6.3.6 Storage of marinated fish products

The storage life of marinated fish products depends on their salt and vinegar concentration. Storage life may be up to 6–12 months under certain conditions. Fish used in the marinating process are fatty fish; therefore, marinated fish products need to be vacuum packed inhibit lipid oxidation induced by oxygen. Temperature is one of the most important storage conditions. Thus, marinated fish products need to be stored at chilled temperatures of 0–3 °C to maintain their quality and inhibit the growth of microorganisms.

6.4 Smoking

6.4.1 Introduction

Smoking is one of the oldest methods used to process and preserve fish and meat (Dore, 1993). Smoking introduces flavour, taste and preservative ingredients into fish muscle by exposing fish to the smoke produced by burning or smouldering plant materials, most often wood. The volatile compounds in smoke penetrate into the fish muscle (Ward, 1995). Smoking usually extends the shelf life of fish due to the combination effects of (i) salting, which lowers water activity resulting in reduced microbial growth, (ii) elevated temperature drying, which provides a physical surface barrier to the passage of microorganisms, and (iii) deposition of antimicrobial and antioxidant compounds, such as aldehydes, carboxylic acid and phenols, which delays microbial growth and rancidity development (Efiuvwevwere and Ajiboye, 1996; Leroi and Joffraud, 2000). Moreover, fish muscle exposed to smoke in combination with a high temperature can effectively limit harmful enzymatic reactions (FAO, 1992). Nowadays, the main purpose of smoking has shifted to enhancing the sensory quality of the product rather than for its preservative effects. Depending on how the smoke is delivered into the fish muscle and the smoking temperature, smoking can be defined as hot smoking, cold smoking, liquid smoking and electrostatic smoking (Wheaton and Lawson, 1985). However, fish is also smoked by the combined method of hot and cold smoking. The smoking process is usually characterized by an integrated combination of salting, drying, heating and smoking steps in a smoking chamber/smokehouses (Alcicek and Atar, 2010) (Figure 6.4). Vacuum, modified or controlled atmosphere packaging or canning can increase the shelf life of the smoked fish products (Bannerman and Horne, 2001).

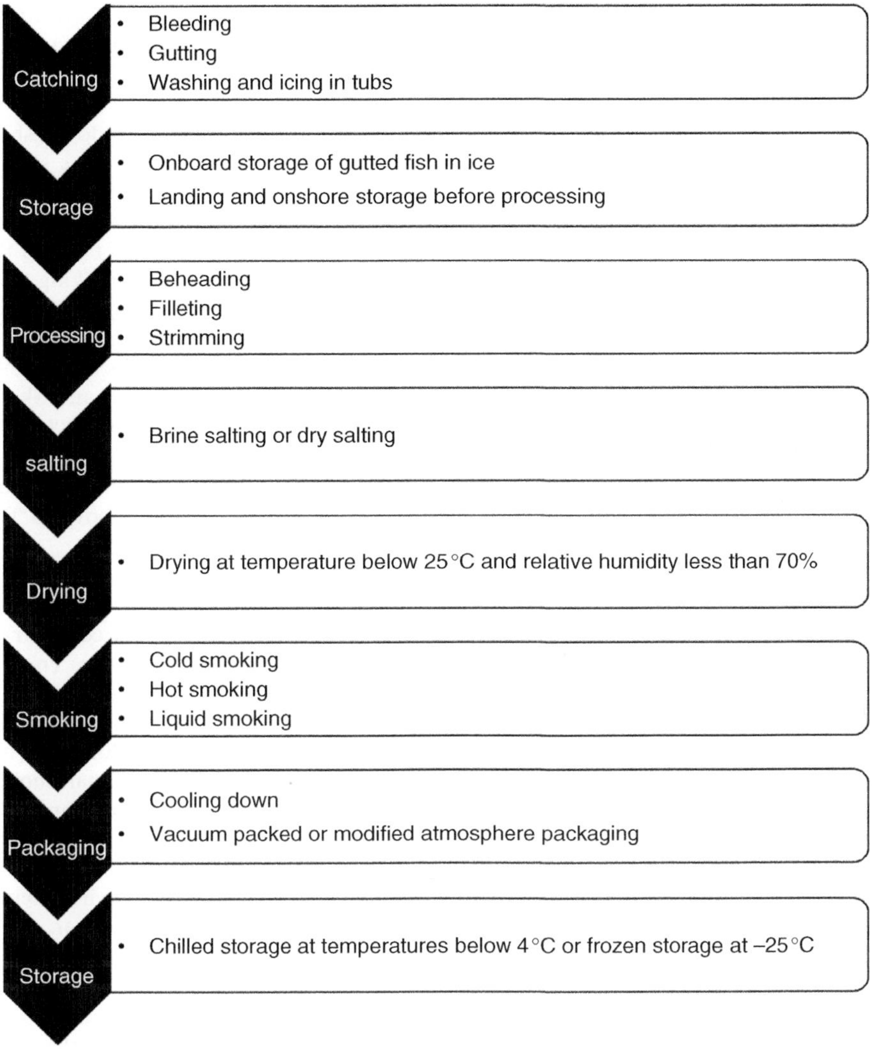

Figure 6.4 Smoking process of fish from catch to final product

6.4.2 Smoking method

Hot smoking Hot smoking is the traditional smoking method using both heat and smoke. The fish is smoked at temperatures that are high enough and for a sufficient period of time to obtain heat coagulation of the protein throughout. Generally, the temperature should be maintained above 30 °C and the normal range is 70–80 °C. In hot smoking, fish is completely cooked and consumer can eat it without further cooking. FDA guidelines suggest that safe vacuum or modified atmosphere packaged hot-smoked fish products require at least 3.5% (w/v) water phase salt and must have reached a centre temperature of at least 62.8 °C for 30 min. This prevents the production of toxins by *Clostridium botulinum* (FDA, 2001). Furthermore, hot-smoked fish products must have a water activity (a_w) of 0.85 or less to give a stable shelf life.

Cold smoking This involves smoking fish at temperatures where the products do not show any signs of heat coagulation of the proteins. In cold smoking, the temperature should be maintained below 30 °C and it is used to impart aroma and flavour in the fish muscle (Doe *et al.*, 1998). The RH should be kept at 75–85%. During cold smoking, the fish will not be cooked; therefore it has to be cooked before consumption, except for fish in the Salmonidae family (e.g. salmon, trout and arctic char). Compared to hot smoking, cold smoking takes longer, has a higher yield and retains the original textural properties much better than hot-smoked products. Cold smoking has been applied to different fish species such as salmon, rainbow trout, cod, tuna, etc. (Dillon and Patel, 1993; Lyhs *et al.*, 1998; Birkeland *et al.*, 2003, 2004a, 2008; Emborg and Dalgaard, 2006; Birkeland and Skara, 2008). Cold-smoked salmon and trout dominate the market of smoked fish products.

Combined method of hot and cold smoking In this method, fish is first smoked below 30 °C for a few hours and finally hot smoked by increasing the temperature above 30 °C.

Liquid smoking Liquid smoking extract is prepared by the dry distillation of wood and then concentrated. Concentrated smoke is dissolved in a solvent, such as water of oil (Maga, 1988) and can be used directly on products. Fish is dipped into a concentrate of a liquid that has been used to absorb smoke (Wheaton and Lawson, 1985). The liquid concentrate transfers the aroma and flavour of smoke into the fish muscle. Liquid smoking has several advantages. It is quick and much easier to achieve a uniform smoke flavour than the traditional cold- and hot-smoking processes (Varlet *et al.*, 2007a). It has a higher processing yield and similar textural properties compared to traditional smoked salmon (Birkeland and Skara, 2008). Undesirable compounds (e.g. polycyclic aromatic hydrocarbons, PAHs) in natural smoke can be removed (Chen and Lin, 1997; Hanne *et al.*, 2007) and antioxidants and antimicrobial activity can be easily determined (Suñen *et al.*, 2003). In addition, liquid smoking has lower operation costs, less environmental pollution and is less time consuming than other smoking methods (Simon *et al.*, 2005; Muratore *et al.*, 2007).

Electrostatic smoking This is a method of smoking fish with infrared radiation. Fish are treated with smoke in an electrical field (usually positively charged) and at the same time fish are negatively charged. The electrical field acts on the ionised smoke particles, accelerating the smoking process, thereby shortening the smoking period. Electrostatic smoking is fully mechanized; therefore it could lower labour and production costs while maintaining high-quality final products.

6.4.3 Changes in fish muscle during smoking

The smoking process leads to weight loss due to both dehydration of the fish muscle and leaching of lipids from the fish muscle. Depending on the raw material, the final product characteristics and the smoking method, the weight loss due to dehydration in the smoking process is about 10–25% (Howgate, 1979). Moreover, the size and shape of the fish affect the processing yield (Rørå *et al.*, 1998). The pH of fish muscle decreases during the smoking process, mainly due to the absorption of acid from the smoke, the loss of moisture and the reaction of phenols, polyphenols and carbonyl compounds with protein and protein constituents (e.g. amino acids, -SH groups) (Hassan, 1988). The higher the temperature used in smoking the lower the pH in the fish muscle, so fish smoked using the electrostatic method shows the least decrease in pH compared to other methods (Espe *et al.*, 2002).

The increase in the salt content of fish muscle during the salting step leads to changes in protein structure (i.e. protein swelling or aggregation). This results in changes in WHC, texture and microstructure of the fish muscle (Thorarinsdottir *et al.*, 2004). These changes vary depending on the composition of the food itself and the conditions under which either type of smoking is undertaken (Toth and Potthast, 1984). In traditional hot smoking, texture changes are mainly due to the denaturation of protein by heat (Gill *et al.*, 1992). In contrast, changes in fish texture during liquid smoking are mainly the consequence of protease enzyme activities that hydrolyse protein, giving good conditions for salt-induced protein conformational changes (Lund and Neilsen, 2001; Hultmann *et al.*, 2004). According to Sigurgisladottir *et al.* (2001), the fish muscle shrinks during the salting and smoking process and a number of fat globules are dispersed among the muscle fibres after smoking. In addition, the components of the smoke used also influence the textural properties of the smoked products (Daun, 1972). The texture of smoked fish products also depends on the salting method used before smoking. A higher gaping score and softer texture have been reported for smoked salmon that underwent prior salting by injection compared to dry salting. However, the injection salting technique gives a more even coloured smoked product (Birkeland *et al.*, 2004a, 2004b).

The changes in flavour and taste of the fish muscle during the smoking process are mainly due to the absorption of volatile compounds, particularly phenols, from the smoke (Doe *et al.*, 1998). The smoking temperature and smoking method allow products to be differentiated according to the intensity of smoke odour and flavour (Cardinal *et al.*, 2001). Smoking at higher temperatures increases the deposit of smoke compounds in the food muscle. The flavour formed may also be due to the lipid oxidation that occurs during the smoking process.

6.4.4 Factors affecting the quality of smoked fish products

Raw materials The chemical and physical characteristics of the raw materials strongly affect the yield and overall appeal of the smoked products. Several factors can affect the quality of raw material such as fishing ground, fishing method, bleeding and handling method, as well as the use of farmed or wild fish. Lipid content and lipid distribution have an important influence on the nutritional value, as well as on the assessed quality of the final products in terms of texture, flavour and taste (Rørå *et al.*, 1998). Generally, a higher phenol content in the smoked fish is obtained from fish with a lower lipid content in their muscle. This might be due to the water availability relative to the fat content. It has been stated that high fat content in the fish muscle can degrade the qualitative characteristics of smoked fish (Espe *et al.*, 2002; Birkeland *et al.*, 2004a). Fat content also affects the diffusion of salt and water during the salting step (Gallart-Jornet *et al.*, 2007a).

Salting method Salting is one of the most important stages in the smoking process. Salt acts as a preservative in the fish muscle, prolonging the shelf life of the product by lowering the water activity and inhibiting bacterial growth. Salting also provides the desired firm texture and salty taste of the final smoked products (Doe *et al.*, 1998). Different salting methods are applied in the smoked fish industry, e.g. dry salting, brine salting, injection salting or their combinations (Chiralt *et al.*, 2001; Espe *et al.*, 2002). Salting method is believed to affect the processing yield and the quality of the final smoked products. The muscle fibre of smoked salmon shrinks more if the fish is dry salted before smoking compared to brine salted (Sigurgisladottir *et al.*, 2000) and a better quality and higher process yield is obtained from brine and injection salted fillets (Birkeland *et al.*, 2004b). However, injection salting prior to smoking results in a higher gaping score compared to dry salting (Birkeland *et al.*, 2003).

Drying method Drying is widely known to contribute a preservative effect to the final smoked fish products, mainly by reducing the water activity, which results in a reduction of microbial and enzyme activities. The drying conditions used for fish vary depending on the producers and the fish species. For fatty fish, such as salmon, the temperature is in the range of 15–26 °C, and RH ranges from 55 to 57% (Birkeland *et al.*, 2003, 2004a, 2004b). Unsuitable use of drying conditions may have significant effects on the product quality, such as texture and colour, as well as the process yield of the product (Cardinal *et al.*, 2001). Microbial flora distribution in smoked fish products varies and is largely dependent on the drying temperature and the drying time (Nickelson *et al.*, 2001).

Smoke composition The composition of the smoke is one of the most important factors affecting the quality of the final smoked products. Smoke is produced by burning or smouldering (i.e. burning without flame) wood. Smoke composition varies depending on the type of wood used and the conditions of smoke generation (Conde *et al.*, 2006). The aroma and flavour of smoked fish are mainly attributable to volatile compounds in the smoke. Phenolic compounds, particularly syringaldehyde and coniferaldehyde, are strongly retained by fish muscle, thereby increasing the aroma of the smoked fish (Varlet *et al.*, 2007b). The aromas of smoked fish are generated by two main pathways: (i) the Maillard reactions (i.e. interactions between smoke carbonyl groups and food amino groups) that produce a smoked aroma; and (ii) lipid oxidation products that provide fishy aromas. It is believed that the smoked taste is generated from guaiacol and its derivatives, while syringol and it derivatives are thought to be responsible for the smoked smell (Martinez *et al.*, 2011). In addition, other compounds in the smoke such as phenols, *p*-cresol, *o*-cresol, guaiacol, 4-memethyguaiacol, 4-ethylguaiacol, eugenol, 4-propylguaiacol and isoeugenol, are thought to contribute to the formation of the smoked aroma and flavour of smoked fish products (Serot and Laffïcher, 2003; Varlet *et al.*, 2006). The colour development of the smoked fish depends on the carbonyl content (Martinez *et al.*, 2007), high molecular weight phenols (Daun, 1972) and volatile aldehydes (Clifford *et al.*, 1980). Furthermore, smoke has bactericidal properties due to formaldehyde and acetic acid, and antioxidant properties due to the presence of three main chemicals (2,6-dimethoxyphenol, 2,6-dimethoxy-4-methylphenol and 2,6-dimethoxy-4-ethylphenol) in the smoke. It is also stated that smoked condensates have antioxidative effects, even at low concentrations (Fretheim *et al.*, 1980; Kjallstrand and Petersson, 2001), mainly due to the presence of 2,6-dimethoxyphenols and 2-methoxyphenols. Oxidation during storage of smoked sausage is lower than the oxidation of sausage without smoking (Ernawati *et al.*, 2012). Bower *et al.* (2009) found that smoke-processing imparts antioxidant potential to the extracted oil of smoked pink salmon, resulting in polyunsaturated fatty acids not being destroyed.

Poisonous substances in smoke include PAHs that are believed to have carcinogenic properties. During the smoking process, PAHs accumulate in the fish muscle and they are difficult to remove due to their lipophilic properties. The content of PAHs in the smoked fish obviously depends on the concentrations of these compounds in the smoke, the smoking conditions, as well as the fish species. To eliminate the carcinogenic compounds, smoke is allowed to pass through electrostatic precipitation equipment that can absorb them.

6.4.5 Packaging and storage of smoked fish products

There are a number of ways in which smoked fish products can become contaminated after smoking. Therefore, it is important to reduce the possibility of contamination by using suitable packaging materials and packaging methods. When smoking is finished, the fish should be cooled rapidly to the ambient temperature or lower before it is packaged, otherwise it will turn flabby, moist and sour or mouldy. Cooling too slowly may promote the growth of microorganisms harmful to humans. After packaging, smoked fish products should be immediately refrigerated and maintained at a temperature of about 0 °C or less without interruption until sold to the consumer. One of the potential hazards in smoked fish, particularly in cold-smoked products, is contamination by *Clostridium botulinum* which produces toxic compounds. Appropriate combinations of NaCl and low temperature are sufficient to guarantee that growth of this organism is eliminated during the shelf life of the product under vacuum packaging (Huss, 1980; Dufresne *et al.*, 2000). No toxic compounds were detected in cold-smoked trout stored for 4 weeks at 4 °C or 8 °C at a salt concentration of 1.7% (w/w) (Dufresne *et al.*, 2000). Also, no toxin was detected by Cann and Taylor (1979) at 2.5–3.5% NaCl (w/w) in hot-smoked trout stored at 10 °C. In addition, vacuum packaging is one of the effective methods for inhibition of *Clostridium botulinum* growth. No spores of *C. botulinum* were detected by Heinitz and Johnson (1998) from 201 vacuum-packed smoked fish and shellfish products. However, 1.7% of positive samples were found in cold-smoked salmon with a level of <1 spore/kg (Nielsen and Pedersen, 1967) and 3% of 64 samples of cold-smoked rainbow trout were positive with levels in the range 40–290 spores/kg (Hyytia *et al.*, 1998).

The changes in the microbiological and physicochemical quality of smoked fish during storage depend on the degree of drying and smoking and the packaging method as well as the temperature at which it is subsequently stored. Modified atmosphere packaging can reduce the total volatile basic nitrogen (TVB-N), trimethylamine nitrogen (TMA) and lipid oxidation of smoked mullet compared to that under vacuum packaging during cold storage (Ibrahim *et al.*, 2008). The same results have been reported by Bilgin *et al.* (2008) for cold- and hot-smoked gilthead seabream and by Yanar (2007) for hot-smoked catfish in which TVB-N, TMA and lipid oxidation products increased significantly during storage at 4 °C. The shelf life of hot-smoked garfish packed in aluminium foil is longer when stored under refrigeration at 4 °C than when stored at ambient temperature (Koral *et al.*, 2009).

References

Ackman, R.G. (1990). Seafood lipids and fatty acids. *Food Review International*, 6(4), 617–646.

Akse, L., Gundersen, B., Lauritzen, K., Ofstad, R. and Solberg, T. (1993). *Salt fish*. Report No. 1/1993 by the Norwegian Institute of Fisheries and Aquaculture, Tromsø, Norway.

Alcicek, Z. and Atar, H.H. (2010). The effects of salting on chemical quality of vacuum packed liquid smoked and traditional smoked rainbow trout (*Oncorhyncus mykiss*) fillets during chilled storage. *Journal of Animal and Veterinary Advances*, 9, 2778–2783.

Andersen, H.J., Bertelsen, G. and Skibsted, L.H. (1990). Colour and colour stability of hot processed frozen minced beef. Results from chemical model experiments tested under storage conditions. *Meat Science*, 28, 87–97.

Andres, A., Rodriguez-Barona, S. and Barat, J.M. (2005a). Analysis of some cod-desalting process variables. *Journal of Food Engineering*, 70, 67–72.

Andres, A., Rodríguez-Barona, S., Barat, J.M. and Fito, P. (2002). Mass transfer kinetics during cod salting operation. *Food Science and Technology International*, 8(5), 309–314.

Andres, A., Rodríguez-Barona, S., Barat, J.M. and Fito, P. (2005b). Salted cod manufacturing: influence of salting procedure on process yield and product characteristics. *Journal of Food Engineering*, 69, 467–471.

Asghar, A., Samejima, K. and Yasui, T. (1985). Functionality of muscle proteins in gelation mechanisms of structured meat-products. *Critical Reviews in Food Science and Nutrition*, 22, 27–106.

Bakowski, M., Riewe, J., Borys, A. and Straszewski, T. (1970). Effect of brine volume on salt content of bacon sides and bacon cuts. *Gospodarka Miesna*, 22(8/9), 16–19.

Bannerman, A. and Horne, J. (2001). Recommendations for the production of smoked salmon. Torry Research Station, Torry Advisory Note 5. Aberdeen, UK, Ministry of Agriculture, Fisheries and Food.

Barat, J.M., Gallart-Jornet, L., Andres, A. *et al.* (2006). Influence of cod freshness on the salting, drying and desalting stages. *Journal of Food Engineering*, 73, 9–19.

Barat, J.M., Rodríguez-Barona, S., Andrés, A. and Fito, P. (2002). Influence of increasing brine concentration in the cod-salting process. *Journal of Food Science*, 67(5), 1922–1925.

Barat, J.M., Rodriguez-Barona, S., Andres, A. and Fito, P. (2003). Cod salting manufacturing analysis. *Food Research International*, 36, 447–453.

Barat, J.M., Rodriguez-Barona, S., Andres, A. and Ibanez, J.B. (2004a). Modelling of the cod desalting operation. *Journal of Food Science*, 69(4), 183–189.

Barat, J.M., Rodriguez-Barona, S., Andres, A. and Visquert, M. (2004b). Mass transfer analysis during the cod desalting process. *Food Research International*, 37, 203–208.

Barat, J.M., Rodriguez-Barona, S., Castello, M., Andres, A. and Fito, P. (2004c). Cod desalting process as affected by water management. *Journal of Food Engineering*, 61, 353–357.

Baygar, T., Alparslan, Y., Güler, M. and Okumuş, M. (2010). Effects of pickling solution on maturing and storage time of marinated sea bass fillets. *Asian Journal of Animal and Veterinary Advances*, 5, 575–583.

Beraquet, N.J., Iaderoza, M., Jardim, D.C.P. and Lindo, M.K.K. (1983). Salting of mackerel (*Scomber japonicas*) II. Comparison between brining and mixed salting in relation to quality and salt uptake. *Coletaneado Instituto de Technologia de Alimentos*, 13, 175–198.

Bilgin, S., Unlusayin, M., Izci, L. and Gunlu, A. (2008). The determination of the shelf life and some nutritional components of gilthead seabream (*Sparus aurata* L., 1758) after cold and hot smoking. *Turkish Journal of Veterinary and Animal Sciences*, 32, 49–56.

Birkeland, S., Akse, L., Joensen, S., Tobiassen, T. and Skara, T. (2007). Injection-salting of pre rigor fillets of Atlantic salmon (*Salmo salar*). *Journal of Food Science*, 72, 29–35.

Birkeland, S., Haarstad, I. and Bjerkeng, B. (2004a). Effects of salt-curing procedure and smoking temperature on astaxanthin stability in smoked salmon. *Journal of Food Science*, 69, 198–203.

Birkeland, S., Røra, A.M.B., Skåra, T. and Bjerkeng, B. (2004b). Effects of cold smoking procedures and raw material characteristics on product yield and quality parameters of cold smoked Atlantic salmon (*Salmo salar* L.) fillets. *Food Research International*, 37, 273–286.

Birkeland, S. and Skara, T. (2008). Cold smoking of Atlantic salmon (*Salmo salar*) fillets with smoke condensate-an alternative processing technology for the production of smoked salmon. *Journal of Food Science*, 73, 326–332.

Birkeland, S., Skara, T., Bjerkeng, B. and Rørå, A.M.B. (2003). Product yield and gaping in cold smoked Atlantic salmon (*Salmo salar*) fillets as influenced by different injection-salting techniques. *Journal of Food Science*, 68, 1743–1748.

Bjarnason, J. (1987). *A Manual of Fish Processing*. I. Curing of salted fish (Fisheries and Quatic Sciences No. 5323), Ottawa, Canada Institute for Scientific and Technical Information, National Research Council.

Bjørkevoll, I., Olsen, J.V. and Olsen, R.L. (2004). Rehydration of salt-cured cod using injection and tumbling technologies. *Food Research International*, 37, 925–931.

Boles, J. and Swan, E. (1997). Effects of brine ingredients and temperature on cook yields and tenderness of pre-rigor processed roast beef. *Meat Science*, 45(1), 87–97.

Bondet, V., Cuvelier, M.E. and Berset, C. (2000). Behaviour of phenolic antioxidants in a partitioned medium: Focus on linoleic acid peroxidation induced by iron/ascorbic acid system. *Journal of the American Oil Chemists' Society*, 77, 813–818.

Botta, J.R., Bonnell, G. and Squires, B.E. (1987). Effect of method of catching and time of season on sensory quality of fresh raw Atlantic cod (*Gadus morhua*). *Journal of Food Science*, 52(4), 928–931.

Bower, C.K., Hietala, K.A., Oliveira, A.C.M. and Wu, T.H. (2009). Stabilizing oils from smoked pink salmon (*Oncorhynchus gorbuscha*). *Journal of Food Science*, 74, 248–257.

Boyd, J.W. and Southcott, B.A. (1965). Effects of phosphates and other salts on drip loss and oxidative rancidity of frozen fish. *Journal of the Fisheries Research Board of Canada*, 22, 53.

Cadun, A.S., Cakli, D. and Kisla, D. (2005). A study of marination of deep water pink shrimp (*Parapenaeus longirostris*, Lucas, 1846) and its shelf life. *Food Chemistry*, 90, 53–59.

Caflisch, A. and Karplus, M. (1994). Molecular dynamics simulation of protein denaturation: solvation of the hydrophobic cores and secondary structure of barnase. *Proceedings of the National Academy of Sciences of the USA*, 91, 1746–1750.

Calligaris, S. and Nicoli, M.C. (2006). Effect of selected ions from lyotropic series on lipid oxidation. *Food Chemistry*, 94, 130–134.

Cann, D.C. and Taylor, L.Y. (1979). The control of the botulism hazard in hot-smoked trout and mackerel. *Journal of Food Technology*, 14, 123–129.

Capaccioni, M.E., Casales, M.R. and Yeannes, M.I. (2011). Acid and salt uptake during marinating process of *Engraulis anchoita* fillets influence of the solution: fish ratio and *aggigation*. *Ciência e Tecnologia de Alimentos*, 31, 884–890.

Cardinal, M., Knockaert, C., Torrissen, O. *et al.* (2001). Relation of smoking parameters to yield, colour and sensory quality of smoked Atlantic salmon (*Salmo salar*). *Food Research International*, 34, 537–550.

Chaijan, M. (2011). Physicochemical changes in tilapia (*Oreochromis niloticus*) muscle during salting. *Food Chemistry*, 129, 1201–1210.

Chen, B.H. and Lin, Y.S. (1997). Formation of polycyclic aromatic hydrocarbons during processing of duck meat. *Journal of Agricultural and Food Chemistry*, 45, 1394–1403.

Chiralt, A., Fito, P., Barat, J.M. *et al.* (2001). Use of vacuum impregnation in food salting process. *Journal of Food Engineering*, 49, 141–151.

Clifford, M.N., Tang, S.L. and Eyo, A.A. (1980). Smoking of food. *Process Biochemistry*, (June/July), 8–11.

Codex Stan 150 (1985). *Codex Standard for Food Grade Salt*, Food Agriculture Organization (FAO), Rome, Italy.

Conde, F.J., Afonso, A.M., González, V. and Ayala, J.H. (2006). Optimization of an analytical methodology for the determination of alkyl- and methoxy-phenolic compounds by HS-SPME in biomass smoke. *Analytical and Bioanalytical Chemistry*, 385, 1162–1171.

Curtis, R.A. and Lue, L. (2006). A molecular approach to bioseparations: protein–protein and protein–salt interactions. *Chemical Engineering Science*, 61, 907–923.

Daun, H. (1972). Sensory properties of phenolic compounds isolated from curing smoke as influenced by its generation parameters. *Lebensmittel-Wissenschaft und-Techologies*, 5, 102–105.

Daya, S., Walker, R.B. and Anoopkumar-Dukie, S. (2000). Cyanide-induced free radical production and lipid peroxidation in rat brain homogenate is reduced by aspirin. *Metabolic Brain Disease*, 15, 203–210.

Dillon, R. and Patel, T. (1993). Effect of cold smoking and storage temperatures on *Listeria monocytogenes* in inoculated cod fillets (*Gadus morhua*). *Food Research International*, 26, 97–101.

Doe, P.E., Hashmi, R., Poulter, R.G. and Olley, J. (1982). Isohalic sorption isotherms. I. Determination for dried salted cod (*Gadus morhua*). *Journal of Food Technology*, 17, 125–134.

Doe, P., Sikorski, Z., Haard, N.F., Olley, J. and Pan, B.S. (1998). Basic principles, in *Fish drying and smoking: Production and quality* (ed. P. Doe), Technomic Publishing Company Inc., Lancaster, Pennsylvania, USA, pp. 13–45.

Dore, I. (1993). *The Smoked and Cured Seafood Guide* (Chapter 1), Urner Barry Publications, Inc., New Jersey.

Dufresne, I., Smith, J.P., Liu, J.N. *et al.* (2000). Effect of films of different oxygen transmission rate on toxin production by *Clostridium botulinum* type E in vacuum packaged cold and hot smoked trout fillets. *Journal of Food Safety*, 20, 251–268.

Duyar, H. and Eke, E. (2009). Production and quality determination of marinade from different fish species. *Journal of Animal and Veterinary Advances*, 8, 270–275.

Efiuvwevwere, B.J. and Ajiboye, M.O. (1996). Control of microbiological quality and shelf-life of catfish (*Clarias gariepinus*) by chemical preservatives and smoking. *Journal of Applied Bacteriology*, 80, 465–470.

Emborg, J. and Dalgaard, P. (2006). Formation of histamine and biogenic amines in cold-smoked tuna: An investigation of psychrotolerant bacteria from samples implicated in cases of histamine fish poisoning. *Journal of Food Protection*, 69, 897–906.

Erikson, U., Veliyulin, E., Singstad, T.E. and Aursand, M. (2004). Salting and desalting of fresh and frozen-thawed cod (*Gadus morhua*) fillets: a comparative study using ^{23}Na NMR, ^{23}Na MRI, low-field ^{1}H NMR, and physicochemical analytical methods. *Journal of Food Science*, 69, 107–114.

Ernawati, E., Purnomo, H. and Estiasih, T. (2012). Antioxidant effect of liquid smoke on oxidation stability of catfish (*Clarias gariepinus*) smoke sausage during storage. *Journal Teknologi Pertanian*, 13, 119–124.

Espe, M., Nortvedt, R., Lie, Ø. and Hafsteinsson, H. (2002). Atlantic salmon (*Salmo salar*) as raw material for the smoking industry. II: Effect of different smoking methods on losses of nutrients and on the oxidation of lipids. *Food Chemistry*, 77, 41–46.

FAO (1992) *Small-Scale Food Processing – A Guide for Appropriate Equipment*. London: Intermediate Technology.

FAO (1983). Cured fish: market patterns and prospects. *Fisheries Technical Papers* 233, 1–10.

FAO (2001). *Torry Advisory Note No. 56. Marinades*, Food and Agriculture Organization, Rome, Italy.

FAO (2010). *Yearbook – Fishery and Aquaculture Statistics*, Food and Agriculture Organization, Rome, Italy, pp. 37–52.

FAO (2012). *The State of the World Fisheries and Aquaculture*, Food and Agriculture Organization, Rome, Italy, pp. 14–64. http://www.fao.org/docrep/016/i2727e/i2727e00.htm [accessed 30 May 2013].

FDA (2001). Chapter 13: *Clostridium botulinum* toxin formation, in *Fish and Fisheries Products Hazards and Control Guide*, 3rd edn, Office of Seafood, Food and Drug Administration, Department of Health and Human Services, US Government Printing Office, Washington, DC.

Fennema, O.R. (1990). Comparative water holding properties of various muscle foods. A critical review relating to definitions, methods of measurement, governing factors, comparative data and mechanistic matters. *Journal of Muscle Foods*, 1, 363–381.

Fretheim, K., Granum, P.E. and Vold, E. (1980). Influence of generation temperature on the chemical composition, antioxidative and antimicrobial effects of wood smoke. *Journal of Food Science*, 45, 999–1002.

Fuentes, A., Fermandez-Segovia, I., Serra, J.A. and Barat, J.M. (2007). Influence of the presence of skin on the salting kinetics of European sea bass. *Food Science and Technology International*, 13, 199–205.

Gallart-Jornet, L., Barat, J.M., Rustad, T. *et al.* (2007a). A comparative study of brine salting of Atlantic cod (*Gadus morhua*) and Atlantic salmon (*Salmo salar*). *Journal of Food Engineering*, 79(1), 261–270.

Gallart-Jornet, L., Barat, J.M., Rustad, T. *et al.* (2007b). Influence of brine concentration on Atlantic salmon fillet salting. *Journal of Food Engineering*, 80(1), 267–275.

Gallart-Jornet, L. and Lindkvist, K.B. (2007). The Spanish salt fish market – a challenge to the Norwegian salt fish industry. Proceedings of the Torskefiskkonferanse [Groundfish Conference], Tromsø, 7 November 2007.

Gill, T.A., Chan, J.K., Phonchreon, K.F. and Paulson, A.T. (1992). Effect of salt concentration and temperature on heat-induced aggregation and gelation of fish myosin. *Food Research International*, 25, 333–341.

Hanne, L., Jensen, K.N., Hylding, G., Nielsen, H.H. and Nielsen, J. (2007). Changes in liquid-holding capacity, water distribution and microstructure during chilled storage of smoked salmon. *Journal of the Science of Food and Agriculture*, 87, 2684–2691.

Hassan, I.M. (1988). Processing of smoked common carp fish and its relation to some chemical, physical and organoleptic properties. *Food Chemistry*, 27, 95–106.

Heinitz, M.L. and Johnson, J.M. (1998). The incidence of *Listeria* spp., *Salmonella* spp., and *Clostridium botulinum* in smoked fish and shellfish. *Journal of Food Protection*, 61, 318–323.

Howgate, P. (1979). Fish, in *Food Microscopy* (ed. J.G. Vaughan), Academic Press, London, pp. 343–389.

Hultmann, L., Røra, A.M.B., Steinsland, I., Skara, T. and Rustad, T. (2004). Proteolytic activity and properties of proteins in smoked salmon (*Salmo salar*) – effect of smoking temperature. *Food Chemistry*, 85, 377–387.

Huss, H.H. (1980). Distribution of *Clostridium botulinum. Applied and Environmental Microbiology*, 39, 764–769.

Hyytia, E., Hielm, S. and Korkeala, H. (1998). Prevalence of *Clostridium botulinum* type E in Finnish fish and fishery products. *Epidemiology and Infection*, 120, 245–250.

Ibrahim, S.M., Nassar, A.G. and El-Badry, N. (2008). Effect of modified atmosphere packaging and vacuum packaging method on some quality aspects of smoked mullet (*Mugil cephalus*). *Global Veternaria*, 2, 296–300.

Jarvis, N.D. (1988). Curing and Canning of Fishery Products: A History. *Marine Fisheries Review*, 50, 180–185.

Jittinandana, S., Kenney, P.B., Slider, S.D. and Kiser, R.A. (2002). Effect of brine concentration and brining time on quality of smoked rainbow trout fillets. *Journal of Food Science*, 67(6), 2095–2099.

Kanner, J., German, J.B., Kinsella, J.E. and Hultin, H.O. (1987). Initiation of lipid peroxidation in biological system. *Critical Reviews in Food Science*, 25, 317–364.

Kanner, J., Harel, S. and Jaffe, R. (1991). Lipid oxidation of muscle food as affected by sodium chloride. *Journal of Agricultural and Food Chemistry*, 39, 1017–1021.

Kanner, J., Hazan, B. and Doll, L. (1988). Catalytic 'free' iron ions in muscle foods. *Journal of Agricultural and Food Chemistry*, 36, 412–415.

Kjallstrand, J. and Petersson, G. (2001). Phenolic antioxidants in alder smoked during industrial meat curing. *Food Chemistry*, 74, 85–89.

Koral, S., Kose, S. and Tufan, B. (2009). Investigating the quality changes of raw and hot smoked garfish (*Blone blone*, Gunther, 1866) at ambient and refrigerated temperatures. *Turkish Journal of Fisheries and Aquatic Sciences*, 9, 53–58.

Lauritzsen, K. (2004). Quality of salted cod (*Gadus morhua* L.) as influenced by raw material and salt composition. PhD dissertation, Norwegian College of Fishery Science, University of Tromsø.

Lauritzsen, K., Akse, L., Gundersen, B. and Olsen, R.L. (2004a). Effects of calcium, magnesium and pH during salting of cod (*Gadus morhua* L.). *Journal of the Science of Food and Agriculture*, 84, 683–692.

Lauritzsen, K., Martinsen, G. and Olsen, R.L. (1999). Copper induced lipid oxidation during salting of cod (*Gadus morhua* L.). *Journal of Food Lipids*, 6, 299–315.

Lauritzsen, K., Olsen, R.L., Akse, L. *et al.* (2004b). Physical and quality attributes of salted cod (*Gadus morhua* L.) as affected by the state of rigor and freezing prior to salting. *Food Research International*, 37, 677–688.

Leroi, F. and Joffraud, J.J. (2000). Salt and smoke simultaneously affect chemical and sensory quality of cold-smoked salmon during 5 °C storage predicted using factorial design. *Journal of Food Protection*, 63, 1222–1227.

Lindkvist, K.B., Gallart-Jornet, L. and Stabell, M.C. (2008). The restructuring of the Spanish salted fish market. *The Canadian Geographer/Le Géographe canadien*, 52(1), 105–120.

Love, R.M. and Mackay, E.M. (1962). Protein denaturation in frozen fish. V. Development of the cell fragility method for measuring cold-storage changes in muscle. *Journal of the Science of Food and Agriculture*, 13, 200–212.

Lund, K.E. and Nielsen, H.H. (2001). Proteolysis in salmon (*Salmo salar*) during cold storage; effect of storage time and smoking process. *Journal of Food Biochemistry*, 25, 379–385.

Lyhs, U., Bjorkroth, J., Hyytia, E. and Korkeala, H. (1998). The spoilage flora of vacuum-packaged, sodium nitrite or potassium nitrate treated, cold-smoked rainbow trout stored at 4 °C or 8 °C. *International Journal of Food Microbiology*, 45, 135–142.

Maga, J.A. (1988). *Smoke in Food Processing*. CRC Press, Inc., Boca Raton, Florida.

Mahon, J.H. and Schneider, C.G. (1964). Minimizing freezing damage and thawing drip in fish fillets. *Food Technology*, 18, 117–118.

Martinez, O., Salmeron, J., Guillen, M.D. and Casas, C. (2007). Textural and physicochemical changes in salmon (*Salmo salar*) treated with commercial liquid smoke flavourings. *Food Chemistry*, 100, 498–503.

Martinez, O., Salmeron, J., Guillen, M.D. and Casas, C. (2011). Characteristics of dry- and brined-salted salmon later treated with liquid smoke flavouring. *Agricultural and Food Science*, 20, 217–227.

Martinez-Alvarez, O. and Gomez-Guillen, M.C. (2005). The effect of brine composition and pH on yield and nature of water-soluble proteins extractable from brined muscle of cod (*Gadus morhua*). *Food Chemistry*, 92, 71–77.

Martínez-Alvarez, O. and Gómez-Guillén, M.C. (2006). Effect of brine salting at different pHs on the functional properties of cod muscle proteins after subsequent dry salting. *Food Chemistry*, 94, 123–129.

Masniyom, P., Benjakul, S. and Visessanguan, W. (2005). Combination effect of phosphate and modified atmosphere on quality and shelf-life of refrigerated seabass slices. *LWT-Food Science and Technology*, 38, 745–756.

Matsumoto, J.J. (1979). Denaturation of fish muscle proteins during frozen storage, in *Proteins at low temperature* (ed. O. Fennema), ACS publications, Washington, DC, pp. 205–224.

McCann, A.M. (1988). The Roman Port of Cosa. *Scientific American*, 258, 84–91.

Muratore, G., Mazzaglia, A., Lanza, C.M. and Licciardello, F. (2007). Effect of process variables on the quality of swordfish fillets flavoured with smoke condensate. *Journal of Food Processing and Preservation*, 31, 167–177.

Murray, J. and Burt, J.R. (1969). The composition of fish. Torry Advisory Note No. 38. Torry Research Station, Aberdeen.

Nguyen, M.V., Arason, S., Thorarinsdottir, K.A., Thorkelsson, G. and Gudmundsdottir, A. (2010). Influence of salt concentration on the salting kinetics of cod loin (*Gadus morhua*) during brine salting. *Journal of Food Engineering*, 100, 225–231.

Nguyen, M.V., Thorarinsdottir, K.A., Gudmundsdottir, A., Thorkelsson, G. and Arason, S. (2011). The effects of salt concentration on conformational changes in cod (*Gadus morhua*) proteins during brine salting. *Food Chemistry*, 125, 1013–1019.

Nguyen, M.V., Thorarinsdottir, K.A., Thorkelsson, G., Gudmundsdottir, A. and Arason, S. (2012). Influences of potassium ferrocyanide on lipid oxidation of salted cod (*Gadus morhua*) during processing, storage and rehydration. *Food Chemistry*, 131, 1322–1331.

Nickelson, R.I., McCarthy, S. and Finne, G. (2001). Fish, crustaceans, and precooked seafoods, in *Compendium of Methods for the Microbiological Examination of Foods* (eds F.P. Downes and K. Ito), American Public Health Association, Washington, DC, pp. 497–505.

Nielsen, S.F. and Pedersen, H.O. (1967). Studies of the occurrence and germination of *Cl. botulinum* in smoked salmon, in *Botulism 1966* (eds M. Ingram and T.A. Robers), Chapman and Hall, UK, pp. 66–72.

Offer, G. and Knight, P. (1988). The structural basis of water-holding in meat. Part 1: General principles and water uptake in meat processing, in *Developments in Meat Science* (ed. R. Lawrie), Elsevier Applied Science, London, pp. 63–171.

Offer, G. and Trinick, J. (1983). On the mechanism of water holding in meat: the swelling and shrinking of myofibrils. *Meat Science*, 8, 245–281.

Osinchak, J.E., Hultin, H.O., Zajicek, O.T., Kelleher, S.D. and Huang, C.H. (1992). Effect of NaCl on catalysis of lipid oxidation by the soluble fraction of fish muscle. *Free Radical Biology and Medicine*, 12, 35–41.

Pedro S., A.M., Nunes, M.L. and Bernado, M.F. (2004). Pathogenic bacteria and indicators in salted cod (*Gadus morhua*) and desalted products at low and high temperatures. *Journal of Aquatic Food Product Technology*, 13(3), 39–48.

Raoult-Wack, A.L. (1994). Recent advances in the osmotic dehydration of foods. *Trends in Food Science and Technology*, 5, 255–260.

Rhee, K.S. and Ziprin, Y.A. (2001). Pro-oxidative effects of NaCl in microbial growth-controlled and uncontrolled beef and chicken. *Meat Science*, 57, 105–112.

Rodrigues, M.J., Ho, P., Lopez-Caballero, M.E., Vaz-Pires, P. and Nunes, M.L. (2003). Characterization and identification of microflora from soaked cod and respective salted raw materials. *Food Microbiology*, 20(4), 471–481.

Rørå , A.M.B., Kvåle, A., Mørkøre, T. *et al.* (1998). Process yield, colour and sensory quality of smoked Atlantic salmon (*Salmo salar*) in relation to raw material characteristics. *Food Research International*, 31, 601–609.

Ross, D.A. and Love, R.M. (1979). Decrease in the cold store flavour developed by fillets of starved cod (*Gadus morhua* L.). *Journal of Food Technology*, 14, 115–122.

Sannaveerappa, T., Ammu, K. and Joseph, J. (2004). Protein-related changes during salting of milkfish (*Chanos chanos*). *Journal of the Science of Food and Agriculture*, 84, 863–869.

Schmidt, F.C., Carciofi, B.A.M. and Laurindo, J.B. (2008). Salting operational diagrams for chicken breast cuts: Hydration–dehydration. *Journal of Food Engineering*, 88, 36–44.

Serot, T. and Lafficher, C. (2003). Optimisation of solid-phase microextraction coupled to gas chromatography for determination of phenolic compounds in smoked herring. *Food Chemistry*, 82, 513–519.

Shearer, D.K. (1994). Factors affecting the proximate composition of cultured fishes with emphasis on salmonids. *Aquaculture*, 119, 63–88.

Shomer, L., Weinberg, Z.G. and Vasilever, R. (1987). Structural binding properties of silver carp (*Hypophthalmichthys molitrix*) muscle affected by NaCl and $CaCl_2$ treatments. *Food Microstructure*, 6, 199–207.

Sigurgisladottir, S., Sigurdardottir, M.S., Ingvarsdottir, H., Torrissen, O.J. and Hafsteinsson, H. (2001). Microstructure and texture of fresh and smoked Atlantic salmon, *Salmo salar* L., fillets from fish reared and slaughtered under different conditions. *Aquaculture Research*, 32, 1–10.

Sigurgisladottir, S., Sigurdardottir, M.S., Torrasen, O., Vallet, J.L. and Hafsteinsson, H. (2000). Effects of different salting and smoking processes on the microstructure, the texture, and yield of Atlantic salmon (*Salmo salar*) fillets. *Food Research International*, 33, 847–855.

Simon, R., de la Calle, B., Palme, S., Meier, D. and Anklam, E. (2005). Composition and analysis of liquid smoke flavouring primary products. *Journal of Separation Science*, 28, 871–882.

Skjerdal, O.T., Pedro, S. and Serra, J.A. (2002). Improved quality and shelf life of desalted cod, an easy-to-use product of salted cod. Final Report to the European Commission. FAIR Project CT98–4179.

Sotelo, C.G., Piñeiro, C. and Pérez-Martín, R.I. (1995). Denaturation of fish protein during frozen storage: role of formaldehyde. *Zeitschrift fuer Lebensmittel Untersuchung und - Forschung*, 200, 14–23.

Stefansson, G. and Hultin, H.O. (1994). On the solubility of cod muscle proteins in water. *Journal of Agricultural and Food Chemistry*, 42, 2656–2664.

Stewart, K.M. (1994). Early hominid utilization of fish resources and implications for seasonality and behaviour. *Journal of Human Evolution*, 27, 229–245.

Suñen, E., Aristimuno, C. and Fernandez-Galian, B. (2003). Activity of smoke wood condensates against *Aeromonas hydrophila* and *Listeria monocytogenes* in vacuum-packed, cold-smoked rainbow trout stored at 4 °C. *Food Research International*, 36, 111–116.

Szymczak, M. (2011). Comparison of physicochemical and sensory changes in fresh and frozen herring (*Clupea harrengus*) during marinating. *Journal of the Science of Food and Agriculture*, 91, 68–74.

Thorarinsdottir, K.A. (2010). The influence of salting procudures on the characteristics of heavy salted cod. PhD dissertation. Lund University, Sweden.

Thorarinsdottir, K.A., Arason, S., Bogason, S.G. and Kristbergsson, K. (2001). Effects of phosphate on yield, quality, and water-holding capacity in the processing of salted cod (*Gadus morhua*). *Journal of Food Science*, 66, 821–826.

Thorarinsdottir, K.A., Arason, S., Bogason, S.G. and Kristbergsson, K. (2004). The effects of various salt concentrations during brine curing of cod (*Gadus morhua*). *International Journal of Food Science and Technology*, 39, 79–89.

Thorarinsdottir, K.A., Arason, S., Geirsdottir, M., Bogason, S.G. and Kristbergsson, K. (2002). Changes in myofibrillar proteins during processing of salted cod (*Gadus morhua*) as determined by electrophoresis and differential scanning calorimetry. *Food Chemistry*, 77(3), 377–385.

Thorarinsdottir, K.A., Arason, S., Sigurgisladottir, S. *et al.* (2011a). The effects of salt-curing and salting procedures in the microstructure of cod (*Gadus morhua*) muscle. *Food Chemistry*, 126, 109–115.

Thorarinsdottir, K.A., Arason, S., Sigurgisladottir, S., Valsdottir, T. and Tornberg, E. (2011b). Effects of different pre-salting methods on protein aggregation during heavy salting of cod fillets. *Food Chemistry*, 124, 7–14.

Thorarinsdottir, K.A., Arason, S., Thorkelsson, G., Sigurgisladottir, S. and Tornberg, E. (2010a). The effects of pre-salting methods from injection to pickling on the yields of heavily salted cod (*Gadus morhua*). *Journal of Food Science*, 75(8), 7–14.

Thorarinsdottir, K.A., Bjørkevoll, I. and Arason, S. (2010b). Production of salted fish in Nordic countries. Variation in quality and characteristics of the salted fish products. Report no. 46–10 by the Matis-Icelandic Food Research and Biotech RandD, Reykjavik, Iceland.

Tokur, B. and Korkmaz, K. (2007). The effects of an iron-catalyzed oxidation system on lipids and proteins of dark muscle fish. *Food Chemistry*, 104, 754–760.

Toth, L. and Potthast, K. (1984). Chemical aspects of the smoking of meat and meat products. *Advances in Food Research*, 29, 87–158.

Toussaint-Samat, M. (2009). *A History of Food*, 2nd edn, Blackwell Publishers Ltd., Massachusetts, USA.

Troller, J.A. and Christian, J.H.B. (1978). *Water Activity and Food*, Academic Press, New York.

van Klaveren, F.W. and Legrende, R. (1965). Salted cod. In *Fish as Food*, vol. III (ed. G. Borgstrom), Academic Press, New York, pp. 133–160.

Varlet, V., Knockaert, C., Prost, C. and Serot, T. (2006). Comparison of odour-active volatile compounds of fresh and smoked salmon. *Journal of Agricultural and Food Chemistry*, 54, 3391–3401.

Varlet, V., Prost, C. and Serot, T. (2007a). Volatile aldehydes in smoked fish: analysis methods, occurrence and mechanisms of formation. *Food Chemistry*, 105, 1536–1556.

Varlet, V., Serot, T., Knockaert, C. *et al.* (2007b). Organoleptic characterization and PAH content of salmon (*Salmo salar*) fillets smoked according to four industrial smoking techniques. *Journal of the Science of Food and Agriculture*, 87, 847–854.

Wang, D., Tang, J. and Correia, L.R. (2000). Salt diffusivities and salt diffusion in farmed Atlantic salmon muscle as influenced by rigor mortis. *Journal of Food Engineering*, 43, 115–123.

Ward, A.R. (1995). Fish smoking in the tropics: a review. *Tropical Science*, 35, 103–112.

Wheaton, F.W. and Lawson, T.B. (1985). *Processing Aquatic Food Products*. John Wiley & Sons Canada Ltd, Toronto, Canada.

Yanar, Y. (2007). Quality changes of hot smoked catfish (*Clarias gariepinus*) during refrigerated storage. *Journal of Muscle Foods*, 18, 391–400.

Yao, Z. and Le Maguer, M. (1996). Osmotic dehydration: an analysis of fluxes and shrinkage in cellular structure. *Transactions of the ASAE*, 39, 2211–2216.

Yashoda, K.P., Rao, R.J., Mahendrakar N.S. and Rao, D.N. (2005). Marination of sheep muscles under effect on meat texture quality. *Journal of Muscle Foods*, 16, 184–191.

Yeannes, M.I. (1991). HACCP in anchovy salting and marinating.Expert Consultation on Quality Assurance in the Fish Industry Conference, Lyngby, Denmark.

Zhang, Q., Xiong, S., Liu, R., Xu, J. and Zhao, S. (2011). Diffusion kinetics of sodium chloride in Grass carp muscle and its diffusion anisotropy. *Journal of Food Engineering*, 107, 311–318.

7

Drying of Fish

Minh Van Nguyen,[1,2] Sigurjon Arason[2,3] and Trygve Magne Eikevik[4]

[1]*Faculty of Food Technology, Nha Trang University, Nha Trang, Vietnam*
[2]*Matís-Icelandic Food and Biotech R & D, Reykjavík, Iceland*
[3]*Faculty of Food Science and Nutrition, University of Iceland, Reykjavik, Iceland*
[4]*Norwegian University of Science and Technology – NTNU, Department of Energy and Process Engineering, Trondheim, Norway*

7.1 Introduction

Drying is one of the thermal treatments applied in many food industries, and the use of dried foods is expanding rapidly (Senadeera *et al.*, 2005). Drying processes can be broadly classified as thermal drying, osmotic dehydration and mechanical and chemical dewatering. Drying is the removal of water content to safe levels that can slow down the actions of enzymes, bacteria, yeasts and moulds (Janjai and Bala, 2012). The effects of heating on the activity of microorganisms and enzymes are also important when food is dried (Rahman, 2006). Research in to the drying process in the food industry has three goals: economic considerations, environmental concerns and product quality aspects (Okos *et al.*, 1992). In many agricultural countries, large quantities of food products are dried to improve shelf life, reduce packaging costs, lower shipping weights, enhance appearance, encapsulate original flavour and maintain nutritional value (Sokhansanj and Jayas 1987; Chou and Chua 2001). The energy used in drying in most industrialized countries accounts for 7–15% of the total industrial energy used, often with a relatively low thermal efficiency ranging from 25 to 50% (Dincer 1998; Chua *et al.*, 2001). Some highly industrialized countries use over one-third of their primary energy for drying operations (Dirk and Markus, 2004).

7.2 Principles of drying

7.2.1 Mass and heat transfer during drying

Drying is a complex process involving simultaneous coupled transient heat, mass and momentum transport in which heat penetrates into the fish muscle and water is removed

Seafood Processing: Technology, Quality and Safety, First Edition. Edited by Ioannis S. Boziaris.

(Haghi and Amanifard, 2008). The removal of water during drying includes a combination of the evaporation of water vapour from the surface and the movement of water from interior layers to the surface layer. The removal of water as vapour from the surface layer depends on the drying temperature, air humidity and air velocity as well as the exposed surface and pressure (Mujumdar, 2006). The movement of water from the inner layers to the surface layer is thought to be a function of the physicochemical properties of fish muscle and temperature. Diffusion is considered the most important mass transfer mechanism responsible for water and vapour transport within fish muscle. Diffusion results from the differences in concentration and pressure caused by protein denaturation/aggregation and muscle shrinkage between the inner layers and outer layers of the fish muscle. Capillary forces are also thought to make a contribution to water migration within the fish muscle. Furthermore, water may be transported inside the fish muscle due to the temperature gradient (thermo-gradient effect) caused by the thermodynamic coupling of heat and mass transport processes (Gavrila *et al.*, 2008). The rate of water evaporation on the surface of the fish muscle should be kept in equilibrium with the rate of water migration from the inner layers to the surface layer. A boundary layer is formed if water in the surface of the fish muscle is vaporized too quickly, which affects the drying process and extends the drying time.

Heat can transfer from the surrounding environment to the fish muscle during the drying process by convection, conduction or radiation or a combination of these. Heat may also be generated internally within the fish muscle by electric resistance (Sahin *et al.*, 2002). The rate of heat flux depends on the air humidity, temperature, and mass and heat transfer coefficients in interaction with the moisture and temperature distribution inside the fish muscle (Gavrila *et al.*, 2008).

7.2.2 Drying kinetics

During the drying process, water is removed from the fish muscle, resulting in decreased water content. The moisture ratio (MR) is calculated by Equation (7.1) (Sacilik *et al.*, 2006):

$$\mathrm{MR} = (W - W_{\mathrm{e}})/(W_{\mathrm{o}} - W_{\mathrm{e}}) \tag{7.1}$$

where MR is moisture ratio; W, W_{e}, and W_{o} are moisture content at any drying time, equilibrium moisture content and initial moisture content of the sample in dry matter, respectively.

The drying rate is obtained from Equation (7.2) (Akpinar *et al.*, 2003):

$$\text{Drying rate} = (W_{t+\mathrm{d}t} - W_t)/\mathrm{d}t \tag{7.2}$$

where $W_{t+\mathrm{d}t}$ and W_t are moisture content at drying time $(t + \mathrm{d}t)$ (kg/kg dry matter) and moisture content at drying time (t) (kg/kg dry matter), respectively.

Second Fick's law is used to calculate the effective moisture diffusivity (D_{eff}) during drying by Equation (7.3) (Crank, 1975):

$$\mathrm{MR} = \frac{W - W_{\mathrm{e}}}{W_{\mathrm{o}} - W_{\mathrm{e}}} = \frac{6}{\pi^2}\sum_{\mathrm{n}=1}^{\infty}\frac{1}{n^2}\exp^{\left(\frac{-\mathrm{n}^2\pi^2 D_{\mathrm{eff}}}{\mathrm{L}^2}t\right)} \tag{7.3}$$

where D_{eff} is the effective moisture diffusivity (m^2/s), L is the thickness of fish fillet and n is the number of terms taken into consideration ($n = 1, 2, 3, \ldots$). When the drying time is long, the number of terms is set as $n = 1$. Therefore, the effective moisture diffusivity is calculated by using Equation (7.4) (Mccabe *et al.*, 1993):

$$\ln(\mathrm{MR}) = \ln\left(\frac{W - W_e}{W_o - W_e}\right) = \ln\left(\frac{6}{\pi^2}\right) - \left(\frac{\pi^2 D_{\mathrm{eff}}}{\mathrm{L}^2}t\right) \quad (7.4)$$

7.2.3 Water activity

Water activity (a_W) is defined as the ratio of the vapour pressure of water in the food muscle (p) and the vapour pressure of pure water at the same temperature (p_o). At the beginning of the drying process (constant rate phase of drying), the water activity remains close to $a_W = 1$, due to saturated air within the boundary layer close to the surface (Doe *et al.*, 1998). However, water activity decreases at later stages of the drying process (falling rate phase of drying).

Water activity is of great importance for food preservation. It has been used to provide a reliable assessment of the microbial growth, lipid oxidation, non-enzymatic and enzymatic activities, and the texture of foods (Rahman and Labuza, 2007; Belessiotis and Delyannis, 2011). It has been stated that water activity is correlated with a deterioration of quality resulting from the growth of microorganisms (Scott, 1957). Rahman and Labuza (2007) noted that pathogenic bacteria cannot grow when the water activity is below 0.85–0.86, but yeast and moulds are more tolerant to a reduced water activity and cannot grow when water activity is 0.62. A Maillard reaction during drying is strongly influenced by water activity in the fish muscle showing a maximum intensity when the water activity is 0.65–0.70 under high temperatures of 80–130 °C (Labuza and Saltmarch, 1981). Different limitations of the water activity of dried fish products are proposed depending on the type of products and the drying method applied. However, to avoid microbial growth, the water activity of dried fish products should be maintained below the critical value of 0.60 (Perera and Rahman, 1997).

7.3 Drying methods

7.3.1 Sun drying

Open sun drying is the most widespread method of drying fish used by artisan fishermen (Kituu *et al.*, 2010). Natural sun drying has been used since ancient times for agricultural products. Sun drying is the most convenient and the cheapest processing technique to preserve fish and fish products (Jain and Pathare, 2007; Immaculate *et al.*, 2012), particularly in tropical and subtropical countries, where solar radiation is abundant, inexhaustible and environmental friendly (Szulmayer, 1971). However, a major problem of open sun drying is the loss of quality due to contamination with dust and excreta from birds and animals (Kituu *et al.*, 2010; Immaculate *et al.*, 2012). Moreover, it is difficult to control the drying process and the drying parameters in the open air due to weather uncertainties and also a large drying area is required (Jain and Pathare, 2007).

During open sun drying, heat is transferred by convection from the surrounding air and by absorption of direct and diffuse radiation on the surface of the fish and fish products. The converted heat is partly conducted to the interior, leading to an increased temperature of the fish and fish products, and is partly used for water and vapour migration from the interior to the surface. The remaining amount of energy is used for evaporation of the water at the surface or lost to the surroundings via convection and radiation. Natural convection supported by wind force removes the evaporated water from the air surrounding the fish and fish products.

7.3.2 Solar drying

Solar dryers are commonly used for drying fish and fish products. Solar drying is an elaboration of sun drying and has proved to be an energy efficient system (Zaman and Bala, 1989; Bala, 1998). Solar drying not only saves energy but also saves a lot of time, occupies a smaller drying area, improves the quality of the final products and makes the process more efficient as well as protecting the environment (VijayaVenkataRaman *et al.*, 2012). It differs from open sun drying because the solar dryer is an enclosed structure that traps heat inside the dryer and uses it efficiently (Immaculate *et al.*, 2012). The main advantages of solar drying over open sun drying are: (i) enhancement of the mass and heat transfer during the drying process, resulting in a shorter drying time; (ii) improvement of the quality of the final products due to a more controlled environment and the use of optimal drying parameters and (iii) reduction of loss during processing.

Various types of solar dryers for drying fish have been designed, developed and tested in different regions of tropical and subtropical countries (Ratti and Mujumdar, 1997; Bala and Debnath, 2012). Two main categories of solar dryers are the natural convection solar dryer (passive dryer) and the forced convection solar dryer (active dryer). In passive dryers, the airflow is driven by buoyancy pressure, whereas, in active dryers, a fan run using electricity, a solar module or fossil fuel provides the airflow. Due to the higher air speed used in an active dryer, drying time is reduced compared to passive dryer. However, a natural convection solar dryer requires a smaller investment and is simpler to operate and maintain compared to a forced convection solar dryer (Soponronnarit, 1995).

7.3.3 Heat pump drying

Heat pump dryers have been in widespread commercial use since the 1970s, particularly in agriculture and the food drying industry. Heat pump drying has the potential to operate more efficiently and at lower temperatures than conventional drying (Filho and Strømmen, 1996). In heat pump drying, both sensible and latent heat can be recovered from the dryer exhaust humid air, improving the overall thermal performance. Heat pump drying can be operated over a wide range of temperatures, providing good conditions for drying heat-sensitive materials (Hawlader *et al.*, 2006). The heat pump presents an efficient and environmentally friendly technology due to its low energy consumption (Strømmen *et al.*, 2000), its high coefficient of performance and the high thermal efficiency of a properly designed dryer (Braun *et al.*, 2002; Dirk and Markus, 2004). In addition, heat pump drying can operate independently of ambient weather conditions and it is environmentally friendly (Perera and Rahman, 1997).

Heat pump drying can improve the quality of fish products due to its low drying temperature and independence from outdoor air. In recent years, interest in applying heat pump drying to foods and biomaterials to enhance the quality of the final products has been a growing (Chou and Chua, 2001). Usually, dried products have a low aroma volatile content, suffer loss of heat-labile vitamins and have a high incidence of colour degradation (Perera and Rahman, 1997).

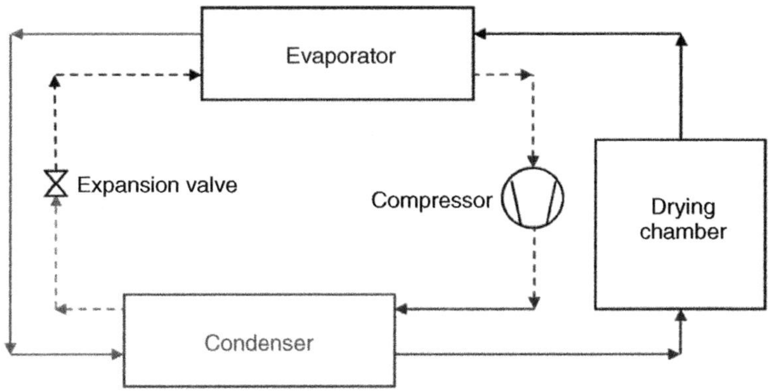

Figure 7.1 Schematic illustration of a heat pump drying system

The heat pump dryer system is a combination of two sub-systems: a heat pump and a dryer (Figure 7.1). In a heat pump system, the working fluid (refrigerant) at low pressure is vaporized in the evaporator by receiving the heat from the dryer exhaust humid air. The compressor raises the enthalpy of the working fluid of the heat pump and discharges it as superheated vapour at high pressure. At the condenser, heat is removed from the working fluid and released to the dried air. The working fluid is then throttled to the low-pressure line (using an expansion valve) and enters the evaporator to complete the cycle. In the dryer system, hot and dried air exiting the condenser passes through the drying chamber where it releases heat to damp material and also extracts the moisture from the damp material by convective transfer to the air. The humid air exiting the drying chamber passes through the evaporator where the moisture in the air will be condensed into water as the air goes below dew point temperature. After passing through the evaporator, the dried air is passed through the condenser to reheat. The reheated air goes to the drying chamber again.

7.3.4 Freeze-drying

Freeze-drying (lyophilization) has been extensively used to dry food since the end of the nineteenth century (Babic *et al.*, 2009). Freeze-drying is an innovative drying process that uses the sublimation of ice as its main drying mechanism; this differentiates the process from the conventional drying methods that rely on the capillary motion and evaporation of liquid water for drying. In freeze-drying, water in products is removed by sublimation at low temperature and pressure below the triple point of water. Firstly, water in products is frozen in the pre-freezing stage, and it is then removed by sublimation of ice in the primary drying stage. The residual water after the primary drying stage is reduced by desorption in the secondary drying stage (Nam and Song, 2007). Freeze-drying provides dried products with a porous structure, small or negligible shrinkage, superior flavour and aroma retention, and improved rehydration capacity compared to the products of other drying methods (Ratti, 2001; Aguilera *et al.*, 2003). It is the best method for preserving materials which are unstable in solution and for dehydrating high-quality and heat-sensitive products such as quality food, pharmaceuticals and biomedical products (Ratti, 2001). Furthermore, because as much as 98% of the water is removed, the food is lightweight, which reduces the cost of transportation and storage. However, note that the disadvantages of freeze-drying are its high costs, both for capital and running costs, and long processing time (Hammami *et al.*, 1999). These costs are usually passed on to the consumers, which makes freeze-dried food products expensive compared to food preserved by some other methods such as canning or freezing. In spite of this, freeze-drying has been used in variety of products from food ingredients and whole ready meals to biological and pharmaceutical products because it gives excellent quality. In the food processing field, freeze-drying has been applied in many products such as fruits (Karathanos *et al.*, 1996), bulbs (Sablani *et al.*, 2007), tuna (Rahman *et al.*, 2002), Atlantic mackerel and horse mackerel (Sarkardei and Howell, 2007) and cod (Eikevik *et al.*, 2005).

7.3.5 Osmotic dehydration

Osmotic dehydration is one of the most effective complementary treatments and food preservation techniques for processing dehydrated food (Alakali *et al.*, 2006; Khan, 2012). It is a common step used in traditional processes such as salting, smoking and marinating, etc. (Collignan *et al.*, 2001). Osmotic dehydration is also widely applied in combination with other drying methods as a seasoning step in order to produce the seasoned dried fish products (Wang *et al.*, 2011; Uribe *et al.*, 2011). The advantages of osmotic dehydration are: (i) reducing the damage of heat to the flavour and colour; (ii) inhibiting the browning by enzymes; (iii) decreasing the energy costs; and (iv) increasing the shelf life of the final products, mainly due to the effects of the chemicals used in the osmotic solution (Petrotos and Lazarides, 2001; Alakali *et al.*, 2006). Osmotic dehydration is carried out by immersing the fish in concentrated osmotic solutions of salt (e.g. brining of fish), sugar, acids (e.g. marinating of fish), etc. During the osmotic dehydration process, the flux of water occurs simultaneously with a counter influx of solutes through the tissue membrane, resulting in dehydration of water out of the fish muscle and increased solute concentration in the fish muscle. Solute diffusion into the fish muscle and dehydration of water lead to changes in physicochemical properties, thereby enhancing the quality attributes of the final products (Raoult-Wack, 1994). Diffusion is the most important mass transfer mechanism taking place during the osmotic dehydration process (Fito, 1994; Barat *et al.*, 2002). Diffusion results from differences in concentration and osmotic pressures between the fish muscle and the osmotic solution (Fito, 1994; Raoult-Wack, 1994) and from concentration gradient within the fish muscle (Erikson *et al.*, 2004). The diffusion rate of water and solutes depends upon various factors such as the temperature and concentration of the osmotic solution, the size and the geometry of the material, the ratio of solution to material mass, and the level of agitation of the solution (Arvanitoyannis *et al.*, 2012). Nowadays, in order to decrease the processing time, vacuum pulse is applied to the osmotic dehydration system. Vacuum pulse can improve the mass transfer rate, mainly due to promoting fast compositional changes by the action of a hydrodynamic mechanism (Fito and Pastor, 1994; Chiralt *et al.*, 1999).

7.4 Changes in fish muscle during drying

7.4.1 Changes in chemical properties of fish muscle

Fish muscle proteins The drying process deeply impacts the final characteristics of the products, mainly in terms of protein denaturation, but protein nitrogen remains (Wu and Mao, 2008). Drying brings about formation of different inter- and intra-molecular bonds. As the result of protein denaturation, protein chains unfold and free carboxylic and amino groups are exposed. Disulfide bonds are formed due to the oxidation of exposed sulfhydryl groups (Sen, 2005). Moreover, the formation of stable covalent linkages and/or Maillard reaction takes place during drying (Eichner and Wolf, 1983; Opstvedt *et al.*, 1984). The changes in protein structure lead to a decrease in protein solubility and digestibility (Wu and Mao, 2008). The denaturation of proteins during drying depends on the drying methods and on the drying parameters such as temperature and air velocity. Raghunath *et al.* (1995) observed that the rate of decrease in protein solubility is higher when fish is dried at higher temperatures and that sulfhydryl groups register a regular and sharp decrease with drying. Protein solubility in grass carp fillets decreases significantly after hot air drying and microwave drying. However, protein solubility of hot air dried fillets is lower than that of microwave dried fillets, mainly due to a higher level of protein denaturation, stable covalent linkage formation and/or a Maillard reaction taking place during hot air drying (Wu and Mao, 2008).

Fish muscle lipids Lipid is an important component, determining both the functionality and the sensory properties of dried fish products. The changes in lipid during drying, such as lipolysis and lipid oxidation in the fish muscle, are associated with quality deterioration. Moreover, lipolysis induced by enzymatic actions (e.g. lipases and phospholipases) results in increased free fatty acids that undergo further oxidation to produce low molecular weight compounds responsible for the rancid off-flavour and taste as well as colour and texture deterioration (Toyomizu *et al.*, 1981; Selmi *et al.*, 2010). Lipid oxidation products are thought to have a hazardous effect on human health. Water activity and oxygen content play a big role in the progress of lipid oxidation during the drying process. Low water activity accelerates lipid oxidation (Sun *et al.*, 2002) and resistance to oxidation is greatest at a water activity around 0.3 (Rahman, 2008). The effect of oxygen on lipid oxidation is associated with the product porosity. Hence, freeze-dried products are more prone to oxidation compared to other dried products during storage, mainly due to their higher porosity (Rahman *et al.*, 2002). The peroxide value (PV) and thiobarbituric acid reactive substances (TBARS) are thought to be the major lipid oxidation indices. These values are significantly different depending on the drying methods and drying parameters. PV and TBARS of *Atherina lagunae* significantly increase after hot air drying and solar drying. Polyunsaturated fatty acids (PUFAs) of polar lipid decrease remarkably during both drying processes with lower levels in solar drying (Selmi *et al.*, 2010). The same results have been found for herring fillets during drying by Shah *et al.* (2009). They reported that PV, carbonyl value and acid value of the lipids from herring fillets increase significantly during the drying period.

7.4.2 Changes in physical properties of fish muscle

Texture of fish muscle Texture is one of the most important sensorial attributes in dried fish products (Telis and Telis-Romero, 2005; Vega-Galvez *et al.*, 2011). Texture is an external reflection of the micro- and macro-structural characteristics of a food product (Aguilera and Stanley, 1999). Therefore, the loss of water and the changes in proteins and lipids of the fish muscle during the drying process result in textural changes. The changes in texture of fish muscle depend on the raw materials and the drying methods as well as drying parameters such as temperature, air velocity and relative humidity (RH) of the air in the drying chamber. The genetic origin of animals, fat composition, pH, protelytic potential, sex and sexual condition of the raw materials have been related to the texture. Vega-Galvez *et al.* (2011) stated that the firmness of osmo-treated jumbo squid increases with increasing air-drying temperature probably due to changes in the food protein matrix. According to Potter and Hotchkiss (1998), the final dried products have more volume and a crust on the surface when a high drying rate is applied. However, a low drying rate gives a uniform and denser product (Brennan, 1994). The texture of dried shrimp obtained by a two-stage drying process (superheated steam dryering followed by heat pump dryering) is less tough and softer than those of samples obtained by single-stage superheated steam dryering (Namsanguan *et al.*, 2004). Tapaneyasin *et al.* (2005) found that dried shrimp obtained using a constant inlet air temperature of 100 °C has lower maximum shear force compared to dried shrimp dried at 70 and 120 °C.

Colour of fish muscle The physical changes and organoleptic qualities are the important characteristics concerned directly with marketing dried fish products. Appearance is one of the most important factors that attracts consumers toward the product. The colour of fish changes markedly when it is dried (Louka *et al.*, 2004). Discolouration of fish products is mainly due to browning reactions that can be classified as enzymatic and non-enzymatic. Maillard and caramelization reactions are known to be the main candidates responsible for yellow-brown discolouration of dried fish products (Louka *et al.*, 2004; Rahman, 2006). In addition, lipid oxidation and degradation of L-methyl-histidine and myoglobin make an important contribution to discolouration of the final products (Pokorny, 1981; Louka *et al.*, 2004; Guizani *et al.*, 2008). The rate of discolouration depends on the drying temperature, drying method, water content and composition of the fish muscle. A higher browning rate is obtained with an intermediate water content compared to very low and very high water contents (Rahman, 2006). Vega-Galvez *et al.* (2011) found that discolouration of osmo-treated jumbo squid is more noticeable at high drying temperatures, mainly due to the combined effects of non-enzymatic browning and protein denaturation/aggregation. Freeze-dried tuna has the highest whiteness and the lowest darkening compared to vacuum-dried and air-dried samples (Rahman *et al.*, 2002). The rate of colour change of sardine sheets during vacuum pulse osmotic dehydration depends on the temperature. Redness and yellowness decrease throughout the dehydration process, while lightness and total colour difference increase (Corzo *et al.*, 2006). Drying fish with a microwave-assisted hot-air dehydration treatment results in a brighter colour and a shift toward red and yellow (Taib and Ng, 2011). Nguyen *et al.* (2011) stated that the lightness of salted cod increases during drying by a heat pump dryer, probably due to water removal and the formation of salt crystals in the fish muscle. Moreover, the increase in the red and yellow colour intensity of salted cod may be due to the oxidation of pigment in the fish muscle by oxygen.

Volume of fish muscle Foods that are dried undergo volume changes either by shrinkage or by pore formation (Namsanguan *et al.*, 2004). The shrinkage of food materials reduces the quality of the dried products and makes a negative impression on the consumer. The shrinkage of the fish muscle is mainly due to the removal of water or by expansion, which is due to the glass transition. The development of pores and shrinkage depend upon the variation of moisture transport mechanism and the external pressure as well as drying method and drying temperature. There are linear relationships between the shrinkage factor and the moisture content and between the shrinkage in volume and the volume of water loss during osmotically dehydration of sardine sheets at temperatures of 30–38 °C (Corzo and Bracho, 2004). There is negligible shrinkage (more pores) in the food muscle if it is dried below the glass transition temperature. Inversely, the food muscle is denser and more shrivelled if dried at a higher glass transition temperature (Rahman, 2001). Therefore, the vacuum dried and freeze-dried food products have higher porosity compared to hot air dried food products (Krokida *et al.*, 1998; Baysal *et al.*, 2003). Namsanguan *et al.* (2004) found that shrinkage of dried shrimp is less when it is processed with a superheated steam dryer followed by heat pump dryer compared with when only a superheated steam dryer is used. Moreover, a lower percentage of shrinkage is observed in shrimp dried at 100 °C compared to that of shrimp dried at 70 and 120 °C (Tapaneyasin *et al.*, 2005). The shrinkage ratio of dried tilapia fillets increases with increasing microwave power and air temperature (Duan *et al.*, 2011). This may be due to the rapid water removal caused by the instantaneous heating of microwaves, resulting in greater changes in the shrinkage ratio (Duan *et al.*, 2005).

7.4.3 Effect of drying on the nutritional properties of fish

Drying is believed to have negative effects on the nutritional properties of food products. The changes in proteins and lipids of the fish muscle consequently decrease the nutritional value of the fish. Protein denaturation/aggregation that takes place during the drying process leads to changes in amino acid composition, protein solubility and protein digestibility. Similarly, lipid oxidation can cause the degradation of fatty acids, particularly PUFAs (e.g. eicosopentaenoic acid (EPA) and docoshexaenoic acid (DHA)). Furthermore, lipid oxidation products could also have a negative effect on human health. The interactions between lipid oxidation products and protein constituents (i.e. amino acids, peptides) cause significant changes in proteins, affecting their solubility and nutritional value (Surono, 1991). Loss of vitamins in the fish muscle also occurrs during drying. Different nutritional components of tilapia muscle undergo different changes at elevated temperatures during conventional smoke kiln drying and electric oven drying. The electric oven dried tilapia has a higher protein quality and lower lipid content compared to smoke kiln dried tilapia (Chukwu, 2009).

7.5 Packing and storage of dried fish products

Dried fish products have a low water content and a low water activity, so the main problems during handling and storage are water adsorption by the product surface in contact with the surroundings, and insect infestation (Ayuba and Omeji, 2006). The rate of water adsorption depends on the RH and the temperature of the surroundings as well as on the water content and water activity of the products. Water adsorption leads to increased water activity and water content of the dried products, promoting enzyme activity and the growth of microorganisms. To prevent water adsorption, dried fish products should be packaged using materials that have a low gas and water vapour permeability. Packaging materials should also be strong enough to protect the packaged products from damage and insect infestation (Davies *et al.*, 2009). In recent years, pouches made of polyester laminated with polythene have been commonly used for consumer packs (Antony, 1993). The temperature and RH of the storage room are also important to prevent water adsorption and to inhibit any deterioration in quality of the product. It is important to maintain the equilibrium between the water activity (a_w) in the dried fish products and the RH in air where the dried fish is stored. Changes in the quality of dried fish products during storage depend on storage conditions (i.e. temperature and RH), packaging methods as well as packaging materials. High density polyethylene (HDPE) woven gusseted bags laminated with 100 gauges low density polyethylene (LDPE) are suitable for packaging dried fish products. HDPE is impervious to microbial and insect infestation (Gopal and Shankar, 2011). The shelf life of dried anchovies packed in LDPE stored at 27–30 °C and RH of 80–90% is limited to 12 weeks. However, the shelf life is extended up to 32 weeks in the plain polyester polythene laminate at the same storage conditions (Gopal *et al.*, 1994). Vacuum packaging and modified atmosphere packaging are popular and effective packaging methods that are accepted by consumers. Note that dried fish needs to be cooled down before packaging to inhibit water vapour condensing on the surface of the food and the packaging material (Sveinsdottir, 1998).

References

Aguilera, J.M. and Stanley, D.W. (1999). *Microstructural principles of food processing and engineering*, 2nd edn, Aspen Publishers, Gaithersburg.

Aguilera, L.M., Chiralt, A. and Fito, P. (2003). Food dehydration and structure. *Trends in Food Science and Technology*, 14, 432–437.

Akpinar, E.K., Bicer, Y. and Midilli, A. (2003). Modelling and experimental study on drying of apple slices in a convective cyclone dryer. *Journal of Food Engineering*, 26, 515–521.

Alakali, J.S., Ariahu, C.C. and Nkpa, N.N. (2006). Kinetics of osmotic dehydration of mango. *Journal of Food Processing and Preservation*, 30, 597–607.

Antony, K.P. (1993). Packaging of dried fish and dried fish products for export and internal markets, in *Fish packaging technology – Materials and methods* (ed. K. Gopakumar), Concept Publishing Company, India, pp. 201–207.

Arvanitoyannis, I.S., Veikou, A. and Panagiotaki, P. (2012). Osmotic dehydration: theory, methodologies, and applications in fish, seafood, and meat products, in *Progress in Food Preservation* (eds R. Bhat, A.K. Alias and G. Paliyath), Wiley-Blackwell, John Wiley & Sons, Ltd., Chichester, pp. 161–189.

Ayuba, V.O. and Omeji, N.O. (2006). Effect of insect infestation on the shelf life of smoked dried fish. *Proceedings of the 21st Annual Conference of the Fisheries Society of Nigeria (FISON)*, 357–359.

Babic, J., Cantalejo, M.J. and Arroqui, C. (2009). The effect of freeze-drying process parameters on broiler chicken breast meat. *LWT-Food Science and Technology*, 42, 1325–1334.

Bala, B.K. (1998). *Solar Drying System: Simulation and Optimization*. Agrotech Academy, India.

Bala, B.K. and Debnath, N. (2012). Solar drying technology: potentials and developments. *Journal of Fundamentals of Renewable Energy and Applications*, 2, 1–5.

Barat, J.M., Rodríguez-Barona, S., Andrés, A. and Fito, P. (2002). Influence of increasing brine concentration in the cod-salting process. *Journal of Food Science*, 67(5), 1922–1925.

Baysal, T., Icier, F., Ersus, S. and Yıldız, H. (2003). Effects of microwave and infrared drying on the quality of carrot and garlic. *European Food Research and Technology*, 218, 68–73.

Belessiotis, V. and Delyannis, E. (2011). Solar drying. *Solar Energy*, 85, 1665–1691.

Braun, J.E., Bansal, P.K. and Groll, E.A. (2002). Energy efficiency analysis of air cycle heat pump dryers. *International Journal of Refrigeration*, 25, 954–965.

Brennan, J.G. (1994). *Food Dehydration: A Dictionary and Guide*. Butterworth-Heinemann Ltd, Oxford.

Chiralt, A., Fito, P., Andres, A. *et al.* (1999). Vacuum impregnation: a tool in minimally processing of foods, in *Processing of Foods: Quality Optimization and Process Assessment* (eds F.A.R. Oliveira and J.C. Oliveira), CRC Press, Boca Raton, FL, pp. 314–356.

Chou, S.K. and Chua, K.J. (2001). New hybrid drying technologies for heat sensitive foodstuffs. *Trends in Food Science and Technology*, 10, 359–369.

Chua, K.J., Mujumdar, A.S., Hawlader, M.N.A., Chou, S.K. and Ho, J.C. (2001). Batch drying of banana pieces-effect of stepwise change in drying air temperature on drying kinetics and product colour. *Food Research International*, 34, 721–731.

Chukwu, O. (2009). Influences of drying methods on nutritional properties of tilapia fish (*Oreochromis nilotieus*). *World Journal of Agricultural Sciences*, 5, 256–258.

Collignan, A., Bohion, P., Deumier, F. and Poligne, I. (2001). Osmotic treatment of fish and meat products. *Journal of Food Engineering*, 49, 153–162.

Corzo, O. and Bracho, N. (2004). Shrinkage of osmotically dehydrated sardine sheets at changes moisture content. *Journal of Food Engineering*, 65, 333–339.

Corzo, O., Bracho, N. and Marval, J. (2006). Effects of brine concentration and temperature on colour of vacuum pulse osmotically dehydrated sardine sheets. *LWT-Food Science and Technology*, 39, 665–670.

Crank, J. (1975). *The Mathematics of Diffusion*, 2nd edn, Clarendon Press, Oxford.

Davies, R.M., Davies, O.A. and Abowei, J.F.N. (2009). The status of fish storage technologies in Niger delta Nigeria. *American Journal of Scientific Research*, 1, 55–63.

Dincer, I. (1998). Moisture transfer analysis during of slab woods. *Heat Mass Transfer*, 34, 17–20.

Dirk, B. and Markus, S. (2004). Heat pump drying (HPD) – how refrigeration technology provides an alternative for common drying challenges. *Ki-luft-und Kaltetechnik*, 40, 140–144.

Doe, P., Sikorski, Z., Haard, N., Olley, J. and Pan, B.S. (1998). Basic principles, in *Fish drying and Smoking: Production and Quality* (ed. P.E. Doe), Technomic Publishing Company, Inc., Pennsylvania, USA, pp. 13–45.

Duan, Z.H., Jiang, L.N., Wang, J.L., Yu, X.Y. and Wang, T. (2011). Drying and quality characteristics of tilapia fish fillets dried with hot air-microwave heating. *Food and Bioproducts Processing*, 89, 472–476.

Duan, Z.H., Zhang, M. and Hu, Q.G. (2005). Characteristics of microwave drying of bighead carp. *Drying Technology*, 23, 637–643.

Eichner, K. and Wolf, W. (1983). Maillard reaction products as indicator compounds for optimizing drying and storage condition, in *The Maillard Reaction in Foods and Nutrition* (eds G.R. Waller and M.S. Feather), ACS Publication, Washington, DC, pp. 317–333.

Eikevik, T.M., Strømmen, I., Alves-Filho, O. and Hemmingsen, A.K.T. (2005). Effect of operating conditions on atmospheric freeze dried cod fish. Paper XIII-3. Proceedings of IADC 2005 – 3rd Inter-American Drying Conference, 21–23 August 2005.

Erikson, U., Veliyulin, E., Singstad, T.E. and Aursand, M. (2004). Salting and desalting of fresh and frozen-thawed cod (*Gadus morhua*) fillets: a comparative study using ^{23}Na NMR, ^{23}Na MRI, low-field ^{1}H NMR, and physicochemical analytical methods. *Journal of Food Science*, 69, 107–114.

Filho, O.A. and Strømmen, I. (1996). The application of heat pump in drying of boimaterials. *Drying Technology*, 14, 2061–2090.

Fito, P. (1994). Modeling of vacuum osmotic dehydration of food. *Journal of Food Engineering*, 22, 313–328.

Fito, P. and Pastor, R. (1994). On some non-diffusional mechanism occurring during vacuum osmotic dehydration. *Journal of Food Engineering*, 21, 513–519.

Gavrila, C., Ghiaus, A.G. and Gruia, I. (2008). Heat and mass transfer in convective drying processes. Proceedings of the 2nd COMSOL Conference, 4–6 November, Hannover.

Gopal, T.K.S., Nair, P.G.V., Kandoran, M.K., Prabhu, P.V. and Gopakumar, K. (1994). Shelf life of dried anchoviella in flexible packaging materials. *Food Control*, 9, 205–209.

Gopal, T.K.S. and Shankar, C.N.R. (2011). Quality and safety of packaging materials for aquatic products, in *Handbook of Seafood Quality, Safety and Health Applications* (eds C. Alasalvar, F. Shahidi, K. Miyashita and U. Wanasundara), Blackwell Publishing Ltd., Oxford, UK, pp. 139–154.

Guizani, N., Obaid Al-Shoukri, A., Mothershaw, A. and Rahman, M.S. (2008). Effects of salting and drying on shark (*Carcharhinus sorrah*) meat quality characteristics. *Drying Technology*, 26, 705–713.

Haghi, A.K. and Amanifard, N. (2008). Analysis of heat and mass transfer during microwave drying of food products. *Brazilian Journal of Chemical Engineering*, 25, 491–501.

Hammami, C., René, F. and Marin, M. (1999). Process-quality optimization of the vacuum freeze drying of apple slices by the response surface method. *International Journal of Food Science and Technology*, 34, 145–160.

Hawlader, M.N.A., Perera, C.O. and Titan, M. (2006). Properties of modified atmosphere heat pump dried foods. *Journal of Food Engineering*, 74, 392–401.

Immaculate, J., Sinduja, P. and Jamila, P. (2012). Biochemical and microbial qualities of *Sardinella fimbriata* sun dried in different methods. *International Food Research Journal*, 19, 1699–1703.

Jain, D. and Pathare, P.B. (2007). Study the drying kinetics of open sun drying of fish. *Journal of Food Engineering*, 78, 1315–1319.

Janjai, S. and Bala, B.K. (2012). Solar drying technology. *Food Engineering Reviews*, 4, 16–54.

Karathanos, V.T., Anglea, S.A. and Karel, M. (1996). Structure collapse of plant materials during freeze drying. *Journal of Thermal Analysis*, 46, 1541–1551.

Khan, M.R. (2012). Osmotic dehydration technique for fruits preservation – a review. *Pakistan Journal of Food Sciences*, 22, 71–85.

Kituu, G.M., Shitanda, D., Kanali, C.L. *et al.* (2010). Thin layer drying model for simulating the drying of tilapia fish (*Oreochromis niloticus*) in a solar tunnel dryer. *Journal of Food Engineering*, 98, 325–331.

Krokida, M.K., Maroulis, Z.B. and Marinos-Kouris, D. (1998). Effect of drying method on physical properties of dehydrated products. Proceedings of the 11th International Drying Symposium (IDS'98), 19–22 August, Halkidiki, Greece, pp. 809–816.

Labuza, T. and Saltmarch, M. (1981). The non-enzymatic browning reaction as affected by water in foods, in *Water Activity: Influence on Food Quality* (eds L.B. Rockland and G.F. Stewart), Academic Press, New York, USA, pp. 605–647.

Louka, N., Juhel, F., Fazilleau, V. and Loonis, P. (2004). A novel colorimetry analysis used to compare different drying fish processes. *Food Control*, 14, 327–334.

Mccabe, W.L., Smith, J.C. and Harriott, P. (1993). *Unit Operations of Chemical Engineering*, McGraw-Hill, Inc., New York, USA.

Mujumdar, A.S. (2006). Principles, classification, and selection of dryers, in *Handbook of Industrial Drying* (ed. A.S. Mujumdar), CRC Press, Boca Raton, FL, pp. 4–31.

Nam, J.H. and Song, C.S. (2007). Numerical simulation of conjugate heat and mass transfer during multi-dimensional freeze drying of slab-shaped food products. *International Journal of Heat and Mass Transfer*, 50, 4891–4900.

Namsanguan, Y., Tia, W., Devahastin, S. and Soponronnarit, S. (2004). Drying kinetics and quality of shrimp undergoing different two-stage drying processes. *Drying Technology: An International Journal*, 22, 759–778.

Nguyen, M.V., Jonsson, A., Gudjonsdottir, M. and Arason, S. (2011). Drying kinetics of salted cod in a heat pump dryer as influenced by different salting procedures. *Asian Journal of Food and Agro-Industry*, 4, 22–30.

Okos, M.R., Narsimhan, G., Singh, R.K. and Weitnauer, A.C. (1992). Food dehydration, in *Handbook of Food Engineering* (eds D.R. Heldman and D.B. Lund), Marcel Dekker, New York, USA.

Opstvedt, J., Miller, R., Hardy, R.W. and Spinelli, J. (1984). Heat induced changes in sulfhydryl groups and disulphide bonds in fish protein and their effect on protein and amino acid digestibility in rainbow trout (*Salmo gairdneri*). *Journal of Agricultural and Food Chemistry*, 32, 929–935.

Perera, C.O. and Rahman, M.S. (1997). Heat pump dehumidifier drying of food. *Trends in Food Science and Technology*, 8, 75–79.

Petrotos, K.B. and Lazarides, H.N. (2001). Osmotic concentration of liquid foods. *Journal of Food Engineering*, 49, 201–206.

Pokorny, J. (1981). Browning from lipid–protein interactions. *Progress in Food Nutrition Science*, 5, 421–428.

Potter, N.N. and Hotchkiss, J.H. (1998). *Food Science*, 5th edn, Chapter 10, Aspen Publishers Inc., Maryland.

Raghunath, M., Sankar, T.V., Ammu, K. and Devadasan, K. (1995). Biochemical and nutritional changes in fish proteins during drying. *Journal of the Science of Food and Agriculture*, 67, 197–204.

Rahman, M.S. (2001). Toward prediction of porosity in foods during drying: a brief review. *Drying Technology: An International Journal*, 19, 1–13.

Rahman, M.S. (2006). Drying of fish and seafood, in *Handbook of Industrial Drying* (ed. A.S. Mujumdar), CRC Press, Boca Raton, FL, pp. 547–559.

Rahman, M.S. (2008). Post-drying aspects for meat and horticultural products, in *Drying Technology in Food Processing* (eds X.D. Chen and A.S. Mujumdar). Blackwell Publishing Ltd, West Sussex, UK, pp. 252–265.

Rahman, M.S., Al-Amri, O.S. and Al-Bulushi, I.M. (2002). Pores and physiochemical characteristics of dried tuna produced by different methods of drying. *Journal of Food Engineering*, 52, 301–313.

Rahman, M.S. and Labuza, T.P. (2007). Water activity and food preservation, in *Handbook of Food Preservation* (ed. M.S. Rahman), CRC Press, Boca Raton, FL, pp. 447–476.

Raoult-Wack, A.L. (1994). Recent advances in the osmotic dehydration of food. *Trends in Food Science and Technology*, 5, 255–260.

Ratti, C. (2001). Hot air and freeze-drying of high-value foods: a review. *Journal of Food Engineering*, 49, 311–319.

Ratti, C. and Mujumdar, A. (1997). Solar drying of food: modelling and numerical simulation. *Solar Energy*, 60, 151–157.

Sablani, S.S., Rahman, M.S., Al-Kuseibi, M.K. *et al.* (2007). Influence of shelf temperature on pore formation in garlic during freeze-drying. *Journal of Food Engineering*, 80, 68–79.

Sacilik, K., Keskin, R. and Elicin, A.K. (2006). Mathematical modelling of solar tunnel drying of thin layer organic tomato. *Journal of Food Engineering*, 43, 231–238.

Sahin A.Z., Dincer I., Yilbas, B.S. and Hussain M.M. (2002). Determination of drying times for regular multi-dimensional objects, *International Journal of Heat and Mass Transfer*, 45, 1757–1766.

Sarkardei, A. and Howell, N.K. (2007). The effects of freeze-drying and storage on the FT-Raman spectra of Atlantic mackerel (*Scomber scombrus*) and horse mackerel (*Trachurus trachurus*). *Food Chemistry*, 103, 62–70.

Scott, W.J. (1957). Water relation of food spoilage microorganisms. *Advances in Food Research*, 72, 83–127.

Selmi, S., Bouriga, N., Cherif, M., Toujani, M. and Trabelsi, M. (2010). Effects of drying process on biochemical and microbiological quality of silverside (fish) *Atherina lagunae*. *International Journal of Food Science and Technology*, 45, 1161–1168.

Sen, D.P. (2005). *Advances in Fish Processing Technology*, Chapter 7. Allied Publishers Pvt. Ltd, New Delhi, India.

Senadeera, W., Bhandari, B.R., Young, G. and Wijesinghe, B. (2005). Modelling dimension shrinkage of shaped foods in fluidised bed drying. *Journal of Food Processing and Preservation*, 29, 109–119.

Shah, A.K.M.A., Tokunaga, C., Kurihara, H. and Takahashi, K. (2009). Changes in lipids and their contribution to the taste of *migaki-nishin* (dried herring fillets) during drying. *Food Chemistry*, 115, 1011–1018.

Sokhansanj, S. and Jayas, D.S. (1987). Drying of foodstuffs, in *Handbook of Industrial Drying* (ed. A.S. Mujumdar), Marcel Dekker, New York, USA, pp. 517–554.

Soponronnarit, S. (1995). Solar drying in Thailand. *Energy for Sustainable Development*, 2, 19–25.

Strømmen, I., Bredesen, A.M., Eikevik, T. *et al.* (2000). Refrigeration, air conditioning and heat pump systems for the 21st century. *Bulletin of the International Institute of Refrigeration*, 2, 2–18.

Sun, Q., Senecal, A., Chinachoti, P. and Faustman, C. (2002). Effect of water activity on lipid oxidation and protein solubility in freeze-dried during storage. *Journal of Food Science*, 67, 2512–2516.

Surono (1991). Nutritional quality changes in salted-dried mackerel (*Scomber scombrus*) and cod (*Gadus morhua*) during processing and storage. Master thesis. Department of Chemical Engineering, Loughborough University, United Kingdom.

Sveinsdottir, K. (1998). The process of fish smoking and quality evaluation. Master thesis. University of Denmark.

Szulmayer, W. (1971). From sun drying to solar dehydration I. Methods and equipment. *Food Technology in Australia*, 23, 440–443.

Taib, M.R. and Ng, P.S. (2011). Microwave assisted hot air convective dehydration of fish slice: Drying characteristics, energy aspects and colour assessment. *International Journal on Advanced Science, Engineering and Information Technology*, 1, 42–45.

Tapaneyasin, R., Devahastin, S. and Tansakul, A. (2005). Drying methods and quality of shrimp dried in a jet-spouted bed dryer. *Journal of Food Process Engineering*, 28, 35–52.

Telis, V. and Telis-Romero, J. (2005). Solids rheology for dehydrated food and biological materials. *Drying Technology*, 23, 759–780.

Toyomizu, M., Hanaoka, K. and Yamaguchi, K. (1981). Effect of release of free fatty acids by enzymatic hydrolysis of phospholipids on lipid oxidation during storage of fish muscle at 5 °C. *Bulletin of the Japanese Society of Scientific Fisheries*, 47, 605–610.

Uribe, E., Miranda, M., Vega-Gálvez, A. *et al.* (2011). Mass transfer modelling during osmotic dehydration of jumbo squid (*Dosidicus gigas*): influence of temperature on diffusion coefficients and kinetic parameters. *Food and Bioprocess Technology*, 4, 320–326.

Vega-Galvez, A., Miranda, M., Claveria, R. *et al.* (2011). Effect of air temperature on drying kinetics and quality characteristics of osmo-treated jumbo squid (*Dosidicus gigas*). *LWT-Food Science and Technology*, 44, 16–23.

VijayaVenkataRaman, S., Iniyan, S. and Goic, R. (2012). A review of solar drying technologies. *Renewable and Sustainable Energy Reviews*, 16, 2652–2670.

Wang, Y., Zhang, M. and Mujumdar, A.S. (2011). Trends in processing technologies for dried aquatic products. *Drying Technology*, 29, 382–394.

Wu, T. and Mao, L. (2008). Influences of hot air drying and microwave drying on nutritional and odorous properties of grass carp (*Ctenopharyngodon idellus*) fillets. *Food Chemistry*, 110, 647–653.

Zaman, M.A. and Bala, B.K. (1989). Thin layer solar drying of rough rice. *Solar Energy*, 42, 167–171.

8

Fish Fermentation

Somboon Tanasupawat[1] and Wonnop Visessanguan[2]

[1]*Department of Biochemistry and Microbiology, Faculty of Pharmaceutical Sciences, Chulalongkorn University, Bangkok, Thailand*
[2]*Food Biotechnology Research Unit, National Center for Genetic Engineering and Biotechnology (BIOTEC), Klong Luang, Pathum Thani, Thailand*

8.1 Definition of the term fermentation in food technology

Fermentation is a natural process that unavoidably affects the human food supply worldwide. The term 'fermentation' is derived from the Latin verb '*fervere*', to boil, which describes the appearance of the action of yeast on extracts of fruit or malted grain during the production of alcoholic beverages. To a microbiologist the word means any process for the production of a product by the mass culture of microorganisms. To a biochemist, however, the word means an energy-generating process in which organic compounds act as both electron donors and acceptors, that is, an anaerobic process where energy is produced without the participation of oxygen or other inorganic electron acceptors (Stanbury *et al*., 1995).

Fermented foods are those whose production involves the action of microorganisms or enzymes which cause desirable biochemical changes and significant modification to the food (Campbell-Platt, 1987). Food fermentation represents one of the oldest known uses of biotechnology. This traditional biotechnology has evolved from 'natural' processes in which nutrient availability and environmental conditions selected particular microorganisms, through the use of starter cultures, strain improvement and most recently, gene technology (Campbell-Platt, 1994).

Fermentation technology plays an important role in food technology. Fermented foods have worldwide acceptance and arise in the human relationship to the microbial environment. Fermented foods such as yoghurt, beer, wine and bread are commonly used as foods. The microbial ecology of fermented foods, such as beer, cheese, bread, yogurt and other foods constitute microbial ecosystems that live in human households across the globe. The production of fermented food does not require knowledge of the biologically mediated nature of fermentation, because the biota that carry out fermentation are present worldwide. Fermentation organisms are ambient in the environment, whether humans

Seafood Processing: Technology, Quality and Safety, First Edition. Edited by Ioannis S. Boziaris.

make use of them or not. The relationships embodied in preserved foods can be treated as microbial ecosystems, sitting on counter tops and stored in food cellars around the world. The main components of the fermentation ecosystem include: microbes (yeasts, moulds and bacteria), organic materials to be fermented, a solution in which the fermentation takes place, a vessel with a controlled gate and various tools which may be used to develop and monitor the fermentation. Fermented foods are preserved by microbes that live in food storage vessels (Scott and Sullivan, 2008). More than anything else, humans have known about the use of microbes for preparation of food products for thousands of years and all over the world a wide range of fermented foods and beverages have contributed significantly to the diets of many people.

Many fermented foods are a less likely vehicle for food infection or intoxication than fresh foods due to the competitive activity and the metabolites of the starter microorganisms. However, risk factors, such as the use of previously contaminated raw materials, lack of pasteurization, the use of poorly controlled natural fermentations, or of sub-optimum fermentation starter cultures, inadequate storage or maturation conditions enabling survival of pathogens, or growth and toxin production and consumption without prior heating, are of importance (Nout, 1994).

8.2 Fermented foods worldwide

Fermentation is one of the oldest and most widely used food preservation methods in households and small-scale food industries as well as in large enterprises. Indigenous food fermentations are one of the oldest forms of biotechnology, although the potential of the associated microorganisms has probably not yet been fully realised. Many food fermentative microorganisms are highly domesticated, although few studies have attempted to assess the extent of the genetic resource which is available, or which strains should be preserved for the future. Many bacterial and fungal fermented foods are based on the utilization of solid substrates. Fundamental studies on these model systems could provide the necessary information to enable a more efficient exploitation of solid substrates in modern biotechnology. Opportunities exist for the use of traditional fermentative microorganisms in biotechnology, either directly or following genetic manipulation using conventional or molecular techniques (Cook, 1994).

Fermented foods are regarded as our staple diet in which the main substrates used for commercial production are milk, meat, soybeans, fish and plant products. The highest production of fermented food occurs in Europe, North America and Africa. In South America, major quantities of beverages and dairy products are fermented. In the Middle East, fermented dairy products are important and in the Indian subcontinent cereals and legumes are fermented in large quantities. In east and south-east Asia, fish and legumes are the most important fermented foods produced, both contributing major protein sources. In these regions also, cereal products may be co-fermented with legumes, as in the use of rice or barley with soybeans in miso production and wheat used with soybeans in soy sauce production. While many fermented foods are consumed where they are made, increasing quantities are found in world trade, ranging from wines, beers and spirits through cheeses to soy sauce and miso (Campbell-Platt, 1994).

In West Africa, indigenous fermented foods and beverages such as fermented starchy foods, *gari*; fermented cereals, *ogi*; alcoholic beverages, *pito*, *burukutu* and *obiolor*; fermented legumes and oil seeds, *dawadawa*, *iru*, *ogiri* and *okpiye*; fermented animals proteins, *nono* and yoghurt are produced in Nigeria (Achi, 2005). In North Africa, a diversity of lactic acid bacteria (LAB) were isolated and have played a role in the process of making traditional Moroccan soft white cheese in different regions (Ouadghiri *et al.*, 2005). Fermented milk products such as *rob* (fermented cow's milk), *zabadi* (yoghurt), *gariss* (fermented camel's milk) *gibna bayda* (white cheese), *gibna mudaffra* (white pickles cheese) and *mish* (fermented milk product with spices) are produced in Sudan (Mohammed Salih *et al.*, 2011).

In East Asia, most fermented foods are non-dairy products featuring various other food raw-materials such as cereals, soybeans, fruits and vegetables, as well as fish and other marine products (Rhee *et al.*, 2011). Traditional clear alcoholic beverages which are generally called *shaosingjiu* in China, *cheongju* in Korea and *sake* in Japan, contain around 15% alcohol and are designated as rice-wine, whereas turbid beverages, *takju* (or *maggolli*) in Korea and *tapuy* in the Philippines, contain less than 8% alcohol along with suspended insoluble solids and live yeasts, are referred to as rice-beer. Fermented vegetable, *kimchi*, is the Korean method of preserving the fresh and crispy texture of vegetables with ingredients such as garlic, red pepper, green onion, ginger as well as added salt (Rhee *et al.*, 2011).

In South and Southeast Asia, indigenous carbohydrate-based fermented foods include rice wine, fermented palm sap and *tapai*. Such products are *bangla mad*, *cholai* and *tari* in Bangladesh; *ara* in Bhutan; *arrack*, *desi sharab*, *tari*, *tharra*, *toddy and fenny* in India; *palm wine* in Indonesia; *raksi*, *tadi* and *chayang tomb* in Nepal; *toddy* and *arrack* in Sri Lanka; *oou*, *sato and krachae* in Thailand; *tuak*, *tapai* Sabah, *arrack* and *samsu* in Malaysia; and *tapuy* and *ruou nep than* in Philippines (Blandino *et al.*, 2003; WHO 2004; Aidoo *et al.*, 2006; Chiang *et al.*, 2006; Law *et al.*, 2011).

8.3 Lactic acid fermentation

LAB are Gram-positive, non-spore-forming rods or cocci, catalase-negative when growing without a heme source, usually non-motile and occasionally nitrate reducers. They use glucose for fermentation and may be either homofermentative, producing more than 85% lactic acid from glucose, or heterofermentative, producing lactic acid, CO_2, ethanol and/or acetic acid in equimolar amounts (Dworkin *et al.*, 2006). Homofermenters use the Embden–Meyerhof–Parnas pathway to generate two moles of lactate per mole of glucose and derive approximately twice as much energy per mole of glucose compared with heterofermenters. The optical isomers of lactic acid produced are only L (+) and D (−) or mixtures of L (+) and D (−) in different proportions. Both L (+) and D (−) isomers are absorbed from the intestinal tract and L (+) is used as an energy source for respiration. D-Lactic acid is not metabolised by humans and is not recommended for infants and young children (Nakazawa and Hosono, 1992; Caplice and Fitzgerald, 1999). The strains in the genera *Lactobacillus, Enterococcus, Lactococcus, Tetragenococcus, Aerococcus, Vagococcus, Pisciglobus, Carnobacterium* produced L-lactic acid; *Leuconostoc* and some *Lactobacillus* strains produced D-lactic acid while some *Lactobacillus, Pediococcus* and *Weissella* strains produced DL-lactic acid (Axelsson, 2004; Tanasupawat *et al.*, 2011b).

Most fermented foods including the major products that are common in the western world, as well as many of those from other sources that are less well characterized, are dependent on LAB to mediate the fermentation process. The end-products of carbohydrate catabolism by these bacteria contribute not only to preservation but also to the flavour, aroma and texture. LAB are involved in many fermentation processes for milk, meats, cereals and vegetables because of their unique metabolic characteristics. The role of LAB in making cheese, yoghurt, bread, *gari, idli, mahewu, ogi,* soy sauce, *nan*, sausages, *kimchi*, sauerkraut and pickles is important for fermented foods from various countries. Although much fermentation is traditionally dependent on inoculation from a previous batch, starter cultures are available for many commercial processes – such as cheese manufacture – thus ensuring consistency of process and product quality (Caplice and Fitzgerald, 1999). Lactic acid fermentation, which is related to and often accompanied by salting, and alcoholic fermentation are good for preservation and also tend to destroy the disease-producing bacteria and parasites that have predominated in food since humans began to use their own excrement as fertilizer (Steinkraus, 1993). In addition, the specific antimicrobial mechanisms of LAB exploited in the biopreservation of foods include the production of hydrogen peroxide, carbon dioxide, diacetyl, broad-spectrum antimicrobials such as reuterin and the production of bacteriocins (De Vuyst and Vandamme, 1994; Pakdeeto *et al.*, 2003).

Lactic acid fermentation is widely used in tropical countries as a low-cost method of enhancing food quality, safety and shelf life. The fermentation occurs during the preparation of a wide variety of foods made from raw materials of plant and animal origin (Nout and Sarkar, 1999). LAB are the most widespread and desirable microorganisms in the low salt food fermentations. They convert most available carbohydrates to lactic acid, with small amounts of acetic acid, resulting in a lower pH (Campbell-Platt, 1987). Homofermentative strains of *Lactobacillus pentosus, Lb. plantarum, Lb. sakei, Lb. farciminis, Lb. acidipiscis*, other *Lactobacillus* sp., *Pediococcus pentosaceus, P. acidilactici, Tetragenococcus halophilus* and *T. muriaticus* are found in a variety of fermented foods and they produce lactic acid as major products (Tanasupawat and Komagata, 2001). Heterofermentative strains of *Lb. vaccinostercus, Lb. fermentum, Lb. brevis*, other *Lactobacillus* sp., *Weissella confusa, W. thailandensis* and *Leuconostoc* sp. are also found; however, they produce a small amount of lactic acid. The DL-lactic acid producing *Lb. pentosus* and *Lb. plantarum* strains are the predominant rod-shaped LAB in fermented foods. *P. pentosaceus* strains are the major coccal bacteria, although *T. halophilus* strains occur in products containing high concentrations of salt. Recently, L-lactic acid producing *Lb. farciminis, Lb. acidipiscis, T. halophilus* and *T. muriaticus* as well as D-lactic acid producer, *W. thailandensis*, were found in salt-fermented fish and soy sauce mash (Table 8.1) (Tanasupawat *et al.*, 1998, 2000, 2002).

8.4 Traditional salt/fish fermentation

The term 'fermented fish products' is used to describe the products of freshwater and marine finfish, shellfish and crustaceans that are processed by the combined action of fish enzymes and bacterial enzymes with salt to cause fermentation and thereby to prevent putrefaction (Saisithi, 1994; Ruddle and Ishige, 2010). The enzymes produced textural change and assisted in the production of flavour while the bacteria involved developed the aroma and flavour for fish fermentation (Beddows, 1998). Fish preservation occurs in areas where there is a seasonal abundance of local fish. The freshwater fish species naturally occurring in local hydrological systems have been fermented. The fermentation process first developed among sedentary farmers and only where salt was easily obtained. Fish fermentation developed in a focal region(s) of continental East Asia and Southeast Asia (Ruddle and Ishige, 2010). Such a wide range of fermented fish is produced in East Asia that a strict classification by product type must be limited to individual countries or linguistic groups. They are classified based on the nature of the final product and

Table 8.1 Isomer of lactic acid produced by LAB from fermented fish and soy sauce mash

Species	strain	Ratio of isomer (%)		Type of isomer	Fermented products
		D	L		
T. halophilus	S21-3	10.1	89.1	L	Soy sauce mash
T. halophilus	S25-1	15.8	84.2	L	Soy sauce mash
T. halophilus	IS20-3	0.3	99.7	L	Fish sauce
T. halophilus	KS11-1	0.4	99.6	L	Fish sauce
T. muriaticus	KS87-14	0.6	99.4	L	Fish sauce
Lb. acidipiscis	SR3-1	4.3	95.7	L	Soy sauce mash
Lb. farciminis	SR18-1	13.5	86.5	L	Soy sauce mash
Lb. pentosus	SR4-1	42.5	57.5	DL	Soy sauce mash
Lb. plantarum	SR62-1	45.6	54.4	DL	Soy sauce mash
Lb. acidipiscis	FS60-1	4.5	95.5	L	*Pla-ra*
Lb. acidipiscis	FS56-1	4.1	95.9	L	*Pla-ra*
Lb. acidipiscis	FS111	4.3	95.7	L	*Pla-chom*
Lb. farciminis	FS73-1	6.7	93.7	L	*Pla-ra*
Lb. farciminis	FS47-3	11.8	88.2	L	*Pla-chom*
Lb. pentosus	FS42-1	59.2	40.8	DL	*Kung-chom*
Lb. farciminis	FS67-1	13.5	86.5	L	*Kung-chom*
W. thailandensis	FS61-1	93.8	6.2	D	*Pla-ra*
W. thailandensis	FS45-1	93.4	6.6	D	*Pla-ra*

the preparation method used. The products combined fish and salt that preserved the shape of the original raw fish material such as *shiokara* in Japan, *prahok* in Cambodia, *yujiang* in China, *bakasam* in Indonesia, *jeot* in Korea, *pa daek* in Laos, *ngapi-gaung* in Myanmar, *bagoong* in Philippines, *pla-ra* in Thailand and *ca mam* in Vietnam. If no vegetable ingredients are added, the salt–fish mixture yields fish sauce, a liquid used as a pure condiment. If cooked vegetable ingredients (carbohydrates, normally just rice) are added to the fish and salt mixture, it becomes *narezushi* as in Japan (Ruddle and Ishige, 2010). This foodstuff occurs only in Southeast and Northeast Asia such as *pla-som* and *pla-ra* in Thailand, *phaak* in Cambodia, *som paa* in Loas, *nga ngapi* in Myanmar, *mam chau* in Vietnam, *burong isda* in Philippines, *pekasam* in Malaysia, *wadi, bakasem* and *ikan masim* in Indonesia and *shikhe* in Korea. Fermented shrimp products, such as shrimp paste, are produced without salt in some areas such as Bangladesh, Myanmar, Indonesia and the Philippines, while some products have a very low salt content or a high salt content, such as *ka-pi* in Thailand (Phithakpol *et al.*, 1995). Two techniques, fermentation and sun drying, are used to preserve epipelagic shrimp. Sun drying is the simplest, but it does not overcome the problem of the coarse texture of the carapace and hence the need for comminuting. However, salting this kind of paste enhances both the taste and shelf life (Ruddle and Ishige, 2010).

8.4.1 Classification of fermented fish

Fermented fish with a large amount of salt Salt has been used for centuries to increase the shelf life of foodstuffs and remains a frequently used component to create an unfavourable environment for food spoilage microorganisms and pathogens (Stringer and Pin, 2005).

In addition, the use of salt allows for a lower preservative content (e.g. sulphur dioxide, benzoic acid), a lower degree of acidity or a milder treatment (e.g. heating), which may have a positive impact on the nutritional or organoleptic quality of the product. Fishery products in south-east Asian countries, especially in Thailand, can be categorized according to the main processing techniques and ingredients (Phithakpol *et al.*, 1995). *Nam-pla* (fish sauce, fish and salt), *ka-pi* (shrimp paste, small shrimp and fish) and *bu-du* (fish and salt) are produced from fish with a large proportion of salt (19.4–40.1%, w/w) which the whole fish are hydrolysed. *Tai-pla* (fermented fish viscera) and *hoi-dong* (fermented sea mussel meat) are made with salt but at lower proportions (11.4–13.5%, w/w) than the products referred to earlier. Most products are made from marine fish in the coastal provinces, especially the east coast of Thailand (Phithakpol *et al.*, 1995).

Fermented fish with salt and carbohydrate There are many kinds of Thai fermented fish with salt (2.3–11.0%, w/w) and carbohydrate such as *pla-ra* (fish, salt and roasted rice powder), *pla-som* (whole fish, salt, cooked rice and garlic), *pla-chao* (fish, salt and *khao-mak), som-fak* (minced fish, salt, cooked rice and garlic) and *pla-chom* (fish, salt, garlic and roasted rice powder). This is used mainly for freshwater fish and also freshwater shrimp (Phithakpol *et al.*, 1995). However, the addition of carbohydrates is intended for different purposes (Saisithi, 1994). The rice is a source of carbohydrate for the microorganisms involved in fermentation, mainly LAB which give a characteristic flavour to the products. The roasted rice gives a brown colour and not only provides flavour but also absorbs excess moisture and prevents the product from sticking together (Saisithi, 1994). There are two types of fermented rice used, *khao-mak*, which is mould and yeast-fermented, and *ang-kak* rice, which is fermented using the mould *Monascus purpureus* that gives a red colour in *pla-paeng-daeng* (Phithakpol *et al.*, 1995).

8.4.2 World fermented fish products

Fermented fish in Europe In Europe, traditional pickled fish products such as *rakeorret* in Norway, *maatjes* in Netherlands and gravad fish in Scandinavia and Scotland are made with a low salt content, while *surströmming* in Sweden is produced with a medium salt content. Other products include a high salt content, salted herring from Scandinavia, Canada and Russia and a sugar-salted herring product from Scandinavia and Finland (Knochel, 1993; Köse, 2012). *Rakeorret* is made from gutted fish placed in containers and either covered with 15% (w/v) brine (optionally 5% sugar) or dry-salt with 5% (w/v) NaCl and which is then allowed to draw brine at 8–10 °C for 2–3 months. *Maatjes-herring* is produced from herring with 5–10% (w/w) salt and fermented at 5 °C for a week (Knochel, 1993). Gravad fish, also known as *gravadlax* and *kryddersild*, is made from salmon, trout or Greenland halibut including mackerel and herring (Knochel, 1993; Köse, 2012). The fillets are combined with sugar, pepper and layered together meat-side to meat-side under light pressure for 1–4 days at 5 °C. The final product contains about 2–9% (w/v) salt in the liquid phase. In vacuum-packed samples, *Lactobacillus* and *Brochotrix thermosphacta* strains dominate (Knochel, 1993). *Surströmming* is made from whole herring pre-salted with a strong brine for 1–2 days and salted in barrels with about 17% (w/v) brine (11–12% final salt in aqueous phase and fermented at 15–18 °C for 2–3 weeks). The products are canned without heat treatment for further distribution but should be kept at a low temperature. Sugar-salted herring is made from fatty herring (17–25% w/w fat), between 90 and 100 kg of fish mixed with 16 kg of NaCl and 6 kg sucrose and spices in barrels. After the fish have drawn brine, the barrels are topped with saturated brine, closed and fermented at 4–8 °C for 0.5–2.0

years. Halophilic Gram-negative bacteria, pedicococci, micrococci and yeast are found in the fermentation (Knochel, 1993). The acidified products, marinated herring in Scandinavia and Germany and *gaffelbitar* or *tidbits* in England, are also produced by placing the fillets in brine with 4–5% (v/v) acetic acid and 12–15% (w/w) NaCl for 2 weeks as the first step (Knochel, 1993). They are marinated in the retail pack in weaker brine with a mixture of sugar and spices for 6 months (Hall, 2002). LAB including *Pediococcus, Lactobacillus* and *Leuconostoc* spp. which are used in fermenting fish, are reported to form histamine (Leisner *et al.*, 1994; Hall, 2002). *Gravadlax* contained 23.1–529.3 ppm, *surströmming* contained 9.3–79.8 ppm and *kryddersild* contained 6.2–9.0 ppm of histamine, while other biogenic amines, tryptamine, phenylethylamine, putrescine, cadaverine, tyramine, spermidine and spermine ranged from 1.4 to 969.4 ppm (Köse *et al.*, 2012).

Fermented fish in Africa Fermented fish products in Africa are habitually whole or cut in pieces, such as *lanhouin* and *benin* and in Togo; *momone, koobi, kako* and *ewule* in Ghana; *guedj* in Gambia; *tambadiang, yet* and *guedj* in Senegal; *djege* and *jalan* in Mali; *fessiekh, kejeick, terkeen* and *mindeshi* in Sudan; *dagaa* in Uganda; *gyagawere, adjonfa* and *adjuevan* in Ivory Coast; and *salanga* in Chad (Anihouvi *et al.*, 2012). *Lanhouin, momone* and *guedj* are produced from fresh fish by scaling, gutting and removal of gills, washing with seawater or well water and then ripening overnight (10–15 h) in a container or buried in the ground. The fish is salted using 20–35% (w/w) salt initially, then salted for a second time using 15–25% salt and then fermented for 3–8 days. After fermentation, the fish product is lightly washed to remove the part of the salt still remaining on its body. At this time, some drops of fuel may be added to the water to repel the flies during drying. The product is sun dried for 2 or 4 days to reduce its water content (Anihouvi *et al.*, 2012). *Momone* and *guedj* are produced using the same procedure as *lanhouin* but have a different fermentation time and salt ratio (Anihouvi *et al.*, 2012). Strains of *Bacillus* spp. and *Staphylococcus* spp. are dominant in *lanhouin*; *Bacillus, Lactobacillus, Pediococcus, Staphylococcus Pseudomonas, Klebsiella, Debaryomyces, Hansenula* and *Aspergillus* strains are involved in *momone*; and *Proteus* spp., *Shewanella putrefaciens and Bacillus* spp. are found in *guedj* (Anihouvi *et al.*, 2012).

Fermented fish in Asian countries Fish is the most important source of animal protein for people in developing countries. Fresh fish is usually preserved by salting, smoking and sun-drying (Saisithi, 1994). Fermented fish and fish products are highly appreciated traditional diets in Asia, especially fish sauce, which is popular as a condiment in some areas and social classes of the region.

South Asian Fermented Fish North-east India is the centre of a diverse food culture comprising fermented and non-fermented ethnic foods and alcoholic beverages (Tamang *et al.*, 2012). Fermented and smoked fish products – *ngari, hentak, tungtap, gnuchi, suka ko maacha, sidra, sukuti, karati, bordia* and *lashim* – are produced and consumed by Meitei, Khasi, Lepcha, Lepcha, Gorkha and Assamese people in this region. *Ngari* is fermented fish (*Puntius sophore* Hamilton) prepared by rubbing the fish with salt, drying them in the sun for 3–4 days, washing and spreading them on to bamboo mats and pressing them tightly an earthen pot. A layer of mustard oil is applied to the inner wall of the pot before filling it with fish. The pot is sealed to make it airtight and then the fish are fermented at room temperature for 4–6 months. *Hentak* is a fermented fish paste prepared from a mixture of sun-dried fish (*Esomus danricus* Hamilton) powder and the petioles (*Alocasia macrorhiza*) of aroid plants. A mixture like thick paste is made into a ball, kept in an earthen pot, tightly sealed and fermented for 7–9 days. *Tungtap* is a fermented fish paste which is made from sun-dried fish (*Danio* spp.) mixed with salt. It is kept in an airtight earthen pot and fermented for 4–7 days.

Gnuchi is a smoked fish (*Schizothorax Acrossocheilus* spp., *Channa* sp.) made from degutted fish mixed with salt and turmeric powder. The larger fish are selected and spread upside down on 'sarhang' and kept above the earthen-oven in a kitchen. Small fish are hung along a bamboo strip above the earthen-oven for 10–14 days. *Gnuchi* is kept at room temperature for 2–3 months. *Suka ko maacha* is a smoked fish product made from hill river fishes (*dothay asala* (*Schizothorax richardsoni* Gray) and *chuchay asala* (*Schizothorax progastus* McClelland)) which are collected in a bamboo basket, degutted, washing and mixed with salt and turmeric powder. Degutted fishes are hooked on a bamboo string and hung above the earthen-oven in a kitchen for 7–10 days. It can be preserved for 4–6 months.

Sidra is a sun-dried fish (*Puntius sarana* Hamilton) made from fish that are collected, washed, dried in the sun for 4–7 days and stored at room temperature for 3–4 months. *Sukuti* is a sun-dried fish (*Harpodon nehereus* Hamilton) made from fish that are collected, washed, rubbed with salt and dried in the sun for 4–7 days and stored for 3–4 months. *Karati, bordia* and *lashim* are sun-dried and salted fish products made from fish that are washed, rubbed with salt and dried in the sun for 4–7 days. The sun-dried fish products are stored at room temperature for 3–4 months before consumption. *Karati, bordia* and *lashim* are prepared from *Gudusia chapra* Hamilton, *Pseudeutropius atherinoides* and *Cirrhinus reba* Hamilton, respectively. Microbial flora found in these fish products are *Lactococcus lactis* subsp. *cremoris, Lc. plantarum, Enterococcus faecium, E. faecalis, P. pentosaceus, Lb. fructosus, Lb. amylophilus, Lb. corynifomis, Lb. plantarum, Leuconostoc mesenteroides, Weissella confusa, Bacillus subtilis, B. pumilus* and *Micrococcus* strain and the yeasts, *Candida* spp. and *Saccharomycopsis* spp.

In Sri Lanka, *jaadi*, a high-salt, fermented fish product consists of partially hydrolysed fish flesh and organs immersed in liquid exudates from fish with the ripe pod of the goraka plant (*Garcinia gamboges*). This product was developed to use *Tilapia* (*Oreochromis niloticus*) (Lakshmi *et al.*, 2010). Fermented shrimp paste, *nappi* in Bangladesh, is fermented using a non-planktonic organism such as juveniles of *Penaeus indicus, Metapenaeus monoceros* and *Perapaeniopsis stylifera*. The shrimp is comminuted with salt and sun-dried for a day, then moistened and comminuted, crumbed and sun-dried (1 day) and sometimes blended to obtain uniform colour, and then packed (Ruddle, 1990). *Chepa shutki*, a semi-fermented fish product, is prepared from a small fish (Indian minor carp/minnow, *Puntius stigma* and *P. ticto*) (Nayeem *et al.*, 2010). *Lona ilish*, a salt-fermented fish, is made from Indian shad (*Tenualosa ilisha*, Hambuch, 1822). This product is widely consumed in North-east India and Bangladesh (Majumdar and Basu, 2010).

In Myanmar, fermented fish, *nga-pi*, is obtained by pounding or grinding fish or shrimp with salt and partially drying the mixture in the sun for 3–4 days. The mixture is stored for 3–6 months after pressing it tighly into earthen jars or concrete vats. There are two types of *nga-pi* made from different raw materials: *nga nga-pi* is prepared from fish and *seinsa nga-pi* or *hmyin nga-pi* is a pink to reddish paste produced from shrimp. There are three types of *nga nga-pi*: *nga-pi gaung* made from fairly large whole fish; *yegyo nga-pi*, prepared from medium and small fish; and *damin nga-pi*, a fermented fish paste made from different kinds of small fish. *Ngan-pya-ye* (fish sauce) is a reddish brown liquid produced by fermenting *yegyo nga-pi* for 4–6 months (Tyn, 1993). *Bacillus* spp. and *Micrococcus* spp. strains are isolated in the products (Tyn, 2004). In addition, fermented products made with boiled rice and fish such as tinfoil barb (*ngachin*), speckled prawn (*pazun-chin*), rohu (*ngagyin-chin*) and featherback fish (*ngaphae-chin*) have a pH of 3.8–4.8 and contain 3.8–8.6% (w/v) NaCl and 0.2–193.4 mg/100 g γ-aminobutyric acid (GABA).

Lb. plantarum, Lb. pentosus and *Lb. farciminis* strains are isolated and they produce GABA with different activities (Thwe *et al.*, 2011).

East Asian fermented fish In Japan, *shottsuru* in Akita Prefecture is prepared from sandfish, sardine, anchovy, horse mackerel, pacific mackerel and mysid. Fish are mixed with salt in the ratio of 3 : 1 or 7 : 2 and fermented either in earthenware jars or in wooden or cement tanks for 12–18 months. The mixture is stirred well and more salt is added several times during fermentation. The liquid is extracted, boiled and filtered through sea sand. After a few months of settling, the supernatant liquor is decanted off and improved by the addition of caramel, soy sauce or seasoning made from the chemical hydrolysate of vegetable protein to the fish before subsequent extraction. Fish sauce, *ikanago* in Kagawa Prefecture, is produced by mixing two parts of sand eel to one part of soy sauce and allowing it to ferment for over 100 days. Production of *ika-shoyu* in Hokkaido Prefecture and *ishiru* in Noto of Ishikawa Prefecture is the same as for *ika-shiokara* (fish paste) except that the fish flesh is allowed to further disintegrate into liquid (Itoh *et al.*, 1993).

In Korea, fish and shellfish sauces are prepared from fish and shellfish such as anchovy, large-eyed herring, many small fishes, oyster, blue mussel, shrimp and so on. They are produced from a mixture of fish or shellfish with over 20% salt (w/w) that is allowed to ferment for 6–12 months (Park *et al.*, 2005). Fermented fish in Korea is characterized by *yunhae* fermentation method which uses only salt as the curing agent, and the *sikhae* method which is used for minor delicacy preparation. *Jeot-kal* is a high-salt fermented fish product that contains the partially hydrolysed tissues of fish organs immersed in the liquid exudates of the fish. The specific name of *jeot-kal* is based on the raw materials used, such as *sae-jeo* for fermented small shrimp and *myeolchi-jeot* for fermented anchovy. *Sikhae* is a low-salt fermented product that combines cereals and is fermented by LAB. *Achromobacter, Bacillus, Corynebacterium, Brevibacterium, Flavobacterium, Halobacterium, Leuconostoc, Micrococcus, Pediococcus, Pseudomonas, Alcaligenes, Aeromonas, Moraxella* and *Sarcina* strains are isolated from *jeot-kals* (Lee *et al.*, 1993). In Taiwan, people prefer fresh fish or shellfish to processed marine food products; however, some fish paste, *yu-lu* (liquid supernatant), marine sauce (fermented fresh fish or shrimp mixed with 30% salt), oyster meat or shrucked oyster and krill sauce are produced (Lee *et al.*,1993). In southern and eastern China, a traditional fermented fish sauce, *yu-lu*, is made from anchovies (or small marine fish) with salt (fish : salt, 3 : 1). Some halotolerant and halophilic microorganisms have been isolated and glutamic acid, lysine, leucine and valine are prominent in the product (Jiang *et al.*, 2007).

South-east Asian fermented fish In the Philippines, traditional fermented fish include fermented cooked rice and shrimp *(balao-balao)*, fermented cooked rice and fish *(burong isda)*, fermented fish paste *(bagoong)*, fermented fish sauce *(patis)*, fermented small shrimp *(alamang)* and *Visayan* fermented fish *(tinabat)* (Sanchez, 1999).

Balao-balao is made from shrimp *(Penaeus indicus* or *Macrobrachium* spp.) and salt (20%) which stands for 2 h before the mixture is drained. Cooked cooled rice is then mixed with the shrimps at a ratio of 1.0 : 4.8 (weight of unsalted shrimp to rice). Salt (3%) is mixed in thoroughly before the mixture is packed in wide-mouth glass jars. The mixture is allowed to ferment for 7–10 days at 28 °C. Changes in acidity during fermentation at 28 °C showed that from 0 to 10 days the pH decreased from 7.52 to 3.80, titratable acidity (as lactic acid) increased from 0.12 to 2.0% and volatile acid (as acetic acid) increased from 0.03 to 0.1% (Sanchez, 1999). A similar product in Thailand is known as *kung-chao* or *kung-som*. It is consumed as a sauce or main dish after sautéing in vegetable oil with garlic and onion. *Leuconostoc mesenteroides, Pediococcus cerevisiae* and *Lactobacillus plantarum* strains were isolated. *Burong isda* is made from different species of fresh fish, *Ophicephalus striatus* (*burong dalag*), *Tilapia plumbeus* (*burong tilapia*), *Chanos chanos* (*burong bangus*), *Therapon plumbeus* (*burong ayungin*) with or without the addition of red rice or *angkak* (*Monascus purpureus*). The fish is scaled, the gills and fins are removed, the fish is split into a butterfly fillet and then washed, packed alternately with a mixture of cooked rice and 2.5% -20-salt at the ratio of 35:65 (salted cooked rice to fish) and fermented for 7 days at 28–30 °C. *Burong isda* is a popular traditional food in the central part of Luzon particularly in Pampanga, Bulacan, Cavite and Nueva Ecija. It is consumed as a sauce or as a main dish after sautéing in vegetable oil with garlic and onion. In other South-east Asian countries similar products are produced such as *phaak* or manchao in Cambodia, *pla-ra* in Thailand, *pekasam* and *cencalok* in Malaysia, *sihhae* in Korea and *narezushi* in Japan. *Bacillus subtilis, B. cereus, Lb. brevis, Streptococcus faecalis. Leuconostoc mesenteroides, P. cerevisiae, Lb. fermentum* and *Lb. plantarum* were isolated during the fermentation.

Bagoong and *patis* is prepared in a similar manner, except that in the latter, fermentation is allowed to proceed until the flesh of the fish disintegrates into a liquid state. In the fermentation mixture, the solid state makes up the *bagoong* while the liquid phase is the *patis*. *Bagoong* is a reddish brown salty paste with a slightly cheese-like odour made from *Dilis* (*Stolephorous* spp.), *tamban* (*Sardinella fimbriate*), and *galonggong* (*Decapterus* spp.) fish. The fish is washed thoroughly, drained and mixed with salt at proportions varying from 1:3 to 2:7 (salt: fish) depending on fish size. The mixture is ferments in a warm (40 °C) place for several months or longer until it develops the characteristic flavour and aroma of the product. Usually, preparation of *patis* takes longer than preparation of *bagoong*. The chemical composition of six-month-old *patis* showed that it has a pH of 5.85, 26.10% sodium chloride, 2.26% total nitrogen, 1.31% formal nitrogen, 0.31% ammonia nitrogen, 1.0% amino nitrogen and 0.84% lactic acid content. *Patis* contained 15 amino acids with no cystine and proline. The dominant amino acids are lysine, histidine and glutamic acid (Sanchez, 1999).

Similar products are the *terasi* of Indonesia, *belachan* of Malaysia, *kapi* of Thailand, *ngapi* of Myanmar and *nam-ca* of Vietnam. *Patis* is clear straw-yellow to amber depending on the type of fish used. The product is made by the slow digestion or fermentation of salted fish and the subsequent separation of the solid from the liquid portion of the hydrolyzate. The endogenous enzymes from *Bacillus subtilis* and *B. coagulans* in addition to those in the intestines of the fish caused protein hydrolysis. The bacterial enzymes were found to be mainly responsible for the deamination and decarboxylation of amino acids to form lower fatty acids and amides that produce the characteristic flavour of the product. *B. pumilus* is the dominant species throughout the fermentation process. Other bacteria responsible for early-stage fermentation were *B. coagulans, B. megaterium* and *B. subtilis*, while later stage fermentation was by *B. licheniformis, Micrococcus colpogenes, M. roseus, M. varians* and *Staphylococcus* species.

Indonesian fermented fish are salted fermented fish and shrimp (*Pakasam, Peda, Pindang, Terasi and Wadi*). *Pakasam* is made from *Gabus* fish *(Ophiocephalus striatus)* or *sepat siam (Trichogaster pectoralis)* mixed with salt (10–20%), and ground roasted rice that is fermented in a sealed container for 3–7 days. *Lb. plantarum-pentosus, Lb. plantarum, Lb. acidophilus, Lb. fermentum, Lactobacillus* sp., *P. acidilactici, Streptococcus thermophilus* and *Weissella* strains are isolated (Rahayu, 2003).

Peda is made from *Kembung (Rastreliger neglectus)* mixed with salt (20–30%) and fermented in a sealed container for 3 days then washed, followed by a second fermentation (1–3 weeks) after mixing with salt for the development of flavour and taste. *Lb. plantarum, Lb. curvatus, Lb. murinus* and *S. thermophilus* strains are isolated (Rahayu, 2003).

Pidang is made from fish trimmed, cleaned, washed and placed in brine followed by boiling for 1–2 h. *Lb. plantarum, Lb. acidophilus, Lb. fermentum, Lactobacillus* sp., *P. acidilactici, S. thermophilus* and *Enterococcus faecium* strains are isolated (Rahayu, 2003).

Terasi is made from shrimp or small fish mixed with salt (2–5%) and sun-dried for 1–3 days. *T. muriaticus* strains were found in this product and were reported to produce histamine in fermented fish in Japan and Thailand (Kobayashi *et al.*, 2000, 2003). *Wadi* is made from Betok/papuyu fish (*Anabas testudineus*) mixed with salt (10–20%) and fermented for 3–7 days.

In Malaysia, *budu, pekasam, belacam and cincaluk* are popular fermented fishery products. *Budu* is made from small fish anchovies (*Anchoviella commerson* and *A. indica*). The mixture of fish and salt (2 : 1 or 3 : 1 ratio) is placed in huge concrete vats or glazed earthenware containers and left to ferment for 3–12 months at ambient temperature. Mature budu is a dark brown liquid extract containing ground fish residue. The product is rich in salt and soluble nitrogen compounds and has a distinct odour and flavour. *Kican ikan* is fish sauce produced from fish other than anchovies species, but the process is similar to that for *budu*. *Pekasam* is made from freshwater fish or marine fish mixed with roast rice, tamarind and salt and fermented for 2–4 weeks. *Belacam* (shrimp paste) is made from small shrimps (*Acetes* and *Mysid*) with 10–15% salt (w/w). The salted shrimp mixtures are spread out onto straw mats for sun-drying for 5–8 h to reduce their moisture and then minced or pounded to paste in a wooden tub and allowed to ferment for one week. The paste is broken into small pies and sun-dried again for a further 5–8 h. This is followed by a second mincing and the paste is again tightly packed into a ball or other desired shape and allowed to ferment for about one month. *Cincaluk* is made from small shrimps (*Acetes* known as *udang geragau, udang bubok, udang bari*) with salt and cooked rice and fermented for 20–30 days. The product is pale pinkish, has a strong flavour and is salty. It is eaten with rice, used as a dip or sauce or can be added to give flavour to certain food items (Ismail and Karim, 1993). In Vietnam, fish sauce (*nuoc-mam*) is made from fish (*Stolephorus* spp., *Ristrellinger* spp., *Engraulis* spp., *Decapterus* spp., *Dorosoma* spp., *Clupea* spp.) with salts (3 : 1 to 3 : 2) and fermented for 4–12 months (Beddows, 1998). The moderately halotolerant bacteria, *Bacillus vietnamensis* strains are isolated from the products (Noguchi *et al.*, 2004).

Fermented fish in Thailand Fermented fish products play an important role in the Thai diet and in the life of the Thai people. Fish sauce (*nam-pla*), *bu-du*, shrimp paste (*ka-pi*), salt-fermented fish (*pla-ra*) and other fermented fish (*pla-som, som-fak, pla-chom* and *kung-chom*) have been consumed daily by Thai people. The fermentation processes vary from simple to complicated; however, some products have similar raw materials, processing methods and composition to products found in other Asian countries (Phithakpol *et al.*, 1995).

Pla-som *Pla-som* is made from freshwater fish (*Puntius gonionotus* (*ta-pian*)) with salt and minced cooked rice that is fermented for 5–7 days (Figure 8.1a). This product is eaten as it is or with vegetables such as lettuce, ginger and shallots. The product contains 62.1–69.6% moisture, 16.0–19.8% protein, 1.9–8.8% fat, 0–0.6% fibre, 4.3–12.4% ash, 4.0–10.7% NaCl, trace to 4.5% of total invert sugar (mg), 0.8–2.7% lactic acid and has a pH 4.0–5.8 (Phithakpol *et al.*, 1995). Microflora found are *Lb. pentosus, Lb. fermentum, P. pentosaceus, Lb. alimentarius/farciminis, W. confusa, Lb. plantarum, Lc. garviae, E. hirae* and *Candida krusei* (Suzuki *et al.*,1987; Tanasupawat and Komagata, 2001; Paludan-Müller *et al.*, 2002).

Figure 8.1 Fermented fish: (a) *pla-som* (*whole fish*), *som-fak* (fish cake); (b) *pla-chom*; (c) *kung-chom*; (d) *pla-ra*

Som-fak *Som-fak* is made from the minced flesh of freshwater fish (*Channa micropeltes* (*cha-do*), *Notopterus chitala* (*kraai*) or *Notopterus notopterus* (*sa-laad*)) with salt, cooked rice and garlic that is fermented for 3–5 days (Figure 8.1a). It is eaten with lettuce, ginger, shallots, chilli and roasted peanuts. The product contains 62.4–76.3% moisture, 15.9–18.5% protein, trace 4.8% fat, 0.1–1.0% fibre, 4.4–4.7% ash, 3.6–3.9% NaCl, 1.2–1.5% lactic acid and has a pH 4.1–5.2 (Phithakpol *et al.*, 1995). LAB found are *Lb. pentosus, Lb. plantarum, Lb. sakei, Lb. brevis, P. pentosaceus,* and *E. faecalis* (Tanasupawat and Komagata, 2001).

Pla-chom *Pla-chom* is made from fresh anchovy (*Danio* sp., *Rasbora* sp. (*siu*)), ground roasted rice, salt and garlic that is fermented for 3–5 days (Figure 8.1b). It is eaten with cooked rice or mix with lemongrass, chopped shallots and chilli or eaten with cooked rice. The product contains 63.2–66.7% moisture, 9.0–29.0% protein, 2.9–14.0% fat, 0.2–1.0% fibre, 4.9–13.2% ash, 2.6–10.2% NaCl, 2.0–7.8% lactic acid and has a pH 4.4–6.1 (Phithakpol *et al.* 1995). It contains *Tetragenococcus halophilus, Lb. acidipiscis, Lb. pentosus, Lb. farciminis, Enterococcus hirae, Halobacillus* spp. and *Staphylococcus carnosus* strains (Tanasupawat *et al.*, 1998, 2000; Tanasupawat and Komagata, 2001; Taprig *et al.*, 2013) and *Saccharomyces cerevisiae* strains (Suzuki *et al.*,1987).

Kung-chom *Kung-chom* is made from small shrimp, ground roasted rice, salt and garlic that is fermented for 3–5 days (Figure 8.1c). It is eaten with cooked rice and either mixed with lemongrass, chopped shallots and chilli or used as a condiment for dipping vegetables in. The product contains 60.7–73.2% moisture, 6.8–14.2% protein, 0.9–2.8% fat, trace to 1.8% fibre, 5.1–12.2% ash, 3.2–9.4% NaCl, 1.1–3.8% lactic acid and has a pH 3.9–4.3 (Phithakpol *et al.*, 1995). It contains LAB, *T. halophilus, Lb. pentosus, Lb. plantarum* and *Lb. farciminis*; and *S. piscifermentans* strains and yeast, *S. cerevisiae*, strains (Suzuki *et al.*, 1987; Tanasupawat *et al.*, 1998, 2000; Tanasupawat and Komagata, 2001).

Pla-ra *Pla-ra* is made from freshwater fish (*Crossocheilus* sp. (*soi*), *Channa striatus* (*chorn*), *Cyclocheilichthys* sp. (*takok*), *Labiobarbus leptocheilus (soi), Puntius gonionotus* (*ta-pian*), *Trichogaster* sp. (*kra-dee*)) fermented with salt and roasted rice/paddy or unroasted rice bran for 6–10 months (Figure 8.1d). The product contains 28.9–68.3% moisture, 7.9–20.3% protein, trace to 6.3% fat, trace to 5.6% fibre, 12.5–28.0% ash, 11.5–23.9% NaCl, trace to 7.3% of total invert sugar (mg), 0.5–2.0% lactic acid and has a pH 4.1–6.9 (Phithakpol *et al.*, 1995). LAB found are *T. halophilus, Lb. acidipiscis, Lb. farciminis, E. faecalis* and *W. thailandensis* (Tanasupawat *et al.*, 1998, 2000; Tanasupawat and Komagata, 2001). Halophilic bacteria involved are *B. vietnamnensis, Halobacterium piscisalsi, T. halophilus, Salinivibrio siamensis, Piscibacillus salipiscarius, Gracilibacillus thailandensis, Oceanobacillus iheyensis, Virgibacillus dokdonensis, V. halodenitrificans, V. marismortui, V. siamensis, Halobacillus* spp., *Chromohalobacter salexigens* and *S. piscifermentans* (Tanasupawat and Komagata, 2001; Tanasupawat *et al.*, 2007b, 2010; Chamroensaksri *et al.*, 2008, 2009, 2010; Yachai *et al.*, 2008; Taprig *et al.*, 2013).

Shrimp paste Fermented shrimp paste (*ka-pi-koong*) is made from planktonic and semi-planktonic saltwater or brackish water shrimp (*Acetes erythraeus* (*koei-dtaa-daeng*), *Acetes* sp. (*koei-malet-khao-saan-som-oh*), *Mesopodopsis orientalis* (*koei-dtaa-dam*)) fermented with salt for 4–6 months. The product is pinkish purple or dark purple (Figure 8.2a). It is used in various spicy soups or flavouring dishes as a condiment called Nam-prik. It is also cooked with fried rice or with coconut milk to make a vegetable dip (Phithakpol *et al.*,

(a) (b)

Figure 8.2 Shrimp paste (a, *ka-pi*); Muslim sauce (b, *bu-du*)

1995). The product contains 26.6–55.0% moisture, 7.7–27.0% protein, 0.1–3.9% fat, trace to 1.9% fibre, 17.5–48.7% ash, 14.0–40.1% NaCl, trace to 1.8% acidity and has a pH 4.0–8.2 (Phithakpol *et al.*, 1995).

Fermented fish paste is made from fish (*Kowala coval* (*ka-tak*) or *Stolephorus indicus* (*ma-li, gluay*)) fermented with salt for 3–4 months. The fish paste contains 35.6–57.3% moisture, 16.8–29.9% protein, 1.6–5.3% fat, 0.1–0.5% fibre, 12.3–25.2% ash, 10.9–19.8% NaCl, 0.1–0.5% acidity and has a pH 6.2–6.6 (Phithakpol *et al.*, 1995).

Halophilic bacteria involved are *T. halophilus, T. muriaticus, Lentibacillus kapialis, (Ln. kapialis), Salinicoccus siamensis, Alkalibacterium kapii, Virgibacillus halodenitrficans, V. dokdonensis, Oceanobacillus kapialis* and *Halobacillus* spp. (Pakdeeto *et al.*, 2007a, 2007b; Ishikawa *et al.*, 2009; Namwong *et al.*, 2009; Tanasupawat *et al.*, 2011a; Taprig *et al.*, 2013) and *Debaryomyces hansenii* strain is isolated (Suzuki *et al.*, 1987).

***Budu* (Muslim sauce)** *Budu* is made from seawater or brackish water fish (*Stolephorus indicus* (*sai-ton*), *Clupeoides* sp. (*ka-tak*), *Sardinella* sp. (*lang-khieo*)) fermented with a large amount salt for 3–6 months. The product is a viscous brown liquid with grey colloidal fish flesh separated from the fish bones suspended in it and it has a salty taste (Figure 8.2b). It is eaten as it is, added during cooking or used to make a dipping sauce. The product contains 45.0–60.2% moisture, 12.1–13.2% (w/v) protein, 1.3–1.8% fat, 0.2–0.4% fibre, 21.6–23.2% ash, 19.4–20.6% NaCl, 0.2% trace of total invert sugar (mg), 0.4–0.9% (w/v) lactic acid and has a pH 5.4–6.5 (Phithakpol *et al.*, 1995). *T. halophilus, S. piscifermentans, Bacillus subtilis, B. lichenformis* strains are isolated from this product (Choorit and Prasertsan, 1992; Tanasupawat and Komagata, 1995).

Fish sauce Fish sauce is a traditional liquid product made from brackish water, seawater and freshwater fish ((*Corica argentius* (*siu-kaew*), *C. soborna, Stolephorus indicus, S. commersonii* (*sai-ton*) and *Clupeoides* sp. (*ka-tak*)) or any other kind of fish fermented in jars with a large amount of salt for 5–18 months. The fish can be found in bountiful supply in the coastal waters of the Gulf of Thailand and the South China Sea. Generally, cleaned fish is mixed with salt in a 2:1 or 3:1 ratio (fish:salt) (w/w), depending on the area of production (Figure 8.3). The product is clear, yellow or brownish yellow, salty and has a fishy aroma. After 12–18 months, when the liquid is a clear reddish brown and has a fragrant aroma, it is removed from the jars, preferably through a spigot at the bottom of the jars so that it passes through the layers of the fish remains. Any sediments are strained out with a clean cloth. Clean jars are filled with the filtered fish sauce which is allowed

(a) (b)

Figure 8.3 Fish sauce fermentation pool

to air out in the sun for a couple of weeks to dissipate the strong fish odours. It is then ready for bottling. The finished product is 100%, top-grade, genuine fish sauce. Second and third grade fish sauces are made by adding salt water to cover the fish remains for 2–3 months, then filtering before bottling. Finally, the fish remains are boiled with salt water to produce the lowest grade fish sauce. Because the flavour is substantially reduced with each fermentation, top-grade fish sauce is frequently added to the lower grades to improve their flavour. In fact, many manufacturers do not market top-grade sauce. For this reason, 100% fish sauce is mixed with second and third grade sauces to produce larger quantities to sell which can still qualify as genuine fish sauce (Saisithi, 1994; Phithakpol *et al.*, 1995; Lopetcharat *et al.*, 2001).

Fish sauce is commonly used as a seasoning for cooking and for dipping seafood and oriental food. It is currently very popular in South-east Asia and with Asian people in Western countries. This hydrolysed fish condiment is produced in many countries under different names such as *nam-pla* (Thailand, Loas), *ketjap-ikan* or *bakasang* (Indonesia), *yu-lu* (China), *patis* (Philippines), *ngan bya yay* (Myanmar), *nuoc-mam* (Vietnam), *teuk trei* (Cambodia), *shotshuru* (Japan), *colombo-cure* (India and Pakistan), *jeot-kal* or *aekjeot* (Korea) and *budu* (Malaysia) (Wood, 1998; Park *et al.*, 2001).

In Thailand, fish sauce is a liquid product with a salty taste that is used as food seasoning. It is divided into three categories by the Thai Ministry of Public Health (2001) (Lopetcharat *et al.*, 2001): (1) genuine fish sauce, which is obtained from the fermentation or digestion of fish or parts of fish, or the residue of fish remaining from the fermentation, by the fish sauce production process; (2) fish sauce made from other animals, which is obtained from the fermentation or digestion of other animals besides fish, or parts of other animals, or the residue of other animals remaining from the fermentation, by the fish sauce production process that includes fish sauce which is made from other animals mixed with genuine fish sauce; (3) mixed fish sauce, which is the fish sauce as stipulated in (1) or (2), in which some other non-hazardous constituents to consumption are added, or is diluted, or for seasoning. Fish sauce should have the following qualities or standards: its colour, odour and flavour are inherent to that specific sauce type; it is clear and is free of sediment, except natural sediment which is not more than 0.1 g/l, as well as salt as sodium chloride which is not less than 200 g/l of fish sauce. When potassium chloride has been used individually or mixed with sodium chloride, the quantity individually or the total quantity of both salts used should not be less than 200 g. Total nitrogen should not be less than 9 g/l of fish sauce. Nitrogen from amino acid should be between 40% and 60% of total nitrogen. Glutamic acid per total nitrogen should be between 0.4% and 0.6%. Only brown sugar or caramel should be added for colour. The use of artificial sweetener should follow the Food Standards of Joint FAO/WHO Codex and may be used alone or in combination with sugar. The factors that influence fish sauce quality are (i) fish species, (ii) type of salt, (iii) the ratio of fish to salt, (iv) the addition of minor ingredients and (v) the fermentation conditions. Fish species also affects the type of proteins that serve as nutrients for microorganisms and substrates for enzymes that hydrolyse proteins into small peptides, free amino acids, ammonia and trimethylamine (TMA), which contribute to the specific aroma and flavour of fish sauce. The pH of raw fish also drops from 7.0–7.2 to 6.6–6.8 after salt is added. The release of free amino acids from proteins and large polypeptides causes the final pH range from 4.8 to 6.7 in commercial fish sauce (Thongsanit, 1999; Lopetcharat and Park, 2002). Fish sauce contains essential (histidine, isoleucine, leucine, lysine, methionine, phenylalanine, tryptophan, valine, arginine, cysteine, glycine, proline and tyrosine) and non-essential (alanine, aspartic acid, glutamic acid and serine) amino acids. Lysine content was 1319 mg per 100 ml of fish sauce from a single batch made from anchovies (*Stolephorus* spp.) after 10 months, fermentation (Lopetcharat *et al.*, 2001; Lopetcharat and Park, 2002; Thongthai and Gildberg, 2005). In addition, every fish contains slightly different fatty acids that are hydrolysed to low molecular weight volatile fatty acids, especially ethanoic and *n*-butanoic acids, which are responsible for the cheesy aroma in fish sauce. The NaCl content of sea salt from Thailand and salt from other countries differ, which affects its diffusion into fish flesh. The chemical composition of salt also affects the type of microbiological flora present during fermentation, which in turn affects the quality of the fish sauce (Lopetcharat *et al.*, 2001).

The fish to salt ratio affects fish sauce quality. Bacterial and enzymatic activity differs at different salt concentrations, resulting in different flavours and digestive products. The NaCl content in fish sauce is 22.2–29.9% (w/v) (Phithakpol *et al.*, 1995; Lopetcharat *et al.*, 2001). Other components present in fish are minerals such as potassium (K), phosphorus (P), sulfur (S), sodium (Na), magnesium (Mg), calcium (Ca), iron (Fe) and vitamins (thiamin, riboflavin, niacin, vitamins B_6 and B_{12}), which contribute to the nutritive value of fish sauce and are essential for a healthy diet. In fish sauce fermentation, the oxygen content is quite high at the surface of the batch, limited under the liquid surface and extremely low at the bottom. Therefore, the fermentation is completed under partial aerobic and anaerobic conditions (Lopetcharat *et al.*, 2001; Park *et al.*, 2001).

During the initial fermentation period (1–6 months), strains of *Tetragenococus halophilus, T. muriaticus, Halobacterium salinarum, Halococcus thailandensis, Hc. saccharolyticus, Natrinema gari, Lentibacillus jurispiscarius, Ln. halophilus, Pisciglobus halotolerans, Chromohalobacter salexigens, Halobacillus* sp., *Virgibacillus* sp., *Filobacillus* sp. and *Marinococcus* sp. are present. The same species are present after six months' fermentation, plus *Ln. salicampi* strains (Table 8.2) (Thongthai *et al.*, 1992; S. Tanasupawat, Chulalongkorn University, 1999, unpublished data; Thongsanit *et al.*, 2002; Hiraga *et al.*, 2005; Namwong *et al.*, 2005, 2006, 2007; Tanasupawat *et al.*, 2006, 2009, 2011b; Sinsuwan *et al.*, 2007, 2010; Tapingkae *et al.*, 2008; Phrommao *et al.*, 2010). *Haloarcula salaria* and *Haloarcula tradensis* strains are also found (Namwong *et al.*, 2011). In fish sauce production in Japan, *Staphylococcus* species dominate during the first month of fermentation, while halophilic and extremly halophilic bacteria increase between 4 and 6 weeks. *T. halophilus* strains are associated with lactic acid production (Fukui *et al.*, 2012).

Role of halophiles in fish fermentation

PROTEASES OF HALOPHILES Fermented fish and fish sauce are basically protein hydrolysates that result from the natural fermentation of fish and salt kept in tanks at an ambient temperature for 12–18 months (Saisithi, 1994). Protein hydrolysis is induced by endogenous proteinases in fish muscle and the digestive tract as well as proteinases produced by halophilic bacteria (Saisithi, 1994; Gildberg and Thongthai, 2001).

The fish sauce strain, *Halobacillus thailandensis* fs-1 produces proteases with a molecular mass of 100, 42 and 17 kDa (Chaiyanan *et al.*, 1999); *Virgibacillus* sp. SK37, 18–81 kDa proteinases (Sinsuwan *et al.*, 2007; Phrommao *et al.*, 2010); *Virgibacillus* sp. SK33, 17, 32 and 65 kDa cell-associated proteinases (Sinsuwan *et al.*, 2010, 2011); *Filobacillus* sp. RF2-5, 49 kDa proteinase (Hiraga *et al.* 2005); and *Halobacillus* sp. SR5-3, 43 kDa proteinase (Namwong *et al.*, 2006). The uniqueness of these enzymes is that their activities were enhanced 2.0–2.5-fold when 10–20 and 25% salt was added (Hiraga *et al.*, 2005; Namwong *et al.*, 2006; Sinsuwan *et al.*, 2011). A subtilisin-like alkaline serine proteinase from *Bacillus licheniformis* RKK-04 and fibrinolytic enzymes from *V. halodenitrificans* SK1-3-7 isolated from fish sauce show high enzymatic activity under high salt conditions (Toyokawa *et al.*, 2010; Montriwong *et al.*, 2012). *V. marismortui* NB2-1 from *pla-ra* produces a substantial level of extracellular proteolytic enzymes which are active in extreme conditions. Using casein as substrate, the protease exhibited maximal activity at pH 10 and 50 °C, in the

Table 8.2 Distribution of halophilic bacteria in Thai fish sauce fermentation

Fermentation days (d)/ months (m)	Isolate no.	Species	Fermentation days (d)/ months (m)	Isolate no.	Species
Initial	IO-1	*Ln. jurispiscarius*	3 m	HDB5-2	*Hc. thailandensis*
	PMO-8	*T. halophilus*		DB5-1B	*Ln. halophilus*
	PWO-15	*T. muriaticus*		K5-51	*T. halophilus*
	CO-1, CO-2	*P. halotolerans*		P5-1, P5-2	*T. muriaticus*
	CBO-1	*L. halophilus*	6 m	HDS6-6 HDS6-1A, HDS6-2	*Hc. thailandensis*
10 d	HIS10-2	*Hc. thailandensis*		DS6-1	*Ln. halophilus*
	HIS10-4	*Hb. salinarum*		HRF6	*Hc. saccharolyticus*
	IS10-5	*Ln. jurispiscarius*		K6-32, K6-35	*T. halophilus*
20 d	HIB20-2	*Hb. salinarum*		P6-1, P6-3	*T. muriaticus*
1 m	HDB1-4	*Hc. thailandensis*	7 m	HDS7-4	*Hc. thailandensis*
	HDB1-1, HDS1-1, HDB1-11, HDB1-31	*Hb. salinarum*		DS7-5, PB7-3	*Ln. halophilus*
				P7-18, P7-37	*T. muriaticus*
	DB1-1, DS1-3B	*Ln. halophilus*	8 m	DB8-4	*Ln. salicampi*
	KF1-1	*Ln. halophilus*		HDB8-2, HDB8-5	*Hc. thailandensis*
	K1-10, K1-31	*T. muriaticus*		DB8-6	*Ln. halophilus*
30 d	PB30-1, PB30-3	*T. halophilus*		K8-1, K8-5	*T. muriaticus*
	HIS30-1	*Hb. salinarum*	9 m	PS9-2	*Ln. jurispiscarius*
35 d	KS35-3	*Hb. salinarum*		DB9-1	*Ln. halophilus*
	HKS35-3	*Hc. thailandensis*		K9-1, K9-2	*T. muriaticus*
40 d	IS40-2, IS40-3	*Ln. jurispiscarius*	10 m	HDB10-5, HIS10-4	*Hb. salinarum*
	HIS40-3	*N. gari*		HDB10-5, HDS10-5	*Hc. thailandensis*
	HIS50-2 (1)	*Hb. salinarum*		DS10-3B,	*Ln. halophilus*
2 m	HDS2-5	*Hb. salinarum*	333 d	KS333-3B	*Ln. salicampi*
	RF2-5	*Filobacillus* sp.		HKS333-2	*Hc. thailandensis*
60 d	PS60-2, PB60-2	*T. halophilus*	11 m	KS11-1	*C. salexigens*
	HIB60-1	*Hb. salinarum*		PS11-2	*Ln. halophilus*
2 m	K2-9, K2-17, C2-1	*T. muriaticus*	12 m	PB12	*C. salexigens*
87 d	HKS87-3	*Hc. thailandensis*	*Bu du*	BN2-2	*Ln. halophilus*
	KS87-5	*C. salexigens*			
3 m	P3-1, P3-5	*T. halophilus*			
	K3-6, K3-26	*T. muriaticus*			
	HDS3-1	*N. gari*			
4 m	HDS4-1	*Hc. thailandensis*			
	DB4-2, DS4-2	*Ln. halophilus*			
	K4-2, K4-16	*T. halophilus*			
	P4-3, P4-5	*T. muriaticus*			

presence of 5% (w/v) NaCl (Chamroensaksri *et al.*, 2008). *V. halodenitrificans* TKNR13-3 isolated from shrimp paste is reported to produce extracellular protease (Tanasupawat *et al.*, 2011a) as previously characterized from other moderately halophilic bacterial strains isolated from fish sauce.

In addition, the extreme halophiles, *Halobacterium salinarum* and *Hc. saccharolyticus* strains show high proteolytic activity at high salt concentrations that will lead them to play their role in fish sauce fermentation (Tanasupawat, 1999; Kanlayakrit *et al.*, 2004; Thongthai and Gildberg, 2005). The *Halobacterium* sp. SP1(1) strain has been used to accelerate fish sauce fermentation (Akolkar *et al.*, 2010).

Halophilic LAB, *T. halophilus* strains utilize fish proteins and oligopeptides as nitrogen sources and some strains increase oligopeptide content, which implies their proteolytic activity for toward fish protein. These bacteria showed high intracellular aminopeptidase activity for toward alanine (2.85–3.67 U/ml) followed by leucine (1.90–2.37 U/ml). Some strains showed the highest aminopeptidase activity toward alanine, while other isolates expressed high activity for methionine or glutamic acid (Udomsil *et al.*, 2010).

Biogenic Amine Formation Biogenic amines are basic nitrogenous compounds occurring in meat and fish products, mainly due to the amino acid decarboxylation activities of certain microbes (Arnold and Brown, 1978). Fish sauce, fish paste and shrimp paste from Southeast Asia and the products sold in the supermarkets in Taiwan were reported to have pH 4.8–6.5, 16.2–45.3% (w/v) of salt, 51–275 mg/100 g of total volatile basic nitrogen, 5.4–53.9 mg/100 g of trimethylamine and 1.0–4.2 log cfu/g of aerobic plate count. The average content for each of eight different biogenic amines in all samples was less than 90 ppm, except for histamine which has an average content of 394, 263 and 382 ppm in fish sauce, fish paste and shrimp paste, respectively. Most of the tested fermented fish products had histamine levels greater than the Food and Drug Administration (FDA) guideline of 50 ppm, while seven (25.9%) contained >500 ppm of histamine. The histamine contents in most fermented fish products exceeded the 50 ppm US FDA guideline value and 7.4% of the tested samples contained >1000 ppm of histamine. Consumption of these products could lead to scombroid poisoning in consumers (Tsai *et al.*, 2006). The unprocessed *budu* samples of Kelantan, Malaysia, contained a high protein concentration of up to 24.88% (w/v) and 58% of all samples analysed are reported to have more than 50 mg/100 g histamine (22.21 mg/100 g to 106.40 mg/100 g sample). A high level of histamine indicates poor handling and processing of fish products, which results in a high degree of contamination (Rosma, *et al.*, 2009). In fish sauce, the most abundant amines are histamine (1220 ppm), putrescine (1257 ppm) and cadaverine (1429 ppm), while tyramine is present in a smaller amounts (1178 ppm). Other amines such as tryptamine, phenylethylamine, spermine and spermidine are minor amines (Zaman *et al.*, 2009). These compounds are generally vasoactive and can cause changes in blood pressure, severe headache, hypertension, renal intoxication, intracerebral haemorrhage and eventually death. The FDA established an advisory level of 500 ppm to be hazardous to human health (FDA, 2011). However, high histamine concentration may not cause severe illness since consumption of fish sauce is less than 23.5 g/day per person (Lopetcharat *et al.*, 2001; Yongsawatdigul *et al.*, 2004).

Histamine was the only major biogenic amine produced by halophilic LAB, *T. muriaticus* (Kimura *et al.*, 2001; Thongsanit *et al.*, 2002). However, in the modified glucose yeast peptone (mGYP) broth, *T. halophilus* strains are reported to form histamine at both 5 and 25% NaCl in the range of 6.62–22.55 and 13.14–20.39 mg/100 ml, respectively. Trace amounts of tyramine (1–5 mg/100 ml mGYP) were also observed. Some strains produced the highest histamine content of 22.55 mg/100 ml at 5% NaCl and 20.39 mg/100 ml at 25% NaCl and produced tyramine slightly at 25% NaCl (Udomsil *et al.*, 2010). According to Yongsawatdigul *et al.* (2004), histamine, cadaverine, putrescine and tyramine were the predominant biogenic amines found in Indian anchovy (*Stolephorus indicus*) left at 35 °C for 16 h and its fish sauce product. Changes of biogenic amines during fish sauce fermentation were subtle, suggesting that raw material, rather than a fermentation process, was a major source of biogenic amines. Use of temperature-abused raw material did not accelerate the fermentation process. Biogenic amines could be used as quality indicators in conjunction with total nitrogen content (Yongsawatdigul *et al.*, 2004).

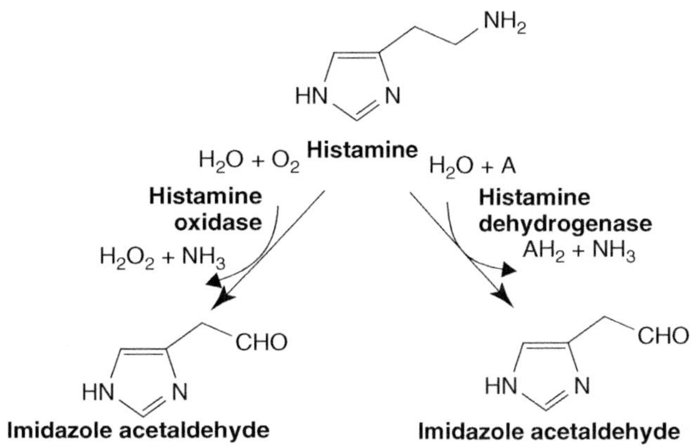

Figure 8.4 Degradation pathways of histamine. A and AH_2 are a two-electron acceptor (or two moles of one-electron acceptor) and its reduced form, respectively. For example; nicotinamide adenine dinucleotide phosphate NAD (P), ferricyanide, flavin adenine dinucleotide (FAD), flavin mononucleotide (FMN), cytochrome c, nitroblue tetrazolium (NBT) and 1-methoxy-5-methylphenazinium methylsulfate (1- methoxy PMS)

HISTAMINE DEGRADATION Many bacteria have been shown to convert primary amines via an oxidative deamination step into products that can be utilized either as a source of carbon and/or energy and as a source of nitrogen, or both (Hacisalihoglu *et al.*, 1997; Levering *et al.*, 1981). Two types of enzymes, amine dehydrogenase (ADH) and amine oxidase (AO), are normally implicated in this form of oxidation. ADHs catalyse the oxidative deamination of primary amines to their corresponding acetaldehyde and ammonia, but AOs generate hydrogen peroxide (H_2O_2) in addition to acetaldehyde and ammonia (Figure 8.4). The potential role of microorganisms in food fermentation with histamine oxidases and histamine dehydrogenases activity has been investigated to prevent or reduce the accumulation of histamine in foods. These findings are useful for the application of microbial degradation of biogenic amines (Ienistea, 1971; Murooka *et al.*, 1979; Umezu *et al.*, 1979a, 1979b; Leuschner *et al.*, 1998; Dapkevicius *et al.*, 2000; Siddiqui *et al.*, 2000; Bozkurt and Erkmen, 2002; Gardini *et al.*, 2002). Many studies have been carried out to evaluate the histamine degrading activity of different microbial species isolated from fermented foods such as *Lactobacillus*, *Micrococcus* and *Staphylococcus* (Ienistea, 1971; Voigt and Eitenmiller, 1977; Murooka *et al.*, 1979; Umezu *et al.*, 1979a, 1979b; Yamashita *et al.*, 1993; Leuschner *et al.*, 1998). Nevertheless, it has been found that the *in vitro* degradation of histamine by microorganisms is not quantitatively reproducible with *in vivo* studies due to the more severe conditions, in particularly the low O_2 tension, pH and temperature.

The presence of histamine-degrading enzymes, either histamine oxidases or histamine dehydrogenases, has been reported in various higher organisms (Pionetti, 1974; Isobe *et al.*, 1980; Rinaldi *et al.*, 1983) as well as in microorganisms (Ienistea, 1971; Yamashita *et al.*, 1993; Siddiqui *et al.*, 2000; Bakke *et al.*, 2005). Therefore, the applications of these microorganisms and/or enzymes might be restricted by unfavourable physiological conditions for enzyme activity such as low stability in the presence of high salt concentration. As none of microorganisms studied were salt tolerant, they are probably not appropriate for the salt-fermented products. The use of fermenting microorganisms to degrade histamine in salt-fermented food is still limited. Therefore, halophilic archaea should be considered as a source of enzyme production, which can degrade histamine and other biogenic amines in fermented foods with a high salt content.

Tapingkae *et al.* (2008, 2010b, 2010a) found that histamine dehydrogenase from *Natrinema gari* BCC 24369, isolated from anchovy fish sauce (*nam-pla*) in Thailand, catalyses the oxidative deamination of histamine to imidazole acetaldehyde and ammonia. The enzyme has the potential to degrade histamine under high salt conditions, high catalytic activity and narrow substrate specificity. Based on histamine assay, the histamine-degrading activity of *N. gari* BCC 24369 was located in the intracellular fraction, required the presence of 1-methoxy-5-methylphenazinium methylsulfate (PMS), an electron carrier and selectively catalysed histamine as a substrate. The histamine-degrading activity was highest at pH 6.5–8, in the presence of NaCl at 3.5–5 M and at 40–55 °C. The activity was found to be stable at pH 6.5–9, in the presence of NaCl above 2.5 M and at temperatures lower than 50 °C. These results suggested a link of the histamine-degrading activity of *N. gari* BCC 24369 to the presence of salt tolerant and thermo-neutrophilic histamine dehydrogenase. The molecular mass of the purified enzyme was 127.5 kDa when analysed using Superose 12 10/300 gel filtration. The enzyme consisted of heterotrimeric subunits with three different molecular masses of 69.1, 29.3 and 27.7 kDa. The enzyme had high affinity and activity toward histamine with V_{max}, K_m and k_{cat} values of 2.5 μmol/min, 57.1 μM and 5.3/s, respectively. Based on unique molecular properties, this enzyme is distinctive from histamine dehydrogenases previously reported in other microorganisms (Siddiqui *et al.*, 2000; Fujieda *et al.*, 2004; Bakke *et al.*, 2005; Sato *et al.*, 2005).

Formation of volatile compounds Various volatile compounds, including acids, carbonyls, nitrogen-containing compounds and sulfur-containing compounds, are formed during fermentation and believed to be responsible for the distinct aroma of fish sauce (Peralta *et al.*, 1996; Shimoda *et al.*, 1996; Fukami *et al.*, 2002). Acetic acid, propanoic acid, 2-methylpropanoic acid, butanoic acid and 3-methylbutanoic acid were detected. Acetic acid appeared to be a predominant acid and 3-methylbutanoic acid was one of potent distinctive volatile fatty acids detected in fish sauce fermentation and contributed to a rancid note. Volatile fatty acids were also associated with cheesy odour and had relatively low threshold values in the range 2.45–7.10 ppb, indicating that the odour of these acids could be easily recognized (Peralta *et al.*, 1996). The fermentation inoculated with *Staphylococcus* sp. SK1-1-5 appeared to have higher levels of volatile fatty acids, particularly butanoic and 3-methyl butanoic acid, than others (Yongsawatdigul *et al.*, 2007). This strain showed lipase activity, which could subsequently promote lipolytic degradation during fermentation. Polyunsaturated fatty acids produced by lipolysis would, in turn, undergo oxidation to form volatile fatty acids. These volatile fatty acids could also be formed through amino acid metabolism of starter cultures.

The addition of starter cultures of *Virgibacillus* sp. and *Stapyloccoccus* sp. to the anchovy hydrolysate increased the degree of hydrolysis during 4 months of fermentation. Total amino acid profiles of fish sauce inoculated by *Virgibacillus* sp. SK33, *Virgibacillus* sp. SK37 and fermented for 4 months were comparable with those of the sample traditionally fermented for 12 months. *Virgibacillus* sp. SK33 also showed histamine-reducing potential during fish sauce fermentation. Volatile compounds detected from the samples inoculated with starter cultures were similar to those of the sample conventionally fermented for 12 months. The sample inoculated by *Staphylococcus* sp. SK1-1-5 was rated to have comparable sensory properties to the commercial fish sauce. *Staphylococcus* sp. SK1-1-5 could potentially accelerate fish sauce fermentation without adversely affecting the sensory characteristics (Yongsawatdigul *et al.*, 2007).

Halophilic LAB produced a higher isopropyl alcohol content when distributed used in fermentation. They contributed to alcohol accumulation in fish broth containing 25% NaCl. Major aldehydes were identified as 2-methylpropanal and benzaldehyde. Branched-chain aldehydes, such as 2-methylpropanal, 2-methylbutanal and 3-methylbutanal, are potent flavour compounds in fermented food such as *ishiru* (Japanese fish sauce), tuna sauce and cheese. Samples inoculated with halophilic LAB showed higher amounts of 2-methylpropanal and contained the highest amount of benzaldehyde; however, smaller amounts of ethyl acetate were found in all samples. Acetone, 2-butanone, 2,3-butanedione and cyclohexanone were found in varied amounts. Sulfur-containing compounds, such as dimethyl sulfide and dimethyl disulfide, which contributed to undesirable faecal notes in fish sauce, were not detected when halophilic LAB were used in the fermentation. *T. halophilus* might be responsible for the flavour/aroma formation of fish sauce. In addition, formation of volatile compounds by *T. halophilus* appeared to be species-dependent (Udomsil *et al.*, 2010). The use of *T. halophilus* strains for fish sauce fermentation improves amino acid profiles and volatile compounds and also reduces the biogenic amine content of the fish sauce product (Udomsil *et al.*, 2011).

8.5 Future trends in fish fermentation technology

Traditional fermented fish have played an essential role in the diet and are consumed either as main dishes or as condiments. Several products undergo similar processing but the ingredients vary, leading to the differences in appearance, flavour and taste as well as in the sensorial properties. On several occasions, this chapter has noted the ubiquity of fermented fish products across the world that reflect the specific combination of raw material, ambient conditions, microbial flora, salt content and dietary traditions which have evolved in each region.

Although the production of traditional fermented fish products is a long-standing industry across the world, its role in securing a sustainable fish processing industry is not developed to its fullest potential. Nevertheless, a sustainable diet in the future might demand a return to more products that use less heat, less energy intense processing, and a safer and healthier diet (e.g. low salt). Owing to the lack of scientific and technological knowledge, fermented fish are generally evaluated on their qualitative attributes such as odour and flavour, which leads to inconsistent quality. Therefore more attention has been paid to the development of the process as well as the use of biotechnology to ensure the quality and safety of the products.

The need for quality improvement and process innovation in the area of fish fermentation have been widely recognized in recent years. Fermented fish contaminated with unacceptable levels of pathogens and chemical contaminants or having other hazardous characteristics, impose substantial health risks to consumers and severe economic burdens on individual communities and nations. Currently, safety is one of the pressing problems for many countries, particularly developing countries with agriculture-based economies. A substantial loss of foreign exchange and reputation is the main result when importing countries reject food containing unacceptable levels of contaminants. As fermented products are mostly uncooked before consumption, they can be a major source of exposure to pathogenic agents, both chemical and biological (viruses, parasites and bacteria).

The prime concern in cured fish fermentation is to keep the salt content as low as possible. During the scale-up of cured fish production from the traditional household level, more salt is added to prevent not only the growth of microorganisms that could pose a health hazard, but also any changes in quality during the long distribution process. However, high salt concentration generally reduces the sensory quality of the products. In addition, excessive salt has adverse health effects. An excessive salt intake can favour the development of high blood pressure, which is a risk factor for cardiovascular disorders. Too much salt is also a risk factor for stomach cancer and osteoporosis. There is a similar requirement to lower the salt content of of lactic acid fermented fish products. The shelf life of lactic acid fermented fish products is primarily dependent on the salt concentration, storage temperature, amount and type of the carbohydrate source and the use of other preservative materials. Many studies have been conducted on the effects of partially replacing NaCl with potassium chloride (KCl) and some of these blends are available on the market (Desmond, 2006). A considerable disadvantage of many salt substitutes is their bitter (metallic) flavour. Alternatives for sodium-containing additives, other than salt, should be considered to reduce the sodium content and to maximize food safety and quality.

Since natural fermentation methods are difficult to control and carry the risk that accompanying microflora may cause spoilage and unsafe products, starter cultures have now become more important in the production of fermented foods. The development of functional starter cultures is also essential for developing a safety control system for fermented products with consistent quality. Due to an increase in microorganisms that are resistant to traditional antibiotics, the search for a natural preservative material that is useful for fermented products has been widely conducted. In particular, novel bacteriocins produced by LAB have received much attention due to their benefits to human health and to food production.

In addition, the consumption of fermented food products has long been associated with good health and longevity, but there is little significant research to back up these claims. It is generally assumed that bacteria from different species and the various microorganisms have their own preferences for growing conditions. An organism that initiates fermentation will grow there until its by-products inhibit further growth and activity. During this initial growth period, other organisms develop which are ready to take over when the conditions become intolerable for the former ones. During fermentation, microorganisms produce a range of bioactive compounds and metabolites, some of which have been associated with health-promoting properties. The most notable of these are enzymes, metabolites and bioactive peptides released from food proteins. Therefore, the promoted health benefits of food fermentation need to be substantiated by scientific data both *in vitro* and *in vivo*.

An accurate reflection of activity *in vivo* that is beneficial to health would greatly facilitate the development of a completely new generation of fermented functional foods tailored to contain clinically proven health-promoting metabolites and/or microorganisms. Future prospects to exploit the potential health-promoting properties of such foods will continue to expand as we gain a greater understanding of the microorganisms concerned and how they, or their metabolites, can directly interact in a positive way with the human host. Besides having a more scientific and technological foundation, products to be marketed should meet consumer's demands and concerns, which are mainly focused on quality, safety and health benefits.

Due to the increase in demand of fermented products in both domestic and international markets, standards for individual products have been established to ensure their authenticity and quality for the customers. Also, the expansion of the market worldwide has lead to strict regulations for exporting, so the specification and quality of products should also comply with the regulations for each country. The development of processes and technologies, such as convenient packaging, need to be carried out to increase productivity and to ensure consistent product quality. Also, the nutritive value and safety of those products need to be standardized to enable expansion of both domestic and international markets. Because of these new developments, the consumption of fermented fish products is again increasing.

References

Achi, O.K. (2005). The potential for upgrading traditional fermented foods through biotechnology. *African Journal of Biotechnology*, 4, 375–380.

Aidoo, K.E., Nout, M.J.R. and Sarkar, P.K. (2006). Occurrence and function of yeasts in Asian indigenous fermented foods. *FEMS Yeast Research*, 6, 30–39.

Akolkar, A.V., Durai, D. and Desai, A.J. (2010). *Halobacterium* sp. SP1(1) as a starter culture for accelerating fish sauce fermentation. *Journal of Applied Microbiology*, 109 (1), 44–53.

Anihouvi, V.B., Kindossi, J.M. and Hounhouigan, J.D. (2012). Processing and quality characteristics of some major fermented fish products from Africa: a critical review. *International Research Journal of Biological Sciences*, 7, 72–84.

Arnold, S.H. and Brown, W.D. (1978). Histamine toxicity from fish products. *Advances in Food Research*, 24, 113–154.

Axelsson, L. (2004). Lactic acid bacteria: classification and physiology, in *Lactic Acid Bacteria, Microbiology and Functional Aspects* (eds S. Salminen, A. von Wright and A. Ouwehand), Marcel Dekker, Inc., New York, pp. 1–66.

Bakke, M., Sato, T., Ichikawa, K. and Nishimura, I. (2005). Histamine dehydrogenase from *Rhizobium* sp.: gene cloning, expression in *Escherichia coli*, characterization and application to histamine determination. *Journal of Biotechnology*, 119, 260–271.

Beddows, C.G. (1998). Fermented fish and fish products, in *Microbiology of Fermented Food*, vol. 2. (ed. B.J.B. Wood), Elsevier Applied Science Publishers, London, pp. 416–440.

Blandino, A., Al-Aseeri, M.E., Pandiella, S.S., Cantero, D. and Webb, C. (2003). Cereal-based fermented foods and beverages. *Food Research International* 36, 527–543.

Bozkurt, H. and Erkmen, O. (2002). Effects of starter cultures and additives on the quality of Turkish style sausage sucuk. *Meat Science*, 61, 149–156.

Campbell-Platt, G. (Ed.) (1987). *Fermented Foods of the World, a Dictionary and Guide.* Butterworths, London.

Campbell-Platt, G. (1994). Fermented foods – a world perspective, *Food Research International*, 27, 253–257.

Chaiyanan, S., Chaiyanan., S., Maugel, T. *et al.* (1999). Polyphasic taxonomy of a novel *Halobacillus, Halobacillus thailandensis* sp. nov. isolated from fish sauce. *Systematic and Applied Microbiology*, 22, 360–365.

Chamroensaksri, N., Akaracharunya, A., Visessanguan, W. and Tanasupawat, S. (2008). Characterization of halophilic bacterium NB2-1 from *pla-ra* and its protease production. *Journal of Food Biochemistry*, 32, 536–555.

Chamroensaksri, N., Tanasupawat, S., Akaracharunya, A. *et al.* (2009). *Salinivibrio siamensis* sp. nov., from fermented fish (*pla-ra*) in Thailand. *International Journal of Systematic and Evolutionary Microbiology*, 59, 880–885.

Chamroensaksri, N., Tanasupawat, S., Akaracharanya, A. *et al.* (2010). *Gracilibacillus thailandensis* sp. nov., from fermented fish (*pla-ra*) in Thailand. *International Journal of Systematic and Evolutionary Microbiology*, 60, 944–948.

Caplice, E. and Fitzgerald, G.F. (1999). Food fermentations: role of microorganisms in food production and preservation. *International Journal of Food Microbiology*, 50, 131–149.

Chiang, Y.W., Chye, F.Y. and Ismail, A.M. (2006). Microbial diversity and proximate pomposition of *Tapai*, A Sabah's fermented beverage. *Malaysian Journal of Microbiology*, 2 (1), 1–6.

Choorit, W. and Prasertsan, P. (1992). Characterization of proteases produced by newly isolated and identified proteolytic microorganisms from fermented fish *(Budu)*. *World Journal of Microbiology and Biotechnology*, 8, 284–286.

Cook, P.E. (1994). Fermented foods as biotechnological resources *Food Research International*, 27, 309–316.

Dapkevicius, M.L.N.E., Nout, M.J.R., Rombouts, F.M., Houbon, J.H. and Wymenga, W. (2000). Biogenic amine formation and degradation by potential fish silage starter microorganisms. *International Journal of Food Microbiology*, 57, 107–114.

Desmond, E. (2006). Reducing salt: a challenge for the meat industry. *Meat Science*, 74, 188–196.

De Vuyst, L. and Vandamme, J.E. (1994). *Bacteriocins of Lactic Acid Bacteria, Microbiology, Genetics and Applications*. Blackie Academic and Professional, London.

Dworkin, M., Falkow, S., Rosenberg, E., Schleifer, K.-H. and Stackebrandt, E. (Eds.) (2006). *The Prokaryotes: A Handbook on the Biology of Bacteria*, vol. 4, 3rd edn, Springer, New York.

FDA (2011) Chapter 7: Scombrotoxin (histamine) formation, in *Fish and Fishery Products Hazards and Controls Guidance*, 4th edn, Department of Health and Human Services Public Health Service Food and Drug Administration Center for Food Safety and Applied Nutrition Office of Food Safety, p. 113.

Fujieda, N., Satoh, A., Tsuse, N., Kano, K. and Ikeda, T. (2004). 6-S-Crysteinyl flavin mononucleotide-containing histamine dehydrogenase from *Nocardiodides simplex*: molecular cloning, sequencing, overexpression and characterization of redox centers of enzyme. *Biochemistry*, 43, 10800–10804.

Fukami, K., Ishiyama, S., Yaguramaki, H. *et al.* (2002). Identification of distinctive volatile compounds in fish sauce. *Journal of Agricultural and Food Chemistry*, 50, 5412–5416.

Fukui, Y., Yoshida, M., Shozen, K.-I. *et al.* (2012). Bacterial communities in fish sauce mash using culture-dependent and -independent methods. *Journal of General and Applied Microbiology*, 58, 273–281.

Gardini, F., Martuscelli, M., Crudele, M.A., Paparella, A. and Suzzi, G. (2002). Use of *Staphylococcus xylosus* as a starter culture in dried sausages: effect on the biogenic amine content. *Meat Science*, 61, 275–283.

Gildberg, A. and Thongthai, C. (2001). The effect of reduced salt content and addition of halophilic lactic acid bacteria on quality and composition of fish sauce made from sprat. *Journal of Aquatic Food Product Technology*, 10, 77–88.

Hacisalihoglu, A., Jongejan, J.A. and Duine, J.A. (1997). Distribution of amine oxidases and amine dehydrogenases in bacteria grown on primary amines and characterization of the amine oxidase from *Klebsiella oxytoca*. *Microbiology*, 143, 505–512.

Hall, G.M. (2002). Safety and quality issues in fish processing, in *Lactic Acid Bacteria in Fish Preservation* (ed. H.A. Bremner), CRC Press, New York, pp. 330–349.

Hiraga, K., Nishikata, Y., Namwong, S. *et al.* (2005). Purification and characterization of serine proteinase from a halophilic bacterium, *Filobacillus* sp. RF2-5. *Bioscience Biotechnology and Biochemistry*, 69, 38–44.

Ishikawa, M., Tanasupawat, S., Nakajima, K. *et al.* (2009). *Alkalibacterium thalassium* sp. nov., *Alkalibacterium pelagium* sp. nov., *Alkalibacterium putridalgicola* sp. nov. and *Alkalibacterium kapii* sp. nov., slightly halophilic and alkaliphilic marine lactic acid bacteria isolated from marine organisms and salted foods collected in Japan and Thailand. *International Journal of Systematic and Evolutionary Microbiology*, 59, 1215–1226.

Ienistea, C. (1971). Bacterial production and destruction of histamine in foods and food poisoning caused by histamine. *Nahrung*, 15, 109–113.

Ismail, M. and Karim, A. (1993). Feremented fish products in Malaysia, in *Fish Fermentation Technology* (eds C.H. Lee, K.H. Steinkraus, and P.J.A. Reilly), UNU Press, Tokyo, pp. 95–106.

Isobe, Y., Ohara, K., Kosaka, M. and Aoki, K. (1980). Relationships between hypothalamic catecholamines, blood pressure and body temperature in spontaneously hypertensive rats. *Japanese Journal of Physiology*, 30, 805–810.

Itoh, H., Tachi, H. and Kikuchi, S. (1993) Fish fermentation technology in Japan, in *Fish Fermentation Technology* (eds C.H. Lee, K.H. Steinkraus, and P.J.A. Reilly), UNU Press, Tokyo, pp. 177–186.

Jiang J.J., Zeng Q.X., Zhu Z.W. and Zhang L.Y. (2007) Chemical and sensory changes associated yu-lu fermentation process – a traditional Chinese fish sauce. *Food Chemistry*, 104, 1629–1634.

Kanlayakrit, W., Bovornreungroj, P., Oka, K. and Goto, M. (2004) Production and characterization of protease from an extremely halophilic *Halobacterium* sp. PB407. *Kasetsart Journal: Natural Science*, 38, 15–20.

Köse, S., Koral, S., Tufan, B. *et al.* (2012). Biogenic amine contents of commercially processed traditional fish products originating from European countries and Turkey. *European Food Research Technology*, 235, 669–683.

Kimura, B., Konagaya, Y. and Fujii, T. (2001) Histamine formation by *Tetragenococcus muriaticus*, a halophilic lactic acid bacterium isolated from fish sauce. *International Journal of Food Microbiology*, 70, 71–77.

Knochel, S. (1993) Processing and properties of North European pickled fish products, in *Fish Fermentation Technology* (eds C.H. Lee, K.H. Steinkraus, and P.J.A. Reilly), UNU Press, Tokyo, pp. 213–230.

Kobayashi, T., Kajiwara, M., Wahyuni, M. *et al.* (2003). Isolation and characterization of halophilic lactic acid bacteria isolated from '*terasi*' shrimp paste: a traditional fermented seafood product in Indonesia. *Journal of Geneneral and Applied Microbiology*, 49, 279–286.

Kobayashi, T., Kimura, B. and Fujii, T. (2000). Differentiation of *Tetragenococcus* populations occurring in products and manufacturing processes of puffer fish ovaries fermented with rice bran. *International Journal of Food Microbiology*, 56, 211–218.

Lakshmi, W.G.I., Prassanna, P.H.P. and Edirisinghe, U. (2010). Production of *Jaadi* using tilapia (*Oreochromis niloticus*) and determination of its physcio-chemical and sensory properties. *Sabaramuwa University Journal*, 9, 57–63.

Law, S.V., Abu Bakar, F., Mat Hashim, D. and Abdul Hamid, A. (2011). Popular fermented foods and beverages in Southeast Asia. *International Food Research Journal*, 18, 475–484.

Lee, C.H., Steinkraus, K.H. and Reilly, P.J.A. (eds.) (1993). *Fish Fermentation Technology*, UNU Press, Tokyo.

Leisner, J.J., Millan, J.C., Huss, H.H. and Larsen, L.M. (1994). Production of histamine and tyramine by lactic acid bacteria isolated from vacuum-packed sugar-salted fish. *Journal of Applied Bacteriology*, 76, 417–423.

Lopetcharat, K., Choi, Y.J., Park, J.W. and Daeschel, M.A. (2001). Fish sauce products and manufacturing: a review. *Food Reviews International*, 17, 68–88.

Lopetcharat, K. and Park, J.W. (2002). Characteristics of fish sauce made from pacific whiting and surimi by-products during fermentation storage. *Journal of Food Science*, 67, 511–516.

Leuschner, R.G., Heidel, M. and Hammes, W.P. (1998). Histamine and tyramine degradation by food fermenting microorganisms. *International Journal of Food Microbiology*, 39, 1–10.

Levering, P.R., van Dijken, J.P., Veenhius, M. and Harder, W. (1981). *Arthrobacter* P1, a fast growing versatile methylotroph with amine oxidase as a key enzyme in the metabolism of methylated amines. *Archives of Microbiology*, 129, 72–80.

Majumdar, R.K. and Basu, S. (2010). Characterization of the traditional fermented fish product *Lona ilish* of Northeast India. *Indian Journal Traditional Knowledge*, 9, 453–458.

Mohammed Salih, A.M., El Sanousi, S.M. and El Zubeir, I.E.M. (2011). A review on the Sudanese traditional dairy products and technology. *International Journal of Dairy Science*, 6, 227–245.

Montriwong, A., Kaewphuak, S., Rodtong, S., Roytrakul, S. and Yongsawatdigul, J. (2012). Novel fibrinolytic enzymes from *Virgibacillus halodenitrificans* SK1-3-7 isolated from fish sauce fermentation. *Process Biochemistry*, 47, 2379–2387.

Murooka, Y., Doi, N. and Harada, T. (1979). Distribution of membrane bound monoamine oxidase in bacteria. *Applied and Environment Microbiology*, 38, 565–569.

Nakazawa, Y. and Hosono, A. (eds) (1992). *Functions of Fermented Milk: Challenges for the Health Sciences*, Elsevier Science Publishers LTD, New York.

Namwong, S., Tanasupawat, S., Smitinont, T. *et al.* (2005). Isolation of *Lentibacillus salicampi* strains and *Lentibacillus juripiscarius* sp. nov. from fish sauce in Thailand. *International Journal of Systematic and Evolutionary Microbiology*, 55, 315–320.

Namwong, S., Hiraga, K., Takada, K. *et al.* (2006). A halophilic serine proteinase from *Halobacillus* sp. SR5-3 isolated from fish sauce: purification and characterization. *Bioscience Biotechnology and Biochemistry*, 70, 1395–1401.

Namwong, S., Tanasupawat, S., Visessanguan, W., Kudo, T. and Itoh, T. (2007). *Halococcus thailandensis* sp. nov., from fish sauce in Thailand. *International Journal of Systematic and Evolutionary Microbiology*, 57, 2199–2203.

Namwong, S., Tanasupawat, S., Lee, K.C. and Lee, J.-S. (2009). *Oceanobacillus kapialis* sp. nov., from fermented shrimp paste in Thailand. *International Journal of Systematic and Evolutionary Microbiology*, 59, 2254–2259.

Namwong, S., Tanasupawat, S., Kudo, T. and Itoh, T. (2011), *Haloarcula salaria* sp. nov. and *Haloarcula tradensis* sp. nov. from salt in Thai fish sauce. *International Journal of Systematic and Evolutionary Microbiology*, 61, 231–236.

Nayeem, M.A., Pervin, K., Reza, M.S. *et al.* (2010). Quality assessment of traditional semi-fermented fishery product (*cheap shutki*) of Bangladesh collected from the value chain. *Bangladesh Research Publications Journal*, 4, 41–46.

Noguchi, H., Uchino, M., Shida, O. *et al.* (2004). *Bacillus vietnamensis* sp. nov., a moderately halotolerant, aerobic, endospore-forming bacterium isolated from Vietnamese fish sauce. *International Journal of Systematic and Evolutionary Microbiology*, 54, 2117–2120.

Nout, M.J.R. (1994). Fermented foods and food safety. *Food Research International*, 21, 291–298.

Nout, M.J.R. and Sarkar, P.K. (1999). Lactic acid food fermentation in tropical climates. *Antonie van Loeuwenhoek*, 76, 395–401.

Ouadghiri, M., Amar, M., Vancanneyt, M. and Swings, J. (2005). Biodiversity of lactic acid bacteria in Moroccan soft white cheese (Jben). *FEMS Microbiology Letters*, 251, 267–271.

Pakdeeto, A., Naranong, N. and Tanasupawat, S. (2003). Diacetyl of lactic acid acid bacteria from milk and fermented foods in Thailand. *Journal of General and Applied Microbiology*, 49, 301–307.

Pakdeeto, A., Tanasupawat, S., Thawai, C. *et al.* (2007a). *Lentibacillus kapialis* sp. nov., from fermented shrimp paste in Thailand. *International Journal of Systematic and Evolutionary Microbiology*, 57, 364–369.

Pakdeeto, A., Tanasupawat, S., Thawai, C. *et al.* (2007b). *Salinicoccus siamensis* sp. nov., isolated from fermented shrimp paste in Thailand. *International Journal of Systematic and Evolutionary Microbiology*, 57, 2004–2008.

Paludan-Müller, C., Madsen, M., Sophanodora, P., Gram, L. and Møller, P.L. (2002). Fermentation and microf lora of *plaa-som*, a Thai fermented fish product prepared with different salt concentrations. *International Journal of Food Microbiology*, 73, 61–70.

Park, J.-N., Fukumoto, Y., Eriko Fujita, E. *et al.* (2001). Chemical composition of fish sauces produced in southeast and east asian countries. *Journal of Food Composition Analysis*, 14, 113–125.

Park, P.-J., Je, J.-Y. and Kim, S.-K. (2005). Amino acid changes in the Korean traditional fermention process for blue mussel, *Mytilus edulis*. *Journal of Food Biochemistry*, 29, 108–116.

Peralta, R.R., Shimoda, M. and Osajima, Y. (1996). Further identification of volatile compounds in fish sauce. *Journal of Agricultural and Food Chemistry*, 44, 3606–3610.

Phithakpol, B., Varanyanond, W., Reungmaneepaitoon, S. and Wood, H. (eds) (1995). *The Traditional Fermented Foods of Thailand*, Institute of Food Research and Product Development, Kasetsart University, Bangkok, 157 pp.

Phrommao, E., Rodtong, S. and Yongsawatdigul, J. (2010). Identification of novel halotolerant bacillopeptidase F-like proteinases from a moderately halophilic bacterium, *Virgibacillus* sp. SK37. *Journal of Applied Microbiology*, 110, 191–201.

Pionetti J.M. (1974). Analytical-band centrifugation of the active form of pig kidney diamine oxidase. *Biochemical and Biophysical Research Communications*, 58, 495–498.

Paludan-Müller, C., Madsen, M., Sophanodora, P., Gram, L. and Møller, P.L. (2002b). Fermentation and microf lora of *plaa-som*, a Thai fermented fish product prepared with different salt concentrations. *International Journal of Food Microbiology*, 73, 61–70.

Rahayu, E.S. (2003). *Lactic acid* bacteria in fermented foods of Indonesian origin. *Agritech*, 23, 75–84.

Rhee, S.J., Lee, J.-E. and Lee, C.-H. (2011). Importance of lactic acid bacteria in Asian fermented foods. *Microbial Cell Factories*, 10 (Suppl 1): S5. Available from: http://www.microbialcellfactories.com/content/10/S1/S5 [accessed 2 June 2013].

Rinaldi, A., Floris, G., Sabatini, S., Finazzi-Agr, A., Giartosio, A. and Rotilio, G. (1983). Reaction of beef plasma and lentil seedlings Cu-amine oxidase with phenylhydrazine. *Biochemical and Biophysical Research Communications*, 115, 841–848.

Rosma, A., Afiza, T.S., Wan Nadiah, W.A., Liong, M.T. and Gulam, R.R.A. (2009). Microbiological, histamine and 3-MCPD contents of Malaysian unprocessed 'budu'. *International Food Research Journal*, 16, 589–594.

Ruddle, K. (1990). Fish processing technologies for entrepreneurial development in Bangladesh, in *Transfer of Technology for Entrepreneurial Development in Bangladesh* (eds Anon.), UN Economic and Social Commission for Asia and the Pacific, Bangkok.

Ruddle, K. and Ishige, N. (2010). On the origins, diffusion and cultural context of fermented fish products in Southeast Asia, in *Globalization, Food and Social Identities in the Asia Pacific Region* (ed. J. Farrer), Sophia University Institute of Comparative Culture, Tokyo. http://icc.fla.sophia.ac.jp/global%20food%20papers/ [accessed 2 June 2013]

Saisithi, P. (1994). Traditional fermented fish: fish sauce production. Ch. 5, in *Fisheries Processing: Biotechnological Applications* (ed. A.M. Martin), Chapman and Hall, London, pp. 111–129.

Sanchez, P.C. (1999). Microorganisms and technology of Philippine fermented foods. *Japanese Journal of Lactic Acid Bacteria*, 10, 19–28.

Sato, T., Horiuchi, T. and Nishimura, I. (2005). Simple and rapid determination of histamine in food using a new histamine dehydrogenase from *Rhizobium* sp. *Analytical Biochemistry*, 346, 320–326.

Scott, R. and Sullivan W.C. (2008). Ecology of fermented foods. *Human Ecology Review*, 15, 25–31.

Shimoda, M., Peralta, R.R. and Osajima, Y. (1996). Headspace gas analysis of fish sauce. *Journal of Agricultural and Food Chemistry*, 44, 3601–3605.

Siddiqui, J.A., Shoeb, S.M., Takayama, S., Shimizu, E. and Yorifuji, T. (2000). Purification and characterization of histamine dehydrogenase from *Nocardioides simplex* IFO 12069. *FEMS Microbiology Letters*, 15, 183–187.

Sinsuwan, S., Rodtong, S. and Yongsawatdigul, J. (2007). NaCl-activated extracellular proteinase from *Virgibacillus* sp. SK37 isolated from fish sauce fermentation. *Journal of Food Science*, 72, 264–269.

Sinsuwan, S., Rodtong, S. and Yongsawatdigul, J. (2010). A NaCl-stable serine proteinase from *Virgibacillus* sp. SK33 isolated from Thai fish sauce. *Food Chemistry*, 119, 573–579.

Sinsuwan, S., Rodtong, S. and Yongsawatdigul, J. (2011). Evidence of cell-associated proteinases from *Virgibacillus* sp. SK33 isolated from fish sauce fermentation. *Journal of Food Science*, 76, C413–C419.

Stanbury, P.., Whitaker, A. and Hall, S.J. (eds) (1995). *Principles of Fermentation Technology*, 2nd Edn, Pergamon Press, Oxford.

Steinkraus, K.H. (1993). Comparison of fermented food of East and West, In *Fish Fermentation Technology* (eds C.H. Lee, K.H. Steinkraus, and P.J.A. Reilly), UNU Press, Tokyo, pp. 1–12.

Stringer S.C. and Pin C. (2005). Microbial risks associated with salt reduction in certain foods and alternative options for preservation. Technical report, April 2005. Institute of Food research, Norwich, UK, 50 pp. http://www.food.gov.uk/multimedia/pdfs/acm740a.pdf [accessed 2 June 2013].

Suzuki, M., Nakase, T., Daengsubha, W., Chaosangket, M., Suyanadana, P. and Komagata, K. (1987). Identification of yeasts isolated from fermented foods and related materials in Thailand. *Journal of General and Applied Microbiology*, 33, 205–220.

Tamang, J.P., Tamang, N., Thapa, S. *et al.* (2012). Microorganisms and nutritional value of ethnic fermented foods and alcoholic beverages of North East India. *Indian Journal of Traditional Knowledge*, 11, 7–25.

Tanasupawat, S., Chamroensaksri, N., Kudo, T. and Itoh, T. (2010). Identification of moderately halophilic bacteria from Thai fermented fish (*pla-ra*) and proposal of *Virgibacillus siamensis* sp. nov. *Journal of General and Applied Microbiology*, 56, 369–379.

Tanasupawat, S. and Komagata, K. (1995). Lactic acid bacteria in fermented foods in Thailand. *World Journal of Microbiology and Biotechnology*, 11, 253–256.

Tanasupawat, S. and Komagata, K. (2001). Lactic acid bacteria in fermented foods in Southeast Asia, in *Microbial Diversity in Asia: Technology and Prospects* (eds B.H. Nga, H.M. Tan, and K. Suzuki), World Scientific Publishing Co. Pte. Ltd, Singapore, pp. 43–59.

Tanasupawat, S., Namwong, S., Kudo, T. and Itoh, T. (2007b). *Piscibacillus salipiscarius* gen. nov., sp. nov., a moderately halophilic bacterium from fermented fish (*pla-ra*) in Thailand. *International Journal of Systematic and Evolutionary Microbiology*, 57, 1413–1417.

Tanasupawat, S., Namwong, S., Kudo, T. and Itoh, T. (2009). Identification of halophilic bacteria from fish sauce (*nam-pla*) in Thailand. *Journal of Culture Collections*, 6, 69–75.

Tanasupawat, S., Okada, S. and Komagata, K. (1998). Lactic acid bacteria found in fermented fish in Thailand. *Journal of General and Applied Microbiology*, 44, 193–200.

Tanasupawat, S., Pakdeeto, A., Namwong, S. *et al.* (2006). *Lentibacillus halophilus* sp. nov., from fish sauce in Thailand. *International Journal of Systematic and Evolutionary Microbiology*, 56, 1859–1863.

Tanasupawat, S., Shida, S., Okada, S. and Komagata, K. (2000). *Lactobacillus acidipiscis* sp. nov. and *Weissella thailandensis* sp. nov. isolated from fermented fish in Thailand. *International Journal of Systematic and Evolutionary Microbiology*, 50, 1479–1485.

Tanasupawat, S., Taprig, T., Akaracharanya, A. and Visessanguan, W. (2011b). Characterization of *Virgibacillus* strain TKNR13-3 from fermented shrimp paste (*ka-pi*) and its protease production. *African Journal of Microbiological Research*, 5, 4714–4721.

Tanasupawat, S., Thongsanit, J., Okada, S. and Komagata, K. (2002). Lactic acid bacteria isolated from soy sauce mash in Thailand. *Journal of General and Applied Microbiology*, 48, 201–209.

Tanasupawat, S., Thongsanit, J., Thawai, C., Lee, K.C. and Lee, J.-S. (2011a). *Pisciglobus halotolerans* gen. nov., sp. nov., isolated from fish sauce in Thailand. *International Journal of Systematic and Evolutionary Microbiology*, 61, 1688–1692.

Tapingkae, W., Parkin, K.L., Tanasupawat, S. *et al.* (2010b). Whole cell immobilisation of *Natrinema gari* BCC 24369 for histamine degradation. *Food Chemistry*, 120, 842–849.

Tapingkae, W., Tanasupawat, S., Itoh, T. *et al.* (2008). *Natrinema gari* sp. nov., a halophilic archaeon isolated from fish sauce in Thailand. *International Journal of Systematic and Evolutionary Microbiology*, 58, 2378–2383.

Tapingkae, W., Tanasupawat, S., Parkin, K.L., Benjakul, S. and Visessanguan, W. (2010a). Degradation of histamine by extremely halophilic archaea isolated from high salt-fermented fishery products. *Enzyme and Microbial Technology*, 46, 92–99.

Taprig, T., Akaracharanya, A., Sitdhipol, J., Visessanguan, W. and Tanasupawat, S. (2013). Screening and characterization of protease-producing *Virgibacillus, Halobacillus* and *Oceanobacillus* strains from Thai fermented fish. *Journal of Applied Pharmaceutical Science*, 3, 025–030.

Thai Ministry of Public Health (2001). The Notification of fish sauce (No. 203), in the *Government Gazette* vol.118, special part 6 Ngor.

Thongsanit, J. (1999). DNA–DNA hybridization in the identification of *Tetragenococcus* species isolated from fish sauce fermentation, Master Thesis. Department of Microbiology, Faculty of Science, Chulalongkorn University.

Thongsanit, J., Tanasupawat, S., Keeratipibul, S. and Jatikavanich, S. (2002). Characterizationand identification of *Tetragenococcus halophilus* and *Tetragenococcus muriaticus* strains from fish sauce (nam-pla). *Japanese Journal of Lactic Acid Bacteria*, 13, 46–52.

Thongthai, C. and Gildberg, A. (2005). Asian fish sauce as a source of nutrition, in *Asian Functional Foods* (eds J. Shi, C-T., Ho and F. Shahidi), Marcel Dekker/CRC Press, New York, pp. 215–265.

Thongthai, C., McGenity, T.J., Suntinanalert, P. and Grant, W.D. (1992). Isolation and characterization of an extremely halophilic archaeobacterium from traditional fermented Thai fish sauce (*nam-pla*). *Letters in Applied Microbiology*, 14, 111–114.

Thwe, S., Kobayashi, T., Luan, T. *et al.* (2011). Isolation, characterization and utilization of γ-aminobutyric acid (GABA)-producing lactic acid bacteria from Myanmar fishery products fermented with boiled rice. *Fisheries Science*, 77, 279–288.

Toyokawa, Y., Takahara, H., Reungsang, A. *et al.* (2010). Purification and characterization of a halotolerant serine proteinase from thermotolerant *Bacillus licheniformis* RKK-04 isolated from Thai fish sauce. *Applied Microbiology and Biotechnology*, 86(6), 1867–1875.

Tsai, Y.H., Lin, C.Y., Chien, L.T. *et al.* (2006). Histamine content of fermented fish products in Taiwan and isolation of histamine-forming bacteria. *Food Chemistry*, 98, 64–70.

Tyn, M.T. (1993). Trends of fermented fish technology in Burma, in *Fish Fermentation Technology* (eds C.H. Lee, K.H. Steinkraus and P.J.A. Reilly), UNU Press, Tokyo, pp. 129–153.

Tyn, M.T. (2004). Industrialization of Myanmar fish paste and sauce fermentation, in *Industrialization of Indigenous Fermented Foods*, 2nd edn, revised and expanded. Marcel Dekker, New York, USA, pp. 737–759.

Udomsil, N., Rodtong, S., Tanasupawat, S. and Yongsawatdigul, J. (2010). Proteinase-producing halophilic lactic acid bacteria isolated from fish sauce fermentation and their ability to produce volatile compounds. *International Journal of Food Microbiology*, 141, 186–194.

Udomsil, N., Rodtong, S., Choi, Y.J., Hua, Y. and Yongsawatdigul, J. (2011). Use of *Tetragenococcus halophilus* as a starter culture for flavour improvement in fish sauce fermentation. *Journal of Agricultural and Food Chemistry*, 59, 8401–8408.

Umezu, M., Shibata, A. and Umegaki, M. (1979a). Oxidation of amines by nitratereducing bacteria and lactobacilli in Saké brewing. *Journal of Fermentation Technology*, 57, 56–60.

Umezu, M., Shibata, A. and Umegaki, M. (1979b). Production and oxidation of amines by hiochi bacteria. *Journal of Fermentation Technology*, 57, 505–511.

Voigt, M.N. and Eitenmiller, R.R. (1977). Production of tyrosine and histidine decarboxylase in dairy-related bacteria. *Journal of Food Protection*, 40, 241–245.

Wood, B.J.B. (1998). *Microbiology of Fermented Foods*, vol.1, Blackie Academic and Professional, London.

WHO (2004). *Global status report on alcohol, 2004*. Geneva, World Health Organization. http://www.who.int/substance_abuse/publications/global_status_report_2004_overview.pdf [accessed 2 June 2013].

Yachai, M., Tanasupawat, S., Itoh, T. *et al.* (2008). *Halobacterium piscisalsi* sp. nov., from fermented fish (*pla-ra*) in Thailand. *International Journal of Systematic and Evolutionary Microbiology*, 58, 2136–2140.

Yamashita, M., Sakaue, M., Iwata, N., Sugino, H. and Murooka, Y. (1993). Purification and characterization of monoamine oxidase from *Klebsiella aerogenes*. *Journal of Fermentation and Bioengineering*, 76, 289–295.

Yongsawatdigul, J., Choi, Y.S. and Udomporn, S. (2004). Biogenic amines formation in fish sauce prepared from fresh and temperature-abused Indian anchovy (*Stolephorus indicus*). *Journal of Food Science*, 69(4), FCT312–FCT319.

Yongsawatdigul, J., Rodtong, S. and Raksakulthai, N. (2007). Acceleration of Thai fish sauce-fermentation using proteinases and bacterial starter cultures. *Journal of Food Science*, 72(9), M382–M390.

Zaman, M.Z., Abdulamir, A.S., Bakar, F.A., Selamat, J. and Bakar, J. (2009). A Review: microbiological, physicochemical and health impact of high level of biogenic amines in fish sauce. *American Journal of Applied Sciences*, 6, 1199–1211.

9

Frozen Surimi and Surimi-based Products

Emiko Okazaki[1] and Ikuo Kimura[2]

[1]*Tokyo University of Marine Science and Technology, Department of Food Science and Technology, Tokyo, Japan*

[2]*Laboratory of Food Engineering, Faculty of Fisheries, Kagoshima University, Shimoarata, Kagoshima, Japan*

9.1 Fish material for frozen surimi

Frozen surimi is a tasteless fish meat protein concentrate which keeps the gel-forming ability of the myofibrillar protein, and is further processed as an intermediate ingredient into surimi-based products such as *kamaboko* and crab-meat analogues.

In the 1960s, 'frozen surimi' processing technology was developed in Japan. At that time, Alaska pollock was used as the only material for frozen surimi production. However, in subsequent years, fish other than Alaska pollock such as Pacific whiting, blue whiting, Menhaden, mackerel, hake, southern blue whiting, hoki, croaker, golden threadfin bream and pike conger, etc., were also used to make frozen surimi (Figure 9.1). In 2012, global production of frozen surimi was about 600–700 thousand metric tonnes per year, pushing the production of surimi-based products to about 1.5 million metric tonnes annually (Pascal, 2012). Therefore, frozen surimi has become an important intermediate raw material for food production all over the world.

9.2 Principles and process of frozen surimi production

Frozen surimi production requires the processing of high amounts of fish materials daily and thus the use of large-scale equipment is required. This is because migratory fish species are usually used for surimi production, so large seasonal catches have to be processed. Because of the large-scale processing, there must be effective use of by-products (e.g. fish meal and fish oil) and the full utilization of the fish without waste. The general production process for frozen surimi is shown in Figure 9.2.

Seafood Processing: Technology, Quality and Safety, First Edition. Edited by Ioannis S. Boziaris.

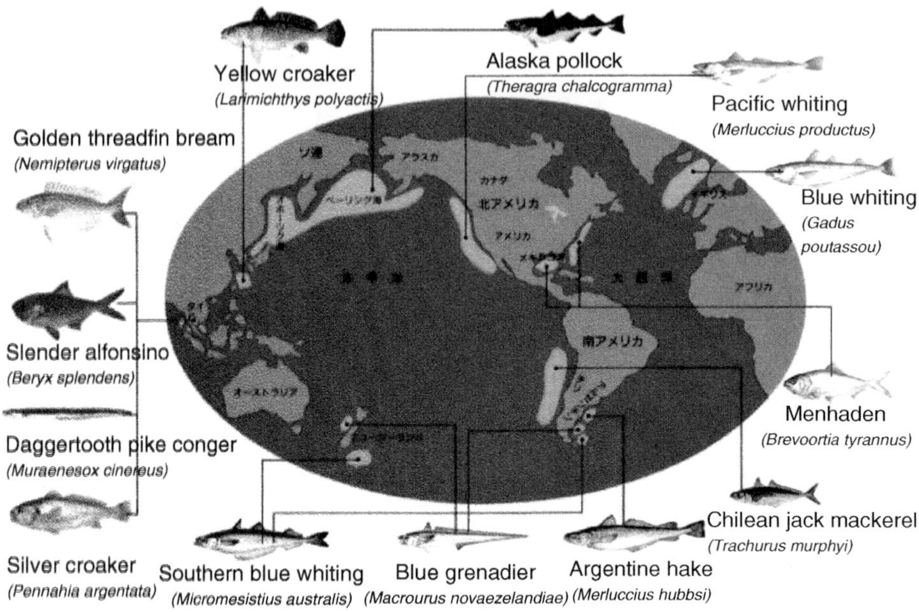

Figure 9.1 Fish materials for frozen surimi around the world

9.2.1 Fish material

The quality and characteristics of frozen surimi are affected by various factors such as the fish species used, fishing season and fish freshness. Among fish species, differences in habitat temperature, salt solubility and thermostability of muscle protein, protease and transglutaminase activity significantly affect the quality of frozen surimi, especially in terms of its gel-forming ability.

The freshness of the materials used has a marked influence on the manufacturing process and quality of frozen surimi. Efficient cooling and low-temperature maintenance is very important for maintaining the freshness of the fish material. Freshness affects surimi gel strength. The dependency of surimi gel strength on the freshness of Alaska pollock is shown in Table 9.1 (Uno and Nakamura, 1958). Freshness is the most important requirement for the raw material of surimi, regardless of the fish species used. Controlling the freshness of fish material minimizes variations in the quality of surimi.

The quality of surimi is also affected by the season and the size of fish used. The surimi quality from fish in the feeding season is generally low.

9.2.2 Washing and scaling of fish

It is important to remove the scales, slime and dirt from the surface of the fish material before processing to obtain good surimi quality. The ease of scale removal is affected by fish freshness. Although scale removal is difficult immediately after the fish are caught, it becomes easier as fish freshness decreases.

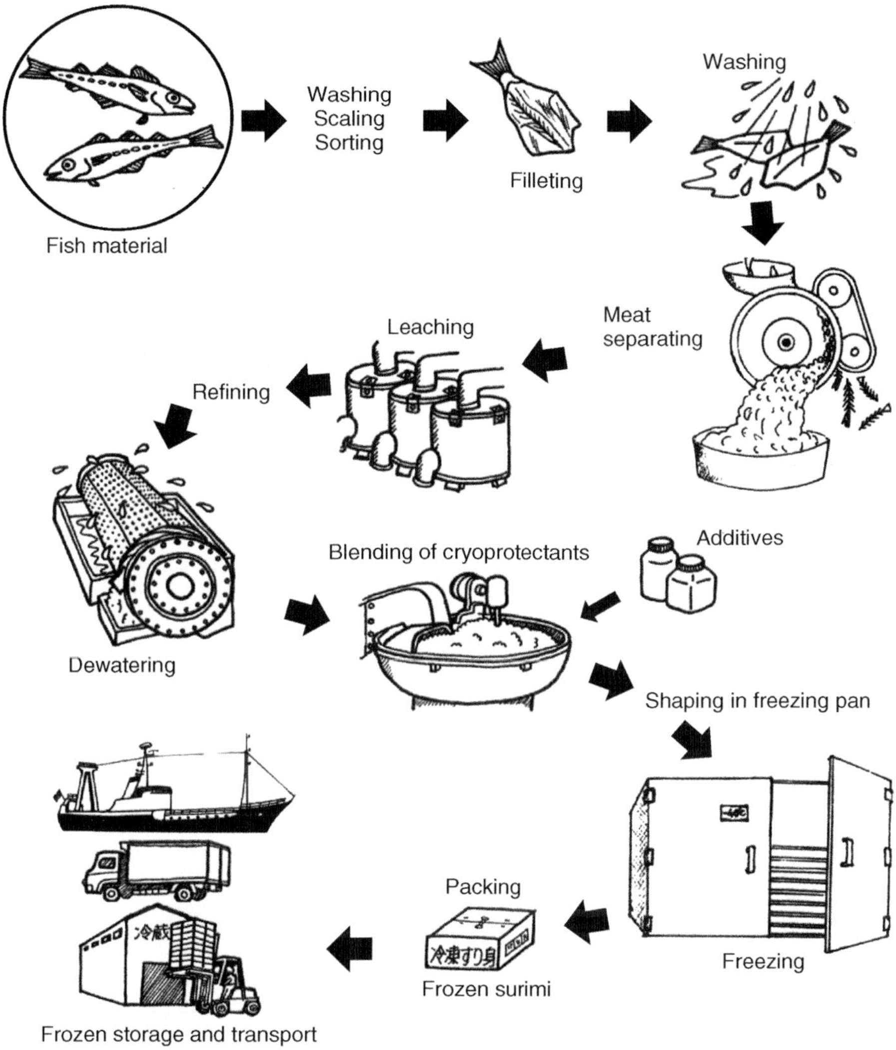

Figure 9.2 The frozen surimi manufacturing process

Table 9.1 Effect of fish freshness (Alaska pollock) on surimi gel strength (Uno and Nakamura, 1958)

	Fish condition (days of storage at 5 °C)			
	Extremely fresh (0 days)	Very fresh (2 days)	Fairly fresh (4 days)	Not fresh (6 days)
	Gel strength (g · cm)			
Unleached surimi	1100	600	350	150
Leached surimi	1200	850	650	400

9.2.3 Sorting of fish

The fish are sorted according to size. This process is needed especially when filleting is carried out using filleting machines to facilitate fish processing and increase fillet yield. Sorting can be performed automatically using a roller or a caterpillar-track-type apparatus.

9.2.4 Filleting of fish

This step affects both the quality and quantity of minced fish meat obtained from subsequent deboning. Filleting may be carried out manually or automatically using a filleting machine. Fish can be supplied to a filleting machine manually or automatically using equipment that positions the fish in a head-to-tail direction and dorsal-to-ventral orientation.

9.2.5 Mechanical separation of fish

The deboning operation separates fish meat from residual bones, fins, or skin. The fish meat separators used are generally of the belt-drum type, in which fish pass between a belt and a perforated drum, with the belt held tightly to the drum by a pressure roller. Relatively soft flesh is forced through holes in the drum to the inside of the drum. Bones, skin, and flesh adhering to the outer surface of the drum are scraped off. The diameter of the holes chosen for the drum markedly affects the subsequent leaching and dewatering processes as well as the yield and quality of surimi. Generally, the hole diameters range from 4 to 7 mm and are chosen in accordance with the size and freshness of the fish.

9.2.6 Leaching

Leaching is an important process not only for removing undesirable substances such as fat, blood, pigments and odorous substances, but also for imparting a better gel forming ability by removing water-soluble proteins (WSPs; sarcoplasmic proteins) that impede gel formation of surimi-based products such as *kamaboko* and crab meat analogues. In this process, the water-soluble components of fish meat, such as WSP, minerals, extractable taste components and organic acids, are removed.

Leaching is performed on a continuous system, which combines a leaching tank with a rotary screen developed for surimi processing. The leaching tank is equipped with paddles that are automatically activated when the tank is filled to a predetermined level. Leached fish meat is intermediately dewatered through the rotary screen prior to the next washing cycle. This step is repeated depending on the number of washing cycles required.

Leached water used for washing fish meat contains not only water-soluble components, but also fat and small meat particles that can cause environmental pollution if not disposed of properly. Therefore, wastewater must be treated to meet local standards. Although scum obtained after wastewater treatment is mainly discarded, its effective utilization is desirable from the viewpoint of conserving food resources. Methods of collecting and utilizing the fat, fish meat particles and soluble proteins mentioned earlier for animal or human consumption have also been tested (Okazaki, 1994).

Table 9.2 Changes in gel forming ability of fish meat by leaching (Okada, 1981)

	Times of leaching	0	2	10
Component of leached meat (%)	Moisture	76.3	82.7	85.5
	Minerals	1.02	0.26	0.19
	Fat	6.9	3.5	0.7
	Protein	15.8	13.4	13.6
	Water soluble nitrogen component	0.64	0.12	0.02
Property of the gel	Gel quality	Brittle	Elastic	Fine Elastic
	Drip (%)	53.5	40.9	33.6

Concentration of myofibrillar protein and removal of water-soluble components The change in the gel-forming ability of fish meat following leaching is shown in Table 9.2. Water-soluble components, minerals and fat are removed from fish meat by leaching, which correspondingly changes the state of the gel from being brittle to elastic (Okada, 1981). Leaching is an effective method of enhancing the gel-forming ability of a raw material that has a weak gel-forming ability. From the measured amounts of water-soluble and myofibrillar proteins in Alaska pollock meat before and after leaching, the former decreased from 20 to 2–4%, and the latter increased from 80 to about 96–98% of the total protein content (Kimura, 2003).

The primary reason for the improvement of the gel-forming ability with leaching is the concentration of myofibrillar proteins consisting mainly of myosin, which is the main protein for the gel formation of fish meat. Generally, WSP has a lower molecular weight than myosin (MW, 480 000), as well as a low gel-forming ability due to the difficulty in forming its network structure. Gel-forming ability is markedly reduced when a myofibrillar protein sol is heated after the addition of the WSP fraction separated by leaching fish meat. On the other hand, the gel-forming ability of myofibrillar proteins is not affected when they are heated after the addition of a heat-coagulated WSP fraction or extractable ingredients excluding heat-coagulated WSP (Shimizu and Nishioka, 1974). During heat processing, it is assumed that denatured WSP, which aggregates with denatured myosin, prevents the formation of fine network structures with myosin molecules.

Active 'suwari' enzyme in surimi Suwari is a very important process that markedly affects the rheological properties of surimi-based products. When salted meat paste is kept at a constant low temperature below approximately 40 °C for several tens of minutes to several hours prior to heating at 80–90 °C, its rheological properties change from sol to gel, and *kamaboko* gels of higher elasticity and higher water holding capacity are produced. This phenomenon is called suwari.

All surimi-based products may be affected to a greater or lesser degree by suwari during their manufacture. Accordingly, control of the suwari process is very important for surimi quality control. In frozen surimi, WSP content is a small percentage of the total protein. Table 9.3 shows the results of the changes in the gel-forming ability of the surimi prepared from blue grenadier with several washings and the effects of extracted WSP on the gel properties of suwari following the addition of WSP and Ca to washed surimi. The gel strength of suwari gel prepared with thoroughly washed surimi was remarkably low (Kimura *et al.*, 1991).

Table 9.3 The effect of WSP and Ca ion on the suwari of the washed surimi (Kimura *et al.*, 1991)

No.	$CaCl_2$ (mM)	WSP (mg)	Gel	JS (g · cm)	MHC (MHC/A)
1	–	–	C	250	0.81
2	–	–	S	214	0.77
3	–	–	C	275	0.81
4	–	–	S	793	0.27
5	5	200	C	237	0.96
6	5	200	S	1289	0.06

C, boiled at 90 °C for 40 min. S, placed in a water bath at 30 °C for 1 h and then boiled at 90 °C for 40 min. JS, jelly strength. No.: 1, 2, washed surimi alone; 3, 4, washed surimi and 5 mM $CaCl_2$; 5, 6, washed surimi, 5 mM $CaCl_2$ and WSP.
MHC myosin HC to actin by measuring SDS-PAGE of kamaboko gel.

Suwari gel strength did not recover even with the addition of calcium ions. The gel strength of a suwari gel prepared from washed surimi with WSP at the same concentration as that in actual surimi was almost the same as that of a suwari gel prepared from original surimi. Thus, it was shown that the suwari process of surimi gel formation can be controlled by adjusting the WSP fraction in surimi. Since the suwari process depends on Ca ions and the cross-linking of ε-(γ-glutamyl)-lysine isopeptides occurs in a suwari gel, direct proof of the participation of transglutaminase (TGase) in suwari gel formation can be obtained, (Kimura *et al.*, 1991).

Since leaching both removes WSPs, most of which inhibit surimi gelation, and maintains TGase, leaching conditions are important in surimi production. It is suggested that thorough washing of fish meat weakens suwari gels. Figure 9.3 shows the relationships between the TGase and Ca-ATPase activities of surimi prepared from Alaska pollock, golden threadfin bream, white croaker and southern blue whiting under different processing

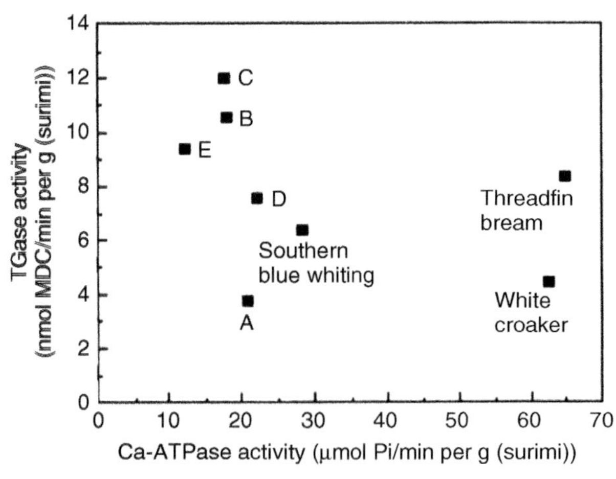

Figure 9.3 The relationship between TGase activity and Ca-ATPase activity of surimi (Seki and Nozawa, 2001). A–E, Alaska pollack surimi

Table 9.4 Proximate and free amino acids composition of Alaska pollock flesh (Kimura, 2003)

Component	Flesh	Surimi
Moisture (%)	80.3	73.4
Fat (%)	1.0	0.9
Protein (%)	17.4	16.8
Minerals (%)	1.3	0.6
Free amino acid (g/100 g)	0.3190	0.0408
Phosphoserine	0.0011	–
Taurine	0.1511	0.0179
Aspartic acid	0.0037	–
Threonine	0.0057	0.0008
Serine	0.0072	0.0009
Glutamic acid	0.0120	0.0012
Glycine	0.0254	0.0057
Alanine	0.0230	0.0033
Citrulline	0.0011	–
Valine	0.0035	–
Methionine	0.0030	0.0014
Cystathionine	0.0012	–
Isoleucine	0.0027	0.001
Leucine	0.0043	0.0011
Tyrosine	0.0023	–
Phenylalanine	0.0012	–
β-Alanine	0.0129	–
Ammonia	0.0064	0.0018
Ornitine	0.0014	0.0004
Lysine	0.0122	0.0025
L-methylhistidine	0.0227	0.0018
Histidine	0.0014	–
Arginine	0.0037	0.0008
Proline	0.0058	–

conditions. Because the TGase activities differed by about threefold among different types of Alaska pollock surimi, it is suggested that leaching conditions affect TGase activity (Seki and Nozawa, 2001).

Removal of extractable compounds by leaching During leaching, taste components such as free amino acids, are removed (Table 9.4) (Kimura, 2003). The amount of these taste components is reduced to less than 20%. Among the extractable components, trimethylamine oxide (TMAO), which is supposed to be involved in osmotic pressure regulation, decomposes during frozen storage, thereby generating dimethylamine (DMA) and formaldehyde (FA). However, TMAO-ase, which decomposes TMAO, does not act under frozen storage; thus TMAO decomposition progresses via a chemical reaction. TMAO decomposition in Alaska pollock meat during frozen storage is shown in Table 9.5 (Tokunaga, 1964). It is shown that about 85 mg/100 g TMAO is contained in Alaska pollock meat along with generated formaldehyde, which supported the progression of

Table 9.5 Freezing-related generation of dimethylamine and formaldehyde (Tokunaga, 1964)

Species	Storage period	TMAO-N (mg/100 g)	DMA-N (mg/100 g)	FA (mg/100 g)
Allaska pollock	0	85.2	0.1	0.2
	7		1.1	2.5
	28		3.7	5.3
	90		11.0	12.7
	180		24.5	20.7
Northern blenny	180	100.7	(±)	(−)
Arabesque greenling	0	81.3	0.15	(±)
	45		0.16	(±)
Bastard halibut	0	78.4	0	(−)
	120		0	(−)

protein denaturation during frozen storage. These components are removed by leaching (Tokunaga, 1965). The DMA generation from leached fish meat during frozen storage is also affected by the frozen temperatures. Although DMA is generated during frozen storage because a little TMAO remains in leached meat, TMAO decomposition and DMA generation are suppressed during frozen storage at below −20 °C.

Inorganic substances are also removed by leaching. It has been shown that about 60% of the inorganic substance is removed, although iron, calcium and magnesium are difficult to remove completely. In addition, the elution level of inorganic substances depends on the components of the leaching water used (Kimura, 2003).

Effect of ionic strength of leaching water Various water-soluble components are removed by leaching, causing muscle protein to swell. Leaching reduces ionic strength which affects water holding capacity. The water holding capacity becomes minimal at ionic strengths as low as 0.05–0.1. It increases at ionic strengths lower than 0.05 and higher than 0.1 (Okada, 1981). With repeated leaching, the ionic strength of washed fish meat decreases, so the meat swells by water absorption. This makes dehydration of washed meat difficult. For efficient leaching, ingenuity to maintain leaching conditions and leaching water quality is necessary. pH is another important factor affecting the hydration of myofibrillar protein. Water holding capacity becomes minimal at pH values near the isoelectric point of myofibrillar protein. The isoelectric point of myofibrillar protein is near pH 5.5; thus, the rate of hydration is lowest at this pH, while the rate of water absorption is higher below or above pH 5.5 (Figure 9.4) (Okada, 1981).

The highest stability of myofibrillar protein occurs at pH values around 7.2, as shown in Figure 9.5. Myofibrillar protein is easily denatured above or below pH 7.2 (Hashimoto and Arai, 1978). Leaching conditions and leaching water quality are very important factors affecting gel-forming ability, remaining water-soluble components, colour of frozen surimi and preservability during frozen storage.

Alkaline saline leaching (see also 9.3.1) Dark-fleshed fatty fish species such as sardine and mackerel are also used as raw materials. The leaching process for fish meat of these species is different from that for white fish meat. Water-soluble components can be

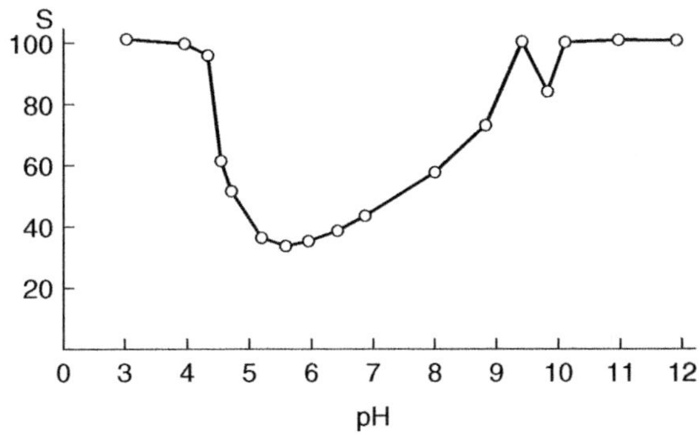

Figure 9.4 Effect of pH on the water holding capacity of horse mackerel meat (Okada, 1981). 'S' is the sedimentation volume of the sample (fish meat : water = 1 : 3) after centrifugation

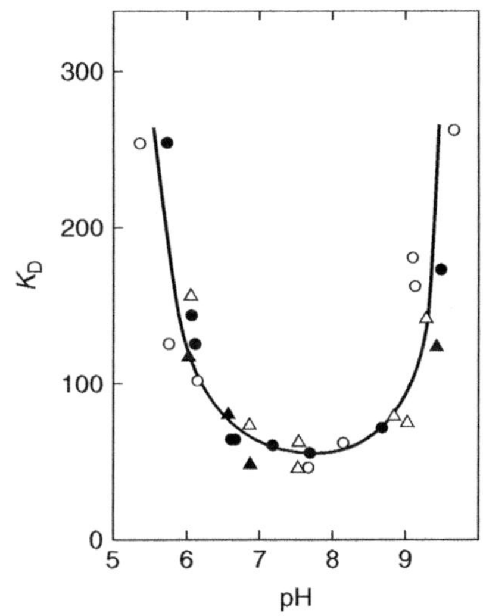

Figure 9.5 Influence of pH on the denaturation of myofibril of various fish species (Hashimoto and Arai,1978). K_D, first order rate constant of inactivation of myofibrillar Ca-ATPase. Incubation of myofibrils were conducted at 30 °C for atka-mackerel (○), at 35 °C for ray (△), chub mackerel (•) and mackerel pike (▲), respectively

efficiently removed by adjusting the pH of fish meat from acidic to neutral by alkaline saline leaching.

9.2.7 Refining

After leaching, the partially dewatered minced fish meat is refined to remove connective tissues, skin, scales, or other undesirable inclusions. Temperature control at this stage is very important because an increase in the temperature of the dehydrated minced fish meat during this process often results in protein denaturation and loss of protein functionality. There are few studies of the effects of connective tissues on surimi quality; however, it is established that connective tissues are generally removed from leached minced fish meat because they lead to brittle surimi gels whose main protein is thermo-reversible collagen (Mizuta, 2001).

9.2.8 Dewatering

Refined meat is dewatered mechanically using a screw press, reducing the water content of the fish meat to about 80–84%. Dewatering efficiency is affected by the type of screw press used and the processing conditions up to the dewatering step. To maintain excellent surimi quality, the control of surimi moisture content is important, necessitating an on-line moisture measurement system applicable to factory operation.

To achieve this, a non-destructive moisture determination using near-infrared spectroscopy was investigated and revealed that the protein and moisture content of surimi can be measured accurately (Uddin and Okazaki, 2008). The practical use of this technology is expected in the future.

9.2.9 Blending of cryoprotectants

Cryoprotectants such as sugar, sorbitol (8–9%), and polyphosphates (0.2–0.3%) are added and mixed to the dewatered fish meat to stabilize fish proteins against denaturation during frozen storage. A silent cutter equipped with a cooling unit is generally used since the temperature of fish meat tends to increase, which affects the quality of the final products. It is important to maintain a low temperature as much as possible to prevent denaturation of fish meat proteins during processing.

9.2.10 Freezing

Blended meat with cryoprotectants is packed into polyethylene bags, which are then placed in freezing pans and quickly frozen by subjecting to high pressure in a contact freezer. To suppress denaturation under freezing, it is necessary to reduce the size of ice crystals in surimi as much as possible. After complete freezing, two surimi blocks are packaged in one box with the following printed information: fish material, type of additive used, date of manufacture, manufacturer's name, and surimi quality specifications.

9.2.11 Frozen storage and transport

The temperature and temperature fluctuation during frozen storage have serious effects on the storage stability of frozen surimi. Repeated temperature fluctuations induce freezing denaturation accompanied by the growth of ice crystals in surimi. The relationship between storage temperature and frozen surimi quality, as reported by Okada, is shown in Figure 9.6. Although the quality of frozen surimi is fairly stable at −20 °C, it is desirable to store surimi at temperatures below −25 °C (Toyoda and Fujita, 1991). The changes in surimi quality during frozen storage have also been confirmed by measuring myosin ATPase activity (Kimura *et al.*, 1977).

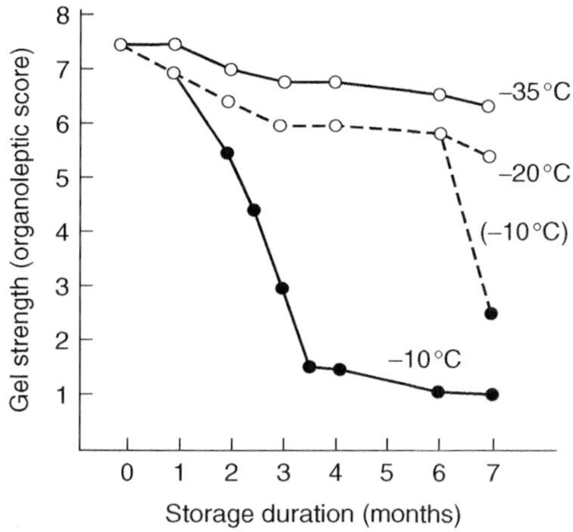

Figure 9.6 Effect of frozen storage temperature on the gel-forming ability of Alaska pollock surimi (Kimura, 2003. The original data was provided by Okada)

9.3 Characteristics of fish material and manufacturing technology

In the previous section, the production method of frozen surimi using mainly Alaska pollock and the basic principles related to surimi quality were introduced. This section will describe the characteristics of the fish material and the surimi manufacturing process.

9.3.1 Surimi from dark-fleshed fatty fish species

Although the basic production technology of frozen surimi from dark-fleshed fatty fish species is almost the same as that for white-flesh fish, methods to deal with the specific characteristics of these fish species were developed. The characteristics of dark-fleshed fish species are the following: (i) high fat content with high seasonal variation (Figure 9.7); (ii) high ratio of dark muscle to ordinary muscle (Table 9.6); (iii) high myoglobin content in ordinary muscles, which is dark red, (iv) rapid drop of fish flesh pH after death; (v) generally small body size compared with white-flesh fish species.

Post-mortem changes of fish muscle protein and strategy for surimi production The post-mortem change of dark-fleshed fish species is characteristically accompanied by a significant pH decrease immediately after death. As shown in Figure 9.5, the denaturation rate constant of myofibrillar Ca-ATPase is rapidly destabilized with a shift to an acidic pH, although myofibrillar Ca-ATPase is most stable at approximately pH 7.2 (Hashimoto and Arai, 1978) . Adequate pH control is very important since protein denaturation of low-pH fish meat cannot be sufficiently prevented even if it is stored at lower temperatures (Arai, 1981) (Figure 9.8).

The relationship between the freshness of the fish material and the gel-forming ability of surimi is shown in Figure 9.9. It is shown that the decrease in the freshness of the material greatly affects surimi quality. For dark-fleshed fatty fish, their freshness affects

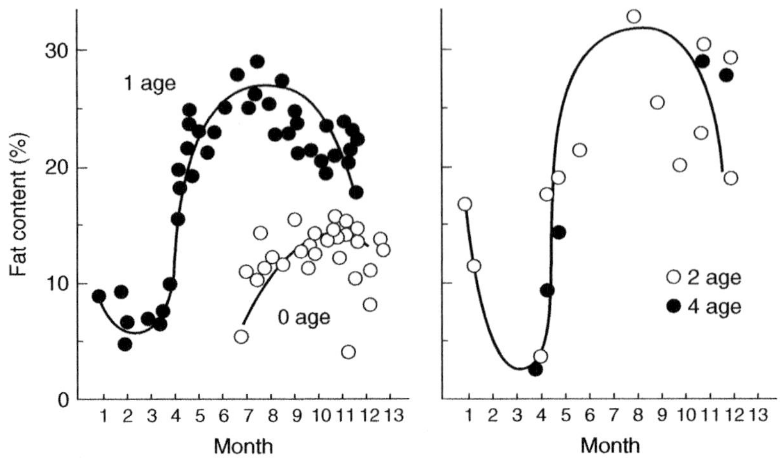

Figure 9.7 Seasonal variation of the fat content of sardine meat (Hasegawa, 1977)

Table 9.6 Ratio of dark muscle in various fish species (Obatake and Heya, 1985)

Species	Dark muscle ratio (%)	Species	Dark muscle ratio (%)
Sardine	20.7	Grunt	5.4
Saury pike	18.6	Japanese barracuda	5.2
Frigate mackerel	14.5	Slipmouth	4.7
Round herring	13.9	Deep-sea smelt	3.8
Gizzard shad	13.6	Goatfish	2.9
Common mackerel	12.0	Golden thread	2.8
Horse mackerel	8.6	Yellow sea bream	2.2
Tigerfish	6.0	Sillago	1.6

surimi quality much more than the case for white-flesh fish. The promotive effect of alkaline saline leaching on the gel-forming ability of surimi is shown in Figure 9.10. In the manufacture of surimi from dark-fleshed fish, the immediate cooling of fish materials after a catch, rapid surimi processing and adjustment to neutral pH by alkaline saline leaching treatment with sodium hydrogen carbonate and/or phosphate solution are essential.

Removal of fat from dark-fleshed fatty fish species The purpose of leaching fish meat of dark-fleshed fatty fish species is not only to adjust pH but also to remove the pigment components (haeme proteins) and fat from the muscles. During leaching, the lipid components in the muscles are extracted into the leaching water. It is desirable to remove the lipids in the muscles as much as possible, since remaining lipids in surimi not only decrease the gel-forming ability of frozen surimi but also cause lipid oxidation during frozen storage, which affects the flavour of the final product (Kochi, 1981).

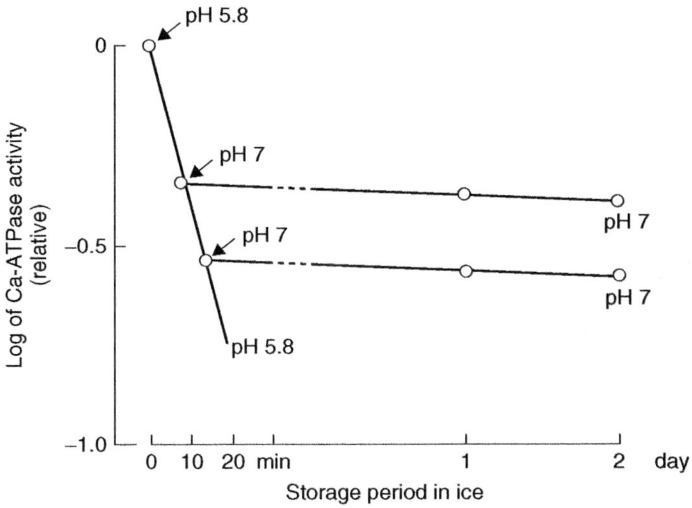

Figure 9.8 Stabilization of sardine myofibrillar protein by pH neutralization (Arai, 1981)

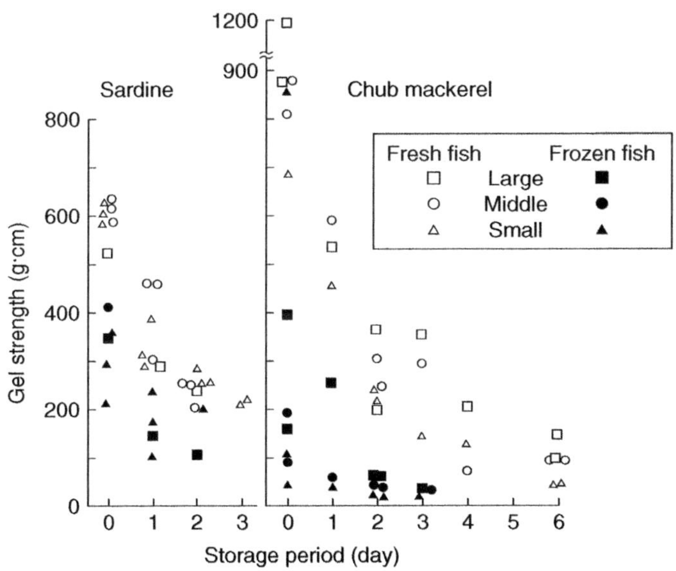

Figure 9.9 Gel strength of *kamaboko* prepared from fresh and frozen fish with different levels of freshness (Fujii, 1981)

The use of a continuous centrifugal dewatering system using a decanter after grinding the fish meat into fine particles in leaching water has been developed and proposed for efficiently removing fat from fatty fish meat (Figure 9.11).

On the other hand, fish oil composed mainly of triglycerides improved the gel-forming ability of surimi by emulsification. The degree of such improvement increases with the

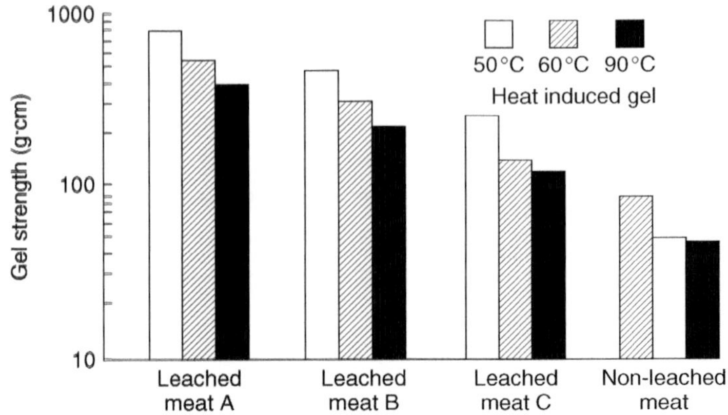

Figure 9.10 Improvement of the gel-forming ability of mackerel meat by alkaline saline leaching (Kimura, 2003. The original data was provided by Shimizu.). Leached meat A, alkaline saline leaching with 0.025 M $NaHCO_3$ and 0.025 M NaCl; B, leaching with water with adjustment to pH 7.0; C, leaching with water without pH adjustment

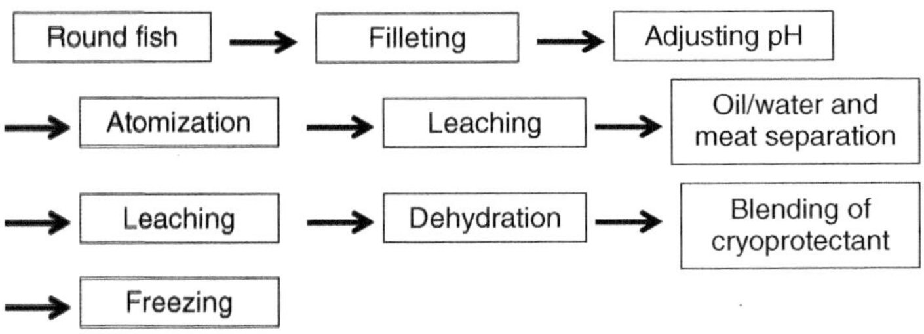

Figure 9.11 Manufacturing process for frozen surimi using a method involving grinding into fine particles and dewatering with a decanter

decrease in the size of the emulsified oil particles (Okazaki *et al.*, 2002) (Figure 9.12). Moreover, emulsified oil is highly stable against oxidation (Azuma *et al.*, 2009); this feature has also been confirmed with emulsified surimi (Okazaki *et al.*, 2002). It is already established that docosahexaenoic acid (DHA) and icosapentaenoic acid (IPA) in marine resources have high functionality. Therefore, intermediate processing materials such as surimi emulsified with the original DHA and IPA content are promising for use as functional materials in the future.

9.3.2 Surimi production from fish species with high protease activity in the muscle

Mucus sporozoans (Myxosporidia) thrive as parasites in the muscle of Pacific whiting, a white-flesh hake of the genus *Merluccius* migrating off the west coast of the United States from Canada. When heated slowly, the fish meat with the mucus sporozoans becomes mushy, even dissolving in extreme cases (Okada *et al.*, 1981).

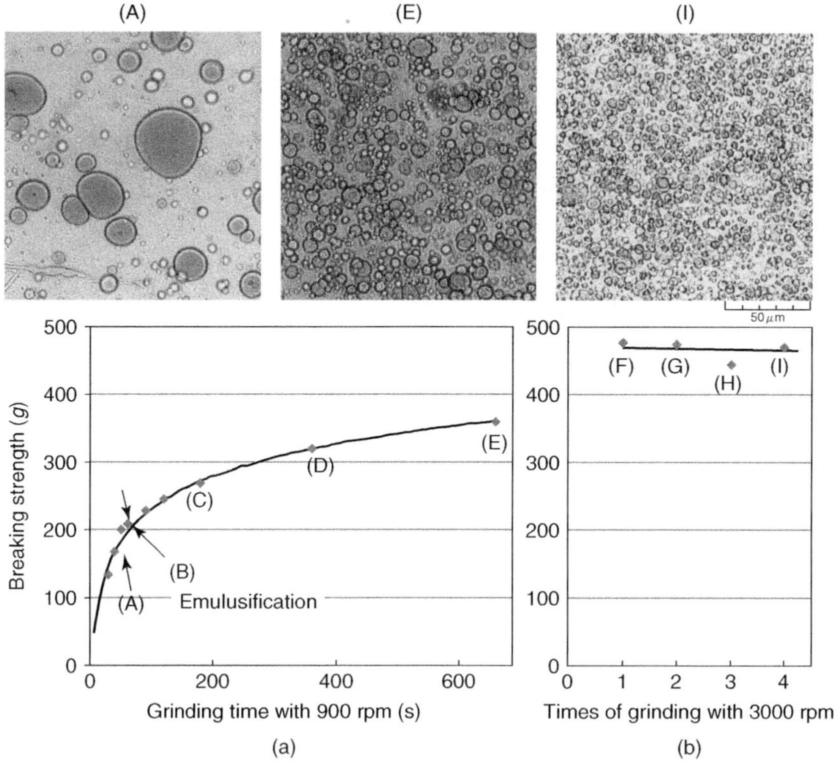

Figure 9.12 Improvement of gel-forming ability of surimi by emulsification of fish oil (Okazaki *et al.*, 2002). (A)–(I): Microscopic observations of fish oil droplets in emulsified surimi obtained by mild and vigorous agitation conditions. Changes in heat-induced gel-forming ability of emulsified surimi obtained by (a) mild (900 rpm) and (b) vigorous (3000 rpm) agitation conditions

The change in the SDS-PAGE pattern of Pacific whiting surimi during heating is shown in Figure 9.13. The myosin heavy chain is degraded almost completely, with the degradation of actin. This indicates that a very strong protease remains in the surimi. This protease has been identified to be the cathepsin L-type enzyme (Toyohara *et al.*, 1993). Egg white is mainly used as a protease inhibitor (Shigeoka, 1981).

9.4 Denaturation of fish protein by freezing and its prevention

There are many factors affecting the denaturation of surimi protein by freezing: the stability of fish protein itself, environmental conditions such as salt concentration surrounding the protein, the concentration of the denaturation-promoting components in the muscles, the concentration and functionality of sugars and polyphosphate as suppressive agents against denaturation, freezing rate and storage conditions.

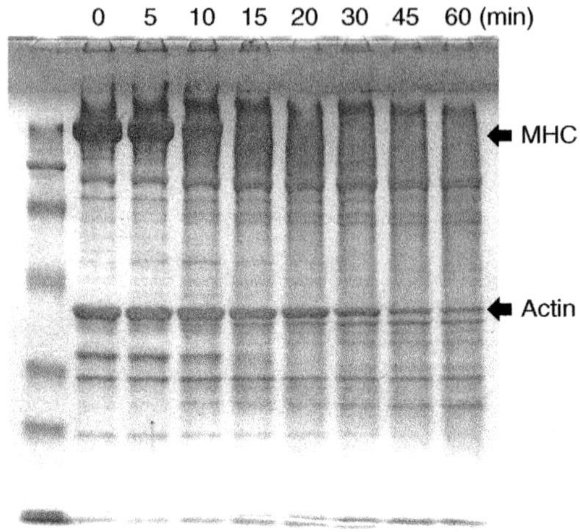

Figure 9.13 Changes in the SDS-PAGE pattern of Pacific whiting surimi during heating at 60 °C (Kimura, 2003)

9.4.1 Stability of fish protein

Myosin, which contributes mainly to the gel-forming ability of surimi, is a unique protein with a high molecular weight and diverse properties such as solubility at high salt concentration, enzymatic action as an ATPase, gel-forming ability of aggregated filaments at low salt concentrations, the coupling action with actin. Therefore, when discussing the stability of fish protein, it is effective to consider using these properties as indices.

The denaturation rate of myofibrillar protein during frozen storage can be represented using the denaturation rate constant K_D of Ca-ATPase. A study using several fish species during frozen storage at several temperatures revealed that denaturation varies depending on species, and that it is greatly affected by storage temperature (Fukuda, 1986), (Figure 9.14). In addition, it has been shown that the protein denaturation of fish meat during frozen storage is greatly affected by storage temperature and pH (Figure 9.15).

Considering the conditions for manufacturing surimi from various fish species as raw materials and studying the effects of additives on them, is very important for understanding the material characteristics of each fish.

9.4.2 Substances promoting protein denaturation during frozen storage

The components of fish meat promoting protein denaturation during frozen storage are acidic components such as organic acids (which lower the pH of fish meat), various types of salts, formaldehyde (degradation product of TMAO) and oxidative lipid products. Calcium chloride is sometimes used to increase the efficiency of dewatering leached fish meat (Figure 9.16) (Saeki *et al.*, 1985). It has been reported that calcium chloride below 15 mM does not promote denaturation of the Ca-ATPase activity of myofibrillar protein (Saeki *et al.*, 1985). However, the denaturation of myofibrillar protein is strongly induced by the addition of calcium at high salt concentrations during *kamaboko* gel preparation.

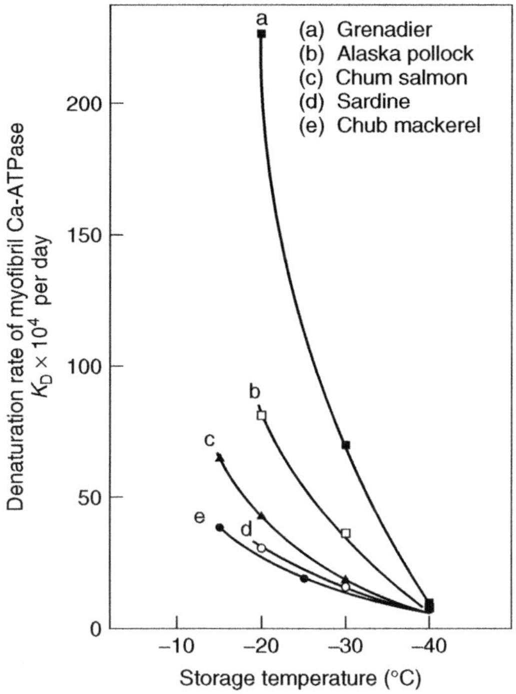

Figure 9.14 Changes in myofibril Ca-ATPase of five fish species during frozen storage (Fukuda, 1986)

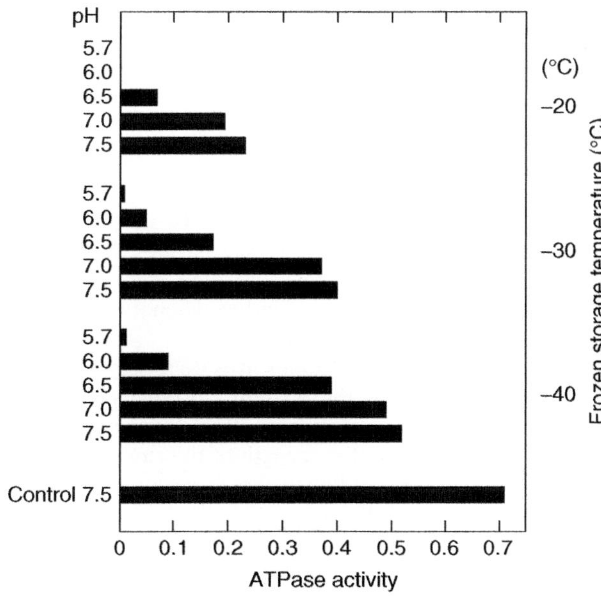

Figure 9.15 Effect of pH and frozen storage temperature on fish protein denaturation (Okazaki, 2009. The original data was provided by Fukuda)

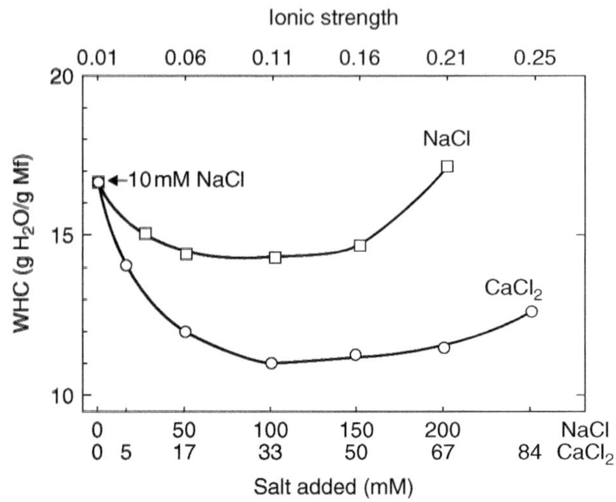

Figure 9.16 Effect of calcium chloride on the water holding capacity (WHC) of carp myofibrillar protein (Saeki *et al.*, 1985)

9.4.3 Cryoprotectants and their mechanism of action

Generally, frozen surimi is manufactured mainly from white-fleshed fish species inhabiting cold-water regions, such as Alaska pollock, southern blue whiting and blue grenadier. The myofibrillar protein of these fish species has an unstable structure, and its denaturation progresses rapidly with freezing and preservation.

Many substances, such as sugars, sugar alcohols and polyphosphates, have been investigated as cryoprotectants against the freezing-induced denaturation of myofibrillar protein (Table 9.7).

To discuss the suppressive effects of additives on the freezing-induced denaturation of myofibrillar protein, it is necessary to measure the degree of denaturation quantitatively.

A positive correlation between the molar concentration (M) of the additives and the logarithmic value of K_D was found (Ooizumi *et al.*, 1981).

The suppressive effects of sugars and sugar alcohols on the heat denaturation of myofibrillar protein are shown in Figure 9.17. Polysaccharides with a high number of OH residues tend to have a highly suppressive effect (Ooizumi *et al.*, 1981). The concentration of polysaccharides required to suppress freezing-induced denaturation of surimi was about 8% (w/w) of those to suppress thermal denaturation.

To evaluate the suppressive effect of additives, it is necessary to know their solubility at low temperatures. In addition, sugars that exhibit a protective effect against freezing-induced denaturation also exhibit an inhibitory effect during pre-heating (suwari) in *kamaboko* manufacture. The change in the strength of the suwari gel pre-heated at 30 °C prior to heating at 90 °C, prepared by adding sugars at various concentrations to Alaska pollock surimi is shown in Figure 9.18 (Kimura, 2003). The suwari rate was suppressed by increasing sugar concentration. Cryoprotectants also affect protein denaturation and gel formation rate during the processing of surimi-based products.

Table 9.7 Preventive effects of various substances on protein denaturation during frozen storage (Noguchi and Matsumoto, 1971; 1975; 1976)

Group	More effective	Effective	Minimally effective or ineffective
Sugar	Xylitol, sorbitol, glucose, galactose, fructose, lactose, sucrose, maltose	Glycerin	Propylene, glycol, starch
Amino acid	Aspartic acid, glutamic acid, cysteine	Lysine, histidine, serine	Glycine, leucine, isoleucine, valine, phenylalanine, tryptophan, threonine, glutamine, asparagine, ornithine
Carboxylic acid	Malonic acid, methylmalonic acid, maleic acid, glutaric acid, lactic acid, L-malic acid, tartaric acid, gluconic acid, citric acid, γ-aminobutyric acid	D,L-malic acid, Adipic acid	Fumaric acid, succinic acid, oxalic acid, pimelic acid, suberic acid
Others	Ethylenediaminetetraacetic acid	Triphosphate acid, Pyrophosphate acid	Ethylenediaminine, creatinine

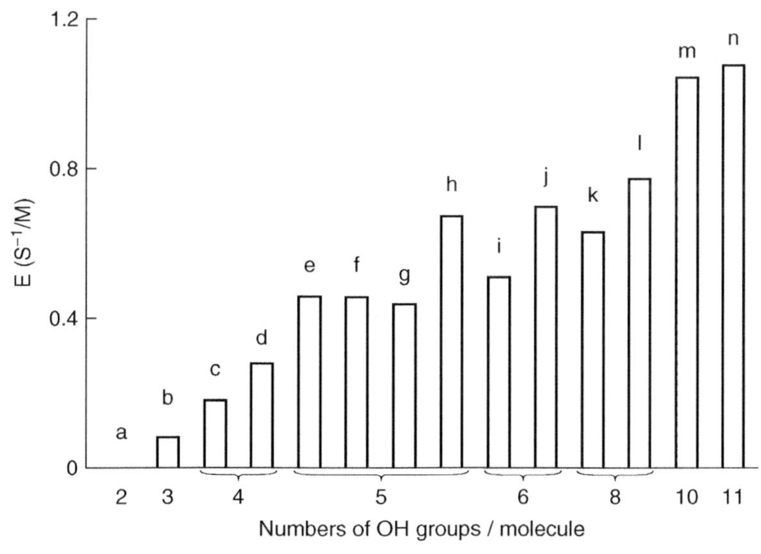

Figure 9.17 Relationship between the protective effects of various sugars on the thermal denaturation of Ca-ATPase activity of mackerel myofibrils and the number of OH residues (Ooizumi *et al.*, 1991). E, the protective effect against denaturation; a, sorbide; b, glycerol; c, sorbitan; d, xylose; e, xylitol; f, fructose; g, mannose; h, glucose; i, mannitol; j, sorbitol; k, maltose; l, sucrose; m, lactitol; n, maltotriose

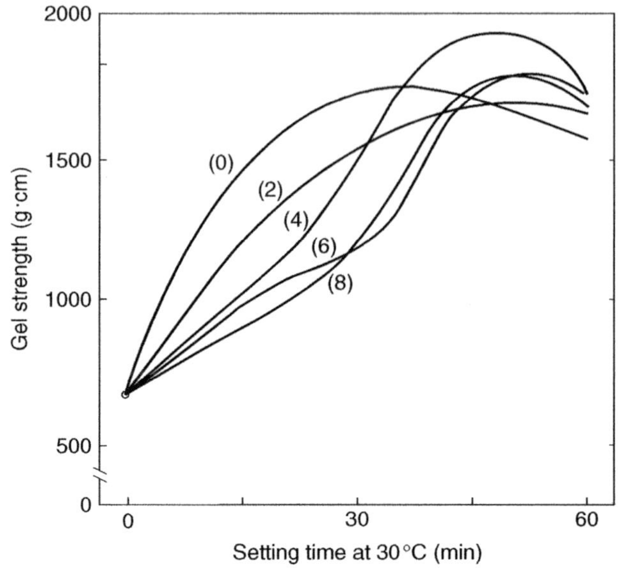

Figure 9.18 Effect of sugar concentration on the rate of suwari progress (Kimura, 2003). Sugar concentration (%)(w/w) in parentheses

9.4.4 Effects of polyphosphates

The polyphosphates generally used in surimi manufacture are sodium pyrophosphate and a mixture of sodium tri-polyphosphate and sodium pyrophosphate. Although the amount of polyphosphate added depends on the type of surimi, 0.2 to 0.3% (w/w) polyphosphates is generally added to surimi. Polyphosphates are supposed to function as follows: (i) adjust (increase) pH, (ii) increase ionic strength, (iii) chelate Ca and Mg ions, and (iv) dissociate myosin and actin. The effect of polyphosphates on surimi noticeably appears during salt grinding and enhances the gel strength of surimi-based products (Figure 9.19). Degradation of polyphosphates may be observed under unfavourable surimi preservation conditions.

9.5 Evaluation of surimi quality

Frozen surimi is generally not consumed directly, but undergoes further processing. This means that the quality of frozen surimi is measured by both its compositional properties and its functional properties for surimi-based products. Therefore, it is strongly recommended to inspect such functional properties, such as the following quality attributes, that are different from those for other fishery products.

It is most important to evaluate the following primary test attributes: moisture content, pH and objectionable matter of raw surimi and gel strength, deformability, and colour of cooked surimi gel. Other secondary attributes may be measured as desired (Box 9.1).

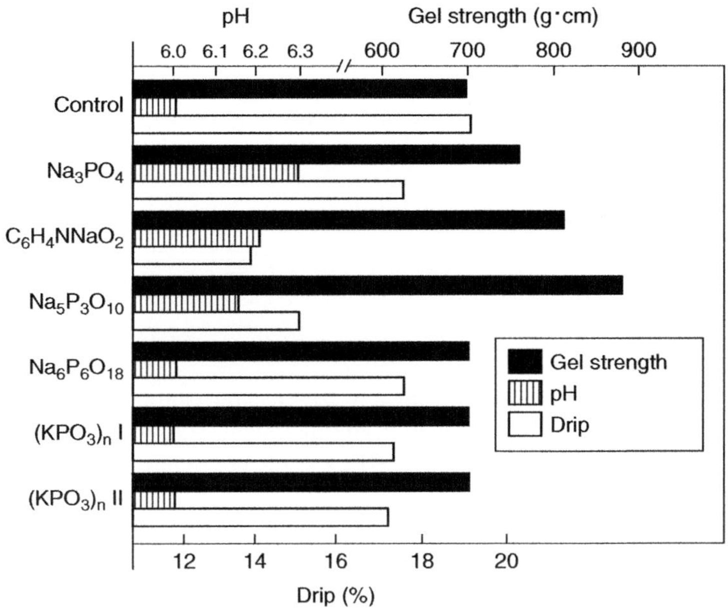

Figure 9.19 Enhancing effect of phosphates on the gel strength of surimi-based products (Okada, 1981)

Box 9.1 Items of quality evaluation of frozen surimi prescribed by the Codex standards

Raw surimi tests	Cooked surimi gel tests
1. Primary quality attribute Moisture content pH Objectionable matter	1. Primary quality attribute Gel strength and deformability (puncture test or torsion test) Color
2. Secondary quality attributes Objectionable matter (scales) Crude protein content Sugar content Crude fat content Color and whiteness Pressure induced drip	2. Secondary quality attributes Preparation of test sample water-added surimi gel: starch-added surimi gel suwari (setting)-treated surimi gel Test method Whiteness Expressible moisture Folding test: Sensory (biting) test

The international standards for the quality evaluation of frozen surimi is presented in CODEX standard (Code of practice for fish and fishery products, CAC/RCP 52–2003; http://www.codexalimentarius.net/input/download/standards/10273/CXP_052e.pdf).

There have been prepared in two ways for gel strength and deformability of cooked surimi gel test and several kinds of secondary quality attributes, the item will be decided between buyers and sellers.

A part of the puncture test, which is most commonly used as a method of measuring the gel strength of frozen surimi, is shown in Box 9.2.

Box 9.2 The procedure of puncture test as a cooked surimi gel test

Preparation of test sample

Put 2–10 kg of frozen surimi in a polyethylene bag, seal the bag, and temper the surimi at room temperature (20 °C) or below so that the temperature of the surimi rises to approximately −5 °C. Do not soften the surface of the test sample.

Preparation of surimi gel for testing: Surimi gel not containing added starch

A. Comminution. Sample volume necessary for surimi paste preparation depends on the capacity of mixing instrument used. Use of 1.5 kg or more is necessary to represent the property of 10 kg of block. Regarding that enough amount of surimi is necessary for consistency of testing, equipment of large capacity which can mix surimi of 1.5 kg or more must be installed in laboratory. A larger size of the equipment also requires an adequate amount of surimi in accordance with equipment to secure enough texture of surimi paste. Crush 1.5 kg or more of the test sample with a silent cutter, then add 3% of salt, and further grind and mash it for 10 min or more into homogenized meat paste. Remember to keep the temperature of the material to be tested at 10 °C or less.
Desirable timing for adding salt is at −1.5 °C.
Desirable temperature of the test material is 5–8 °C.

B. Stuffing. Stuff a polyvinylidene chloride tube of 48 mm width (30 mm diameter), when flat, with approximately 150 g (to make it approximately 20 cm long) of the meat paste using a stuffer with an 18 mm diameter stuffing tube, and tie the both ends of the tube.

C. Heating. Heat the test material in hot water at 84–90 °C for 30 min.
When the test material is put in, the drop in temperature should not exceed 3 °C.

D. Cooling. Immediately after finishing the heat treatment, put the test material in cold water and cool it fully, and then leave it at the room temperature for 3 h or longer.

Test method

Between 24 and 48 h after cooking, carry out the following measurements on the prepared inspection sample of surimi gel which has a temperature that equilibrates to the room temperature, and record the temperature of the sample at the time of measurement.

Measure the gel strength and deformability of the inspection sample of surimi gel with a squeeze stress tester (rheometer). Use a spherical (plunger) with diameter 5 mm and speed 60 mm/min.

Remove film off the inspection sample of surimi gel, cut it into 25 mm long test specimen, and place test specimen on the sample deck of the tester so that the centre of the test specimen comes just under the plunger. Apply load to the plunger, and measure the penetration force in *g* and the deformation in mm at breakage.
Record the values obtained for penetration and deformation in *g* by integral number. Record the value obtained for deformation in mm to the first decimal place.
Prepare six or more test specimens from the same inspection sample of surimi gel, and test each of them. Record the average values thereby obtained.

9.6 Surimi-based products

9.6.1 The production of surimi-based products in the world

The global production of surimi-based products is estimated at about 1.5 million tonnes (2010) (Pascal, 2012). The largest producer is Japan (37.5%), followed by China (22.4%) and Korea (9.8%) South-East Asia (10.5%) and the European Union (7.2%), etc. The surimi-based products in Japan are mainly traditional products such as *kamaboko, chikuwa* and fried *kamaboko* as well as fish sausage and crab meat-analogue. In South-east Asian countries, fish ball is very popular. Seafood analogues such as crab meat-analogue were developed in Japan in 1975, and have since become very popular all over the world.

9.6.2 General processing techniques of surimi-based products

Surimi-based products are generally processed by the following procedures: thawing the surimi, grinding with 2–3% salt and other ingredients, shaping and heating (steaming, baking, frying, retort-heating, ohmic heating, etc.). The main characteristic of a surimi-based product is its resilient texture. The essential element in the formation of this texture is actomyosin derived from myofibrils in surimi. Grinding surimi with salt is required for the formation of heat-induced gels. Here, salt is not used as seasoning, but for increasing the ionic strength of fish meat to solubilize the actomyosin from surimi into sol form. Heating actomyosin sol until it reaches a certain temperature will result in a network structure with a resilient texture. The general process of manufacturing surimi-based product is shown in Figure 9.20.

9.6.3 Recent technological changes in the production of surimi-based products

Ohmic heating and extrusion cooking Ohmic heating, which is generated by flowing electric current through the food directly is becoming popular in the surimi-based production industry. The fish paste can be heat-processed in a short time with excellent energy efficiency. In addition, the fish paste is cooked uniformly regardless of the size. As the

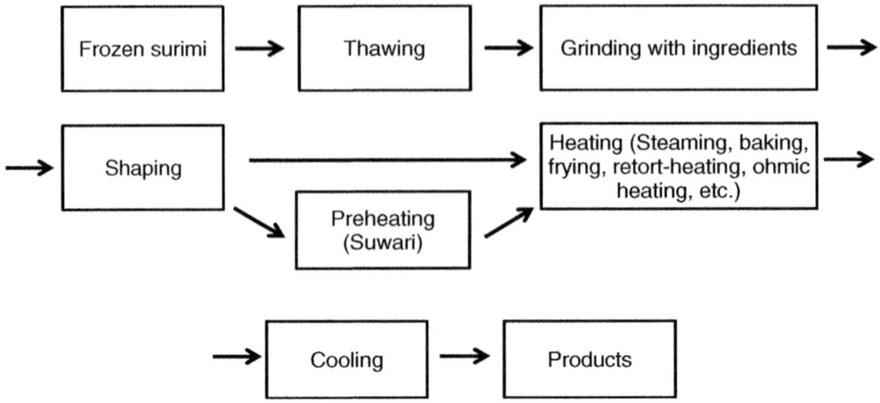

Figure 9.20 General processing procedures of surimi-based products

degradation of the gel network structure at 60–70 °C is minimized, surimi-based products with high gel strength can be produced by ohmic heating rather than by conventional heating methods (Shiba, 1992). Ohmic heating is also used as a pre-cooking or suwari treatment of thin surimi-based products such as fried *kamaboko* and crab-meat analogue.

By using twin-screw extrusion cooking, new products such as seafood analogues with unique textures have been developed.

New ingredients The suwari process is important to give an elastic texture to products. As a transglutaminase preparation of bacterial origin has recently become available in the market, many manufacturers are applying the enzyme to reinforce gel strength. Proteinase inhibitors such as egg white are used for low-grade surimi such as Pacific whiting infected by myxosporidian parasites to prevent the enzymatic degradation of the surimi gel occurring during the heating process.

Recent trends of surimi-based products Many manufacturers have been trying to expand their market by reducing production costs and developing new products to fit consumer trends. In general, the consumers, especially the young ones, prefer softer textured foods that are beneficial to health. Enriched products with functional components such as docosahexanoic acid or calcium as well as low-salt content products are becoming popular.

9.7 Future prospective

Frozen surimi has evolved the manufacturing of fish-paste products which were a traditional seafood in Japan and South-east Asian countries into a modern food industry. Frozen surimi has the following advantages: (i) any kind and any size of fish can be utilized as raw materials; (ii) large-scale production targeting the migratory fish species that allows for large seasonal catches is possible; (iii) it is possible to process the surimi-based products minimizing the wastewater treatment or the disposal treatment; (iv) standardization of the raw material is possible, etc. Owing to these advantages, surimi production has a strong demand all over the world.

In the future, approaches from the following viewpoints will be needed: (i) development of methods for producing surimi in accordance with the characteristics of a variety of fish species with various characteristics; (ii) development for rationalization of the production process (e.g. the methods to minimize the leaching of the fish meat and the addition of additives, etc.)

On the other hand, surimi-based products have great advantages such as: (i) manufacturers can easily obtain the surimi as raw material through the year, and its quality is stable so that the quality control of the product is easy; (ii) any kind of food materials such as seasonings and functional components can be freely mixed according to consumer preferences; (iv) it can be shaped to any form; (v) the products are edible without cooking, etc.

Owing to these advantages, surimi-based products, especially crab meat-analogue, are now produced in many countries. In the future, new types of surimi-based products which are very practical and meet the tastes of many people are expected to be produced.

References

Arai, K. (1981). Characteristics of ordinary muscle, in *More Efficient Utilization of Sardine and Mackerel* (in Japanese), The Japanese Society of Fisheries Science, Koseisya Koseikaku Publishing Co., Tokyo, pp. 20–32.

Azuma, G., Kimura, N., Hosokawa, M. and Miyashita, K. (2009). Effect of droplet size on the oxidative stability of soybean oil TAG and fish oil TAG in oil-in-water emulsion. *Journal of Oleo Science*, 58, 329–338.

Fujii, Y. (1981). Processing of jelly products, in *More Efficient Utilization of Sardine and Mackerel* (in Japanese), The Japanese Society of Fisheries Science, Koseisya Koseikaku Publishing Co., Tokyo, pp. 76–88.

Fukuda, Y. (1986). Effect of freezing condition on quality of frozen fish meat. *The Refrigeration* (in Japanese), 61, 18–29.

Hasegawa, K. (1977). A materials study of sardine. *Bulletin of Ibaraki Prefectural Institute of Fisheries* (In Japanese), 37, 1–5.

Hashimoto, A. and Arai, K. (1978). The effect of pH and temperature on the stability of myofiblillar Ca-ATPase from some fish species. *Nippon Suisan Gakkaishi* (in Japanese), 44, 1389–1393.

Kimura, I. (2003). Chapter 2. Frozen surimi, in *Kamaboko* (in Japanese) (eds Y. Fukuda, N. Seki and M. Yamazawa), Koseisya Koseikaku Publishing Co., Tokyo), pp. 107–147.

Kimura, I., Murozuka, T. and Arai, K. (1977). Comparative studies on biochemical properties of myosin from frozen muscle of marine fishes. *Bulletin of Japanese Society of Fisheries Science*, 43, 315–321.

Kimura, I., Sugimoto, M., Toyoda, K. and Fujita, T. (1991). A study on the cross-linking reaction of myosin in kamaboko 'suwari' gels. *Nippon Suisan Gakkaishi*, 57, 1389–1396.

Kochi, M. (1981). Lipids, in *More Efficient Utilization of Sardine and Mackerel* (in Japanese), The Japanese Society of Fisheries Science, Koseisya Koseikaku Publishing Co., pp. 45–59.

Mizuta, S. (2001). Collagen, in *Roles of Muscle Proteins Other than Myosin Ashi Formation of Kamaboko – Contribution of Myosin, Other Muscle Proteins and Enzymes* (in Japanese) (eds N. Seki and Y. Itoh), Koseisha Koseikaku Publishing Co., Tokyo, pp. 86–96.

Noguchi, S. and Matsumoto, J.J. (1971). Studies on the control of the denaturation of the fish muscle proteins during frozen storage–II. Preventive effect of amino acids and related compounds. *Nippon Suisan Gakkaishi*, 37, 1115–1122.

Noguchi, S. and Matsumoto, J.J. (1975). Studies on the control of the denaturation of the fish muscle proteins during frozen storage–III. Preventive effect of some amino acids, peptides, acetylamino acids and sulfur compounds. *Nippon Suisan Gakkaishi*, 41, 243–249.

Noguchi, S. and Matsumoto, J.J. (1976). Studies on the control of the denaturation of the fish muscle proteins during frozen storage–IV. Preventive effect of carbohydrates. *Nippon Suisan Gakkaishi*, 42, 77–82.

Obatake, A. and Heya, H. (1985). A rapid method to measure dark muscle content in fish. *Nippon Suisan Gakkaishi* (in Japanese), 51, 1001–1004.

Okada, M., Nicanor, A.T. and Edward, Y.Y. (1981). Myxosporidian infestation of peruvian hake. *Nippon Suisan Gakkaishi*, 47, 229–238.

Okada, M. (1981). The theory of manufacture and current situation, in *Shinpan Gyoniku Neriseihin* (in Japanese) (eds M. Okada, T. Kinumaki and M. Yokoseki), Koseisya Koseikaku Publishing Co., Tokyo, pp. 169–224.

Okazaki, E. (2009). Freezing of seafood, in Refrigeration and Freezing Technology of Food (in Japanese), Japan Society of Refrigerating and Air Conditioning Engineers, pp. 82.

Okazaki, E. (1994). A study on the recovery and utilization of sarcoplasmic protein of fish meat discharged during the leaching process of surimi processing. *Bulletin of the National Research Institute of Fisheries Science* (in Japanese), 6, 79–160.

Okazaki, E., Yamashita, Y. and Omura, Y. (2002). Emulsification of fish oil in surimi by high-speed mixing and improvement of gel-forming ability. *Nippon Suisan Gakkaishi* (in Japanese), 68, 547–553.

Ooizumi, T., Hashimoto, K., Ogura, J. and Arai, K. (1981). Quantitative aspect for protective effect of sugar and sugar alcohol against denaturation of fish myofibrils. *Nippon Suisan Gakkaishi* (in Japanese), 47, 901–908.

Pascal, G. (2012). Surimi Market update 2012. The 12th Surimi Industry Forum, pp. 40–42, Oregon State University.

Saeki, H., Ozaki, H., Nonaka, M. *et al.* (1985). Effect of $CaCl_2$ on water-holding capacity of carp myofibrils and on their thermal stability. *Nippon Suisan Gakkaishi* (in Japanese), 51, 1311–1317.

Seki, N. and Nozawa, H. (2001). Transglutaminase, in *Roles of Muscle Proteins Other than Myosin Ashi Formation of Kamaboko – Contribution of Myosin, Other Muscle Proteins and Enzymes* (in Japanese) (eds N. Seki and Y. Itoh), Koseisya Koseikaku Publishing Co., Tokyo, pp. 98–107.

Shiba, M. (1992). Properties of kamaboko gels prepared by using a new heating apparatus. *Nippon Suisan Gakkaishi* (in Japanese), 58, 903–907.

Shigeoka, R. (1981). Development of surimi processing with protease inhibitor. United States Patent 4282653.

Shimizu, Y. (1981). The theory of manufacture and current situation, in *Shinpan Gyoniku Neriseihin* (in Japanese) (eds M. Okada, T. Kinumaki and M. Yokoseki), Koseisya Koseikaku Publishing Co., Tokyo, pp. 220.

Shimizu, Y. and Nishioka, F. (1974). Interactions between horse mackerel actomyosin and sarcoplasmic proteins during heat coagulation. *Nippon Suisan Gakkaishi* (in Japanese), 40, 231–234.

Tokunaga, T. (1964). Studies on the development of dimethylamine and formaldehyde in Alaska pollack muscle during frozen storage. *Bulletin of the Hokkaido Regional Fisheries Research Laboratory* (in Japanese), 29, 108–122.

Tokunaga, T. (1965). Studies on the development of dimethylamine and formaldehyde in Alaska pollack muscle during frozen storage –II– factors affecting the formation of dimethylamine and formaldehyde. *Bulletin of the Hokkaido Regional Fisheries Research Laboratory* (in Japanese), 30, 90–97.

Toyoda, K. and Fujita, T. (1991). Modification in frozen surimi, in *Denaturation of fish muscular protein by processing and its control.* (in Japanese) (ed. K. Arai), Koseisya Koseikaku Publishing Co., Tokyo, pp. 64–72.

Toyohara, H., Kinoshita, M., Kimura, I., Satake, M. and Sakaguchi, M. (1993). Cathepsin-L-like protease in pacific hake muscle infected by myxosporidian parasites. *Nippon Suisan Gakkaishi*, 59, 1101.

Uddin, M. and Okazaki, E. (2008). Fish and related products, in *Infrared Spectroscopy for Food Quality Analysis and Control* (ed. D. Sun), Academic Press, Netherlands, pp. 215–240.

Uno, T. and Nakamura, M. (1958). Studies on the characteristic qualities of fish meat – I. on 'kamaboko'-(steamed fish cake) forming ability. *Hokkaido Regionals Fisheries Reasearch Laboratory Bulletin* (in Japanese), 18, 45–53.

10

Packaging of Fish and Fishery Products

Bert Noseda,[1] An Vermeulen,[1,2] Peter Ragaert[1,2] and Frank Devlieghere[1,2]

[1] *Laboratory of Food Microbiology and Food Preservation, Member of Food2know, Department of Food Safety and Food Quality, Ghent University, Ghent, Belgium*
[2] *Pack4Food, Department of Food Safety and Food Quality, Ghent University, Ghent, Belgium*

10.1 Introduction

The recent increase in the use of different packaging materials has been rapid and encompassed a change from their primary function of containment towards a wide range of properties from acting as a barrier to gases, water and flavours to offer convenience for consumers and as a surface for communicating with the consumers. Furthermore, the use of a diverse range of plastic packaging materials in conjunction with metals (such as aluminium), glass and paper/cardboard has led to multi-component packaging concepts that have an increasing role in the sustainable and social aspects of today's society. Fishery products are highly perishable food products, so adequate preservation techniques must be used to maintain the desired levels of quality and safety throughout storage.

The preservation function of packaging as the final barrier before the distribution and storage of seafood becomes more important as the use of additives are reduced and milder processing techniques (e.g. pasteurization instead of sterilization) are used. The use of vacuum and modified atmosphere packaging (MAP) to increase the shelf life of seafood has therefore increased; however, both techniques – which will be elaborated in the next section – require materials with good gas barrier properties as well as appropriate sealing characteristics.

Another function of packaging that has become more important is convenience which can be seen in the development of portion packaging, microwaveable packaging, easy-to-open packaging and re-closable packaging. Often, these extra properties have to be combined with appropriate gas and water barrier properties to apply them as modified atmosphere packaging (MAP). Finally, packaging attract consumers to buy the packaged product by acting as an important communication interface using a combination of well-designed shapes and well-selected colours. Future developments in food packaging related to these topics will be elaborated in the final section of this chapter.

Seafood Processing: Technology, Quality and Safety, First Edition. Edited by Ioannis S. Boziaris.

10.2 MAP principles and importance for packaging fresh fish

Because of the perishable nature of fishery products, even during chilled storage, the quality of aerobically stored fresh fish deteriorate rapidly and limits the shelf life of the product. The degradation of the product is the result of physico-chemical, autolytical and microbiological processes. Usually, microbiological processes are responsible for spoilage. MAP is a well-known, generally used technique to prolong the shelf life of perishable food products. MAP can be defined as the enclosure of food in a package in which the atmosphere (% CO_2, % CO, % O_2, % N_2) inside the package is modified to provide an optimal atmosphere for retarding microbiological growth on the food product and the accompanying deteriorative microbiological and chemical processes (Young *et al.*, 1988). In the fish industry, two packaging techniques are currently widely used as a preservation strategy: vacuum packaging and packaging with a modified gas composition in the headspace.

10.2.1 Principles of MAP

Vacuum packaging Vacuum packaging can be considered as a specific case of MAP packaging. Removing the gas phase from inside the packaging before it is sealed, reduces the oxygen concentration present inside. So, aerobic growth and the accompanying aerobic metabolism of microorganisms is inhibited, potentially resulting in a longer shelf life of the product. This can be the case for packaging freshwater fish products, but vacuum packaging marine fish products is not recommended due to the enhancing effect of anaerobic conditions on trimethylamine formation (see 10.4.2). Vacuum packaging is generally used for packaging preserved (especially smoked) fish products. It is also used for packaging frozen fish products to avoid the formation of rancid odours and flavours that result from oxidation in the products and to prevent dehydration during frozen storage. Important advantages of vacuum packaging besides an increase of the shelf life are: (i) product volumes are minimized, which is interesting from a logistical point of view; (ii) potential gas leakages in the packaging are easily detectable by simple visual inspection.

Gas packaging A modified atmosphere can be obtained by changing the gas composition (% CO_2, % CO, % O_2, % N_2) inside the packaging. The application of this type of MAP has been widely introduced in the fish industry during the 2000s. The basic effects of the gases used in MAP on fishery products are highlighted here.

Carbon dioxide (CO_2) This is the most important gas used in MAP fish. The dissolution of CO_2 leads to a certain equilibrium of dissolved CO_2 concentration ($[CO_2]_{diss}$) in the aqueous phase of the fishery product according to the following equations (Sivertsvik *et al.*, 2002):

$$CO_{2(gas)} + H_2O_{(liquid)} \rightleftarrows CO_{2(liquid)} + H_2O_{(liquid)}$$

$$CO_{2(liquid)} + H_2O_{(liquid)} \rightleftarrows H_2CO_{3\,(liquid)} \rightleftarrows HCO_3^- + H^+{}_{(liquid)}$$

$$\rightleftarrows CO_3^{2-} + 2H^+{}_{(liquid)}$$

Table 10.1 Example of the concentration of dissolved CO_2 ($[CO_2]_{diss}$) in the water phase of two fishery products packaged under modified atmospheres at 7 °C (Devlieghere and Debevere, 2000)

Fishery product	Water content (g/100 g)	MA (% CO_2/% O_2/% N_2)	$[CO_2]_{diss}$ (mg/L) Gas : product = 1 : 1	Gas : product = 2 : 1
Cooked shrimp	83.6	35/0/65	310 ± 11	561 ± 6
Cod	81.1	50/30/20	417 ± 16	623 ± 45

The solubility of CO_2 greatly increases with decreased temperature and is dependent on the product's water and fat content. Devlieghere *et al.* (1998a, 1998b) have demonstrated that the inhibition of growth of spoilage microorganisms in a modified atmosphere is determined by the concentration of dissolved CO_2 in the product. It is therefore of utmost importance that a sufficient level of CO_2 with a suitable gas/product ratio (G/P) is used in the headspace of the package to achieve the antimicrobial effect. An example of $[CO_2]_{diss}$ depending on the modified atmosphere condition and G/P applied is given in Table 10.1 for cod and cooked shrimp.

Devlieghere and Debevere (2000) demonstrated the antimicrobial effect of $[CO_2]_{diss}$ on typical gram-negative (*Pseudomonas fluorescens, Photobacterium phosphoreum, Shewanella putrefaciens, Aeromonas hydrophila*) and gram-positive (*Lactobacillus sake, Brochothrix thermosphacta, Bacillus circulans*) spoilage bacteria at 7 °C. A linear relationship between $[CO_2]_{diss}$ and the maximum specific growth rate μ_{max} as well as between $[CO_2]_{diss}$ and the inverse of the lag phase λ was established (Figure 10.1a, b).

The growth parameters (λ as well as μ_{max}) of gram-negative bacteria were more influenced by $[CO_2]_{diss}$ than the growth parameters of gram-positive bacteria. The acidic effect of dissolved CO_2 leads to a drop of product surface pH in MAP fish, but this depends on the buffering capacity of the fish product. This decrease in pH could not entirely explain the bacteriostatic effect of CO_2. Other inhibiting mechanisms of CO_2 were also identified and described (Daniels *et al.*, 1985; Sivertsvik *et al.*, 2002). Carbon dioxide can penetrate the bacterial membranes of the cell, leading to a decrease in intracellular cytoplasmic pH, causing an inhibition of cytoplasmic enzymes and enzyme production, but also altering cell membrane functions including nutrient uptake and absorption. Again, as demonstrated by Devlieghere and Debevere (2000), not all spoilage microorganisms on fish are sensitive to CO_2, for example some lactic acid bacteria (LAB; Debevere and Boskou, 1996; Dalgaard *et al.*, 1997).

Oxygen (O_2) Oxygen concentrations in MAP food products are generally kept as low as possible with the primary intention of inhibiting the growth of aerobic spoilage bacteria and to avoid oxidative rancidity. Whether O_2 in MAP may promote oxidative rancidity during storage is not always apparent (Chen *et al.*, 1984; Sivertsvik *et al.*, 2002) (see Section 10.3.2). However, O_2 can also have quality improving properties. Therefore, for certain situations, O_2 can be used:

1. MAP packaging of living shellfish to enhance survival of the species inside the package and thus the coinciding shelf life of the product (Pastoriza *et al.*, 2004; Goncalves *et al.*, 2009).

2. High oxygen concentrations may be used to enhance the sensorial quality (colour and odour) of the product (see Section 10.3.1).
3. High O_2 concentrations in combination with CO_2 are used to package lean marine fish species. O_2 will directly inhibit the reduction of trimethylamine oxide (TMAO) to trimethylamine (TMA) (Boskou and Debevere, 1997) resulting in a slightly longer sensorial perceived shelf life (see Section 10.4.2).
4. It is also known that high O_2 concentrations have an inhibiting effect on certain psychrotrophic spoilage organisms when applied with CO_2 (e.g. *Acinetobacter-Moraxella*, lactic acid bacteria, *Photobacterium phosphoreum*) (Dalgaard *et al.*, 1997; Amanatidou *et al.*, 1999; Jacxsens *et al.*, 2001). Applied elevated O_2 in the package leads to intracellular generation of reactive oxygen species (ROS) such as O_2^-, H_2O_2, OH•, which affect vital cell components and reduce cell viability (Halliwell and Gutteridge, 1984; Fridovich, 1986; Amanatidou *et al.*, 1999). Evidently, microorganisms have developed strategies such as induction of O_2-decomposing enzymes (catalase, peroxidase, superoxide dismutase) or radical scavengers (e.g. glutathione) to avoid lethal damage by oxygen. However, an inhibiting effect of oxygen on the growth of certain bacterial species could be found (Condon, 1987; Fu and Mathews, 1999).

Carbon monoxide (CO) Low levels of CO can be included in MAP of red fish meat (e.g. tuna fillets) to give the fillets a stable bright red colour due to the strong binding of CO to myoglobin and the formation of carboxymyoglobin. In the EU, however, the use of this gas in food packaging is not allowed. For aquatic food, in particular for histidine-rich fishes such as tuna, the use of CO provides an additional risk of histamine formation (Droghetti *et al.*, 2011). Since 2002, the Food and Drug Administration (FDA) has accepted CO as generally recognized as safe (GRAS). It can be used in a gas mixture at a maximum level of 0.4%. The other gases are then CO_2 and N_2. Schubring (2008) reported the successful application of small amounts of CO (0.2–1.0%) in gas mixtures to avoid discoloration of salmon gills. Another possibility is the use of CO for smoking fish.

Nitrogen (N_2) This is an inert and tasteless gas, mostly used as filler gas to complement the desired CO_2 and O_2 concentrations and to prevent the 'snug down' effect of the packages, because of its low solubility in water and fat. Nitrogen has no antimicrobial activity and no demonstrated direct effect to any chemical or biochemical reaction in the fish.

10.2.2 Importance of MAP

The industrial application of MAP has resulted in a 28–52% increase in shelf life for retail packaged cod and haddock fillets (Sivertsvik *et al.*, 2002). Longer shelf lives positively affected the entire market in this sector. In the past there has been a shift in the marketing of fishery products in European retail. Fresh fishery products offered on ice were being replaced by pre-packaged fishery products in combination with a modified atmosphere to extend product shelf life and for consumer convenience. A larger market was reached and a broader variety of products was offered. In Europe, MAP fish by end use already represented 8% of the market in 1986 and 12% in 1990 (Lopez-Galvez *et al.*, 1995) and this segment has continued to grow in the late 2000s and the early 2010s. From data obtained between 2008 and 2011, we can state that in Belgium the percentage of MAP fishery products in retail amounts to 62–68% of the fresh market. For the total fresh fish volume

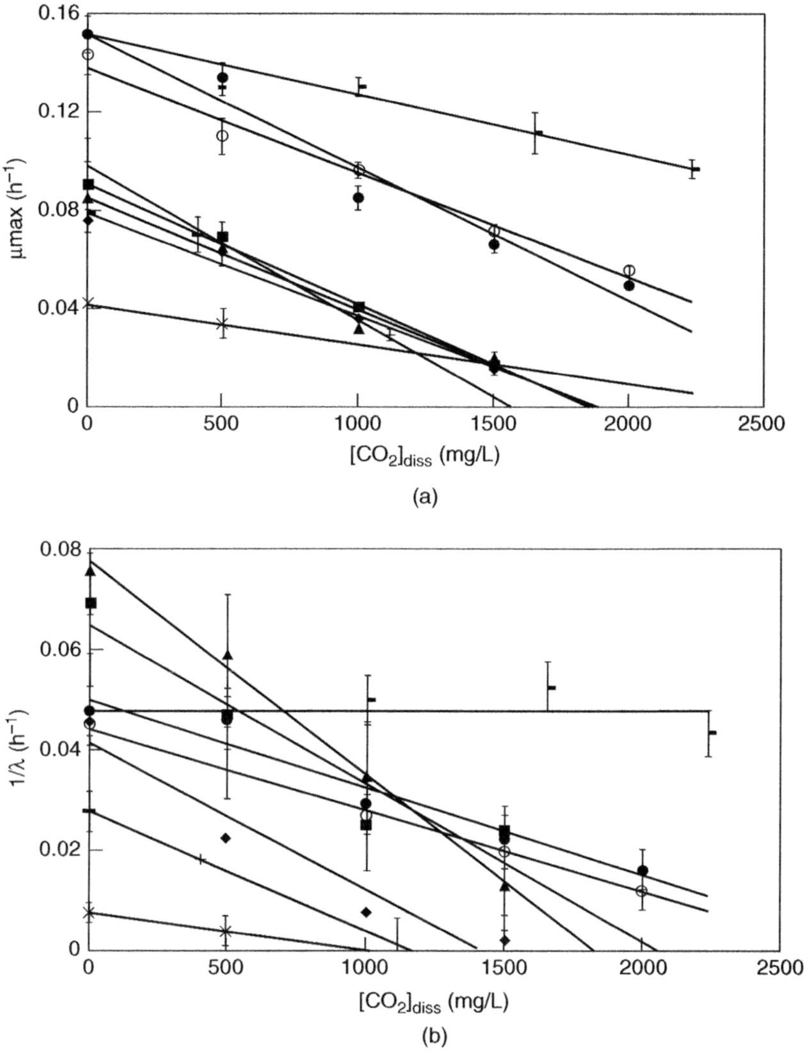

Figure 10.1 (a) Influence of $[CO_2]_{diss}$ on maximum specific growth rate μ_{max} (h^{-1}) at 7 °C of different spoilage organisms. (b) Influence of $[CO_2]_{diss}$ on the inverse of the lag phase $1/\lambda$ (h^{-1}) at 7 °C of different spoilage organisms (◆ *P. fluorescens*, ■ *P. phosphoreum*, ▲ *S. putrefaciens*, + *A. hydrofila*, × *B. circulans*, • *B. thermosphacta* (aerobic), ○ *B. thermosphacta* (anaerobic), - *L. sake*) (Devlieghere and Debevere, 2000)

as end product, thus besides retail and also including special trade, this share still amounts to 40–44%. In the Netherlands the total share of MAP fish in retail exceeds 80% of the market. In France, this is 18–19% because a lot of big retailers are still offering fishery products in cooled counters on ice. It is expected that within the next decade, European consumers will increasingly look for good quality, portion-sized fish, boneless, skinless fresh fish and fishery products (FAO, 2010). These products are mostly fresh, minimally processed foods and ready-to-cook products, which are time-saving, safe, healthy and nutritious products. Although convenience costs money, European consumers are apparently increasingly willing to pay extra for that convenience. In both product groups, MAP plays an important role in meeting the demand for products with longer shelf lives, thus it is probable that the MAP market segment for fishery products will continue to increase.

10.3 Non-microbial effects of MAP

10.3.1 Effect on sensorial quality

MAP gas composition may affect the colour of the fishery product. Carbon dioxide in the MAP may cause bleaching of the fishery product. Fish species which contain carotenoid pigments (e.g. salmon) may bleach during prolonged storage in atmospheres with high carbon dioxide concentrations (Han, 2005). As already suggested, O_2 also plays a role in affecting the colour of packaged red fish meat (e.g. tuna, yellowtails). Fresh tuna fillets are highly perishable due to haemoglobin oxidation, which causes detrimental colour changes. However, tuna fillets are often packaged under super atmospheric O_2 concentrations as this can reduce these colour changes (Torrieri *et al.*, 2011). Carbon monoxide (CO) is also often used to obtain an agreeable colour as it forms a very stable red colour complex with myoglobin (Mb), which is more resistant to oxidation than oxymyoglobin.

Texture and exudate losses may also be affected when using MAP. Increasing the CO_2 concentrations in MAP enhances exudation by acidifying fish muscle, so that the water holding capacity of fish proteins is reduced (Pastoriza *et al.*, 1998). In turn, this may have an effect on the texture of the product (Sivertsvik *et al.*, 2002). This problem may be minimized by dipping the fishery product in a sodium chloride solution (Pastoriza *et al.*, 1998).

The MAP gas composition may also have an effect on the perceived odour and flavour of the fishery product. It is assumed that oxygen concentrations higher than 50% also provide a fresh odour to the packaged product (Hotchkiss, 1988; Dixon and Kell, 1989).

10.3.2 Effect on oxidative rancidity

Generally, a distinction is made between lean fish and fatty fish regarding the occurrence of oxidative rancidity of MAP fresh fish. Oxygen is preferentially excluded from the atmosphere when packaging fatty fishes (e.g. salmon, trout) in order to inhibit oxidative rancidity. However, according to Haard (1992), the development of rancid off-odours in O_2 containing MAP fatty fish species has not been registered as a problem. Giménez *et al.* (2002) found that O_2 concentration in the MAP gas mixtures definitely influenced the oxidation level of MAP rainbow trout fillets (fatty fish species) during storage at $1 \pm 1\,°C$. Significant differences were obtained in thiobarbituric acid (TBA) values towards the end of shelf life, between vacuum packaged fillets and those packaged with 10% O_2, but also between these latter samples and samples packaged with gas mixtures containing 20 and 30% O_2. Vacuum packages presented the lowest TBA reactive substances (TBARS) values, because of the absence of O_2 in the package. The TBARS values increased with rising O_2 concentrations in the gas mixture. Yet, samples with high O_2 concentrations scored better in the sensorial analysis indicating that the oxidation products were not causing rejection of the products (Gimenez *et al.*, 2002). For lean fish, the use of O_2 in

MAP could or should be used for the reasons previously discussed (see Section 10.2). Pastoriza *et al.* (1998) examined the influence of oxygen on MAP stored hake fillets (lean fish species) at 2 ± 1 °C. Packages containing 30% O_2 showed significantly higher TBA values compared to samples packaged with 10% O_2, but again, this did not produce off-odours or off-flavours so the product was not rejected. This implies that the analysis of oxidation products (e.g. TBARS) cannot be used as a good shelf-life indicator.

10.4 Effects of MAP on fish spoilage

A lot of data are available in the literature about fish stored under ice. However, a change in atmosphere and in storage temperature from 0 °C (under ice) up to 4–7 °C (in MAP), implies a change in the microbiological spoilage of fishery products.

10.4.1 Effect of MAP on the spoilage microbiota

The establishment of the specific spoilage organism (SSO) concept has contributed significantly to the understanding of seafood spoilage (Gram and Huss, 1996). The SSOs are typically present in very low numbers and constitute a very small fraction in newly processed seafood (Gram and Dalgaard, 2002). The growth of these SSOs is correlated to spoilage. Once a SSO has reached 10^7–10^8 cfu/g, fish stored in ice start to spoil (Dalgaard *et al.*, 2002). Yet, one cannot just assume that spoilage is attributed to one specific species. Jørgensen *et al.* (2000) introduced the term 'metabolic spoilage association' to describe situations where two or more microbial species contribute to spoilage through exchange of nutrients and/or metabolites. This scenario could be covered by the SSO concept, but note that a consortium of microorganisms interact to spoil the product (Gram *et al.*, 2002). The use of modified atmospheres will cause a change in the dominating microbiota depending on the gas composition used.

Effect of MAP on the spoilage microbiota of marine fish *Shewanella putrefaciens* and *Pseudomonas* spp. are considered to be the main SSOs for iced marine fish regardless of the origin of the fish (Gram and Huss, 1996). These species are able to outcompete other microbiological groups or microorganisms when the fish is stored in air. *Photobacterium phosphoreum* has also been found in spoiled aerobically stored fish species, but high levels of other microorganisms made it unlikely that *P. phosphoreum* was the primary cause of spoilage (Dalgaard *et al.*, 1997). An overview of identified SSOs recovered from marine fish species under different (modified) atmospheric conditions is shown in Table 10.2. On the one hand, the use of CO_2 in MAP inhibits the growth of *Pseudomonas* spp. and *Shewanella spp*. On the other hand, it favours the growth of gram-positive bacteria (e.g. LAB, *Brochothrix thermosphacta*) or the CO_2 resistant *P. phosphoreum* (Gram and Huss, 1996). This latter species has been suggested as the general SSO when CO_2 is applied to packages of temperate water marine fish. Spoilage starts once *P. phosphoreum* counts reach 10^7 cfu/g of fish (Dalgaard 1995a, 1995b).

Although the presence of oxygen in the MAP gas mixture reduced the growth rate of *P. phosphoreum*, it is still the dominant spoilage microflora of cod fillets stored in oxygen-containing modified atmospheres (Dalgaard 1995a, 1995b). Furthermore, O_2 inside the headspace of the packaged fishery product inhibits TMAO reduction into TMA (Boskou

Table 10.2 Overview of SSOs recovered from marine fish species under different atmosphere conditions and different temperatures

Atmosphere	Product	Temp (°C)	MAP % $CO_2/O_2/N_2$	Identified SSOs	Reference
Air	Plaice Pollock, haddock	0		*S. putrefaciens*	Karl and Meyer, 2007
	Haddock	0–15		*S. putrefaciens*, *Pseudomonas* spp. *P. phosphoreum*	Olafsdottir *et al.*, 2006a
	Cod	0.5		*S. putrefaciens*, *Pseudomonas* spp. *P. phosphoreum*	Olafsdottir *et al.*, 2006b
	Pearl spot	0		*Shewanella* spp. *Pseudomonas* spp. *Aeromonas* spp.	Lalitha *et al.*, 2005
	Sea bass	0		*S. putrefaciens* *Pseudomonas spp.*	Poli *et al.*, 2006
(0–50% CO_2) without O_2	Cod	0–5	0/0/100	*P. phosphoreum*	Dalgaard, 1995b
	Little tuna	0	50/0/50	*P. phosphoreum*	Dalgaard *et al.*, 1997
	Sea bream	0	50/0/50	*P. phosphoreum*	Dalgaard *et al.*, 1997
(51–100% CO_2) without O_2	Cod	0–5	100/0/0	*P. phosphoreum*	Dalgaard, 1995b
	Cod Saithe Redfish	0	60/0/40	*P. phosphoreum*	Dalgaard *et al.*, 1997
	Sea bass	0	60/0/40	Lactic acid bacteria	Poli *et al.*, 2006
(0–50% CO_2) with O_2	Pearl spot	0	40/60/0 50/50/0	*B. thermosphacta* *S. putrefaciens*	Lalitha *et al.*, 2005
	Hake	2	40/12/48	*B. thermosphacta* Lactic acid bacteria	Ordonez *et al.*, 2000
	Mullet	4	20/10/70 40/10/70	*S. putrefaciens* Lactic acid bacteria *Enterobacteriaceae*	Pournis *et al.*, 2005
	Cod	0	45/5/50	*P. phosphoreum*	Dalgaard *et al.*, 1997

(*continued overleaf*)

Table 10.2 (*continued*)

Atmosphere	Product	Temp (°C)	MAP % $CO_2/O_2/N_2$	Identified SSOs	Reference
(51–100% CO_2) with O_2	Pearl spot	0	60/40/0	*B. thermosphacta* *S. putrefaciens*	Lalitha *et al.*, 2005
	Mullet	4	60/10/70	*S. putrefaciens* Lactic acid bacteria *Enterobacteriaceae*	Pournis *et al.*, 2005
	Cod	0	60/40/0	*P. phosphoreum*	Dalgaard *et al.*, 1997
Vacuum	Cod	0–4		*S. putrefaciens* *P. phosphoreum*	Gram and Huss, 1996
	Cod	5		*S. putrefaciens* *P. phosphoreum*	Venugopal, 2006

and Debevere, 1997). This might partly explain the competitiveness of *P. phosphoreum* in MAP containing elevated oxygen concentrations, shown in Table 10.2. However, an underestimation of the number of *P. phosphoreum* in the studies used to compile Table 10.2 is also plausible because *P. phosphoreum* is temperature sensitive. Quantification should be made by streak-plating on a salt-containing medium (1% NaCl) followed by incubation which is optimal at 15–20 °C. This sensitivity to high temperature also explains why *P. phosphoreum* is a typical SSO for MAP marine fish from cold and temperate waters and not from fish originating from tropical waters (Dalgaard *et al.*, 1997).

Next to *P. phosphoreum*, MAP also favours the development of LAB in fresh fish. Numerous studies on MAP of fatty and lean fish species have shown that total LAB counts were relevant in fresh MAP fish: Banks *et al.* (1980) and Lannelongue *et al.* (1982) in ocean perch, Oberlender *et al.* (1983) in swordfish, Stenström (1985) in cod, Ordonez *et al.* (2000) in hake, Rudi *et al.* (2004) in salmon, Lalitha *et al.* (2005) in pearl spot and Leroi (2010) on herring fillets showed that preservation under 100% CO_2 leads to a predominance of *Lactobacillus* spp. in the microbiota, while under air *Shewanella* spp. and *Pseudomonas* spp. are more common (Molin *et al.*, 1983; Molin and Stenstrom, 1984); however, a high CO2 level results in poor sensory quality due to the development of a carbonated flavour (Leroi, 2010). However, in some cases MAP does not seem to significantly increase the number of LAB (Debevere and Boskou, 1996). Leroi (2010) reported that LAB probably have been disregarded in seafood for a long time as some genera, especially *Carnobacterium*, grow poorly in the presence of acetate contained in the de Man Rogosa (MRS) medium, which is classically used to enumerate LAB. Since less selective media such as Nitrite Actidione Polymixine agar or Elliker agar have been used, presence of LAB in processed fishery products has been highlighted (Leroi, 2010). LAB also play an important role in MAP thawed marine fish and MAP preserved fish (e.g. smoked salmon) (Leroi, 2010). Emborg *et al.* (2002) showed that *P. phosphoreum* was the dominant bacterial species in fresh salmon MAP fillets (60% CO_2, 40% N_2). However, when the fillets were frozen for 4 weeks at −20 °C and thawed before MAP, *P. phosphoreum*, being a very cold-sensitive species, was eliminated enabling *Carnobacterium* spp. to become dominant (Emborg *et al.*, 2002; Leroi, 2010).

Brochothrix thermosphacta has been associated with the dominant microflora in CO_2 and CO_2–O_2 enriched atmospheres and plays an important role in the spoilage of MAP meat and fish (Pin *et al.*, 2002; Mejlholm *et al.*, 2005). This species is more associated with MAP marine fish from tropical waters (>18 °C) (Mejlholm *et al.*, 2005).

Effect of MAP on the spoilage microbiota of freshwater fish *Pseudomonas* spp., *Aeromonas* spp. and some Enterobacteriaceae are considered to be the general spoilage bacteria of iced and aerobically stored tropical freshwater fish (Gram *et al.*, 1990; Gram and Huss, 1996). *S. putrefaciens* has been isolated from tropical freshwater, but did not appear to be important in the spoilage process (Gram *et al.*, 1990; Gram and Huss, 1996). This may be due to the occurrence of very low numbers and the inability of the organism to compete with high numbers of antagonistic pseudomonads (Gram and Huss, 1996; Gram *et al.*, 2002). Carbon dioxide and vacuum packaging of fish caught in freshwater results in gram-positive bacteria, mainly LAB, dominating the microflora (Gram and Huss, 1996; ICMSF, 2005). LAB and *B. thermosphacta* have been identified as SSOs of MAP freshwater fish (Ahvenainen, 2003; Leroi, 2010). Noseda *et al.* (2012) studied the effect of MAP on the microbiological growth and metabolism occurring on thawed *Pangasius hypophthalmus* fillets during storage at 4 °C. The change from air storage to storage with 50% CO_2 caused a shift in dominating flora from *Serratia* spp. and *Pseudomonas* spp. towards LAB and *B. thermosphacta*. The LAB appear to originate from the gastrointestinal tract of the fish (Ringo and Gatesoupe, 1998). However, the effects of MAP on the spoilage microbiota and spoilage processes has so far been studied mainly on marine fish species, leaving a gap in the scientific data for fresh MAP freshwater fish.

10.4.2 Effect of MAP on the spoilage mechanism

Bacterial spoilage of a fish product occurs when the metabolism of the microbiota on the fish causes a reduction in the quality of the fish. This can be perceived sensorially as a change in odour and flavour from neutral or non-specific to off-odorous and undesired. Spoilage reaction mechanisms have already been described in the literature. There are various substrates for the production of volatiles: TMAO, carbohydrates (e.g. lactate), amino acids, nucleotides and other non-protein nitrogen (NPN) molecules (Gram and Huss, 1996). The NPN fraction includes water-soluble, low molecular weight, nitrogen-containing compounds of a non-protein nature. The major components in this fraction are bases such as ammonia and TMAO, creatine, free amino acids, nucleotides, purine bases and, for cartilaginous fish, also urea (Huss, 1995). The off-odour that develops on a fish is dependent on the composition of the fish species and its origin (Gram and Huss, 1996). This is because the chemical composition (e.g. NPN fraction) varies for different fish species. TMAO has, for instance, an osmoregulatory function in marine fish species and the amount of TMAO present in the muscle tissue depends on the species, season and the marine environment (Kelly and Yancey, 1999). The concentration of TMAO in the muscle will also depend on the depth at which the fish live; the greater the depth, the higher the TMAO concentration in the muscle tissue (Kelly and Yancey, 1999). Some gadoid fish species are able to reduce TMAO in an autolytic pathway to dimethylamine (DMA) and formaldehyde (Castell *et al.*, 1973; Lundstrom *et al.*, 1982). Cartilaginous fish such as shark and ray have generally higher concentrations of TMAO in their muscle than bony fish, but also store high concentrations of urea in the cartilaginous skeleton (Huss, 1995).

Bacterial spoilage of marine temperate-water fish stored under ice is characterized by the development of off-odour and flavours caused by proteolytic breakdown, which produces volatile bases and sulfur compounds. This sensory impression is distinctly different from some tropical and freshwater fish, where sweet, fruity, sulfide off-odours and flavours are more typical (Gram and Huss, 1996). TMA is an especially important spoilage component for marine seafood. In air, primarily aerobic organisms will grow using carbohydrates and lactate as the energy source and producing CO_2 and H_2O. This initial metabolic process gradually decreases the oxido-reduction potential on the surface of the fish. Oxygen becomes less available for the surface bacteria, which favours the growth of facultative anaerobic bacteria. Facultative anaerobic bacteria such as *S. putrefaciens, P. phosphoreum, Aeromonas* spp., *Vibrio* spp. and *Enterobacteriaceae* are able to use TMAO instead of oxygen as the final electron acceptor in the metabolic pathway. These bacteria all contain the enzyme TMAO reductase that catalyses the reduction of TMAO into TMA, which has been associated with an unpleasant fishy odour. This reduction is accompanied with the production of acetic acid as a main component. Debevere *et al.* (2001), suggested the following reaction mechanism:

$$\underset{\text{lactic acid}}{CH_3CHOH\ COOH} + \underset{\textbf{TMAO}}{\mathbf{(CH_3)_3NO}} \rightleftharpoons \underset{\text{pyruvate}}{CH_3CO\ COOH} + \underset{\textbf{TMA}}{\mathbf{(CH_3)_3N}} + H_2O$$

$$\underset{\text{pyruvate}}{CH_3CO\ COOH} + \underset{\textbf{TMAO}}{\mathbf{(CH_3)_3NO}} + H_2O \rightleftharpoons \underset{\text{acetic acid}}{CH_3COOH} + \underset{\textbf{TMA}}{\mathbf{(CH_3)_3N}} + CO_2 + H_2O$$

The use of MAP gases will not only have an effect on the selection of the spoilage microbiota, they will also affect this spoilage reaction mechanism. As TMAO reductase is pH sensitive, using CO_2 inside the package indirectly inhibits TMA production. Easter *et al.* (1982) reported an optimum TMAO-reductase activity in *Shewanella* spp. at pH 6.8. According to Boskou and Debevere (1997), the presence of 30% CO_2 (G:P ratio 4:1, 7 °C) in the package already significantly inhibited the TMA production by *Shewanella* spp. Packaging for marine fish that contains a low concentration of or no oxygen, enhances this metabolism (Boskou and Debevere, 1997).

Other typical spoilage off-odours are caused by the degradation products of proteins and amino acids (Herbert and Shewan, 1975; Gram and Huss, 1996; Boskou, 1998). Protein and amino acid substrates are utilized once the small amount of carbon sources present are depleted (Dainty, 1996). Proteolysis is often a result of a collaboration of the bacterial community. Proteins cannot migrate through the bacterial cell membrane, so they are initially hydrolysed to free amino acids by bacterial proteases. Free amino acids are further broken down by oxidative deamination [1], reductive deamination [2] and/or decarboxylation reactions [3]:

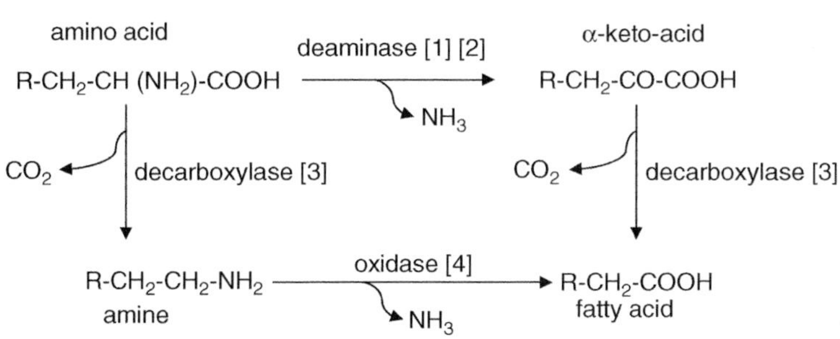

These amino acids are particularly important substrates for the formation of ammonia (Gram and Huss, 1996; Boskou, 1998). Ammonia can also be formed in elasmobranch fish species by the conversion of urea into ammonia, but whether this has a microbial cause is currently not clear (Mugica *et al.*, 2008; Broekaert, 2012). The production of biogenic amines (histamine, putrescine, cadaverine, ...) originates from the decarboxylation of amino acids. This decarboxylation may occur during the spoilage process and has been related to a bacterial stress response (e.g. acid stress). Most bacteria identified as SSO on fish produce one or several volatile sulfides from either L-cysteine or methionine (Herbert and Shewan, 1975; Whitfield, 1998).

The effect of the MAP gases on the amino acid breakdown metabolism of the microbiota still remains unclear because of the interaction between different parameters such as the chemical composition of the fish product, the dominating microorganisms and their metabolism. The microorganisms associated with the production of volatile sulfides in fish are described by Whitfield (1998). Interestingly, neither *Pseudomonas* spp. nor *P. phosphoreum* can produce significant concentrations of hydrogen sulfide during chilled storage. It has been claimed that hydrogen sulfide is a spoilage characteristic for aerobic chilled fish but is not detected in CO_2 packaged spoiled cod (Dalgaard *et al.*, 1993; Gram and Huss, 1996; Whitfield, 1998). Ammonia production is also found to be lower in MAP fish products compared with aerobically stored fish products. There are some reports of the effect of MAP gases on biogenic amine formation. High CO_2 concentrations in MAP appear to have a repressing effect on biogenic amine production. Santos (1996) reported that histidine decarboxylase activity is inhibited in atmospheres of 80% CO_2. The presence of oxygen also appeared to have a significant effect on biogenic amine synthesis. *In vitro* studies demonstrated that *Enterobacter cloacae* only produced half as much putrescine in anaerobic conditions than in aerobic conditions and *Klebsiella pneumoniae* synthesized significantly less cadaverine under anaerobic conditions, but was then able to produce putrescine. On the other hand, Halasz *et al.* (1994) reported that reducing the redox potential stimulated histamine production, and histidine decarboxylase activity seemed to be inactivated or destroyed in the presence of oxygen.

As suggested, a modified atmosphere does not only have an effect on microbiological growth but also on the spoilage metabolism of the microflora. This effect on the metabolism can be illustrated by the results obtained by Pin *et al.* (2002), who studied the effect of MAP on the metabolism of *B. thermosphacta*. When the O_2 level was higher than the solution of the equation [$\% \ O_2 = 10.17 + 0.6717 * \% \ CO_2$] for a given CO_2 concentration (%), the metabolism of *B. thermosphacta* was found to be predominantly aerobic, and it was predominantly anaerobic when the oxygen level was lower than the solution of this equation. The main metabolites of the consumption of glucose by *B. thermosphacta* under anaerobic conditions were lactic acid and ethanol, but no acetoin, and only small or no amounts of short-chain fatty acids have been detected. Under aerobic conditions, *B. thermosphacta* produces acetoin, acetic acid, isobutyric, 2-methylbutyric and isovaleric acids; and 3-methyl-1-butanol (Pin *et al.*, 2002).

10.5 Effects of MAP on the microbial safety of fish products

Pathogenic microorganisms can (i) be naturally present in the fish (*Clostridium botulinum* type E, B and F, *Vibrio* spp., *Aeromonas hydrophila, Plesiomonas shigelloides*), (ii) come from the environment (*Listeria monocytogenes, C. botulinum* types A and B, *Clostridium perfringens, Bacillus* spp), or (iii) originate from an animal/human contamination

(*Salmonella, Shigella, Escherichia coli* and *Staphylococcus aureus*) (Sivertsvik *et al.*, 2002). For most of these pathogens, no differences in survival are observed between air-packed or MAP-packed fish. Kimura and Murakami (1993) proved that the risk caused by *E. coli, S. aureus, V. parahaemolyticus* and *C. perfringens* stored under MAP (40% CO_2/60% N_2) or 100% N_2 was minimal when the packages were stored under chilled conditions. For *Salmonella*, even a CO_2 dependent bacteriostasis was found at chilled temperatures (Slade and Davies, 1997). Consequently, *C. botulinum* and *L. monocytogenes* are the pathogens of major concern and will be discussed in detail.

10.5.1 *Listeria monocytogenes*

Listeria monocytogenes is a vegetative food infectant which implies that metabolically active cells need to be digested to cause illness. Therefore, it is important to distinguish between fresh or minimally processed fish which can be eaten without heat treatment (e.g. cold smoked salmon, sushi, cooked shellfish) and products that undergo sufficient heat treatment before consumption to inactivate the pathogen ($P_{70} = 2\,\text{min}$) (e.g. most fresh fish). Growth control of *L. monocytogenes* is particularly important in the first category.

The prevalence of *L. monocytogenes* (presence in 25 g) in lightly preserved fish products is highly variable but relatively elevated. The contamination is partly due to their presence in the raw material but mostly due to processing (Leroi, 2010). However, the potential growth of the pathogen is more important than its initial contamination of the product as this is mainly below 1 cfu/g. Therefore, good strategies to delay or avoid growth, such as storage at refrigerated temperatures and MAP should be applied. However, as *L. monocytogenes* can grow under aerobic and anaerobic conditions, the effect of vacuum packaging and elevated CO_2 concentrations is limited. Even for products packed under 100% CO_2, complete growth inhibition did not occur but growth delay could be observed in comparison with air-packed products. Also, elevated CO_2 levels in combination with an oxygen absorbent, did not completely inhibit the growth of *L. monocytogenes* (Lyver *et al.*, 1998c). It is generally known that the effectiveness of MAP increases as the storage temperature of the packed product decreases. The strict temperature control of MAP seafood is therefore of utmost importance (Rutherford *et al.*, 2007). Additional barriers to microbial growth (e.g. chemical and/or natural preservatives) are desirable if temperature control ($< 3\,°C$) is difficult during transportation, retail and storage at the consumer's location (Mejlholm *et al.*, 2005).

Vacuum packaging or MAP with enhanced CO_2 concentrations will delay the growth of *L. monocytogenes* but is not sufficient to control the growth of this pathogen completely. However, according to the EU legislation 2073/2005 (EU, 2005), growth of *L. monocytogenes* can be tolerated as long as the food business operator (FBO) can demonstrate that the cell count of *L. monocytogenes* will not exceed the limit of 2 log cfu/g during its shelf life. Consequently the barriers to delay *L. monocytogenes* growth should always be related to the influence of this technique to spoilage phenomena (see earlier). *L. monocytogenes* must not be allowed to exceed the limit of 2 log cfu/g or to reach infectious doses in unspoiled products.

10.5.2 *Clostridium botulinum*

Non-proteolytic *Clostridium botulinum* is a strict anaerobic food intoxicant which can grow and produce toxins at refrigerated temperatures as low as 3 °C (Graham *et al.*, 1997). Also, these bacteria are widely distributed in freshwater, estuarine and marine environments and consequently can be found on fishery products harvested in these areas (Sivertsvik *et al.*, 2002). If fishery products are packed under modified atmosphere, the normal bacterial flora can be inhibited, which can permit non-proteolytic *C. botulinum* to grow and produce toxins during refrigerated storage and/or temperature abuse. This implies that these products may become toxic while their odour and appearance remain acceptable to the consumer (Sivertsvik *et al.*, 2002). Dodds and Austin (1997) demonstrated that toxin production generally takes weeks at low temperature. The results on challenge tests with *C. botulinum* on fishery products are inconsistent; in some studies toxin was detected in MAP fish prior to the spoilage of the product, while in other studies the products were spoiled before or at the same time as toxin production started (Sivertsvik *et al.*, 2002). As one of the first attempts to inhibit the growth of the anaerobic pathogen *C. botulinum*, O_2 was introduced in the package; however, this did not give satisfactory results. Smoked herring, inoculated with type E spores and packed under air or in 100% O_2, can also be toxic after 6–9 days storage at 15 °C. Dufresne *et al.* (2000) showed that neither the inclusion of O_2 in the headspace nor the use of packaging material with a high OTR (oxygen transmission rate) inhibited or delayed growth of *C. botulinum*. This latter was also proved by Post *et al.* (1985) for *C. botulinum* on cod, whiting and flounder fillets. This implies that the presence of O_2 in MAP does not enhance the safety of fish products and may be misleading. Even under these conditions, reduced oxygen conditions or anaerobic conditions can occur on the surface of the fish depending on the degree of contamination with oxygen-consuming microorganisms. Gibson *et al.* (2000) demonstrated that 100% CO_2, on the other hand, could inhibit the growth of *C. botulinum* at refrigerated temperatures. However, on flounder fillets packed under 100% CO_2, toxin formation occurred much earlier than spoilage phenomena, which can compromise food safety. The safest MAP gas mixture for reducing toxin production was a mixture of equal parts of O_2 and CO_2 (Sivertsvik *et al.*, 2002). An atmosphere of 40% CO_2 and 60% O_2 was most effective for refrigerated tuna, as it enhances colour stability and delays microbial growth (Lopez-Galvez *et al.*, 1995).

Extra barriers that can be combined with MAP to inhibit the growth of this pathogen are, for example, decreasing the pH level, increasing the salt concentration, adding competitive flora (e.g. *Bacillus* species), addition of preservatives (e.g. nisin) and treatment with ozone (Taylor *et al.*, 1990; Lyver *et al.*, 1998a, 1998b; Gibson *et al.*, 2000; Bono and Badalucco, 2012).

Also the type of fish, and more specifically the fat content, can influence the safety of MAP products. For leaner fish (e.g. cod, tilapia), the time between sensorial deterioration and toxin production was longer than for fattier fish (Reddy *et al.*, 1999).

10.6 Application of MAP on fish and fishery products

The search for a suitable gas composition to optimize the shelf life of a fishery product often involves some trial and error, because the choice for a certain gas composition depends on some intrinsic and extrinsic factors. Intrinsic factors such as the fish species, its origin (tropical/ temperate/ cold waters, marine/ brackish/ fresh waters), chemical composition and the microbiota present influence the shelf life of MAP fish. Also, extrinsic factors such as temperature and the G : P ratio are involved. The storage temperature of MAP fish ranges between 0 and 4–7 °C and the G : P ratio between 0.5 and 3. Higher ratios are possible but are in practice incidental for logistical reasons. However, with this knowledge, well-considered choices for the initial gas composition can be made. Once MAP conditions are optimized for a specific product, shelf life extensions of 30–50% are plausible at low temperatures.

10.6.1 Fresh fish

The use of O_2 to package lean marine fish species can extend shelf life because of the inhibitory effect of O_2 on the formation of TMA. However, the potential stimulating effect of O_2 towards the growth of certain spoilage microorganisms need to be considered. Generally, combinations of 40–60% CO_2 and 10–50% O_2, supplemented with N_2 are reported. Debevere and Boskou (1996) indicated that a modified atmosphere consisting of 60/30/10 (% CO_2, % O_2, % N_2) inhibited TMA production during the storage of cod fillets, but at the expense of CO_2 and its antimicrobial activity. Stenström (1985) suggested 50% CO_2/50% O_2 as modified atmosphere for retail packages of cod. Lopez-Caballero *et al.* (2002) reported a delay in microbiological growth and some biochemical indexes during storage of MAP whole hake, when gas mixtures containing CO_2 in combination with 20–60% O_2 were used.

Because of their very low TMAO content, the formation of TMA in non-marine lean fish species is of minor importance. An extended shelf life is mainly achieved by the addition of 40–60% CO_2. Yet, Noseda *et al.* (2012) found that a superatmospheric concentration of 50% O_2 had an additional antimicrobial effect towards LAB in combination with a high CO_2 concentration (50% CO_2) for MAP of defatted *Pangasianodon hypophthalmus* fillets stored at 4 °C. Reddy *et al.* (1995) evaluated the effect of CO_2-enriched atmospheres 75/0/25, 50/0/50 and 25/0/75 (% CO_2, % O_2, % N_2) on the shelf life of *Tilapia* spp. fillets at 4 °C. The authors observed that *Tilapia* fillets packaged in 75/0/25 showed an increased shelf life of 25 days compared with storage in air and during the shelf life sensorial characteristics remained acceptable (Reddy *et al.*, 1995).

Fish products with a high fat content should be packaged in an O_2-free environment to avoid enhancing oxidative rancidity. As an example, for Atlantic salmon, CO_2 is recommended with a partial pressure ranging between 0.4 to 0.6 (Hansen *et al.*, 2009; Schirmer *et al.*, 2009; Pettersen *et al.*, 2011; Mace *et al.*, 2012).

As already mentioned above, O_2 is important for red colour formation in tuna meat. After cutting the fish, the tuna meat surface is exposed to oxygen, which reacts with the myoglobin forming a bright red pigment (oxymyoglobin). However, due to the continued exposure to oxygen, the red colour will gradually change to a brownish colour during storage. This is caused by the oxidation of oxymyoglobin (Fe^{2+}) to metmyoglobin (Fe^{3+}) (Tajima and Shikama, 1987). In a recent study, Torrieri *et al.* (2011) showed a significant decrease in O_2 concentration of around 15% due to oxidation processes in tuna fillet that was air packaged or under super atmospheric conditions (60% O_2 and 40% CO_2). The CO_2 concentration remained almost constant (38% CO_2) in the MAP conditions even if a collapse of the packaging occurred due to solubilization in the product. Results on the protective effect of high O_2 concentrations on myoglobin oxidation are inconsistent. Lopez-Galvez *et al.* (1995) reported a protective effect on refrigerated tuna steaks, while Torrieri *et al.* (2011) demonstrated no difference between tuna fillets packaged at high oxygen concentration and air-packaged tuna fillets.

10.6.2 Fresh crustaceans

Crustaceans generally have a low fat content, so the recommendations made earlier for lean fish are also valid for fresh crustaceans. Lopez-Caballero *et al.* (2002) studied the effect of MAP conditions 40/30/30 and 45/5/50 (% CO_2, % O_2, % N_2) on packaged deepwater pink shrimp (*Parapenaeus longirostris*) during storage at $1.6 \pm 0.4\,°C$, where the applied MAP delayed microbiological growth, TMA and total volume base nitrogen (TVB-N) production. The authors suggested the involvement of *P. phosphoreum* in the shrimp spoilage both in air as under MAP (Lopez-Caballero *et al.*, 2002).

Many crustacean products are cooked and peeled before cooled storage under MAP. Mejlholm *et al.* (2005) performed storage trials with naturally contaminated cooked and peeled MAP shrimps (*Pandalus borealis*) with a gas composition of 50/20/30 (% CO_2, % O_2, % N_2) and stored at 2, 5 and 8 °C. *B. thermosphacta* and *C. maltaromaticum* were responsible for sensory spoilage of the cooked and peeled MAP shrimps (Mejlholm *et al.*, 2005). Chen and Xiong (2008) investigated the MAP condition 80/10/10 (% CO_2, % O_2, % N_2) to prolong the shelf life of cooked and peeled red claw crayfish *(Cherax quadricarinatus)*. In this study, *Carnobacterium* spp. were suggested as spoilage microorganisms (Chen and Xiong, 2008). Jaffrès *et al.* (2009) highlighted the main bacterial genera and species related to the spoilage of cooked and peeled tropical shrimps, packaged in MAP of 50/0/50 and 100/0/0 (% CO_2, % O_2, % N_2). LAB, mainly represented by the genera *Carnobacterium, Vagococcus* and *Enterococcus*, were found to dominate the microbiota of the studied shrimps, followed by *B. thermosphacta* and *S. liquefaciens* (Jaffrès *et al.*, 2009, 2011). Noseda *et al.* (2012) investigated the effect of different O_2–CO_2 enriched atmospheres on microbiological growth and volatile metabolite production in MAP cooked and peeled grey shrimp (*Crangon crangon*). In this research, the combination of 50% CO_2 and 50% O_2 was the best gas composition since microbial growth was significantly inhibited. Furthermore, a diminishing effect on the production of metabolites was also observed for this MAP condition, especially for amines and sulfur compounds, which were the biggest contributors to offensive odour (Noseda *et al.*, 2012).

10.6.3 Fresh molluscs

Another effect of MAP is its effectiveness in keeping molluscs alive, thereby prolonging the marketable period. The author's research on packaging conditions for fresh mussels (*Mytilus edulis*) during storage at 4 °C, indicated that using a superatmospheric oxygen concentration kept the mussels alive and prolonged shelf life (Figure 10.2). Mussels were packaged under three atmospheric conditions: air, 0/0/100 and 0/90/10 (% CO_2, % O_2, % N_2). Among the three conditions, there was a clear difference in quality which was related to the survival of these molluscs (Figure 10.2). A possible explanation is that the O_2 is more available for respiration at a gas composition of 0/90/10 (% CO_2, % O_2, % N_2) compared to the other conditions. Similar results were obtained by Bernardez and Pastoriza (2011). They investigated the effect of high oxygen concentrations on the quality of live packaged Mediterranean mussels (*Mytilus galloprovincialis*) during storage at 2 °C. Packaging mussels in atmospheres with richer O_2 concentrations (75–85% O_2) decreased mortality of the mussels and under these conditions, packaged mussels tended to produce less metabolic products.

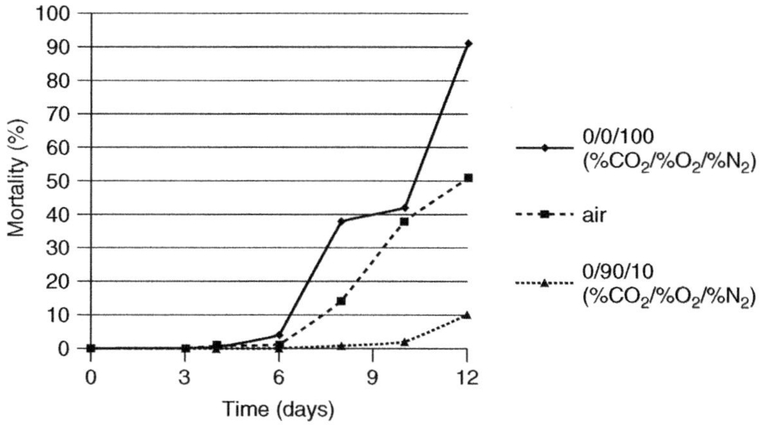

Figure 10.2 Influence of the atmosphere inside the packaging on the mortality (%) of mussels (*Mytilus edulis*) during storage at 4 °C

10.6.4 Smoked fish products

Smoked fishery products are generally packaged under vacuum or MAP in such way that O_2 is limited from the packaging. Anaerobic microbiota such as LAB typically predominate the microflora. However, growth of other organisms, including Enterobacteriaceae, *B. thermosphacta*, yeasts and *P. phosphoreum* may also occur (Gonzalez-Rodriguez *et al.*, 2002). Both techniques (vacuum or MAP) can be used to extend the shelf life of smoked fishery products.

10.7 Packaging materials and future developments

The earlier sections have clearly shown the added value of MAP for seafood products. There are, however, several conditions that need to be fulfilled to apply MAP successfully. An appropriate combination of the gases O_2, CO_2 and N_2 needs to be selected for extending the shelf life of seafood products. Next packaging materials as well as the filling equipment also need to be considered.

The packaging materials need to have, a good gas and water barrier in combination with good sealing characteristics. Such barriers can be achieved by using multilayer packaging materials, mostly consisting of different polymer structures. Examples are: PP/EVOH/PP, PA/PE/EVOH/PE, PE/EVOH/PE in which the polyolefin-type polymers (polyethylene, PE; polypropylene, PP) exhibit good water barriers, while ethylene vinyl alcohol (EVOH) is a polymer with good gas barriers. The degree of barrier needed for the MAP application is dependent on several factors including the sensitivity of the seafood products towards O_2 and the desired shelf life period. Taking the example of packaging consisting of a tray with a topfoil, a monolayer tray could be sufficient for maintaining the desired gas mixture in the headspace for rather short shelf-life periods as, in this case, the thickness of the tray would provide an adequate gas and water barrier.

In many MAP applications, however, barrier materials are necessary, especially taking into consideration the trend towards thinner and/or lightweight packaging. The use of EPS as a foam-structured tray in MAP applications becomes possible if the tray is coated with a thin barrier film.

In the next section, specific topics regarding the packaging of seafood will be highlighted, focussing on trends and developments in packaging materials used for MAP.

10.7.1 Barrier materials

In seafood packaging, polyamide (PA) and polyethylene terephthalate (PET) as moderate and EVOH as good gas barrier material are frequently used for different types of seafood products. The gas barriers PA and EVOH are dependent on the relative humidity, which a decrease in barrier properties at increasing relative humidity. In contrast, polyvinylidene chloride (PVdC) is a polymer having both good gas and water barrier properties and that could be applied as a coating on packaging material. Aluminium is another coating material that is, frequently applied on polymers by a metallization process using vacuum deposition. Next to metalized materials (e.g. MPET), an aluminium layer can also be applied as a laminate in multilayer packaging materials. Next to the excellent gas and water barrier properties of metals, it provides the packaging materials (when applied sufficiently thickly) with a good protection against UV and visible light (Robertson, 2006). In addition to these polymers and metals, components blended in packaging material could also improve the barrier properties. Much research in this area focuses on the use of nanoparticles, homogenously dispersed in the packaging material that cause a torturous path inside the material to slow down the permeation of gases through these materials (Brody *et al.*, 2008).

10.7.2 Active and intelligent packaging

As mentioned before, seafood products are highly susceptible to microbiological processes. Chemical processes such as oxidation reactions with the fat fraction could also influence the quality of certain seafood products. For packing highly O_2-sensitive seafood products, the use of oxygen scavengers in the packaging concept could be considered. These scavengers remove residual oxygen which is still present after sealing and oxygen that permeates the packaging material. Oxygen scavengers belong to the group of active packaging because the conditions of the packaged product are changed after the sealing step (Ozdemir and Floros, 2004). This is also the case for water absorbers, which are frequently used at the bottom of a tray to absorb possible drip losses from the fish. There is an increasing interest for both systems to integrate them in packaging materials instead of applying them as separate sachets in the packs (Restuccia *et al.*, 2010).

The increased use of MAP for seafood products has also resulted in an increased interest for monitoring the product quality without opening the packages. Such types of packaging give information of the quality of the packaged product and are called intelligent packaging (Restuccia *et al.*, 2010). In general, three groups of intelligent packaging concepts can be distinguished that are relevant to seafood products packaged under modified atmosphere.

- Time temperature indicators: these are mostly applied as labels attached to the outside of the packaging, changing colour depending on the time–temperature conditions to which the packaging has been exposed (Mendoza *et al.*, 2004).

- Leak indicators: these are present on the inside of the packaging, changing colour depending on the presence or absence of certain gas components (e.g. O_2 entering the packaging) (Mills *et al.*, 2008).
- Quality indicators: contrary to the two earlier indicators, these are a direct measure of the quality of the packaged product by monitoring specific volatile components in the headspace of the pack.

Most research and the few commercial applications on these quality indicators have focused on colour changes due to the presence of some volatile components, causing a pH change on the indicator attached to the inside of the packaging (Pacquit *et al.*, 2006). There is, however, an increasing interest in other detection systems including micro-electronics and optical sensors (Alimelli *et al.*, 2007; Puligundla *et al.*, 2012).

10.7.3 New resources for packaging materials

There has been an increasing interest in sustainable packaging materials that can be reused, recycled and/or originate from renewable bio-based resources (Brody *et al.*, 2008; Siracusa *et al.*, 2008). Packaging materials produced from the latter belong to the group of bio-plastics: examples are bio-PE, polylactic acid (PLA), cellulose, starch and polyhydroxybutyrate (PHB). By combining different bio-plastics materials in multilayer structures, it is also possible to achieve appropriate gas and water barriers required for a certain shelf life period.

References

Ahvenainen, R. (2003). *Novel food packaging techniques.* Woodhead Publishing Limited, Cambridge.

Alimelli, A., Pennazza, G., Santonico, M. *et al.* (2007). Fish freshness detection by a computer screen photoassisted based gas sensor array. *Analytica Chimica Acta*, 582, 320–328.

Amanatidou, A., Smid, E.J. and Gorris, L.G.M. (1999). Effect of elevated oxygen and carbon dioxide on the surface growth of vegetable-associated microorganisms. *Journal of Applied Microbiology*, 86, 429–438.

Banks, H., Nickelson, R. and Finne, G. (1980). Shelf-life studies on carbon-dioxide packaged finfish from the Gulf of Mexico. *Journal of Food Science*, 45, 157–162.

Bernardez, M. and Pastoriza, L. (2011). Quality of live packaged mussels during storage as a function of size and oxygen concentration. *Food Control*, 22, 257–265.

Bono, G. and Badalucco, C. (2012). Combining ozone and modified atmosphere packaging (MAP) to maximize shelf-life and quality of striped red mullet (*Mullus surmuletus*). *Lwt-Food Science and Technology*, 47, 500–504.

Boskou, G. (1998). Reduction of bacterial spoilage associated with the trimethylamineoxide-dependent respiration in fish fillets packed under modified atmosphere. PhD dissertation. Faculty of Applied Biological Sciences, Ghent University, Ghent.

Boskou, G. and Debevere, J. (1997). Reduction of trimethylamine oxide by *Shewanella* spp. under modified atmospheres *in vitro*. *Food Microbiology*, 14, 543–553.

Brody, A.L., Bugusu, B., Han, J.H., Sand, C.K. and Mchugh, T.H. (2008). Innovative food packaging solutions. *Journal of Food Science*, 73, R107-R116.

Broekaert, K. (2012). Molecular identification of the dominant microbiota and their spoilage potential of Crangon crangon and Raja sp. PhD dissertation. Faculty of Applied Biological Sciences, Ghent University, Ghent.

Castell, C.H., Neal, W.E. and Dale, J. (1973). Comparison of changes in trimethylamine, dimethylamine, and extractable protein in iced and frozen gadoid fillets. *Journal of the Fisheries Research Board of Canada*, 30, 1246–1248.

Chen, G. and Xiong, Y.L. (2008). Shelf-stability enhancement of precooked red claw crayfish (*Cherax quadricarinatus*) tails by modified CO2/O-2/N-2 gas packaging. *Lwt-Food Science and Technology*, 41, 1431–1436.

Chen, H.M., Meyers, S.P., Hardy, R.W. and Biede, S.L. (1984). Colour stability of astaxanthin pigmented rainbow-trout under various packaging conditions. *Journal of Food Science*, 49, 1337–1340.

Condon, S. (1987). Responses of lactic-acid bacteria to oxygen. *FEMS Microbiology Reviews*, 46, 269–280.

Dainty, R.H. (1996). Chemical/biochemical detection of spoilage. *International Journal of Food Microbiology*, 33, 19–33.

Dalgaard, P. (1995a). Modeling of microbial activity and prediction of shelf-life for packed fresh fish. *International Journal of Food Microbiology*, 26, 305–317.

Dalgaard, P. (1995b). Qualitative and quantitative characterization of spoilage bacteria from packed fish. *International Journal of Food Microbiology*, 26, 319–333.

Dalgaard, P., Buch, P. and Silberg, S. (2002). Seafood spoilage predictor – development and distribution of a product specific application software. *International Journal of Food Microbiology*, 73, 343–349.

Dalgaard, P., Gram, L. and Huss, H.H. (1993). Spoilage and shelf-life of cod fillets packed in vacuum or modified atmospheres. *International Journal of Food Microbiology*, 19, 283–294.

Dalgaard, P., Mejholm, O., Christiansen, T.J. and Huss, H.H. (1997). Importance of *Photobacterium phosphoreum* in relation to spoilage of modified atmosphere-packed fish products. *Letters in Applied Microbiology*, 24, 373–378.

Daniels, J.A., Krishnamurthi, R. and Rizvi, S.S.H. (1985). A review of effects of carbon-dioxide on microbial-growth and food quality. *Journal of Food Protection*, 48, 532–537.

Debevere, J. and Boskou, G. (1996). Effect of modified atmosphere packaging on the TVB/TMA-producing microflora of cod fillets. *International Journal of Food Microbiology*, 31, 221–229.

Debevere, J., Devlieghere, F., van Sprundel, P. and De Meulenaer, B. (2001). Influence of acetate and CO_2 on the TMAO-reduction reaction by *Shewanella baltica International Journal of Food Microbiology*, 68, 115–123.

Devlieghere, F. and Debevere, J. (2000). Influence of dissolved carbon dioxide on the growth of spoilage bacteria. *Lebensmittel-Wissenschaft Und-Technologie-Food Science and Technology*, 33, 531–537.

Devlieghere, F., Debevere, J. and van Impe, J. (1998a). Concentration of carbon dioxide in the water-phase as a parameter to model the effect of a modified atmosphere on microorganisms. *International Journal of Food Microbiology*, 43, 105–113.

Devlieghere, F., Debevere, J. and van Impe, J. (1998b). Effect of dissolved carbon dioxide and temperature on the growth of *Lactobacillus sake* in modified atmospheres. *International Journal of Food Microbiology*, 41, 231–238.

Dixon, N.M. and Kell, D.B. (1989). The inhibition by CO_2 of the growth and metabolism of microorganisms. *Journal of Applied Bacteriology*, 67, 109–136.

Dodds, K.L. and Austin, J.W. (1997). Clostridium botulinum, in *Food Microbiology – Fundamentals and Frontiers* (eds M.E. Doyle, L.R. Beuchat, and T.J. Montville), ASM Press, Washington, pp. 288–304.

Droghetti, E., Bartolucci, G.L., Focardi, C. *et al.* (2011). Development and validation of a quantitative spectrophotometric method to detect the amount of carbon monoxide in treated tuna fish. *Food Chemistry*, 128, 1143–1151.

Dufresne, I., Smith, J.P., Liu, J.N. *et al.* (2000). Effect of headspace oxygen and films of different oxygen transmission rate on toxin production by *Clostridium botulinum* type E in rainbow trout fillets stored under modified atmospheres. *Journal of Food Safety*, 20, 157–175.

Easter, M.C., Gibson, D.M. and Ward, F.B. (1982). A conductance method for the assay and study of bacterial trimethylamine oxide reduction. *Journal of Applied Bacteriology*, 52, 357–365.

Emborg, J., Laursen, B.G., Rathjen, T. and Dalgaard, P. (2002). Microbial spoilage and formation of biogenic amines in fresh and thawed modified atmosphere-packed salmon (*Salmo salar*) at 2 degrees C. *Journal of Applied Microbiology*, 92, 790–799.

EU (2005). Commission regulation (EC) N° 2073/2005 of 15 November 2005 on microbiological criteria for foodstuff.

FAO (2010) *The State of World Fisheries and Aquaculture*, Fisheries and Aquaculture Department, Rome.

Fridovich, I. (1986). Biological effects of the superoxide radical. *Archives of Biochemistry and Biophysics*, 247, 1–11.

Fu, W.G. and Mathews, A.P. (1999). Lactic acid production from lactose by *Lactobacillus plantarum*: kinetic model and effects of pH, substrate, and oxygen. *Biochemical Engineering Journal*, 3, 163–170.

Gibson, A.M., Ellis-Brownlee, R.C.L., Cahill, M.E. *et al.* (2000). The effect of 100% CO_2 on the growth of nonproteolytic *Clostridium botulinum* at chill temperatures. *International Journal of Food Microbiology*, 54, 39–48.

Giménez, B., Roncales, P. and Beltran, J.A. (2002). Modified atmosphere packaging of filleted rainbow trout. *Journal of the Science of Food and Agriculture*, 82, 1154–1159.

Goncalves, A., Pedro, S., Duarte, A. and Nunes, M.L. (2009). Effect of enriched oxygen atmosphere storage on the quality of live clams (*Ruditapes decussatus*). *International Journal of Food Science and Technology*, 44, 2598–2605.

Gonzalez-Rodriguez, M.N., Sanz, J.J., Santos, J.A., Otero, A. and Garcia-Lopez, M.L. (2002). Numbers and types of microorganisms in vacuum-packed cold-smoked freshwater fish at the retail level. *International Journal of Food Microbiology*, 77, 161–168.

Graham, A.F., Mason, D.R., Maxwell, F.J. and Peck, M.W. (1997). Effect of pH and NaCl on growth from spores of non-proteolytic *Clostridium botulinum* at chill temperature. *Letters in Applied Microbiology*, 24, 95–100.

Gram, L. and Dalgaard, P. (2002). Fish spoilage bacteria – problems and solutions. *Current Opinion in Biotechnology*, 13, 262–266.

Gram, L. and Huss, H.H. (1996). Microbiological spoilage of fish and fish products. *International Journal of Food Microbiology*, 33, 121–137.

Gram, L., Ravn, L., Rasch, M. *et al.* (2002). Food spoilage – interactions between food spoilage bacteria. *International Journal of Food Microbiology*, 78, 79–97.

Gram, L., Wedellneergaard, C. and Huss, H.H. (1990). The bacteriology of fresh and spoiling lake Victorian Nile perch (*Lates-niloticus*). *International Journal of Food Microbiology*, 10, 303–316.

Haard, N.F. (1992). Technological aspects of extending prime quality of seafood: a review. *Journal of Aquatic Food Product Technology*, 1, 9–27.

Halasz, A., Barath, A., Simonsarkadi, L. and Holzapfel, W. (1994). Biogenic-amines and their production by microorganisms in food. *Trends in Food Science and Technology*, 5, 42–49.

Halliwell, B. and Gutteridge, J.M.C. (1984). Lipid-peroxidation, oxygen radicals, cell-damage, and antioxidant therapy. *Lancet*, 1, 1396–1397.

Han, J.J. (2005). *Innovations in Food Packaging*, Elsevier Academic Press, London.

Hansen, A.A., Morkore, T., Rudi, K. *et al.* (2009). Quality changes of prerigor filleted Atlantic salmon (*Salmo salar* L.) packaged in modified atmosphere using CO_2 emitter, traditional MAP, and vacuum. *Journal of Food Science*, 74, M242–M249.

Herbert, R.A. and Shewan, J.M. (1975). Precursors of volatile sulfides in spoiling North-Sea cod (*Gadus morhua*). *Journal of the Science of Food and Agriculture*, 26, 1195–1202.

Hotchkiss, J.H. (1988). Experimental approaches to determining the safety of food packaged in modified atmospheres. *Food Technology*, 42, 55.

Huss, H.H. (1995) Quality and changes in fresh fish. FAO Fisheries Technical Paper, 348. Food and Agriculture Organization of the United Nations.

ICMSF (2005). *Microorganisms in Food 6: Microbiological Ecology of Food Commodities*, Kluwer Academic/Plenum Publishers, New York, USA.

Jacxsens, L., Devlieghere, F., Van der Steen, C. and Debevere, J. (2001). Effect of high oxygen modified atmosphere packaging on microbial growth and sensorial qualities of fresh-cut produce. *International Journal of Food Microbiology*, 71, 197–210.

Jaffres, E., Lalanne, V., Mace, S. *et al.* (2011). Sensory characteristics of spoilage and volatile compounds associated with bacteria isolated from cooked and peeled tropical shrimps using SPME-GC-MS analysis. *International Journal of Food Microbiology*, 147, 195–202.

Jaffrés, E., Sohier, D., Leroi, F. *et al.* (2009). Study of the bacterial ecosystem in tropical cooked and peeled shrimps using a polyphasic approach. *International Journal of Food Microbiology*, 131, 20–29.

Jorgensen, L.V., Huss, H.H. and Dalgaard, P. (2000). The effect of biogenic amine production by single bacterial cultures and metabiosis on cold-smoked salmon. *Journal of Applied Microbiology*, 89, 920–934.

Karl, H. and Meyer, C. (2007). Effect of early gutting on shelf life of saithe (*Pollachius virens*), haddock (*Melanogrammus aeglefinus*) and plaice (*Pleuronectes platessa*) stored in ice. *Journal fur Verbraucherschutz und Lebensmittelsicherheit – Journal of Consumer Protection and Food Safety*, 2, 130–137.

Kelly, R.H. and Yancey, P.H. (1999). High contents of trimethylamine oxide correlating with depth in deep-sea teleost fishes, skates, and decapod crustaceans. *Biological Bulletin*, 196, 18–25.

Kimura, B. and Murakami, M. (1993). Fate of food pathogens in gas-packaged jack mackerel fillets. *Nippon Suisan Gakkaishi*, 59, 1163–1169.

Lalitha, K.V., Sonaji, E.R., Manju, S. *et al.* (2005). Microbiological and biochemical changes in pearl spot (*Etroplus suratensis* Bloch) stored under modified atmospheres. *Journal of Applied Microbiology*, 99, 1222–1228.

Lannelongue, M., Hanna, M.O., Finne, G., Nickelson, R. and Vanderzant, C. (1982). Storage characteristics of finfish fillets (*Archosargus probatocephalus*) packaged in modified gas atmospheres containing carbon-dioxide. *Journal of Food Protection*, 45, 440–444.

Leroi, F. (2010). Occurrence and role of lactic acid bacteria in seafood products. *Food Microbiology*, 27, 698–709.

Lopez-Caballero, M.E., Goncalves, A. and Nunes, M.L. (2002). Effect of CO_2/O_2-containing deepwater pink shrimp modified atmospheres on packed (*Parapenaeus longirostris*). *European Food Research and Technology*, 214, 192–197.

Lopez-Galvez, D., Delahoz, L. and Ordonez, J.A. (1995). Effect of carbondioxide and oxygen-enriched atmospheres on microbiological changes and chemical changes in refrigerated Tuna (*Thunnus Alalunga*) steaks. *Journal of Agricultural and Food Chemistry*, 43, 483–490.

Lundstrom, R.C., Correia, F.F. and Wilhelm, K.A. (1982). Enzymatic dimethylamine and formaldehyde production in minced american plaice and blackback flounder mixed with a red hake TMAO-ase active fraction. *Journal of Food Science*, 47, 1305–1310.

Lyver, A., Smith, J.P., Austin, J. and Blanchfield, B. (1998a). Competitive inhibition of *Clostridium botulinum* type E by *Bacillus* species in a value-added seafood product packaged under a modified atmosphere. *Food Research International*, 31, 311–319.

Lyver, A., Smith, J.P., Nattress, F.M., Austin, J.W. and Blanchfield, B. (1998b). Challenge studies with *Clostridium botulinum* type E in a value-added surimi product stored under a modified atmosphere. *Journal of Food Safety*, 18, 1–23.

Lyver, A., Smith, J.P., Tarte, I., Farber, J.M. and Nattress, F.M. (1998c). Challenge studies with Listeria monocytogenes in a value-added seafood product stored under modified atmospheres. *Food Microbiology*, 15, 379–389.

Mace, S., Cornet, J., Chevalier, F., Cardinal, M., Pilet, M.F., Dousset, X. and Joffraud, J.J. (2012). Characterisation of the spoilage microbiota in raw salmon (*Salmo salar*) steaks stored under vacuum or modified atmosphere packaging combining conventional methods and PCR-TTGE. *Food Microbiology*, 30, 164–172.

Mejlholm, O., Boknaes, N. and Dalgaard, P. (2005). Shelf life and safety aspects of chilled cooked and peeled shrimps (*Pandalus borealis*) in modified atmosphere packaging. *Journal of Applied Microbiology*, 99, 66–76.

Mendoza, T.F., Welt, B.A., Otwell, S. *et al.* (2004). Kinetic parameter estimation of time – temperature integrators intended for use with packaged fresh seafood. *Journal of Food Science*, 69, M90–M96.

Mills, A., Tommons, C., Bailey, R.T., Tedford, M.C. and Crilly, P.J. (2008). UV-activated luminescence/colourimetric O_2 indicator. *International Journal of Photoenergy*, article ID 547301.

Molin, G. and Stenstrom, I.M. (1984). Effect of temperature on the microbial flora of herring fillets stored in air or carbon dioxide. *Journal of Applied Bacteriology*, 56, 275–282.

Molin, G., Stenstrom, I.M. and Ternstrom, A. (1983). The microbial flora of herring fillets after storage in carbon dioxide, nitrogen or air at 2 degrees C. *Journal of Applied Bacteriology*, 55, 49–56.

Mugica, B., Barros-Velazquez, J., Miranda, J.M. and Aubourg, S.P. (2008). Evaluation of a slurry ice system for the commercialization of ray (*Raja clavata*): Effects on spoilage mechanisms directly affecting quality loss and shelf-life. *Lwt-Food Science and Technology*, 41, 974–981.

Noseda, B., Goethals, J., De Smedt, L. *et al.* (2012). Effect of O_2-CO_2 enriched atmospheres on microbiological growth and volatile metabolite production in packaged cooked peeled gray shrimp (*Crangon crangon*). *International Journal of Food Microbiology*, 160, 65–75.

Oberlender, V., Hanna, M.O., Miget, R., Vanderzant, C. and Finne, G. (1983). Storage characteristics of fresh swordfish steaks stored in carbon dioxide enriched controlled (flow-through) atmospheres. *Journal of Food Protection*, 46, 434–440.

Olafsdottir, G., Lauzon, H.L., Martinsdottir, E. and Kristbergsson, K. (2006a). Influence of storage temperature on microbial spoilage characteristics of haddock fillets (*Melanogrammus aeglefinus*) evaluated by multivariate quality prediction. *International Journal of Food Microbiology*, 111, 112–125.

Olafsdottir, G., Lauzon, H.L., Martinsdottir, E., Oehlenschlager, J. and Kristbergsson, K. (2006b). Evaluation of shelf life of superchilled cod (*Gadus morhua*) fillets and the influence of temperature fluctuations during storage on microbial and chemical quality indicators. *Journal of Food Science*, 71, S97–S109.

Ordonez, J.A., Lopez-Galvez, D.E., Fernandez, M., Hierro, E. and de la Hoz, L. (2000). Microbial and physicochemical modifications of hake (*Merluccius merluccius*) steaks stored under carbon dioxide enriched atmospheres. *Journal of the Science of Food and Agriculture*, 80, 1831–1840.

Ozdemir, M. and Floros, J.D. (2004). Active food packaging technologies. *Critical Reviews in Food Science and Nutrition*, 44, 185–193.

Pacquit, A., Lau, K.T., McLaughlin, H. *et al.* (2006). Development of a volatile amine sensor for the monitoring of fish spoilage. *Talanta*, 69, 515–520.

Pastoriza, L., Bernaardez, M., Sampedro, G., Cabo, M.L. and Herrera, J.J.R. (2004). Elevated concentrations of oxygen on the stability of live refrigerated mussel stored refrigerated. *European Food Research and Technology*, 218, 415–419.

Pastoriza, L., Sampedro, G., Herrera, J.J. and Cabo, M.L. (1998). Influence of sodium chloride and modified atmosphere packaging on microbiological, chemical and sensorial properties in ice storage of slices of hake (*Merluccius merluccius*). *Food Chemistry*, 61, 23–28.

Pettersen, M.K., Bardet, S., Nilsen, J. and Fredriksen, S.B. (2011). Evaluation and suitability of biomaterials for modified atmosphere packaging of fresh salmon fillets. *Packaging Technology and Science*, 24, 237–248.

Pin, C., De Fernando, G.D.G. and Ordonez, J.A. (2002). Effect of modified atmosphere composition on the metabolism of glucose by *Brochothrix thermosphacta*. *Applied and Environmental Microbiology*, 68, 4441–4447.

Poli, B.M., Messini, A., Parisi, G. *et al.* (2006). Sensory, physical, chemical and microbiological changes in European sea bass (*Dicentrarchus labrax*) fillets packed under modified atmosphere/air or prepared from whole fish stored in ice. *International Journal of Food Science and Technology*, 41, 444–454.

Post, L.S., Lee, D.A., Solberg, M. *et al.* (1985). Development of botulinal toxin and sensory deterioration during storage of vacuum and modified atmosphere packaged fish fillets. *Journal of Food Science*, 50, 990–996.

Pournis, N., Papavergou, A., Badeka, A., Kontominas, M.G. and Savvaidis, I.N. (2005). Shelf-life extension of refrigerated Mediterranean mullet (*Mullus surmuletus*) using modified atmosphere packaging. *Journal of Food Protection*, 68, 2201–2207.

Puligundla, P., Jung, J. and Ko, S. (2012). Carbon dioxide sensors for intelligent food packaging applications. *Food Control*, 25, 328–333.

Reddy, N.R., Solomon, H.M. and Rhodehamel, E.J. (1999). Comparison of margin of safety between sensory spoilage and onset of *Clostridium botulinum* toxin development during storage of modified atmosphere (MA)-packaged fresh marine cod fillets with MA-packaged aquacultured fish fillets. *Journal of Food Safety*, 19, 171–183.

Reddy, N.R., Villanueva, M. and Kautter, D.A. (1995). Shelf-Life of modified atmosphere packaged fresh Tilapia fillets stored under refrigeration and temperature abuse conditions. *Journal of Food Protection*, 58, 908–914.

Restuccia, D., Spizzirri, U.G., Parisi, O.I. *et al.* (2010). New EU regulation aspects and global market of active and intelligent packaging for food industry applications. *Food Control*, 21, 1425–1435.

Ringo, E. and Gatesoupe, F.J. (1998). Lactic acid bacteria in fish: a review. *Aquaculture*, 160, 177–203.

Robertson, G.L. (2006). *Food Packaging. Principles and Practice*, 2nd edn, Taylor and Francis, Boca Raton, FL, USA.

Rudi, K., Maugesten, T., Hannevik, S.E. and Nissen, H. (2004). Explorative multivariate analyses of 16S rRNA gene data from microbial communities in modified-atmosphere-packed salmon and coalfish. *Applied and Environmental Microbiology*, 70, 5010–5018.

Rutherford, T.J., Marshall, D.L., Andrews, L.S. *et al.* (2007). Combined effect of packaging atmosphere and storage temperature on growth of *Listeria monocytogenes* on ready-to-eat shrimp. *Food Microbiology*, 24, 703–710.

Santos, M.H.S. (1996). Biogenic amines: their importance in foods. *International Journal of Food Microbiology*, 29, 213–231.

Schirmer, B.C., Heiberg, R., Eie, T. *et al.* (2009). A novel packaging method with a dissolving CO2 headspace combined with organic acids prolongs the shelf life of fresh salmon. *International Journal of Food Microbiology*, 133, 154–160.

Schubring, R. (2008) Use of 'filtered smoke' and carbon monoxide with fish. *Journal fur Verbraucherschutz und Lebensmittelsicherheit – Journal of Consumer Protection and Food Safety*, 3, 31–44.

Siracusa, V., Rocculi, P., Romani, S. and Dalla Rosa, M. (2008). Biodegradable polymers for food packaging: a review. *Trends in Food Science and Technology*, 19, 634–643.

Sivertsvik, M., Jeksrud, W.K. and Rosnes, J.T. (2002). A review of modified atmosphere packaging of fish and fishery products – significance of microbial growth, activities and safety. *International Journal of Food Science and Technology*, 37, 107–127.

Slade, A. and Davies, A.R. (1997). Fate of foodborne pathogens on modified atmosphere packaged (MAP) cod and trout, in *Seafood from Producer to Consumer, Integrated Approach to Quality* (eds J.B. Luten, T. Borresen and J. Oehlenschläger), Elsevier, Amsterdam, pp. 455–461.

Stenstrom, I.M. (1985). Microbial-flora of cod fillets (*Gadus morhua*) stored at 2-degrees-C in different mixtures of carbon dioxide and nitrogen oxygen. *Journal of Food Protection*, 48, 585–589.

Tajima, G. and Shikama, K. (1987). Autoxidation of oxymyoglobin – an overall stoichiometry including subsequent side reactions. *Journal of Biological Chemistry*, 262, 12603–12606.

Taylor, L.Y., Cann, D.D. and Welch, B.J. (1990). Antibotulinal properties of nisin in fresh fish packaged in an atmosphere of carbon-dioxide. *Journal of Food Protection*, 53, 953–957.

Torrieri, E., Carlino, P.A., Cavella, S. *et al.* (2011). Effect of modified atmosphere and active packaging on the shelf-life of fresh bluefin tuna fillets. *Journal of Food Engineering*, 105, 429–435.

Venugopal, V. (2006). *Seafood Processing: Adding Value through Quick Freezing, Retortable Packaging and Cook Chilling*, CRC Press, Boca Raton.

Whitfield, F.B. (1998). Microbiology of food taints. *International Journal of Food Science and Technology*, 33, 31–51.

Young, L.L., Reviere, R.D. and Cole, A.B. (1988). Fresh red meats – a place to apply modified atmospheres. *Food Technology*, 42, 65–69.

11
Fish Waste Management

Ioannis S. Arvanitoyannis and Persefoni Tserkezou
Laboratory of Food Technology, Department of Agriculture, Ichthyology and Aquatic Environment, University of Thessaly, Volos, Greece

11.1 Introduction

Global fish and shrimp production has increased steadily over the past two decades. Of the estimated 131 million tonnes of fish produced in 2000 worldwide, nearly 74% (97 million tonnes) was used for human consumption. The rest (about 26%) was used for various non-edible products, mainly meal and oil. In 2000, more than 60% of the total fish produced worldwide underwent some form of processing (FAO, 2002). In aquaculture, a large amount of species, rearing techniques and husbandry methods are used. Extensive aquaculture includes farming finfish or shellfish in a 'natural' habitat with no additional food and with a minimum effect on the environment (Midlen and Redding, 1998; OAERRE, 2001). Fish processing generates large amounts of solid and liquid wastes as shown in Figure 11.1. Furthermore, more than half of the raw-material weight is unused, for example only about 15% of the round shrimp becomes a canned shrimp product (El-Beltagy *et al.*, 2005).

Fish industries usually process wastes that are suspected to pose a significant risk to the environment. Technologies to treat these wastes, or even better to recover some useful organic materials before disposal, are necessary to mitigate pollution (Quitain *et al.*, 2001). Fish farm waste affects not only the immediate area but it can also change a wider coastal zone through reduction of biomass, density and diversity of the benthos, plankton and nekton, and by modifying natural food webs (Gowen, 1991; Pillay, 1991).

The main reason for life cycle assessment (LCA) is to associate, using a small number of indicators, the polluting emissions and resources used during the course of the product's life (Payraudeau and van der Werf, 2005) (Table 11.1).

The European Union has brought into force many Directives over the past 30 years for the implementation of national legislation concerning the management of the environmental impacts of aquaculture by establishing appropriate Environmental Quality Objectives (EQOs) and Environmental Quality Standards (EQSs). In the framework of aquaculture, environmental protection measures have been established at three levels: (i) general policy, (ii) specific measures, and (iii) regulations that control specific local conditions (Eleftheriou and Eleftheriou, 2001). The most relevant European Commission

Seafood Processing: Technology, Quality and Safety, First Edition. Edited by Ioannis S. Boziaris.

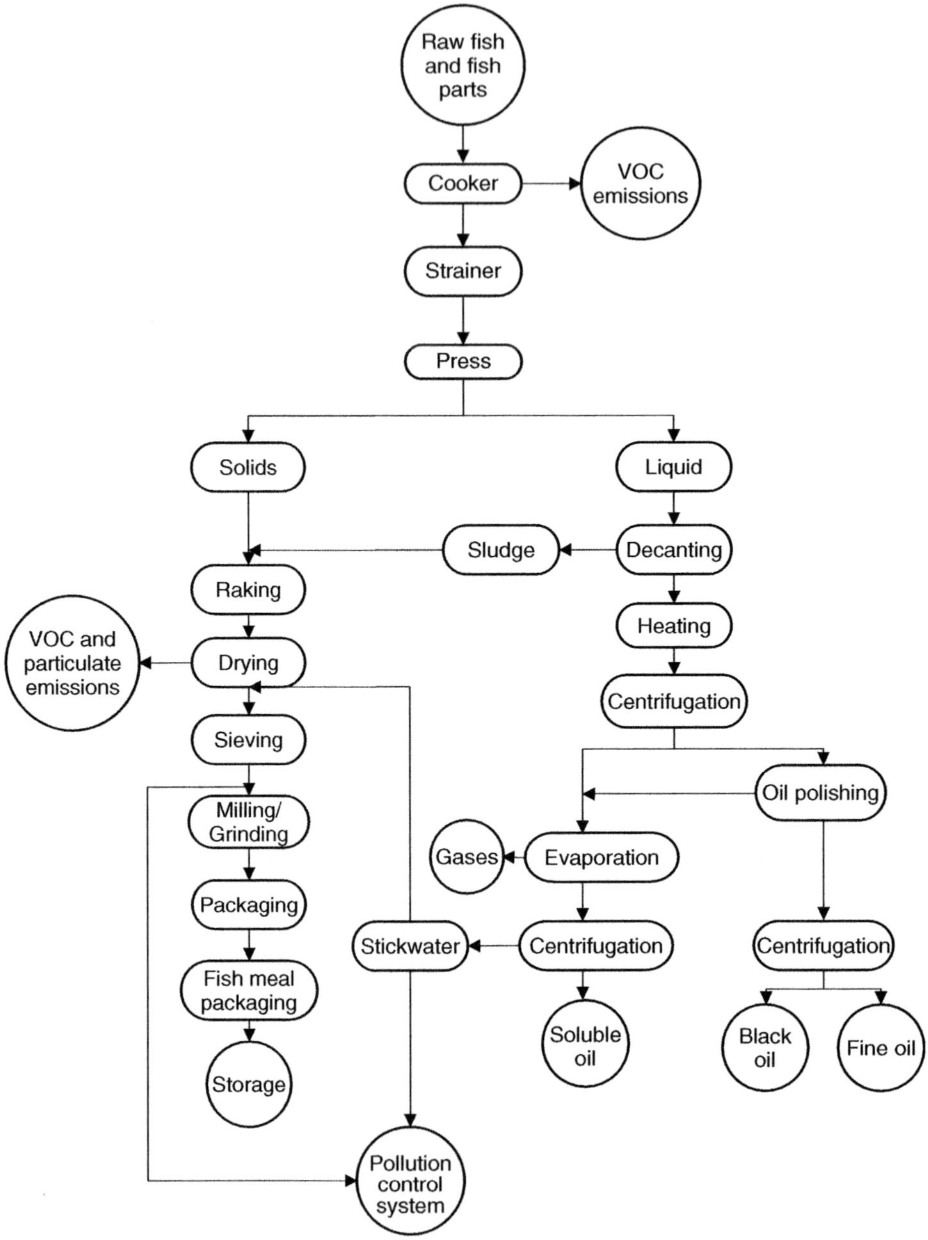

Figure 11.1 Flow diagram of fish meal and fish oil production (*Source*: EPA (US government).)

Table 11.1 Electricity (MJ) and thermal energy (kWh) requirements of canning tuna industry

Process	Thermal energy (MJ)	Electricity (kWh)
Fish reception	–	0.67
Fish thawing	111	
Cutting	–	23.62
Cooking	1385	24.19
Manual cleaning	–	9.69
Liquid dosage and can filling	–	47.19
Sterilisation	2083	45.61
Quality control and packaging	–	33.5
Ancillary activities (wastewater treatment, tinplate production for packaging)	–	64.29

Source: Hospido *et al.*, (2006). Reproduced with permission of Elsevier.

(EC) Directives regarding the management of the environmental impacts are: Directive 76/464/EEC for hazardous substances, Directive 79/923/EEC for quality of shellfish growing waters, Directive 91/492/EEC for shellfish, Directives 85/337/EEC and 97/11/EEC for environmental impact assessment, Directive 2001/42/EC for strategic environmental assessment, Directive 92/43/EEC for species and habitats, Directive 79/409/EEC for wild birds and Directive 2000/60/EC for water quality (Read and Fernades, 2003).

This chapter aims to summarize the current and potential treatment methods for fish waste and treated waste uses by means of informative flow diagrams, tables and the inputs and outputs (energy consumption, wastewater, solid waste) of various fish processes.

11.2 Treatment methods

The methods used for fish waste treatment are hydrolysis, bioremediation, anaerobic treatment, filtration, screening and several other miscellaneous/multifunctional methods.

11.2.1 Hydrolysis

Most of the first steps in fish protein hydrolysis were performed in the 1960s. Emphasis was given to fish protein concentrate as an inexpensive nutritious protein source for developing countries (Kristinsson and Rasco, 2000). Enzymatic hydrolysis of fish protein was used as an alternative technique for converting partially processed fish biomass into edible protein products (Suzuki, 1981; Diniz and Martin, 1996).

Fish meat (a horse mackerel) was mushed using hydrolysis in a reactor under subcritical (10^{-3} kg fish meat, 3.36×10^{-6} cm^3 Milli-Q water) and supercritical (10^{-3} kg fish meat, 3.06×10^{-6} (280 °C, 30 MPa) or 1.76×10^{-6} (400 °C, 30 MPa) cm^3 Milli-Q water) conditions at 200–400 °C (Yoshida *et al.*, 1999, 2003). At 200 °C, 3.35 MPa, and 5-min reaction time, lactic acid (0.03 kg/kg of dry meat), phosphoric acid (0.12 kg/kg of dry meat) and histidine (0.01 kg/kg of dry meat) were formed. At 280 °C, 6.42 MPa and 30 min reaction time, pyroglutamic acid (0.095 kg/kg of dry meat) was produced. In contrast, amino acids (cystine, alanine, glycine, and leucine) were obtained at 270 °C and 5.51 MPa. Oil was extracted from a water-insoluble phase with hexane, and was rich in fatty acids, such as eicosapentanoic acid (EPA) and docosahexianoic acid (DHA) (Yoshida *et al.*, 1999). According to Yoshida *et al.* (2003), at 200–400 °C, and 5-min reaction time, oil, amino acids (cystine, alanine and glycine) and organic acids (pyroglutamic acid) were formed under subcritical conditions.

Gao *et al.* (2006) applied two different acid hydrolysis methods in order to produce lactic acid from fish heads and bony parts. According to process A, minced raw material mixed with water (1:1 ratio) (initial pH 1) were hydrolysed at 121 °C for 20 min and centrifuged at 2706 *g* for 20 min, while the supernatant was kept and used as a nutrient source for lactic acid production. In process B, minced fish wastes were pretreated at 121 °C for 20 min without the addition of acid. The mixture was centrifuged and the residue and the supernatant obtained were treated according to process A (Figure 11.2). Lactic acid production was inhibited when unhydrolysed fish waste was used as the substrate for fermentation, whereas glucose was converted to lactic acid when hydrolysed fish waste was used as the nutrient source. Lactic acid production increased with hydrolysis time, especially up to 40 min. The fish waste hydrolysate produced by process B was rich in nutrients and can be used effectively as a substitute for yeast extract.

The potential production of amino acids and glucosamine by hydrothermal treatment of shrimp shells using high-temperature and high-pressure (45 MPa) water was studied by Quitain *et al.* (2001). Shrimp shells and deionised water (weight ratio = 1:125, dry basis) were placed in a batch reactor apparatus under subcritical and supercritical conditions at various temperatures (90–400 °C) and reaction times (5–60 min). The maximum production of amino acids (70 mg/g dry shrimp) was reported at 250 °C in 60 min, which was about 2.5 times higher than the total amino acids obtained at 90 °C. Glycine (Gly) and alanine (Ala) production increased with temperature, whereas it reduced at temperatures above 250 °C.

11.2.2 Bioremediation

Bioremediation contains all those processes and actions used to biotransform an environment, already altered by contaminants, to its original status (Thassitou and Arvanitoyannis, 2001).

Marinho-Soriano *et al.* (2011) conducted a laboratory study to test the efficacy of the macroalgae *Gracilaria caudata* and the microcrustacean *Artemia franciscana* to remove nutrients from aquaculture effluents. The experiment consisted of three treatments: (1) macroalgae; (2) macroalgae and *Artemia*; and (3) only *Artemia*. The results indicated that treatment 2 was the most efficient in removing the nitrogenated forms. The maximum reduction values were: NH_4 = 29.8%; NO_2 = 100%; NO_3 = 72.4% and DIN = 44.5%. In contrast, nutrient concentration was significantly higher at the end of treatment 3 than for the other treatments. PO_4 showed a significant increase during the experiment, especially in treatments 2 and 3. The results showed that *G. caudata* and *A. franciscana* performed well in absorbing nitrogenated forms. Therefore, the use of these organisms as biofilters has the potential for being an ecologically correct practice and may contribute to improving the water quality of coastal regions.

Sterilized wheat bran (700 g), *Aspergillus awamori* spores (0.7 g) and minced and sterilized sardine fish (2638 g) were combined and placed in a ventilated incubator for 5 days at 40 °C. The fermented product was mixed with corn-soybean mixture at 0.1, 1.0, 5.0 and 10.0% to examine the digestibilities of crude protein (CP) and dry matter (DM). After fermentation, crude protein, crude fat and nitrogen-free extracts were reduced by

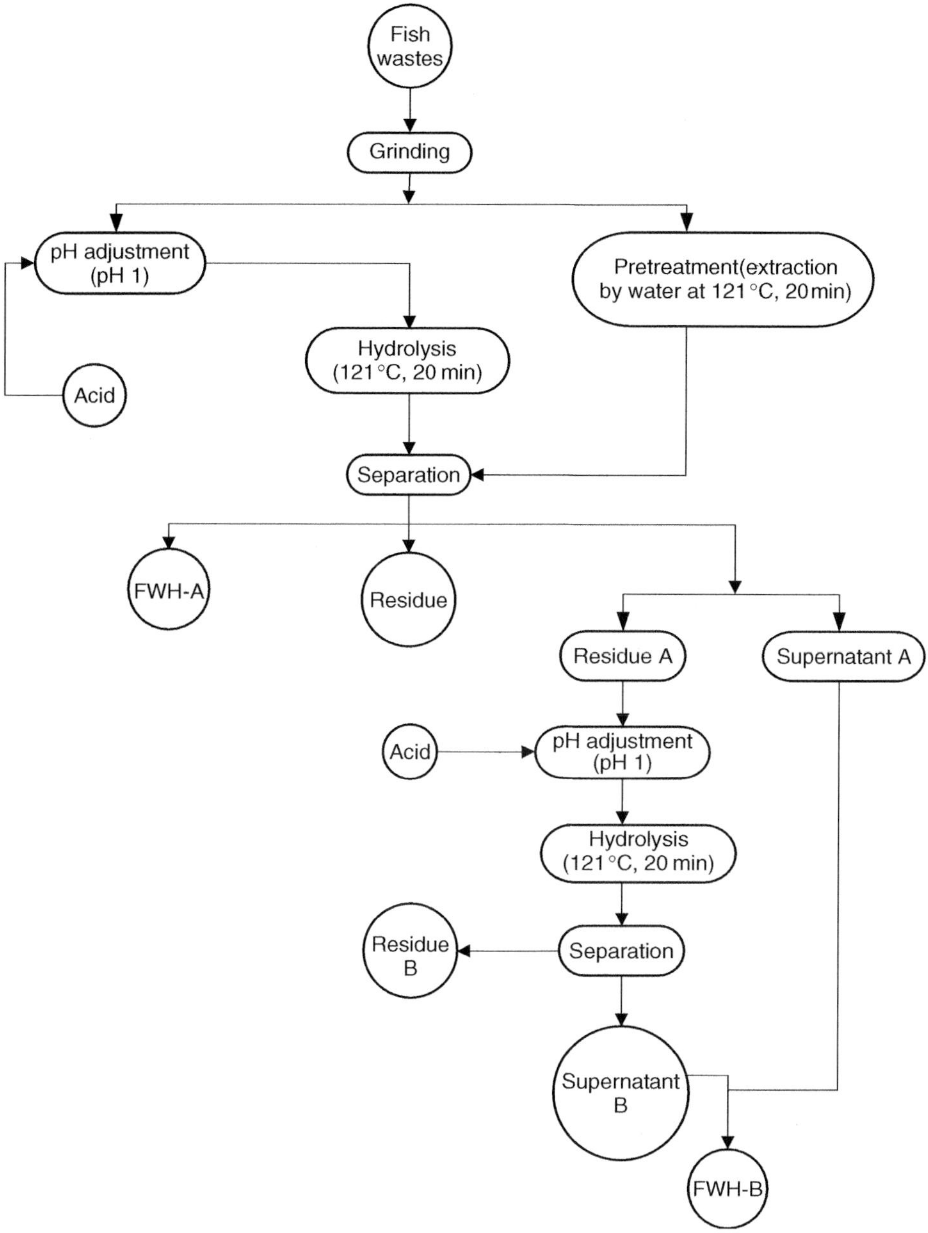

Figure 11.2 Flow diagram of the production of fish waste hydrolysate (FWH) (*Source*: Gao *et al.*, (2006). Reproduced with permission of Elsevier.)

up to 2.6, 22.2 and 25.7%, respectively. On the other hand, an increase up to 16.1% in crude fibre content was documented; therefore showing that crude fibre may be synthesized by *Aspergillus awamori*. Furthermore, various enzymes were produced during fermentation, such as glucoamylase (9.71 U), α-glucosidase (0.21 U), α-amylase (21.69 U) and acidic protease (17 778 U). Fermented fishmeal had no effect on CP digestibility, while an increase in DM digestibility was revealed in spite of the fishmeal levels and it was significant at 5 and 10% (Yamamoto *et al.*, 2005).

Shih and his co-workers (2003) performed 5-month fermentation experiments to produce fish sauce using different groupings of body parts (flesh body, head, fins and viscera) of bonito with or without the addition of enzymes, such as soybean koji (*Aspergilus oryzae*), ang-khak (*Monacus purpureus*) or viscera (*Treatment 1*: blended and sterilized bonito wastes were mixed with viscera (1 : 2, w/w) and a 20% salt solution; *Treatment 2*: blended and sterilized whole bonitos were mixed with viscera (1 : 2, w/w) and a 20% salt solution; *Treatment 3*: blended bonito wastes were mixed with a 20% salt solution; *Treatment 4*: blended whole bonitos were mixed with a 20% salt solution; *Treatment 5*: blended and sterilized bonito wastes were mixed with 38% soybean koji, stored for 2 days at room temperature, while a 20% salt solution was added; *Treatment 6*: blended and sterilized whole bonitos were mixed with 38% soybean koji, stored for 2 days at room temperature, while a 20% salt solution; *Treatment 7*: blended and sterilized bonito wastes were mixed with 38% ang-khak, stored for 2 days at room temperature, while a 20% salt solution was added; *Treatment 8*: blended and sterilized whole bonitos were mixed with 38% ang-khak, stored for 2 days at room temperature, while a 20% salt solution was added was added). The results indicated that treated bonito (whole fish and waste) with soybean koji and ang-khak presented higher total sugar content and amylase activity values ranging from 10.8 to 17.4 mg/l and from 5.0 to 100 U/ml, respectively. On the other hand, total sugar content of treated bonitos without the addition of enzymes was 3.6–5.9 mg/l, while no amylase activity was detected. Moreover, protein and amino acid values of treated whole bonitos were slightly higher than treated fish waste. Volatile basic nitrogen (VBN) and trimethylamine (TMA), two indices for fishiness odour, were found to exceed the thresholds of 6 mg/100 g and 50 mg/100 g. Fish sauces with better qualities were found with whole fish treatment with ang-khak.

Martin (1999) studied the potential for improving the valuable biomass present in fish waste by composting the processing waste from fisheries with turf. A mixture of fish waste and turf (bulking agent) were composted in a passively ventilated composting system. Compost extracts were sterilized and inoculated at a ratio of 5% (v/v) before use as a substrate for the acid-resistant fungus *Scytalidium acidophilum* ATCC *26774* (150 rpm, 25.0 °C, pH 2.0 ± 0.1, 8–10 days). Compost acid extracts showed higher nutrient concentration (lipids, carbohydrates and protein) than the compost. Compost extracts can be successfully used as a substrate for the growth of the fungus *S. acidophilum*, for the biofiltration of liquid and gas effluents, and in the growth of mushrooms.

A mixture of fish processing waste (offal) and a bulking agent (a blend of sawdust and wood shavings, 1 : 1 by volume) (3 : 1 wet weight ratio) was composted for 150 days in static ventilated piles (in-vessel method) under different weather conditions (winter and summer). Temperatures were kept above 45 °C for 60 days, at 55 °C for 9 days and then peaked several times, and at 25 °C for 120–150 days. A significant reduction in electric conductivity (EC), ammonium as nitrogen (NH_4^+-N) and water soluble carbon (WSC) was observed, while after 80 days volatile fatty acids (VFA) reduced to non-detectable levels, with the exception of acetic values. The ratio WSC : TN (total nitrogen), an index of compost maturity, revealed that compost obtained during summer was mature (WSC : TN=0.7) within 150 days, while compost obtained in winter required more days to accomplish maturity (WSC : TN=1.2) (Laos *et al.*, 2002).

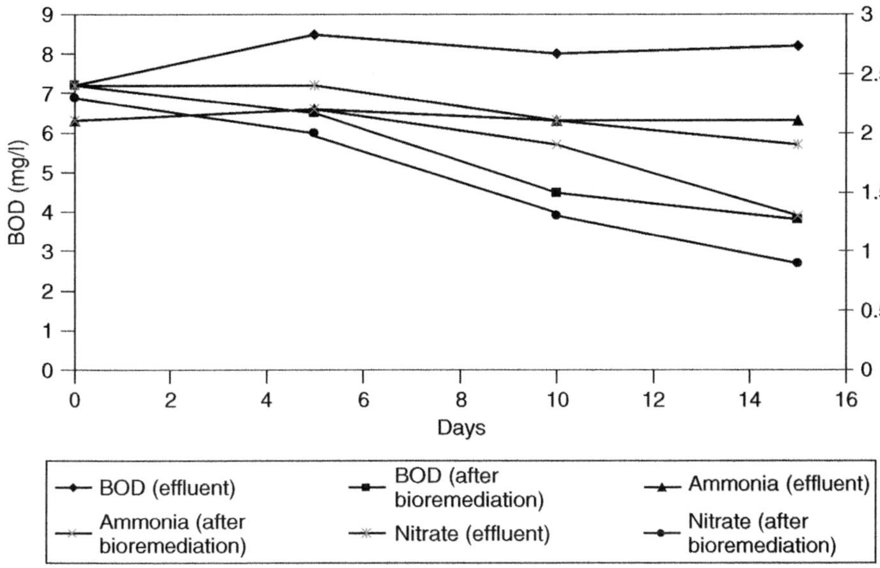

Figure 11.3 Trends in BOD_5 (biochemical oxygen demand of wastewater during decomposition occurring over a 5-day period), ammonia, and nitrate concentration in effluent and after bioremediation with mats (adapted from Paniagua-Michel and Garcia, 2003)

Paniagua-Michel and Garcia (2003) used indigenous microbial groups to create microbial mats for the bioremediation of a semi-extensive shrimp (*Litopenaeus vannamei*) pond. The shrimp hatchery effluent was fed in to three glass reactors at 24–27 °C, pH 7.8–8.0, oxygen concentration 8 mg/l, and hydraulic resistance time 5 days. Each reactor was composed of a retention chamber at the top for the water treatment, three stages with the constructed mats were covered at each stage and had several small holes serving as a sprinkling filter at the bottom. According to the design of the bioreactor, the effluent flows throughout the surface of the first bed which contains mats, and then sprinkles to the second and third bed. After 20 days of treatment, 97% ammonia nitrogen and 95% nitrate nitrogen concentration removal was reported. Furthermore, at the end of the experimental period, 80% BOD_5 reduction occurred (Figure 11.3).

11.2.3 Anaerobic treatment

The anaerobic treatment of wastewater proceeds with degradation of the organic load to gaseous products (mainly methane and carbon dioxide), which constitute most of the reaction products and biomass. Treatment with anaerobic digestion may be a feasible option, incorporating waste utilization and energy production in the form of biogas.

Manufacturing processes in fish canning industries generate a considerable amount of solid waste that can be digested anaerobically. Eiroa *et al.* (2012) studied the biochemical methane potential of different solid fish wastes. For tuna, sardine and needle fish waste, around 0.47 g $COD{-}CH_4$/g COD_{added} was obtained in batch experiments with 1% total solid (TS); whereas for mackerel waste, the methane production attained 0.59 g $COD{-}CH_4$/g COD_{added}. The increase in the waste : inoculum ratio, from 1.1–1.3 to 2.8–3.3 g volatile solids $(VS)_{waste}$/g $VS_{inoculum}$, led to overload due to accumulation of volatile fatty acids and long chain fatty acids. Afterwards, co-digestion assays of fish waste with gorse were undertaken but the biochemical methane potential did not improve.

Anaerobic batch digestion of sisal pulp and fish wastes separately, as well as co-digestion of both substrates was investigated by Mshandete *et al.* (2004). Sisal pulp and fish waste (offal, scales, gills) and washing water, were digested separately at 5–60% wet biomass (v/v) in 30 batch bioreactors for 25 and 29 days, respectively. In the co-digestion experiment, mixtures of fish waste and sisal pulp of different proportions were digested in 15 bioreactors for 24 days at 27 ± 1 °C. The highest methane yield of 0.32 m^3 CH_4/kg VS for sisal pulp and 0.39 m^3 CH_4/kg VS for fish waste at 5% TS was reported after 25 and 29 days, respectively. Co-digestion with 33% fish waste and 67% sisal pulp, at 16.6% TS and C:N ratio 16, resulted in the highest methane yield of 0.62 m^3 CH_4/kg VS, an increase of 59–94% in the methane yield compared to that obtained during the digestion of sisal pulp and fish wastes at 5% TS. Biogas recovery with 60–65% methane content was recorded for anaerobic co-digestion of fish waste and sisal pulp.

Nges *et al.* (2012) studied anaerobic digestion of the original fish waste and the fish sludge remaining after enzymatic pre-treatment to extract fish oil, and fish protein hydrolysate was assessed for potential methane production. The results showed high biodegradability of both fish sludge and fish waste, giving specific methane yields of 742 and 828 m^3 CH_4/t VS added, respectively. However, chemical analysis showed high concentrations of light metals which, together with high fat and protein contents, could be inhibitory to methanogenic bacteria. The feasibility of co-digesting the fish sludge with a carbohydrate-rich residue from crop production was thus investigated, and a full-scale process outlined for converting odorous fish waste to useful products.

Vidotti and his colleagues (2003) studied the amino acid composition of fish silages produced by acid digestion (20 ml/kg formic acid and 20 ml/kg sulfuric acid) and anaerobic fermentation (50 g/kg *Lactobacillus plantarum*, 150 g/kg sugar cane molasses) from different raw materials, such as saltwater fish (SW), freshwater fish (FW) and tilapia filleting residue (TR). The raw materials were separately mixed with 15% sugar cane molasses, 5% *Lactobacillus plantarum* culture media and 0.25% sorbic acid (w/w), to obtain fermented silage. When 2% formic acid (w/v) and 2% sulfuric acid (w/v) were added to this mixture, acid silage was produced (Figure 11.4). The results indicated that acid silages had higher protein content 69.91% (SW), 44.38% (FW) and 39.59% (TR), compared to fermented silages, 59.61% (SW), 42.09% (FW) and 35.84% (TR), respectively. Histidine, threonine and serine levels increased, while levels of valine, isoleucine, and leucine decreased in all products. All products can be used in balanced fish diets, due to their nutritional value.

11.2.4 Filtration/screening

Filtration and screening are recorded under a single heading because they have a common characteristic: separation of particles (cells) that depends upon the difference between the size of the particles and that of the openings (screen) or pores (filter medium). The problem is that the screen or filter medium becomes blocked before a workable ‘cake’ can be accumulated (United Nations Environment Programme, 2005).

Cattaneo-Vietti and his co-workers (2003) used a small underwater device for wastewater processing. The system called MUDS (marine underwater depuration system) was placed at sea over the urban sewage outflow of a submarine pipeline. Total suspended matter (TSM), BOD_5 and coliform-bacteria concentrations were decreased up to 90%, while the porous structure of the filter resulted in the development of protozoan and microbenthic communities, therefore indicating the consumption of a significant amount

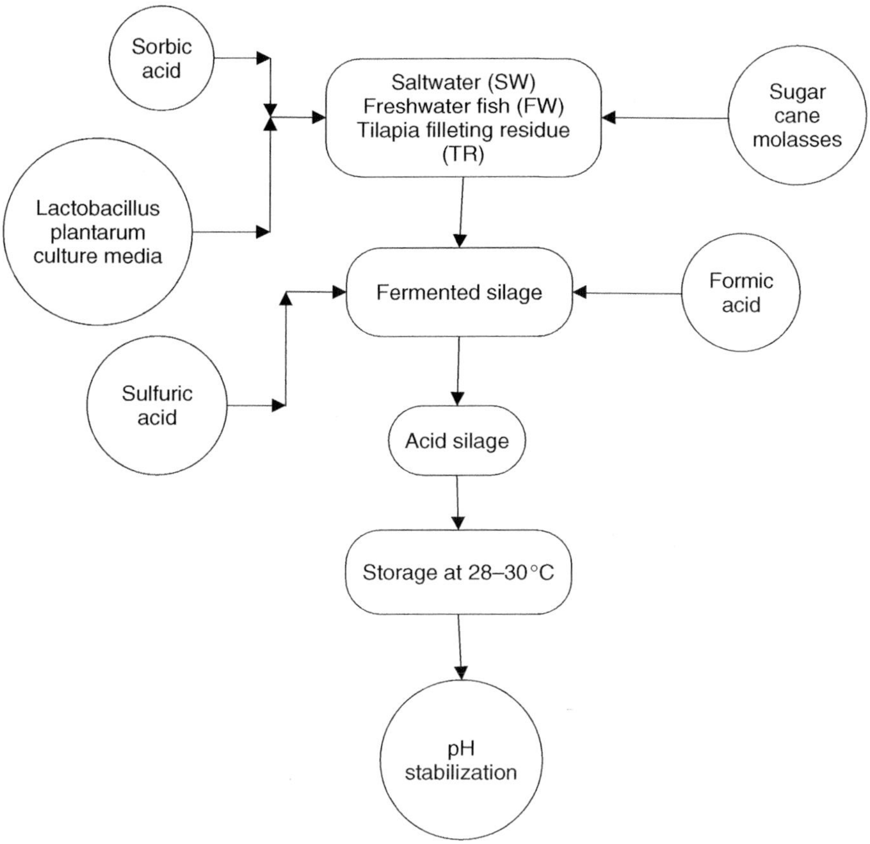

Figure 11.4 Flow diagram of silages production from three different raw materials (adapted from Vidotti *et al.*, 2003; Arvanitoyannis and Kassaveti, 2008)

of organic matter. Moreover, predators (such as detritus feeder invertebrates and benthic and pelagic fishes) were attracted to the filter of the system causing recycling of organic matter. The system favoured the development of fish, the aggregation of fish species and the selection of marine organisms (bacteria, fungi, protozoa) capable of supporting the changes in water salinity and temperature.

Guerdat *et al.* (2011) assessed the organic carbon impact on three different commercially available biological filters at a large scale under actual recirculating aquaculture production conditions. The study was performed at the North Carolina State University Fish Barn, a commercial scale research and demonstration recirculating aquaculture facility operated by the Biological and Agricultural Engineering department. The study was based on a $60\,m^3$ *Tilapia* system with average daily feed rates of 45 kg using a 40% protein feed and an average biomass of 6750 kg. The system was dosed with sucrose ($C_{12}H_{22}O_{11}$) to increase the concentration of biodegradable organic carbon in the system. The effect of elevated organic carbon concentrations on total ammonia nitrogen (TAN) removal rates was assessed and determined based on biofilter media volume. Variability increased substantially in the volumetric TAN removal rate (VTR) for all three filters. VTR for all three filter types was reduced by approximately 50% compared to normal production conditions. The results demonstrate the importance of controlling the concentration of biologically available organic carbon in a recirculating aquaculture system.

The potential use of an ultra-low pressure membrane for aquaculture wastewater treatment was studied by Ali *et al.* (2005). Fish wastewater samples after sand treatment filtration were passed through polyethersulfone (PES) membranes, which were prepared by using a dry/wet phase inversion technique. The results showed that total ammonium and phosphorous were removed up to 85.70 and 96.49%, respectively at pressure 0.4–0.8 MPa.

Degradation rate experiments of solid waste from Atlantic salmon fish farms were performed for 21 days by Buryniuk and his colleagues (2006). Fish feed waste removal rate experiments were also conducted. Frozen samples (−10 °C) of solid waste were placed in tanks filled with saline (25–30% salinity) and distilled water, which were kept in the dark at 8 °C and at oxygen concentration above 80%. After 21 days, C/N and nitrogen free extracts (NFE) decreased from 11 to 8.5, and from 48.6 to 36.5% dry weight, respectively. Fish feed COD reduced from 2.9×10^5 to 1.2×10^6 mg/l·g dry weight, while fish solids COD did not change significantly (from 9.6×10^6 to 7.7×10^6 mg/l·g dry weight). Ammonia concentration increase, bubbles and foam formation, and wastewater decolourisation were reported during the experiment. Furthermore, after 30–40 days in cold and saline water, approximately 50% of the waste matter disappeared.

11.2.5 Miscellaneous/multifunctional methods

Lupatsch and his colleagues (2003) studied the potential of using grey mullet for removing effluent from fish farms. Net-cage enclosures with no bottoms were stocked with bottom-feeding grey mullets (*Mugil cephalus*), which accessed the enclosed organically enriched sediments and the particulate matter (PM) falling from the overlying cages. Fish was fed with a known composition diet (303 g/kg crude protein and 16 MJ/kg gross energy) to estimate the potential growth rate and the energy and protein requirements of the mullet. During the experiment, mullet in the enclosures consumed all the available food in the sediment and increased weight up to 0.78 g/day per fish. The feeding trial for energy and protein requirements for maintenance and growth revealed that mullet removed 4.2 g organic carbon, 0.70 g nitrogen and 7.5 mg phosphorus/kg mullet per m^2/day from the organically enriched sediment. In conclusion, rearing grey mullet on the organically enriched sea floor below commercial fish farms can effectively lead to the improvement of sediment status.

By-products from frozen herring (*Clupea harengus*) were used for crude fish oil production (Aidos *et al.*, 2001). Minced herring by-products were pumped into an insulated scraped-surface heat exchanger indirectly heated by steam. The heated suspension was then separated in a three-phase decanter into a semi-solid phase (protein phase), a water phase (stickwater) and lipid phase (oil) (Figure 11.5). The oil produced was rich in valuable polyunsaturated *n*-3 fatty acids, such as EPA and DHA. Herring oil had very low copper content (<0.1 mg/kg oil) compared to iron concentration (0.8 mg/kg oil). Maatjes herring fillets had a 32% higher copper content than by-products, while by-products had a three times higher iron content than fillets. Furthermore, α-tocopherol and free fatty acid (FFA) concentrations remained stable over time at different storage conditions (light at room temperature, dark at room temperature, dark at 50 °C).

Crude oils had very low copper content (<0.1 mg/kg oil), in comparison to iron content that reached 0.8, 0.1, and 0.03 mg/kg oil, respectively, for oil from maatjes and frozen and fresh by-products. FFA content of maatjes and fresh oil was low and remained almost stable, while an increase in FFA production for frozen oil was observed. Secondary

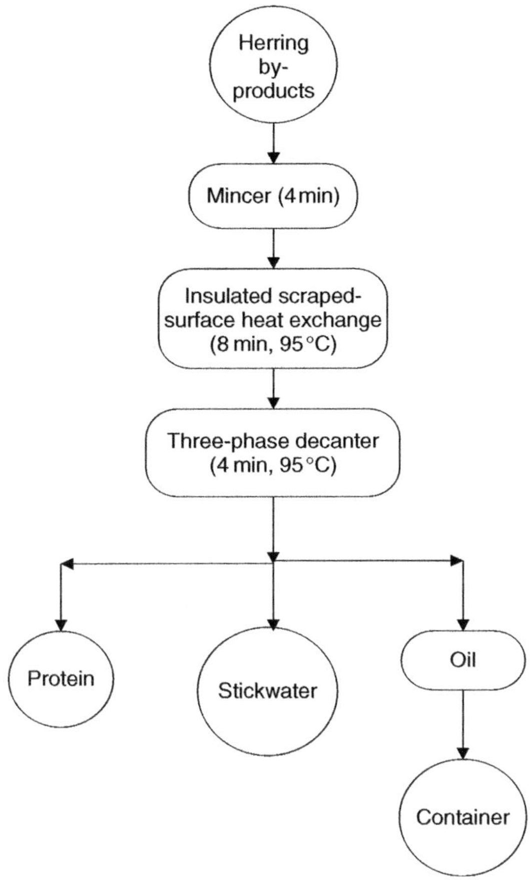

Figure 11.5 Schematic presentation of crude fish oil production by herring by-products (*Source*: Aidos *et al.* (2001). Reproduced with permission of Elsevier.)

oxidation products were detected in fresh oil, while tertiary oxidation products were recorded for the maatjes and frozen by-products oil. During storage, maatjes and frozen oils had intense odour, correlating positively at the end with sensory attributes of train-oil, acidic, marine and fishy. The best oils were produced when fresh and unsalted herring by-products were used for primary oil extraction (Aidos *et al.*, 2003).

Frozen paste from the filleting waste of silver carp (*Hypophthalmichthys molitrix*) was mixed with microbial transglutaminase (3 g/kg), dairy proteins, such as whey protein and sodium caseinate (10 g/kg), and NaCl (0, 10, 20 g/kg) for 2 min at 15 °C. The sample was placed into stainless steel tubes and sprayed with commercial regular vegetable oil. The tubes were submerged in water at 40 °C for 1 h, and afterwards at 90 °C for 20 min. Then the mixture was placed in a water bath and cooled at 4–5 °C for 30 min. Restructured fish products with different levels of salt (0, 1 and 2%) were obtained. The mechanical properties of fish gels increased when the salt level of the samples containing dairy proteins or not increased. Otherwise the mechanical properties of fish gels with 0 and 1% salt content increased only when dairy proteins (sodium caseinate) were used for filleting waste treatment (Uresti *et al.*, 2001).

Five chemical coagulation/flocculation treatments of fish canning wastewater were tested using $FeCl_3$ (0.4 g Fe/l) (treatment 1), $Al_2(SO_4)_3$ (1.2 g Al/l) (treatment 2), $Ca(OH)_2$ (0.5 g Ca/l) (treatment 3), $FeCl_3$ and $Ca(OH)_2$ (0.4 g Fe/l + 0.2 g Ca/l) (treatment 4), $Al_2(SO_4)_3$ and $Ca(OH)_2$ (1.2 g Al/l + 0.3 g Ca/l) (treatment 5). BOD_5, COD, total suspended solids (TSS), and total dissolved solids (TDS) were reduced significantly up to 80–88%, 75–85%, 94–95% and 31–53%, respectively. In treatments 3, 4 and 5, total hardness increased by 61%, 319% and 419%, respectively, while in treatments 1, 2, 3 and 4, ammonia content reduced up to 90%, 56%, 100% and 100%, respectively. Phosphate decreased by 90% in treatments 2 and 3, and was undetectable in treatments 1, 2 and 3. A significant reduction in oil and grease, and total protein amounting to 73–89%, and 70–92%, respectively was observed. Furthermore, the maximum decrease in oil and grease was reported in treatments 1, 2 and 4, while for protein the maximum reduction was observed in treatments 3 and 4. The best method for the treatment of fish canning wastewater was the use of 0.4 g Fe/l and 0.2 g Ca/l (treatment 4) and the dried precipitate obtained contained 40% (by weight) recovered protein and 20% recovered fat (Fahim *et al.*, 2001).

Rainbow trout (*Oncorhynchus mykiss*) raceways effluents were used for the construction of a baffled sedimentation basin. The trout effluent entering the basin consisted of two consecutive sections (first and second), and within the first section, four linear segments of biofiltration media were installed. TSS removal ranged from 71 to 79% during normal conditions, and from 79 to 92% during cleaning/harvesting conditions. On the other hand, doubling the basin volume and surface area caused up to 50% TSS removal. The highest removal of TSS was observed in the first section of the basin (84% and 94%, TSS removal for normal and cleaning/harvesting conditions, respectively). Moreover, dissolved nutrients, such as TAN, nitrate and nitrite, removal was observed, with the exception of orthophosphate (OP). The recorded retention times were shorter by up to 32–37% for the first section, and by 17–27% for the overall basin, compared to theoretical values. Finally, an increase in hydraulic efficiency with an increase in L : W (length : width) ratios was reported (Stewart *et al.*, 2006).

Organic wastes from a fish (cod) enclosure were banded with lipid biomarkers, using sediment traps to collect settling matter. Sediment traps were placed 1.5 m away from the seafloor at distances of 5, 10, 30 and 100 m from the cages in a transect out of the cove. The samples taken from the traps and the feed (whole herring) were frozen, transported to the laboratory and stored at −20 °C in chloroform and under nitrogen until analysis. Lipids were extracted from the samples with a mixture of chloroform : methanol. The major lipid classes detected were: phospholipids (21–10%), acetone mobile polar lipids (17–42%), FFAs (8–41%), and in smaller amounts triacylglycerols (4–9%), sterols (5–10%), wax esters and sterol esters (0–11%), alcohols (0–5%) and hydrocarbons (3–5%). On the other hand, the major fatty acids detected were: 14:0 (6–9%), 16:0 (17–22%), $16{:}1\omega_7$ (12–21%), $18{:}1\omega_7$ (6–7%), $18{:}1\omega_9$ (5–10%), $20{:}5\omega_3$ (7–11%) and 20:3a (0–9%). The three long-chain monounsaturated fatty acids $20{:}1\omega_9$, $22{:}1\omega_9$ and $22{:}1\omega_{11}$ were found at proportions below 3%, and were at high levels close to the cages, and reduced with distance from the cages. The feed was rich in fatty acids $20{:}1\omega_9$ and $22{:}1\omega_{11}$, which were not digested by cod and other fish; thus resulting in their use as biomarkers for the dispersion of organic waste from fish farms (Van Biesen and Parrish, 2005).

Table 11.2 provides a synopsis of the most important treatment methods (parameters, quality control and results). The advantages and disadvantages of various methods for fish waste treatment are presented in Table 11.3.

Table 11.2 Traditional African fish smoking ovens and their characteristics

Kind of waste	Treatment	Parameters	Methodology	Quality control methods	Results	References
Fish meat	Hydrolysis of fish meat under subcritical and supercritical conditions	Organic acids, amino acids, TOC, carbon, nitrogen, hydrogen	Fish meat liquefaction by hydrolysis in a reactor under subcritical (10^{-3} kg fish meat, 3.36×10^{-6} cm^3 Milli-Q water) and supercritical (10^{-3} kg fish meat, 3.06×10^{-6} (280 °C, 30 MPa)	1. Organic acids, amino acids: HPLC 2. TOC: TOC analyser 3. Carbon, nitrogen, hydrogen: CHN corder	1. Formation of lactic acid (0.03 kg/kg of dry meat), phosphoric acid (0.12 kg/kg of dry meat), and histidine (0.01 kg/kg of dry meat) (200 °C, 3.35 MPa, 5 min) 2. Production of pryoglutamic acid (0.095 kg/kg of dry meat) (280 °C, 6.42 MPa, 30 min) 3. Production of amino acids (cystine, alanine, glycine, and leucine) (270 °C, 5.51 MPa) 4. Oil extraction (rich in EPA and DHA) 5. Formation of amino acids (cystine, alanine, and glycine) and organic acids (pyroglutamic acid) (200–400 °C, 5 min)	Yoshida *et al.*, 1999, 2003

(continued overleaf)

Table 11.2 *(continued)*

Kind of waste	Treatment	Parameters	Methodology	Quality control methods	Results	References
Minced salmon muscle	Enzymatic hydrolysis	Protein, moisture content, total mineral content, lipid content, nitrogen recovery, emulsifying capacity, emulsifying stability, fat absorption/oil-holding capacity	Homogenized minced salmon muscle treated with one of four alkaline proteases (Alcalase 2.4 l, Flavourzyme 1000 l, Corolase PN-L, and Corolase 7089) or endogenous digestive proteases (pH 7.5, 40 °C, 7.5% protein).	1. Protein: using a modified $CuSO_4/TiO_2$ mixed catalyst Kjeldahl method according to method 4.2.03 (AOAC, 1990) 2. Moisture content: convectionless oven method 24.003 (AOAC, 1990) 3. Total mineral content: direct ash method 14.006 (AOAC, 1990) 4. Lipid content: modified acid hydrolysis method 948.15 (AOAC, 1990) 5. Nitrogen recovery: calculated as the amount of protein (%N × 6.25) present in the hydrolysates relative to the initial amount of protein present in the reaction mixture 6. Emulsifying capacity: oil titration method similar to that of Webb *et al.* (1970) with some modifications 7. Emulsifying Stability: method of Yatsumatsu *et al.* (1972) as modified by Miller and Groninger (1976) with some additional modifications 8. Fat Absorption/Oil-Holding Capacity: method of Shahidi *et al.* (1995) with some modifications	1. Protein content ranged from 71.7 to 88.4% 2. Low lipid content 3. Nitrogen recovery ranged from 40.6 to 79.9% 4. High hydrolysates solubility at high pH (pH 7) 5. Hydrolysates solubility ranged from 92.4 to 99.7% 6. Fish protein emulsification capacity was 75–299 ml of oil emulsified per 200 mg of protein 7. Fish protein emulsification stability was 50–70% 8. Greater fat absorption at 5 and 10% DH fish protein hydrolysates (3.22–5.90 ml of oil/g of protein) than for 15% hydrolysates	Kristinsson and Rasco, 2000

Fish wastes (heads and bony parts)	Acid hydrolysis	Glucose, lactic acid, total nitrogen, cell nitrogen, dry weight	*Process A* Minced wastes mixed with water (1 : 1 ratio) (pH 1) was hydrolyzed at 121 °C for 20 min and centrifuged at 2706 g for 20 min *Process B* Process A plus fish wastes pre-treatment (an extraction by water at 121 °C for 20 min)	1. Glucose and lactic acid: HPLC with refractive index detector (for glucose) and with a UV/VIS detector at 210 nm (for lactic acid) 2. Total nitrogen: elementary analyser 3. Cell density: spectrophotometer at 660 nm 4. Dry weight: drying in an oven at 104 °C	1. Lactic acid production by both processes 2. Process B increased lactic acid productivity by 22% 3. Fish waste hydrolysate (produced by Process B) can be used as a substitute for yeast extract	Gao *et al.*, 2006
Shrimp shells	Hydrothermal treatment	Amino acids, acetic acid, organic acid	Treatment of shrimp shells at batch reactor apparatus in high temperatures (90–400 °C), high pressure water (45 MPa) and various reaction times (5–60 min)	1. Amino acids: amino acids analyser 2. Acetic acid: organic acid analyser 3. Organic acid: system consists of an ion-exclusion column and an electroconductivity detector	1. Maximum production of amino acids (70 mg /g of dry shrimp shel)l at 250 °C in 60 min 2. Production of Gly and Ala increased with increasing temperature, but decreased for temperatures above 250 °C	Quitain *et al.*, 2001
Fish waste	Fermentation process using *Aspergillus awamori*	Dry matter (DM), crude protein (CP), lipid, ash, fiber content, glucoamylase, acidic protease activity, released tyrosine	Wheat bran (700 g), minced fresh sardines (2638 g) and *A. awamori* spores (0.7 g) mixed together and kept in a ventilated incubator for 5 days at 40 °C	1. DM, CP, lipid, ash and fiber content: proximate chemical analysis (AOAC, 1984) 2. Glucoamylase: using kits 3. Acidic protease activity: using casein as a substrate 4. Released tyrosine: photometrically at 660 nm	1. Decrease of crude protein, fat, and nitrogen-free extracts up to 2.6, 22.2 and 25.7%, respectively 2. 16.1% increase of crude fibre	Yamamoto *et al.*, 2005

(*continued overleaf*)

Table 11.2 *(continued)*

Kind of waste	Treatment	Parameters	Methodology	Quality control methods	Results	References
					3. Production of glucoamylase (9.71 U), α-glucosidase (0.21 U), α-amylase (21.69 U) and acidic protease (17 778 U) 4. CP content remained stable	
Bonito waste (flesh body, head, fins, and viscera)	Fermentation	pH, protein, reducing sugar, total sugar, trimethylamine (TMA), volatile basic nitrogen (VBN), free amino acids, color, volatile compounds	Different combinations of fish body parts with or without the addition of enzymes (viscera, soybean koji or ang-khak) were inoculated into a porcelain vase for fermentation	1. pH and protein: standard methods (AOAC, 1984). 2. Reducing sugar and total sugar: Miller's method (Miller, 1959) and phenol-sulfuric method (Dubois *et al.*, 1956), respectively 3. Trimethylamine (TMA) and volatile basic nitrogen (VBN): Castell *et al.*'s method (Castell *et al.*, 1974), and Cobb *et al.*'s method (Cobb *et al.*, 1973), respectively 4. Free amino acids: amino acid analysanalyser 5. Colour: colour measuring system 6. Volatile compounds: GC	1. Higher total sugar content (10.8–17.4 mg/l) and amylase activities (5.0–100 U/ ml) of soybean koji and ang-khak treated bonitos 2. Lower total sugar content (3.6–5.9 mg/l), and no detection of amylase activities of treated bonitos without the addition of enzymes 3. Volatile basic nitrogen (VBN) and trimethylamine (TMA) exceed thresholds of 6 mg/100 g and 50 mg/100 g (except for treatments 1 and 3)	Shih *et al.*, 2003

Fisheries processing waste	Fermentation	Ash, carbohydrates, lipids, protein, biomass concentration	Composting with peat in a passively aerated composting system. Compost extracts were used as a substrate for *Scytalidium acidophilum* ATCC *26774* (agitation speed of 150 rpm, temperature of 25.0 °C, pH of 2.0 ± 0.1, and cultivation time 8–10 days)	1. Ash: combustion in a muffle furnace at 600 °C 2. Carbohydrates: anthrone reagent method presented by Morris (1948) 3. Lipids: method described by Folch *et al.* (1957) 4. Protein: multiplying the nitrogen, found by AOAC Micro-Kjeldahl Method 47.021 (Anon, 1980), by the factor 6.25 5. Biomass concentration: procedure of Martin *et al.* (1990)	1. Acid extracts of compost had higher concentrations of nutrients (lipids, carbohydrates, and protein) than compost 2. Fish wastes–peat compost: inexpensive nutrient source for fermentation processes 3. Substrate for the growth of acid-tolerant fungus *S. acidophilum* 4. Used for the biofiltration of liquid and gas effluents and in the growth of mushrooms	Martin, 1999
Fish processing waste (offal)	Composting of fish waste with sawdust and wood shavings using static aerated piles	pH, electrical conductivity (EC), ammonium as nitrogen (NH_4^+-N), water soluble carbon (WSC), total organic carbon (TOC), total N (TN) and volatile fatty acids (VFA)	Fish offal was mixed with the bulking agent (a blend of sawdust and wood shavings 1 : 1 by volume) (3 : 1 wet weight ratio). The mixture was placed in the reactors and remixed after 20 days and reloaded in the reactors (temperature above 45 °C, duration 150 days) (the experiment was carried out in winter and summer)	1. Electrical conductivity, pH, NH_4^+-N and WSC: mechanically shaking of the samples with distilled water (1 : 10 ratio) for 2 h and filtering through Whatman N° 1 filter paper 2. Ammonium: indophenol-blue method 3. WSC (as chemical oxygen demand (COD)): wet acid digestion, followed by spectrophotometric determination (APHA, 1992)	1. Decrease of EC, NH_4^+-N, WSC 2. Decrease of VFA to non detectable levels after 80 days, with exception of acetic values 3. WSC/TN = 0.7 (summer): compost is mature in 150 d 4. WSC/TN = 1.2 (winter): additional days would be needed for the compost to achieve maturity	Laos *et al.*, 2002

(*continued overleaf*)

Table 11.2 *(continued)*

Kind of waste	Treatment	Parameters	Methodology	Quality control methods	Results	References
				4. Total N: semi-micro Kjeldahl method 5. TOC: ignition at 550 °C (Navarro *et al.*, 1990) 6. VFA: gas chromatograph equipped with a hydrogen flame detector		
Fish waste (offal, scales, gills, washing water)	Anaerobic batch co-digestion of sisal pulp and fish waste	Partial alkalinity (PA), total alkalinity (TA), pH, TS, VS, total nitrogen, total lipids, organic carbon	Batchwise digestion of sisal pulp and fish waste separately (25 and 29 days, respectively) and with mixtures in various proportions (24 days) in bioreactors at 27 ± 1 °C	1. Partial alkalinity (PA), total alkalinity (TA), pH: TIM titration manager with an ABU 901 Autoburette (Bjornsson *et al.*, 2000) 2. TS, VS: APHA Standard Methods (APHA, 1995) 3. Total nitrogen: Kjeldahl method 4. Total lipids: Soxhelt extraction using petroleum ether solvent extraction (APHA, 1995) 5. Organic carbon: dichromate oxidation method (Nelson and Sommers, 1996)	1. Digestion of sisal pulp and fish waste separately: 0.32 and 0.39 m^3 methane/kg VS, respectively 2. Highest methane yield (33% fish waste and 67% sisal pulp): 0.62 m^3 methane/kg VS 3. Co-digestion sisal pulp and fish waste: biogas production (60–65% methane content)	Mshandete *et al.*, 2004

Marine fish waste, commercial freshwater fish waste, and tilapia filleting residue	Acid digestion (formic acid and sulfuric acid) and anaerobic fermentation (*Lactobacillus plantarum*, sugar cane molasses)	Crude protein, amino acid composition	Saltwater (SW), commercial freshwater fish waste (FW), and tilapia filleting residue (TR) were mixed with 15% sugar cane molasses, 5% *Lactobacillus plantarum* and 0.25% sorbic acid (w/w) (fermented silage), while 2% formic acid (w/v) and 2% sulfuric acid (w/v) (acid silage) were added to the mixture	1. Crude protein: micro-Kjedahl method, according to AOAC-code 981.10 (1990) 2. Amino acid composition: liquid chromatography, using a cationic exchange resin column and ninhidrine post-column derivation in auto-analyser	1. Acid silages: higher protein content 69.91% (SW), 44.38% (FW) and 39.59% (TR) compared to fermented silages, 59.61% (SW), 42.09% (FW) and 35.84% (TR), respectively 2. Increase in histidine, threonine and serine levels, for both processes and all three raw materials used 3. Decrease of valine, isoleucine, and leucine decreased in all products. 4. All products can be used in balanced fish diets – nutritional value	Vidotti *et al.*, 2003
Fish wastewater	Underwater filtration	Total suspended matter (TSM), BOD_5, coliform-bacteria concentrations	Marine underwater depuration system (MUDS) consisted of a percolating filter was placed at sea over an urban sewage outflow of a submarine pipeline	–	1. 90% reduction of TMS, BOD_5 and coliform-bacteria concentrations 2. Development of protozoan and microbenthic communities, increasing the depuration efficiency of the system and the consumption of the organic matter 3. Predators (such as detritus feeder invertebrates and benthic and pelagic fishes) attraction, resulting in organic matter recycling	Cattaneo-Vietti *et al.*, 2004

(*continued overleaf*)

Table 11.2 *(continued)*

Kind of waste	Treatment	Parameters	Methodology	Quality control methods	Results	References
					4. Development of fish, aggregation of fish species 5. Selection of marine organisms (bacteria, fungi, protozoans) causing changes in salinity and temperature	
Aquaculture wastewater	Filtration followed by treatment by polyethersulfone (PES) membranes	Total ammonium, total phosphorous	Aquaculture wastewater obtained after sand filtration treatment passed through polyethersulfone (PES) membranes, which were prepared by using a dry/wet phase inversion technique	Total ammonium and total phosphorous: standard method and digestion technique by using spectrophotometer at 880 and 640 nm, respectively	High removal of total ammonium and total phosphorous up to 85.70 and 96.49%, respectively at pressure 0.4–0.8 MPa	Ali *et al.*, 2005
Solid waste and fish feed	Degradation rate experiments	Dry mass, ash, total carbon, nitrogen, sulphur content, COD, carbohydrates, lipids, protein content	Frozen samples (−10 °C) were placed in tank filled with saline (25–30% salinity) and distilled water, and kept in the dark at 8 °C and oxygen concentration >80%	1. Dry mass and ash: Standard Methods for the Examination of Water and Wastewater (American Public Health Association, 1998) 2. Total carbon, nitrogen, and sulfur content: by the UBC Earth and Ocean Sciences Laboratory with a Carlo Erba N2A-1500 Analyzer 3. COD: pre-packaged mercury-free standard range digestion kit	1. C/N and nitrogen free extracts (NFE) decrease from 11 to 8.5, and from 48.6 to 36.5% dry wet, respectively. 2. Fish feed COD decrease from 2.9×10^5 to 1.2×10^6 mg/l per g dry weight 3. Fish solids COD remained stable (from 9.6×10^6 to 7.7×10^6 mg/l per g dry weight)	Buryniuk *et al.*, 2006

				4. Carbohydrates and lipids: assumed to be the nitrogen free extract (NFE) (sample minus ash and protein) 5. Protein content: calculated from total nitrogen content after multiplying the total nitrogen content by 6.25	4. 50% waste matter degradation after 30–40 days in cold and saline water 5. Ammonia concentration increase, bubbles and foam formation, wastewater decolourisation	
Fish farm effluents	Removal efficiency of fish farm effluents by grey mullets	Nitrogen, dry matter, ash, protein, total phosphorus, total carbohydrate, gross energy, organic carbon	Use of the bottom-feeding grey mullet (*Mugil cephalus*) within benthic enclosures as a means to reduce the benthic impacts of a net cage fish farm	1. Nitrogen: Kjeldahl technique 2. Dry matter: weight loss after 24 h at 105 °C 3. Ash: weight loss after 24 h at 550 °C in a muffle furnace 4. Protein: multiplying N by 6.25 5. Total phosphorus: ashing the samples and using the vanado-molybdate method 6. Total carbohydrate: calculated as the difference 7. Gross energy: combustion in a calorimeter using benzoic acid as the standard 8. Organic carbon: C content of protein (520 g C, 23.6 MJ), lipid 766 g C, 35.0 MJ) and carbohydrate (400 g C, 17.50 MJ)	1. Removal of 4.2 g organic carbon, 0.70 g nitrogen and 7.5 mg phosphorus/kg/mullet m^2 per day from the organically enriched sediment 2. Improvement of the quality of sediments below intensive net-cage fish farms	Lupatsch *et al.*, 2003

(*continued overleaf*)

Table 11.2 *(continued)*

Kind of waste	Treatment	Parameters	Methodology	Quality control methods	Results	References
Fish oils extraction		Fatty acids composition, lipid class distribution, free fatty acids (FFA), peroxide value (PV), anisidine value (AV), fluorescence products (FP), α-tocopherol, lipid content, moisture, protein, salt, copper, iron	Minced herring by-products pumped to an insulated scraped-surface heat exchanger indirectly heated by steam and separated in a three phase decanter into a high solids phase, a water phase and lipid phase (oil)	1. Fatty acids composition: GC with a flame ionisation detector (FID) 2. Lipid class distribution: gravimetrically by solid phase extraction (SPE) (Kaluzny *et al.*, 1985) 3. Free fatty acids (FFA): titration according to the AOCS Official Method Ca 5a-40 4. Peroxide value (PV): AOCS Official Method Cd-8b-90 5. Anisidine value (AV): AOCS Official Method Cd 18–90 6. Fluorescence products (FP): excitation (ex) maximum at 367 nm and an emission (em) maximum at 420 nm were using FIA (Undeland *et al.*, 1998) 7. α-Tocopherol: reversed phase HPLC 8. Lipid content: gravimetrically after extraction 9. Moisture: according to Karl–Fischer method 10. Protein: Kjeldahl digestion method 11. Salt: titration according to Volhard's method (Kolthoff and Sandell, 1952)	1. Oil: rich in polyunsaturated *n*-3 fatty acids (EPA and DHA). 2. Herring oil: low copper content (<0.1 mg/kg oil), iron concentration (0.8 mg/kg oil) 3. Maatjes herring fillets: 32% higher copper content than by-products, whereas by-products had 3 times higher iron content than fillets 4. α-tocopherol and FFA concentrations: remained stable at different storage conditions (light at room temperature, dark at room temperature, dark at 50 °C) 5. At dark, in the absence of oxygen, and at room temperatures: low oil oxidation	Aidos *et al.*, 2001

				13. Copper: with a Perkin-Elmer 5100 graphite furnace atomic absorption spectrometer (AAS) 14. Iron: with a flame atomic absorption spectrometer		
Herring by-products (maatjes, frozen and fresh by-products)	Fish oils extraction	Free fatty acids (FFA), peroxide value (PV), anisidine value (AV), R-tocopherol, total lipid soluble fluorescent lipid oxidation products (FP), fatty acid composition, copper, iron	Minced herring by-products pumped to an insulated scraped-surface heat exchanger indirectly heated by steam and separated in a three phase decanter into a high solids phase, a water phase and lipid phase (oil)	1. Free fatty acids (FFA): titration according to the AOCS Official Method Ca 5a-40 2. Peroxide value (PV): AOCS Official Method Cd-8b-90 3. Anisidine value (AV): AOCS Official Method Cd 18–90 4. α-Tocopherol: slightly modified method (Lie *et al.*, 1994) 5. Total lipid soluble fluorescent lipid oxidation products (FP): excitation (ex) maximum at 367 nm and an emission (em) maximum at 420 nm 6. Fatty acid composition: AOCS Official method Ce 1b-89 7. Copper: with a Perkin-Elmer 5100 graphite furnace atomic absorption spectrometer (AAS) 8. Iron: with a graphite furnace atomic absorption spectrometer	1. Crude oils: very low content of copper (<0.1 mg/kg oil), and iron values (0.03, 0.1, 0.8 mg/kg oil for maatjes, frozen and fresh by-products oil, respectively) 2. Maatjes oil: reduced amount of α-tocopherol 3. Maatjes and fresh oil: low presence of FFA that remained constant over storage 4. Frozen by-products oil: constant increase of FFA during storage 5. Secondary oxidation products were measured for fresh oil, while for the maatjes and frozen by-products oil, tertiary oxidation products were detected 6. Maatjes and frozen oils: intense odour, correlating positively at the end with sensory attributes of train-oil, acidic, marine and fishy	Aidos *et al.*, 2003

(*continued overleaf*)

Table 11.2 *(continued)*

Kind of waste	Treatment	Parameters	Methodology	Quality control methods	Results	References
Filleting waste from silver carp	Treatment with dairy proteins and microbial transglutaminase	Mechanical properties (textural profile analyses, punch test and torsional test analyses and expressible water)	Microbial transglutaminase (3 g/kg), whey protein and sodium caseinate (10 g/kg), and NaCl (0, 10, 20 g/kg) were mixed with the paste. The mixture was sprayed with commercial regular vegetable oil, and immersed in water at 40 °C for 1 h, followed by immersion in water at 90 °C for 20 min. Then placed in a water bath and cooled at 4–5 °C for 30 min.	Mechanical properties: Texturometer	1. Increase of mechanical properties when the salt level was also increased 2. Increase of mechanical properties of non-salted and low-salt products by using dairy proteins 3. Microbial transglutaminase increased expressible water	Uresti *et al.*, 2004
Fish canning waste water	Coagulation/ flocculation treatment	BOD, COD, total suspended solids, total dissolved solids, oil, grease, potassium, ammonia, phosphate, sulfate, sodium, total hardness, sulfide, nitrate, chloride, total protein	Five chemical coagulation/flocculation treatments were tested using $FeCl_3$ (0.4 g Fe L^{-1}), $Al_2(SO_4)_3$ (1.2 g Al L^{-1}), $Ca(OH)_2$ (0.5 g Ca L^{-1}), $FeCl_3$ and $Ca(OH)_2$ (0.4 g Fe L^{-1} + 0.2 g Ca L^{-1}), $Al_2(SO_4)_3$ and $Ca(OH)_2$ (1.2 g Al L^{-1} + 0.3 g Ca L^{-1}).	1. BOD: membrane electrode method 2. COD: open reflux method 3. Total suspended solids: dried residue gravimetric method 4. Total dissolved solids: dried residue gravimetric method 5. Oil and grease: partition gravimetric method 6. Potassium: flame photometry	1. 80–88% BOD_5, 75–85% COD, 94–95% total suspended solids (TDS), and 31–53% total dissolved solids (TDS) decrease 2. In treatments 3, 4, and 5 total hardness increased by 61%, 319% and 419%, respectively	Fahim *et al.*, 2000

7. Ammonia: phenate colorimetric method 8. Phosphate: stannous chloride colorimetric method 9. Sulfate: barium chloride drying of residue gravimetric method. 10. Sodium, total hardness, sulfide, nitrate and chloride: ion selective electrode method 11. Total protein: Biuret reaction colorimetric method	3. In treatments 1, 2, 3, and 4 ammonia content decreased up to 90%, 56%, 100%, and 100%, respectively 4. In treatments 2 and 3 phosphate decreased by 90% 5. 73–89% oil and grease, and 70–92% total protein decrease 6. Best method involved the use of $FeCl_3$ and $Ca(OH)_2$ (0.4 g Fe/l and 0.2 g Ca/l, respectively) (Treatment 4) 7. Dried precipitate averaged 50 g L^{-1} contained 40% by weight recovered protein and 20% recovered fat. 8. Solid was ideal for on-site reprocessing as animal feed. 9. Final effluent safe for controlled use in some irrigation applications or forestry projects at the desert area

(*continued overleaf*)

Table 11.2 *(continued)*

Kind of waste	Treatment	Parameters	Methodology	Quality control methods	Results	References
Rainbow trout raceway effluent	Baffled sedimentation with and without the installation of artificial substrates	Temperature, pH, total suspended solids, (TSS), total organic carbon, (TOC), total ammonia nitrogen, (TAN), nitrite (NO_2-N), nitrate (NO_3-N), orthophosphate, OP (PO_4-P), total phosphorus TP (PO_4-P), dissolved oxygen (DO)	The trout effluent was entered the basin consisted of two consecutive sections (first and second), and within the first section, four linear segments of biofiltration media were installed. Treatment efficiency was determined under normal and cleaning/harvesting operations.	1. Temperature: dissolved oxygen/temperature meter 2. pH: w/electrosylicate pH probe 3. Total suspended solids (TSS): Standard Methods (APHA, 1998) 4. Total organic carbon, (TOC): Persulfate-ultraviolet oxidation method, TOC analyser 5. Total ammonia nitrogen (TAN): Salycilicate method 6. Nitrite (NO_2-N): Dianosation method 7. Nitrate (NO_3-N): Cadmium reduction method 8. Orthophosphate OP (PO_4-P): Ascorbic acid method 9. Total phosphorus TP (PO_4-P): Acid persulfate digestion, ascorbic acid detection method 10. Dissolved oxygen (DO): dissolved oxygen/temperature meter	1. 71–79% TSS removal during normal and 79–92% cleaning and harvesting conditions. 2. Majority of TSS removal in the first section of the basin (84% and 94% TSS removal for normal and cleaning/harvesting conditions, respectively). 3. Dissolved nutrients (TAN, nitrate, and nitrite) removal, with the exception of OP. 4. Shorter retention times for the first section and the overall basin (32–37% and 17–27%, respectively) than the theoretical values 5. Hydraulic efficiency increased with increasing L : W (leangth : width) ratios.	Stewart *et al.*, 2006

Table 11.3 Advantages and disadvantages of fish waste treatment methodologies (adapted from Arvanitoyannis and Kassaveti, 2008)

Treatment	1. Advantages	2. Disadvantages	References
Hydrolysis	1. Improves digestion process by reducing hydraulic digestion time (HRT) or increasing digestion 2. Increases gas production by increasing digestion 3. Low cost solution to digester upgrade	1. High operation cost 2. Not necessarily a benefit to well operated plants 3. Requires pilot work to determine benefits	Harlan, 2006
Bioremediation	1. Natural process – acceptable by the public 2. Hazardous compounds degradation 3. Less expensive than other techniques 4. In situ/ex situ	1. Not rapid 2. Contaminants degradation under favourable conditions 3. Application only for biodegradable compounds 4. Constant monitoring for effectiveness assurance	Vidali, 2001; Sasikumar and Papinazath, 2003
Anaerobic treatment	1. Low energy consumption 2. Biogas production 3. Less waste sludge generation 4. Low investment costs	1. Odour problems 2. High maintenance and power costs 3. It is a pre-treatment method; thus requiring a post-treatment method for waste disinfection	LENNTECH, 2012
Filtration	1. Small particles removal (1–60 microns) 2. Low cost method at low volume flows 3. No chemical and power requirements 4. Simple installation, operation, maintenance	1. Expensive at high volume flows 2. Clogging of filters 3. A pre-treatment step may be required 4. Applied as a single method for wastewaters with low levels of contamination	Miller and Semmens, 2002; United Nations Environment Programme, 2005

(continued overleaf)

Table 11.3 *(continued)*

Treatment	1. Advantages	2. Disadvantages	References
Screening	1. Simple operation 2. Low cost 3. Easy installation 4. Large solids (0.7 mm or larger) removal	1. Clogging of screen	http://www.fao.org/DOCREP/003/V9922E/V9922E05.htm; United Nations Environment Programme, 2005
Evaporation	1. Established and proven technology 2. Simple to operate 3. Widely applicable	1. High energy requirements 2. Multistage counter current rinsing essential 3. Returns impurities to bath 4. Additional treatment might be needed to control impurities	EPA, 1992
Coagulation Flocculation	1. Removal of very fine particles. 2. Increase in agronomic value of the sludge. 3. Low maintenance 4. Reagent consumption optimised 5. Possibility of running in batch mode	1. Expensive process 2. Process may generate toxic sludge 3. Additional management of larger amounts of removed sludge 4. Reagent addition must be carefully controlled to preclude unacceptable concentrations in treatment effluent 5. As discharge standards become more stringent, further treatment may be required	Medrzycka and Tomczak-Wandzel, 2005

11.3 Uses of fish waste

Food industry wastes are an important source of environmental contamination. Research has been carried out to develop methods for converting these wastes into useful products. More than 50% of the material remaining from the total fish capture is not used as food and involves almost 32 million tonnes of waste (Perea *et al*., 1993; Kristinsson and Rasco, 2000; Larsen *et al*., 2000; Guerard *et al*., 2001; Coello *et al*., 2002; Laufenberg *et al*., 2003).

11.3.1 Animal feed

Nowadays, using food waste as animal feed is of interest because it benefits the environment and the public and reduces the cost of animal production (Myer *et al*., 1999; Samuels *et al*., 1991; Westendorf *et al*., 1998; Westendorf, 2000).

Kotzamanis *et al.* (2001) studied the potential uses of trout offal as an ingredient in the diets of gilthead bream *Sparus aurata* (L.). Trout offal (heads, skeletons, tails and intestines) was minced, homogenized and mixed thoroughly with other dietary ingredients to make pellets for experimental diets (diet A (control): 410 g/kg fish meal and 58 g/kg fish oil; diet B: fish meal 338 g/kg and trout heads, skeletons and tails; diet C: similar to diet A, but fish oil was substituted with trout intestines). The microbiological load of the trout offal was low (10^4 cfu/g), while the fatty acid (FA) composition indicates that it is a good lipid source, because of total *n*-3 highly unsaturated fatty acids (HUFA) and arachidonic acid (AA) 20:4*n*-6 levels. However, the high level of 18:2*n*-6 fatty acid of trout intestine, which is not a natural constituent of sea bream lipids, is the main drawback. Differences in liver glycogen content were more pronounced than differences in dietary carbohydrate. Increased haematocrit values with diets (such as B) prompting faster growth have been reported (Barnhart, 1969). Trout offal is an alternative, non-polluting way of using the by-products of fish industry in sea bream diets.

Fish waste has proved to be a valuable source of minerals, protein (58% DM) and fat (19% DM). Fatty acids (monounsaturated acids, palmitic and oleic acids) are plentiful in fish waste; while the high ash content (22% DM) indicates a high percentage of minerals in fish meal. Toxic substances (such as As, Pb, Hg and Cd) were detected in fish waste at low concentrations. Waste digestibility increased with temperature, and consequently temperatures over 105 °C should not be used in the treatment to reduce moisture and to ensure the microbiology quality. Fish waste can be used as alternative feedstuffs in swine diets, partially to meet the protein requirements and to serve as a substitute for common sources of protein (i.e. soybean meal and commercial fish meal) (Esteban *et al*., 2007).

Coward-Kelly *et al.* (2006) treated shrimp head waste (*Penaeus indicus*) with lime at different temperatures (75, 100, 125 °C) and lime : shrimp ratios (0, 0.05, 0.1, 0.2 g $Ca(OH)_2$: g dry shrimp) to determine the repeatability, the temperature effect and the effect of lime loading on solubilizing protein in shrimp head waste. Shrimp heads were hydrolysed in less than 15 min and do not require strong treatment conditions (low temperature, low lime loading and short times). Shrimp head waste contained 20% ash, 10.3% total Kjeldahl nitrogen (TKN) corresponding to 64% crude protein and chitin, 18% lipids and other compounds; while little amino acid degradation occurred. The protein-rich material can be used as a monogastric animal feed supplement, and the residual solid-rich in calcium carbonate and chitin material, can be used to produce chitin and chitosan. The latter has many applications individually or as blends either as a diet product or as edible films for food preservation purposes (Arvanitoyannis *et al*., 1997, 1998).

11.3.2 Biodiesel/biogas

Biogas is derivative of fish oil or the waste itself and is usually produced by anaerobic digestion. A system used for effluents from rainbow trout processing plants consisted of a digester connected to a sedimentation column with an aerobic filter, and final polishing using a Zeolite column. The digester reduced the total, soluble, suspended and volatile solids and the zeolite column reduced total nitrogen and wastewater contaminants (Jayasinghe and Hawboldt, 2012).

Kato *et al.* (2004) assessed the fish waste oil treated with ozone for use as a diesel fuel for transport. Fish oil, a fish-powder by-product, was pretreated by filtration, placed in a reactor with two catalysts (iron oxide and calcium phosphate monobasic) and mixed using ozone bubbling (5 g/h, 16 g/m^3 (about 8000 ppm) for 1 h at room temperature (primary ozone treatment). Then the sample was filtered and treated with ozone at the same conditions for 30 min, but without the presence of catalysts (secondary ozone treatment). The oil produced from fish waste was tested for its density, flash point, pour point, heating value, distillation test and sulfur content. The yield of the fuel produced was 95–96% after filtration and primary and secondary treatments. The oil obtained had suitable properties for use in diesel engines, such as an almost identical higher heating value (HHV) (10 700 kcal/kg) and density (at 15 °C, 0.87 g/cm^3), lower flash and pour points (37 °C and −16 °C, respectively) compared to commercial diesel fuel, no production of sulfur oxides, and lowered or no soot, polyaromatic or carbon dioxide emissions. These properties suggest that the oil obtained has better properties than methyl esterificated vegetable oil waste and is suitable for diesel engines, especially in low temperature areas.

The anaerobic treatment of sludge from a salmon smolt hatchery in a continuous stirred tank reactor (CSTR) at mesophilic temperature (35 °C) and 55–60 days hydraulic retention time (HRT) was investigated by Gebauer and Eikebrokk (2006). The main components of sludge treated with 1.5–3.3% total (dry) solids (TS) were: 32% nitrogen, 8.5% phosphorous, low potassium content, acceptable concentration of heavy metal apart from zinc and high levels of volatile fatty acids (VFA) that may cause phytotoxicity. The treated sludge is in liquid form and can be used as a liquid fertilizer on cultivated land and meadows. Moreover, the methane content in biogas stabilized at 59.4–60.5% vol, methane yield was 0.14–0.15 l/g COD, nitrogen mineralization increased to 70%, with 44.8–53.5% COD removed. The potential of the sludge for energy production was also examined. The net energy production from the biogas was 43–47 MW h/year, and could cover 2–4% of the energy demand in flow-through hatcheries, and at least twice as much in recirculation hatcheries.

11.3.3 Natural pigments

Shrimp waste, such as head and body carapace, were used for carotenoid extraction with various organic solvents (methanol, ethyl methyl ketone, isopropyl alcohol (IPA), ethyl acetate, ethanol, petroleum ether, hexane) and solvent mixtures (acetone and hexane, IPA and hexane) using various extraction conditions (percentage of hexane in the solvent mixture of IPA and hexane, ratio of solvent to waste and number of extractions) (Sachindra *et al.*, 2006). The results indicated that the highest yield of carotenoids (43.9 μg/g waste) was observed when the carotenoids were extracted with a mixture of IPA and hexane, followed by IPA (40.8 μg/g) and acetone (40.6 μg/g); while the lowest carotenoid yield was attained with petroleum ether (12.1 μg/g) and hexane (13.1 μg/g). The optimized conditions for extraction of carotenoids from shrimp waste were 60% hexane in the solvent mixture of IPA and hexane, a solvent-to-waste ratio of 5 in each extraction and three extractions. The recovered carotenoids can be efficiently used instead of synthetic carotenoids in aquaculture feed formulations, and the residue available after extraction can be used for the preparation of chitin/chitosan (Sachindra *et al.*, 2006).

11.3.4 Food industry/cosmetics

Whole shrimp heads from northern pink shrimp (*Pandalus eous*), endeavour shrimp (*Metapenaeus endeavouri*) and black tiger shrimp (*Penaeus monodon*), were used for isolation of shrimp head protein hydrolysates (SHPH). The preparation of hydrolysate was carried out according to Iwamoto *et al.* (1991) with slight modifications, such as: thaw of shrimp heads overnight at 4 °C, pH adjustment to 8.0 by a dilution of 1N NaOH, and pH adjustment to 6.0 using 10% HCl. The SHPH obtained are a good source of protein (90–91%) and amino acids (71–84%), but are low in fat (0.01–0.02%). The preparation of fish myofibrils was conducted according to Katoh *et al.* (1977) with some modifications introduced by Nozaki *et al.* (1991). The SHPH or glutamate obtained was mixed with 5% myofibrils (dry basis) at 5 °C and pH was adjusted to 7.0. Dehydration took place at 5 °C, and when the moisture content reached 10%, further dehydration occurred. The results indicated that SHPH could suppress dehydration-induced denaturation of myofibrillar protein by hydrated water stabilization, decrease Ca-ATPase inactivation, and increased monolayer sorbed water and multilayer sorbed water of myofibrillar protein. Therefore, SHPH can be used as a natural food additive to suppress the denaturation of myofibrillar protein and to maintain moisture in intermediate moisture foods (Ruttanapornvareesakul *et al.*, 2005, 2006).

Aoki *et al.* (2004) partially purified proteases from northern shrimp *Pandalus borealis*, in order to tenderize beef. Shrimp head protein was extracted, partially purified, dissolved in PBS (pH 7.4) and diluted to a final concentration of 10 μg/ml, before adding to raw beefsteaks. The mixture was incubated for 1 h at 10 °C, cooked at 70 °C, cooled and stored at 4 °C for 24 h. Shrimp proteases can be used at an industrial scale in the food industry as they have proved to be effective for beef meat tenderization, were inactive after mild heat treatment and active at low temperatures, thus resulting in energy savings through operation at room temperature.

Khan and his co-workers (2003) prepared fish protein hydrolysate (FPH) from fish scrap of five marine species (white croker *Agryrosomus argentatus*, horse mackerel *Tachurus japonicus*, flying fish *Cypselurus heterurus*, chub mackerel *Scomber japonicus* and sardine *Sardinops melanostictus*). Fish scrap (head, viscera, scale, skin, caudal fin and bone) were treated according to the method suggested by Iwamoto *et al.* (1991) with the following modifications: pH adjustment to 8.0 by 1N NaOH, pH adjustment to 6.0 using DL-malic acid, and pH adjustment to 7.0 by 1N NaOH prior to ultrafiltration. The FPH obtained had the following properties: 82.3–85.8% peptides, 0.2–0.3% lipid, 7.7–9.4% ash, 3.3–4.2% sugar, trace sodium chloride, and amino acids (Glx, Arg, Lys, Ser, Ala, and Leu). Lizard fish surimi was minced and mixed with 5% FPH, while moisture content was adjusted to 82.0%. Then, the samples were stored at −25 °C for 180 days. FPH can be used as a cryoprotectant for the suppression of muscle protein denaturation of lizard fish meat during frozen storage, because it suppressed the decrease of unfrozen water, and maintained a high-gel forming ability and Ca-ATPase activity.

Fish skin, bone and fin (from skipjack tuna *Katsuwonus pelamis*, Japanese sea-bass *Lateolabrax japonicus*, ayu *Plecoglossus altivelis*, yellow sea bream *Dentex tumifrons*, chub mackerel *Scomber japonicus*, bullhead shark *Heterodontus japonicus* and horse mackerel *Trachurus japonicus*) were examined for the potential isolation of collagen. It was found that collagen recovery ranged from 36 to 54%, with the highest value recorded for ayu *Plecoglossus altivelis* bone, and the lowest for Japanese sea-bass *Lateolabrax japonicus* fin. The denaturation temperatures were: skin collagen $25.0 \pm 26.5\,^{\circ}C$, bone collagen $29.5 \pm 30.0\,^{\circ}C$ and fin collagen $28.0 \pm 29.1\,^{\circ}C$, and were lower than porcine collagen (Nagai and Suzuki, 2000).

Squid protein hydrolysate (SPH) was extracted from four squid species (Japanese flying squid *Todarodes pacificus*, bigfin reef squid *Sepioteuthis lessonania*, swordtip squid *Loligo edulis*, and golden cuttlefish *Sepia esculenta*) with hydrolysis using endo-type protease, derived from *Bacillus subtilis*, at 60 °C for 2 h. The enzymatic activity was terminated by increasing the temperature to 90 °C for 30 min. After pH adjustment to 6.0, the sample was hydrolysed by exo-type protease, derived from *Aspergillus oryzae*, and ultra-filtrated. The SPH composition was 84–88% protein, 6–7% ash, 3% sugar, 61–64% hydrophilic amino acids, crude lipids and NaCl trace components. The development of a functional food from squid opens up the potential use of low-cost squid meat in food processing and preservation (Hossain *et al.*, 2003).

Morimura *et al.* (2002) treated fish bone from yellowtail fish for collagen isolation. The pre-treated fish waste (80% fat and 90% inorganic compounds removal from bone) was placed in a reactor with enzyme L and agitated at 200 rpm for 60 min at 60 °C and pH 8.0. The mixture was then centrifuged at 8000 *g* for 10 min and the precipitate weight was determined after drying at 105 °C for 24 h. The results indicated protein recovery up to 53%, high collagen degradation efficiency by enzyme L (85.5%) in comparison to enzyme K (69.2%) and other enzymes (both enzymes L and K originated from *Bacillus* species). Fat and inorganic materials from the pre-treatment step can be used for production of cooking oil and calcium apatite, respectively. The hydrolysate, a composite of oligopeptides, can be used as a food additive due to its anti-radical (antioxidant) activity and for blood pressure reduction. Finally, the solid material from enzymic hydrolysis can be used as fertilizer (Figure 11.6).

Hake (*M. hubbsi*) filleting waste was minced, mixed with water and autolysed at 60 °C. After centrifugation, the supernatant was freeze-dried to obtain solid fish protein hydrolysate (FPH). The FPH produced contained 80% protein and a low content of free amino acids and appeared to have significant nutritional value for supporting the growth of bacteria and archaea. Therefore, FPH can be used as an alternative substrate for culturing microorganisms (*Halobacterium salinarum, Escherichia coli, Bacillus subtilis* and *Staphylococcus epidermidis*) (Martone *et al.*, 2005).

11.3.5 Waste management

Fish bone, acquired from a natural Japanese sea bream, was heated at 600 °C for 24 h or 900 °C for 12 h, and then a removal test was conducted by adding the powder produced to nitrate solutions with aqueous chromium (Cr^{+3}) concentration 3×10^{-4} M. The mixture was stirred and soaked for 2 h, and then held for 6 days without stirring. Chromium concentration in the solution was measured by inductively coupled plasma atomic emission spectrometer. The fish bone heated at 600 °C, showed better removal capacity while raw bone and bone powder heated at 900 °C were of lower and similar activity, respectively. Raw fish bone had low crystallinity, the bone sample heated at

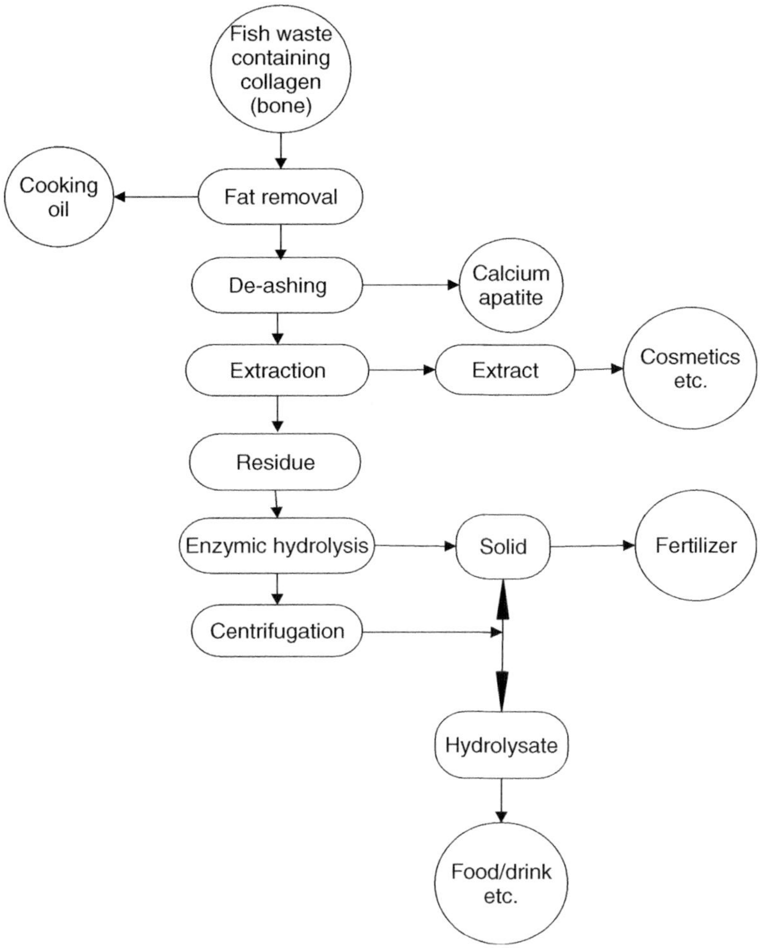

Figure 11.6 Flow diagram of total procedure for extraction of protein and production of peptides by enzymatic hydrolysis from fish bone (*Source*: Morimura *et al.* (2002). Reproduced with permission of Elsevier.)

600 °C was well-crystallized hydroxyapatite, and the sample heated at 900 °C displayed the development of high crystallinity hydroxyapatite. The results confirmed the potential utilization of heated (600° or 900 °C) fish bone for Cr-immobilization (Ozawa *et al.*, 2003).

Stepnowski and his co-workers (2004) performed sorption experiments using dry scales from redfish (*Sebastes marinus*) packed without pressure in 10–70-cm long PE columns supported with a reservoir filled with wastewater from a shrimp factory for 5 h at room temperature. The results showed that 88–95% of astaxanthin in its esterified form was bound to scales, and the maximum loading capacity (362 mg/kg dry wt) was found in 25–35 cm of suspension on top of the scales. Furthermore, slowing down the outflow of wastewater passing through the scales increased the percentage of astaxanthin retention onto scales up to 90–97%; however, the process time was prolonged. Therefore, fish scales can be used as a natural adsorbent for a carotenoid pigment (astaxanthin) from seafood industry wastewater.

11.3.6 Miscellaneous uses

Anderson *et al.* (1999) studied the possibility of benthic *Gracilaria* population grown close (1.5 km and in the waste plume) to a fish-factory waste release site and 3.5 km away (control site) during 1996. In October–November, all the *Gracilaria* at the control site died whereas growth at the fish waste site was 8–10% per day. In November–December control plants grew slightly faster than those from the waste site, in February the reverse occurred, and in March–June growth was similar at both sites. A considerable uptake of the fish-waste nitrogen by *Gracilaria* was reported even at the control site. It was thus concluded that fish waste provided an important source of nitrogen for seaweed cultivation.

Casani *et al.* (2006) examined the potential of water reuse obtained in a shrimp processing line. Iced shrimps, *Pandalus borealis*, stored for maturation, were de-iced, rinsed and steam cooked.The cooked shrimps were then conveyed to a peeling machine and the edible meat was carried to a scrubber, finally brined and packed. The water recovered from peeling was screened, pre-filtrated and treated by means of reverse osmosis (RO). The results revealed that the quality of the treated water was higher than that of the drinking water. Despite the considerable advantages of water reuse, the capital cost for the treatment equipment was too high for routine operation. Table 11.4 provides a synopsis of olive oil waste treatment methods, applied substrates, physicochemical characteristics and final product applications/uses.

11.4 Inputs and outputs in fisheries

Minimization and enforced reduction of pollutant loads from point-source aquatic animal production facilities are increasingly demanded by various segments of the public and the regulatory community (MacMillan *et al.*, 2003).

The solid wastes from fish farms falling to the seabed below fish cages are enriched in carbon, nitrogen and phosphorus relative to the natural sediments, hence fish farming may considerably alter the physicochemical nature of sediments below and adjacent to the operation but this is usually limited to the vicinity (50 m) of the cages. This increase in carbon sedimentation results in an increase in oxygen consumption by bottom-living animals. The sediments will become anoxic (i.e. contain no oxygen) if this additional oxygen demand exceeds oxygen supply, at which point there may be severe consequences for both benthic organisms and the fish farming itself (El-Fadel *et al.*, 1997).

Although smoke and particulates may be a problem, odours are the most objectionable emissions from fish processing plants. The largest odour source in the fish by-products segment is from the fish meal driers. Odorous gases from reduction cookers consist primarily of hydrogen sulfide (H_2S) and trimethylamine ($(CH_3)_3N$) but these are emitted from this stage in appreciably smaller volumes than from fish meal driers. Some odours are also released by the canning processes. Fish cannery and fish by-product processing odours can be controlled using afterburners, chlorinator-scrubbers or condensers (http://www.epa.gov/ttn/chief/ap42/ch09/final/c9s13-1.pdf). Both inputs and outputs of various fish processes are summarized in Table 11.5.

Table 11.4 Fish waste; treatment method and physicochemical characteristics, substrate to be applied and final product uses (adapted from Arvanitoyannis and Kassaveti, 2008)

Substrate to be applied	Treatment method	Physicochemical characteristics	Final product / Uses	References
Trout offal (heads, skeletons, tails and intestines)	Mincing, homogenisation, pellets preparing of experimental diets	Low microbiological load (10^4 cfu/g), good lipid source due to total *n*-3 highly unsaturated fatty acids (HUFA) and arachidonic acid 20: 4*n*-6 (AA) levels, main drawback the high level of 18: 2*n*-6 fatty acid of trout intestine	Dietary ingredient of gilthead bream *Sparus aurata (L.)*	Kotzamanis *et al.*, 2001
Fish waste (mainly heads, bones, skin, viscera and sometimes whole fish and parsley)	Heat treatments at 65, 80, 105 and 150 °C for 12 h for moisture content reduction to 10–12%	High source of minerals, 58% protein, 19% fat, detection of toxic substances (As, Pb, Hg and Cd) at non-problematic concentrations, decrease of waste digestibility with temperature	Feedstuff in swine diets as a protein substitute	Esteban *et al.*, 2007
Shrimp head waste	Lime treatment at different temperatures (75, 100, 125 °C) and lime/shrimp ratios (0, 0.05, 0.1, 0.2 g $Ca(OH)_2$/g dry shrimp)	20% ash, 10.3% TKN corresponding to 64% crude protein and chitin, 18% lipids and other compounds, little amino acid degradation	1. Protein-rich material can be used as a monogastric animal feed supplement 2. Residual solid – rich in calcium carbonate and chitincan can be used to generate chitin and chitosan	Coward-Kelly *et al.*, 2006
Raw fish oil	Filtration pretreatment with or without the presence of two catalysts (iron oxide and calcium phosphate monobasic) and ozone treatment (5 g/h, 16 g/m^3 (about 8000ppm)) at room temperature for 1h and 30 min, respectively	Almost identical HHV (10700 kcal/kg) and lower flash and pour points (37 °C and −16 °C, respectively) compared to commercial diesel fuel, no production of sulfur oxides, lowered or no soot, polyaromatic and carbon dioxide emissions	Biodiesel fuel for transportation	Kato *et al.*, 2004

(*continued overleaf*)

Table 11.4 *(continued)*

Substrate to be applied	Treatment method	Physicochemical characteristics	Final product / Uses	References
Sludge from salmon smolt hatching	Mesophilic anaerobic treatment in a continuous stirred tank reactor (CSTR) at 35 °C and 55–60 days hydraulic retention time (HRT)	1. Treated sludge characteristics: 32% nitrogen, 8.5% phosphorous, low potassium content, acceptable concentration of heavy metal apart from zinc, high levels of volatile fatty acids (VFA) that may cause phytotoxicity, fertilizing value of 1 ton treated sludge 3.2–6.4 kg N and 1.2–2.4 kg P, 44.8–53.5% COD reduction 2. Methane yields 0.14–0.15 l/g COD, net energy production from the biogas 1 million smolts 43–47 MW h/year	1. Treated sludge may cause phytotoxicity – requirements for special means of application 2. Biogas can cover 2–4% energy demands of a flow-through hatchery.	Gebauer and Eikebrokk, 2006
Shrimp waste (head and body carapace)	Extraction of shrimp waste carotenoids using different organic solvents and solvent mixtures at various extraction conditions	Highest carotenoids yield (43.9 μg/g waste) obtained when the carotenoids were extracted with a mixture of IPA and hexane, followed by IPA (40.8 μg/g) and acetone (40.6 μg/g) Lowest carotenoid yield obtained with petroleum ether (12.1 μg/g) and hexane (13.1 μg/g)	1. Carotenoids recovered from shrimp waste may be used instead of synthetic carotenoids in aquaculture feed formulations 2. The residue available after carotenoids extraction may be used for the preparation of chitin/chitosan	Sachindra *et al.*, 2006

Fish myofibrils	Shrimp head protein hydrolysates (SHPH) were obtained according to Iwamoto *et al.* (1991) with some modifications. Fish myofibrils preparation was according to Katoh *et al.* (1977) with some modifications by Nozaki *et al.* (1991).	SHPH characteristics: suppress dehydration - induced denaturation of myofibrillar protein by hydrated water stabilisation, decrease Ca-ATPase inactivation, and increase monolayer sorbed water, multilayer sorbed water of myofibrillar.	Food additive to suppress the denaturation of myofibrillar protein and maintain moisture in intermediate moisture foods.	Ruttanapornvareesakul *et al.*, 2005, 2006
Shrimp heads	Protease extracted from shrimp waste, purified, dissolved in PBS (pH 7.4), added to raw beefsteaks, incubated for 1h at 10 °C, cooked at 70 °C, cooled and stored at 4 °C for 24h, prior shear force analysis.	Overdegradation of meat proteins, efficient at low temperatures - energy minimanisation, enzyme inactivation after mild heat treatment	Meet industry - beef tenderisation	Aoki *et al.*, 2004
Lizard fish surimi	Fish scrap (head, viscera, scale, skin, caudal fin, and bone) were treated according to Iwamoto *et al.* (1991) with some modifications. Minced lizard fish meat was mixed with 5.0% (dry weight/wet weight) FPH, the moisture was adjusted at 82.0%, and stored at −25 °C for 180 days.	FPH components: peptide (11.5–16.3% 0 and other nitrogenous compounds (82.3–85.8%), high gel - forming ability and Ca-ATPase activity, suppression of decrease of unfrozen water content.	Cryoprotectant for the suppression of denaturation of muscle protein of lizard fish meat during frozen storage.	Khan *et al.*, 2003
Fish skin, bone and fin	Collagen isolation	36–54 % collagen recovery and denaturation temperatures of skin collagen (25.0 ± 26.5 °C), bone collagen (29.5 ± 30.0 °C) and fin collagen (28.0 ± 29.1 °C)	Use as alternatives to mammalian collagen in foods, cosmetics and biomedical materials	Nagai and Suzuki, 2000

(continued overleaf)

Table 11.4 (*continued*)

Substrate to be applied	Treatment method	Physicochemical characteristics	Final product / Uses	References
Fish bone wastes	Pre-treated (fat removal, de-ashing), treated enzymatically, agitated at 200 rpm for 60 min at 60 °C and pH 8.0, centrifuged at 8000 × g for 10 min and the precipitate weight was measured after drying at 105 °C for 24 h	50% collagen recovery, high water retention capacity, high anti-radical activity, amino acid composition	1. Extract: used as a cosmetic material, ability to repair rough skin, lack of any odour problem and absence of harmful effects on skin 2. Hydrolysate: used as a food additive, high 12 potential for lowering high blood pressure 3. Fat and inorganic materials from pre-treatment: production of cooking oil and calcium apatite, respectively 4. Solid from extraction process: used as fertiliser	Morimura *et al.*, 2002
Fish bone waste	Heat treatment of raw bone at 600 °C for 24 h or 900 °C for 12 h	Better removal capacity and well-crystallized hydroxyapatite at 600 °C, raw bone showed lower activity and crystallinity, bone sample heated at 900 °C showed developed similar activity with raw bone and developed crystallinity of hydroxyapatite	Cr-immobilisation (when treated at 600 °C)	Ozawa *et al.*, 2003

Fish scales and wastewater	Sorption experiments using dry scales packed without pressure in 10–70 cm long PE columns supported with a reservoir filled with wastewater (5 h, at room temperature)	88–95% bond of asthaxanthin to scales, maximum loading capacity (362 mg kg^{-1} dry wt) at 25–35cm of suspension on top of scales, 90–97% astaxanthin retention onto scales by slowing down the out-flow of wastewater	Natural adsorbent for a carotenoid pigment (astaxanthin) from the seafood industry wastewater	Stepnowski *et al.*, 2004
Fish-factory waste	Release of large quantities of nitrogen-rich fish waste	Significant source of N for cultivated seaweed (*Gracilaria gracilis)*	Nitrogen removal introduced by the fish-factories	Anderson *et al.*, 1999
Shrimp	Wastewater (from shrimp process) recycling with RO membranes	Higher quality than drinking water	Drinking water	Casani *et al.*, 2006

Table 11.5 Inputs and outputs of various fish processes (adapted from Arvanitoyannis and Kassaveti, 2008)

Process	INPUTS		OUTPUTS	
	Fresh or frozen fish (kg)	Energy (kW h)	Wastewater	Solid waste (kg)
White fish filleting	1000	Ice: 10–12 Freezing: 50–70 Filleting: 5	5–11 m^3: BOD 35 kg COD_5 50 kg	Skin: 40–50 Heads: 210–250 Bones: 240–340
Oily fish filleting	1000	Ice: 10–12 Freezing: 50–70 Filleting: 2–5	5–8 m^3: BOD 50 kg COD_5 85 kg Nitrogen 2.5 kg N Phosphate 0.1–0.3 kg P	400–450
Canning	1000	150–190	15 m^3: BOD 52 kg COD_5 116 kg Nitrogen 3 kg N Phosphate 0.1–0.4 kg P	Heads/entrails: 250 Bones: 100–150
Fish meal and fish oil production	1000	Fuel: 49 l Electricity: 32	–	–
Frozen fish thawing	1000	–	5 m^3: COD_5 1–7 kg	–
De - icing and washing	1000	0.8–1.2	1 m^3: COD_5 0.7–4.9 kg	0–20
Granding	1000	0.1–0.3	0.3–0.4 m^3: COD_5 0.4–1.7 kg	0–20
Scaling of white fish	1000	0.1–0.3	10–15 m^3	Scales: 20–40
De - heading of white fish	1000	0.3–0.8	1 m^3: COD_5 2–4 kg	Heads and debris: 270–320
Filleting of de - headed white fish	1000	1.8	1–3 m^3: COD_5 4–12 kg	Frames and offcuts: 200–300
Filleting of un - gutted oily fish	1000	0.7–2.2	1–2 m^3: COD_5 7–15 kg	Entrails, tails, heads and frames: 400
Skinning white fish	1000	0.4–0.9	0.2–0.6 m^3: COD_5 1.7–5.0 kg	Skin: 40
Skinning oily fish	1000	0.2–0.4	0.2–0.9 m^3: COD_5 3.0–5.0 kg	Skin: 40

Table 11.5 (*continued*)

	INPUTS		OUTPUTS	
Process	Fresh or frozen fish (kg)	Energy (kW h)	Wastewater	Solid waste (kg)
Trimming and cutting white fish	1000	0.3–3.0	0.1 m^3	Bones and cut - off: 240–340
Packaging of fillets	1000	5.0–7.5	–	–
Freezing and storage	1000	10.0–14.0	–	–
Unloading fish for canning	1000	3.0	2.0–5.0 m^3: COD_5 27.0–34.0 kg	–
Grading of fish	1000	0.15	0.2 m^3: COD_5 0.35–1.7 kg	0.30
Nobbing and packing in cans	1000	0.4–1.5	0.2–0.9 m^3: COD_5 7.0–15.0 kg	Heads and entrails: 150 Bones and meat: 100–150
Skinning of nobbed fish	1000	–	17.0 m^3: COD_5 3.0–5.0 kg	Skin: 55
Precooking of fish to be canned	1000	0.3–1.1	0.07–0.27 m^3	Inedible parts: 150
Draining of cans containing precooked fish	1000	0.3	0.1–0.2 m^3: COD_5 3.0–10.0 kg	–
Sauce filling	1000	–	–	Spillage of sauce and oil: varies
Can sealing	1000	5.0–6.0	–	–
Washing of cans	1000	7.0	0.04 m^3	–
Sterilisation of cans	1000	230	3.0–7.0 m^3	–
Handling and storage of fish	1000	10.0–12.0	COD_5 130.0–140.0 kg	–

(*continued overleaf*)

Table 11.5 (*continued*)

Process	INPUTS		OUTPUTS	
	Fresh or frozen fish (kg)	Energy (kW h)	Wastewater	Solid waste (kg)
Unloading of fish	1000	3.0	2.0–5.0 m^3: COD_5 27.0–34.0 kg	–
Cooking of fish	1000	90.0	–	–
Pressing the cooked fish	1000	–	750 kg water 150 kg oil	Press cake: 100 kg dry matter
Drying of press cake	1000	340.0	–	–
Fish oil polishing	1000	Hot water	0.05–0.1 m^3: COD_5 5 kg	–
Stickwater evaporation	1000	475.0	–	Concentrated stickwater: 250 Dry matter: 50

References

Aidos, I., Van der Padt, A., Boom, R.M. and Luten, J.B. (2001). Upgrading of maatjes herring byproducts: production of crude fish oil. *Food Chemistry*, 49, 3697–3704.

Aidos, I., Vis-Smit, R.S., Veldman, M. *et al.* (2003). Chemical and sensory evaluation of crude oil extracted from herring by-products from different processing operations. *Food Chemistry*, 51, 1897–1903.

Ali, N., Mohammad, A.W., Jusoh, A. *et al.* (2005). Treatment of aquaculture wastewater using ultra-low pressure asymmetric polyethersulfone (PES) membrane. *Desalination*, 185, 317–326.

Anderson, R.J., Smit, A.J. and Levitt, G.J. (1999). Upwelling and fish-factory waste as nitrogen sources for suspended cultivation of *Gracilaria gracilis* in Saldanha Bay, South Africa. *Hydrobiologia*, 398/399, 455–462.

Aoki, H., Ahsan, Md.N., Matsuo, K., Hagiwara, T. and Watabe, S. (2004). Partial purification of proteases that are generated by processing of the Northern shrimp *Pandalus borealis* and which can tenderize beef. *International Journal of Food Science and Technology*, 39, 471–480.

Arvanitoyannis, I., Kolokuris, I., Nakayama, A., Yamamoto, N. and Aiba, S. (1997). Physico-chemical studies of chitosan-poly (vinyl alcohol) blends plasticized with sorbitol and sucrose. *Carbohydrate Polymers*, 34(1/2), 9–19.

Arvanitoyannis, I., Nakayama, A. and Aiba, S. (1998). Chitosan and gelatin based edible films: state diagrams, mechanical and permeation properties. *Carbohydrate Polymers*, 37, 371–382.

Arvanitoyannis, I.S. and Kassaveti, A. (2008). Treated waste fish: environmental treatment, current and potential uses. *International Journal of Food Science and Technology*, 43, 726–745.

Barnhart, R.A. (1969). Effects of certain variables on hematological characteristics of rainbow trout. *Transactions of the American Fisheries Society*, 98, 411–418.

Buryniuk, M., Petrell, R.J., Baldwin, S. and Lo, K.V. (2006). Accumulation and natural disintegration of solid wastes caught on a screen suspended below a fish farm cage. *Aquacultural Engineering*, 35(1), 18–90.

Casani, S., Leth, T. and Knochel, S. (2006). Water reuse in a shrimp processing line: safety considerations using a HACCP approach. *Food Control*, 17, 540–550.

Cattaneo-Vietti, R., Benatti, U., Cerrano, C. *et al.* (2003). A marine biological underwater depuration system (MUDS) to process waste waters. *Biomolecular Engineering*, 20, 291–298.

Coello, N., Montiel, E., Concepcion, M. and Christen, P. (2002). Optimisation of a culture medium containing fish silage for L-lysine production by *Corynebacterium glutamicum*. *Bioresource Technology*, 85, 207–211.

Coward-Kelly, G., Agbogbo, F.K. and Holtzapple, M.T. (2006). Lime treatment of shrimp head waste for the generation of highly digestible animal feed. *Bioresource Technology*, 97, 1515–1520.

Diniz, F.M. and Martin A.M. (1996). Use of response surface methodology to describe the combined effects of pH, temperature and E/S ratio on the hydrolysis of dogfish (*Squalus acanthias*) muscle. *International Journal of Food Science and Technology*, 31, 419–426.

Eiroa, M., Costa, J.C., Alves, M.M., Kennes, C. and Veiga, M.C. (2012). Evaluation of the biomethane potential of solid fish waste. *Waste Management*, 32, 1347–1352.

El-Beltagy, A.E., El-Adawy, T.A., Rahma, E.H. and El-Bedawey, A.A. (2005). Purification and characterisation of an alkaline protease from the viscera of bolti fish (*Tilapia nilotica*). *Journal of Food Biochemistry*, 29, 445–458.

El-Fadel, M., Findikakis A.N. and Leckie, J.O. (1997). Environmental Impacts of Solid Waste Landfilling. *Journal of Environmental Management*, 50, 1–25

Eleftheriou, A. and Eleftheriou, M. (2001). Legal and regulatory background to best environmental practice, in *The Implications of Directives, Conventions and Codes of Practice on the Monitoring and Regulation of MarineAquaculture in Europe* (eds P.A. Read, T.F. Fernandes, K.L. Miller, *et al.*), Proceedings of the Second MARAQUA Workshop, 20–22 March 2000, Institute of Marine Biology, Crete, Scottish Executive, Aberdeen, UK, pp. 75–83.

EPA (1992). *Guides to Pollution Prevention: The Metal Finishing Industry*, Office of Research and Development, Washington, DC.

Esteban, M.B., Garcia, A.J., Ramos, P. and Marquez, M.C. (2007). Evaluation of fruit-vegetable and fish wastes as alternative feedstuffs in pig diets. *Waste Management*, 27(2), 193–200.

Fahim, F.A., Fleita, D.H., Ibrahim, A.M. and El-Dars, F.M.S (2001). Evaluation of some methods for fish canning wastewater treatment. *Water, Air, and Soil Pollution*, 127, 205–226.

FAO (2002). *FAO Fisheries Statistical Yearbook 2002*, Food and Agricultural Organization of the United Nations, Rome.

Gao, M.-T., Hirata, M., Toorisaka, E. and Hano, T. (2006). Acid-hydrolysis of fish wastes for lactic acid fermentation. *Bioresource Technology*, 97(18), 2414–2420.

Gebauer, R. and Eikebrokk, B. (2006). Mesophilic anaerobic treatment of sludge from salmon smolt hatching. *Bioresource Technology*, 97(18), 2389–2401.

Gowen, R.J. (1991). Aquaculture and the environment, in *Aquaculture and the Environment* (eds N. DePauw and J. Joyce, European Aquaculture Society Special Publications No. 16, Ghent, Belgium, pp.30–38.

Guerard, F., Dufosse, L., De La Broise, D. and Binet, A. (2001). Enzymatic hydrolysis of proteins from yellow fin tuna (*Thunnus albacares*) wastes using Alcalase. *Journal of Molecular Catalysis B: Enzymatic*, 11, 1051–1059.

Guerdat, T.C., Losordo, T.M., Classen, J.J., Osborne, J.A. and DeLong, D. (2011). Evaluating the effects of organic carbon on biological filtration performance in a large scale recirculating aquaculture system. *Aquaculture Engineering*, 44, 10–18.

Harlan, K.G. (2006). Emerging processes in biosolids treatment, 2005. *Journal of Environmental Engineering and Science*, 5(3).

Hospido, A., Vazquez, M.E., Cuevas, A., Feijoo, G. and Moreira, M.T. (2006). Environmental assessment of canned tuna manufacture with a life-cycle perspective. *Resources, Conservation and Recycling*, 47, 56–72.

Hossain, Md. A., Ishihara, T., Hara, K., Osatomi, K., Khan, MD.A.A. and Nozaki, Y. (2003). Effect of proteolytic squid protein hydrolysate on the state of water and dehydration-induced denaturation of lizard fish myofibrillar protein. *Journal of Agricultural and Food Chemistry*, 51, 4769–4774.

Iwamoto, M., Fujiwara, R. and Yokoyama, M. (1991). Immunological effects of BM-2, an enzymatic digestive extract of chub markerel. *Journal of Japan Society for Cancer Therapy*, 26, 936–947.

Jayasinghe, P. and Hawboldt, K. (2012). A review of bio-oils from waste biomass: focus on fish processing waste. *Renewable and Sustainable Energy Reviews*, 16(1), 798–821.

Kato, S., Kunisawa, N., Kojima, T., and Murakami, S. (2004). Evaluation of ozone treated fish waste oil as a fuel for transportation. *Journal of Chemical Engineering of Japan*, 37(7), 863–870.

Katoh, N., Uchiyama, H., Tsukamoto, S. and Arai, K. (1977). A biochemical study on fish myofibrillar ATPase. *Nippon Suisan Gakkaishi*, 43, 857–867.

Khan, M.A.A., Hossain, M.A.A., Hara, K. *et al.* (2003). Effect of enzymatic fish-scrap protein hydrolysate on gel-forming ability and denaturation of lizard fish *Saurida wanieso* surimi during frozen storage. *Fisheries Science*, 69, 1271–1280.

Kotzamanis, Y.P., Alexis, M.N., Andriopoulou, A., Castritsi-Cathariou, I. and Fotis, G. (2001). Utilisation of waste material resulting from trout processing in gilthead bream (*Sparus aurata* L.) diets. *Aquaculture Research*, 32(Suppl. 1), 288–295.

Kristinsson, H.G. and Rasco, B.A. (2000). Fish protein hydrolysates: production, biochemical and functional properties. *CRC Critical Reviews in Food Science and Nutrition*, 40(1), 43–81.

Laos, F., Mazzarino, M.J., Walker, I. *et al.* (2002). Composting of fish offal and biosolids in northwestern Patagonia. *Bioresource Technology*, 81, 179–186.

Larsen, T., Thilsted, S.H., Kongsback K. and Hanse, M. (2000). Whole small fish as a rich calcium source. *British Journal of Nutrition*, 83, 191–196.

Laufenberg, G., Kunz, B. and Nystroem, M. (2003). Transformation of vegetable waste into value added products. *Bioresource Technology*, 87, 167–198.

LENNTECH (2012). Water treatment solutions. http://www.lenntech.com/Pre-fermer.htm [accessed 3 June 2013].

Lupatsch, I., Katz, T., and Angel, D.L. (2003). Assessment of the removal efficiency of fish farm effluents by grey mullets: a nutritional approach. *Aquaculture Research*, 34, 1367–1377.

MacMillan, J.R., Huddleston, T., Woolley, M. and Fothergill, K. (2003). Best management practice development to minimize environmental impact from large flow-through trout farms. *Aquaculture*, 226, 91–99.

Marinho-Soriano, E., Azevedo, C.A.A., Trigueiro, T.G. *et al.* (2011). Bioremediation of aquaculture wastewater using macroalgae and *Artemia*. *International Biodeterioration and Biodegradation*, 65(1), 253–257.

Martin, A.M. (1999). A low-energy process for the conversion of fisheries waste biomass. *Renewable Energy*, 16, 1102–1105.

Martone, C.B., Perez Borla, O. and Sanchez, J.J. (2005). Fishery by-product as a nutrient source for bacteria and archaea growth media. *Bioresource Technology*, 96, 383–387.

Medrzycka, K. and Tomczak-Wandzel, R. (2005). The utilization of sludge generated in coagulation of fish processing wastewater for liquid feed production. *Environmental Science Research*, 59, 215–221.

Midlen, A., and Redding, T. (1998). *Environmental Management for Aquaculture*. Chapman and Hall, London, 223 pp.

Miller, D. and Semmens, K. (2002). Waste management in aquaculture. *Aquaculture Information Series*, AQ02-1.

Morimura, S., Nagata, H., Uemura, Y. *et al.* (2002). Development of an effective process for utilisation of collagen from livestock and fish waste. *Process Biochemistry*, 37, 1403–1412.

Mshandete, A., Kivaisi, A., Rubindamayugi, M. and Mattiasson, B. (2004). Anaerobic batch co-digestion of sisal pulp and fish wastes. *Bioresource Technology*, 95, 19–24.

Myer, R.O., Brendemuhl, J.H. and Johnson, D.D. (1999). Evaluation of dehydrated restaurant food waste products as feedstuffs for finishing pigs. *Journal of Animal Science*, 77, 685–692.

Nagai, T. and Suzuki, N. (2000). Isolation of collagen from fish waste material – skin, bone and fin. *Food Chemistry*, 68, 277–281.

Nges, I.A., Mbatia, B. and Björnsson, L. (2012). Improved utilization of fish waste by anaerobic digestion following omega-3 fatty acids extraction. *Journal of Environmental Management*, 110, 159–165.

Nozaki, Y., Ichikawa, H. and Tabata, Y. (1991). Effect of amino acid on the state of water and ATPase activity accompanying dehydration of fish myofibrils. *Nippon Suisan Gakkaishi*, 57, 1531–1537.

OAERRE (2001). Oceanographic Applications to Eutrophication in Regions of Restricted Exchange [On-line]. http://www.wise-rtd.info/en/info/oceanographic-applications-eutrophication-regions-restricted-exchange-0 [accessed 3 June 2013].

Ozawa, M., Satake, K. and Suzuki, R. (2003). Removal of aqueous chromium by fish bone waste originated hydroxyapatite. *Journal of Materials Science Letters*, 22, 513–514.

Paniagua-Michel, J. and Garcia, O. (2003). Ex-situ bioremediation of shrimp culture effluent using constructed microbial mats. *Aquacultural Engineering*, 28, 131–139.

Payraudeau, S. and Van der Werf, H.M.G. (2005). Environmental impact assessment for a farming region: a review of methods. *Agriculture, Ecosystems and Environment*, 107, 1–19.

Perea, A., Ugalde, U., Rodriguez, I. and Serra, J.L. (1993). Preparation and characterisation of whey protein hydrolysates: application in industrial whey bioconversion processes. *Enzyme Microbial Technology*, 15, 418–423.

Pillay, T.V.R. (1991). *Aquaculture and the Environment*, Blackwell Scientific, London, pp. 26–45.

Quitain, A.T., Sato, N., Daimon, H. and Fujie, K. (2001). Production of valuable materials by hydrothermal treatment of shrimp shells. *Industrial and Engineering Chemistry Research*, 40, 5885–5888.

Read, P. and Fernandes, T. (2003). Management of environmental impacts of marine aquaculture in Europe. *Aquaculture*, 226, 139–163.

Ruttanapornvareesakul, Y., Ikeda, M., Hara, K. *et al.* (2005). Effect of shrimp head protein hydrolysates on the state of water and denaturation of fish myofibrils during dehydration. *Fisheries Science*, 71, 220–228.

Ruttanapornvareesakul, Y., Ikeda, M., Hara, K. *et al.* (2006). Concentration-dependent suppressive effect of shrimp head protein hydrolysate on dehydration-induced denaturation of lizardfish myofibrils. *Bioresource Technology*, 97, 762–769.

Sachindra, N.M., Bhaskar, N. and Mahendrakar, N.S. (2006). Recovery of carotenoids from shrimp waste in organic solvents. *Waste Management*, 26(10), 1092–1098.

Samuels, W.A., Fontenot, J.P., Allen, V.G. and Abazinge, M.D. (1991). Seafood processing wastes ensiled with straw: utilisation and intake by sheep. *Journal of Animal Science*, 69, 4983–4992.

Sasikumar, C.S. and Papinazath, T. (2003). Environmental Management:- Bioremediation of Polluted Environment, in (eds J. Martin, V. Bunch, M. Suresh and T. Vasantha Kumaran), Proceedings of the Third International Conference on Environment and Health, Chennai, India, 15–17 December 2003. Department of Geography, University of Madras, Chennai and Faculty of Environmental Studies, York University, p. 465–469. http://www.yorku.ca/bunchmj/ICEH/proceedings/Sasikumar_CS_ICEH_papers_465to469 .pdf search=%22bioremediation%20%20disadvantages%22 [accessed 15 June 2013].

Shih, I.-L., Chenb, L.-G., Yu, T.-S., Changb, W.-T. and Wang, S.-L. (2003). Microbial reclamation of fish processing wastes for the production of fish sauce. *Enzyme and Microbial Technology*, 33, 154–162.

Stepnowski, P., Olafsson, G., Helgason, H. and Jastor, B. (2004). Recovery of astaxanthin from seafood wastewater utilizing fish scales waste. *Chemosphere*, 54, 413–417.

Stewart, N.T,. Boardman, G.D. and Helfrich, L.A. (2006). Treatment of rainbow trout (*Oncorhynchus mykiss*) raceway effluent using baffled sedimentation and artificial substrates. *Aquacultural Engineering*, 35(2), 166–178.

Suzuki, T. (1981). *Fish and Krill Protein: Processing Technology*, Applied Science, London.

Thassitou, P.K. and Arvanitoyannis, I.S. (2001). Bioremediation: a novel approach to food waste management. *Trends in Food Science and Technology*, 12, 185–196.

United Nations Environment Programme (2005). Solid waste management. http://www.unep .org/ietc/InformationResources/Publications/SolidWasteManagementPublication/tabid /79356/Default.aspx [accessed 3 June 2013].

Uresti, R.M., Tellez-Luis, S.J., Ramirez, J.A. and Vazquez, M. (2001). Use of dairy proteins and microbial transglutaminase to obtain low-salt .sh products from filleting waste from silver carp (*Hypophthalmichthys molitrix*). *Process Biochemistry*, 36, 809–812.

van Biesen, G. and Parrish, C.C. (2005). Long-chain monounsaturated fatty acids as biomarkers for the dispersal of organic waste from a fish enclosure. *Marine Environmental Research*, 60, 375–388.

Vidali, M. (2001). Bioremedation. An overview. *Pure Appli. Chem.*, 73(7), 1163–1172 http://www.iupac.org/publications/pac/2001/pdf/7307x1163.pdf search=%22bioremediation %20advantages%22 [accessed 15 June 2013].

Vidotti, R.M., Viegas, E.M.M., Dalton, V. and Carneiro, D.J. (2003). Amino acid composition of processed fish silage using different raw materials. *Animal Feed Science and Technology*, 105, 199–204.

Westendorf, M.L. (2000). Food waste as animal feed: an introduction, in: *Food Waste to Animal Feed*, (ed. M.L. Westendorf), Iowa State University Press, Ames, pp. 3–16, 69–90.

Westendorf, M.L., Dong, Z.C. and Schoknecht, P.A. (1998). Recycled cafeteria food waste as a feed for swine: nutrient content digestibility, growth, and meat quality. *Journal of Animal Science*, 76, 2976–2983.

Yamamoto, M., Saleh, F., Ohtsuka, A. and Hayashi, K. (2005). New fermentation technique to process fish waste. *Animal Science Journal*, 76, 245–248.

Yoshida, H., Takahashi, Y. and Terashima, M. (2003). A simplified reaction model for production of oil, amino acids, and organic acids from fish meat by hydrolysis under sub-critical and supercritical conditions. *Journal of Chemical Engineering of Japan*, 4, 441–448.

Yoshida, H., Terashima, M. and Takahashi, Y. (1999). Production of organic acids and amino acids from fish meat by sub-critical water hydrolysis. *Department of Chemical Engineering*, 15, 1090–1094.

Electronic Sources

http://www.fao.org/DOCREP/003/V9922E/V9922E05.htm (accessed 2012, 12 December)

http://www.epa.gov/ttn/chief/ap42/ch09/final/c9s13-1.pdf (accessed 2012, 12 December)

12

Fish Processing Installations: Sustainable Operation

George M. Hall[1] and Sevim Köse[2]

[1] *University of Central Lancashire, Centre for Sustainable Development, Preston, Lancashire, UK*
[2] *Department of Fisheries Technology Engineering, Faculty of Marine Sciences, Karadeniz Technical University, Trabzon, Turkey*

12.1 Introduction

All the major fish processing techniques are covered in other chapters in this book, thus, this chapter will concentrate on the sustainability issues surrounding the sector. A limited description of the techniques is given in order to put them in the context of sustainability. Cross referencing to the individual detailed chapters will enhance the reader's deeper understanding of each process and the sustainability parameters discussed here.

12.1.1 Defining sustainability

The concept of sustainability currently in use can be dated to the 1972 United Nations Conference on the Human Environment (the 'Stockholm Conference') which recognized environmental problems such as greenhouse gas emissions and depletion of the ozone layer. Subsequently, the United Nations (UN) set up the Brundtland commission to study the issues of global inequality and resource redistribution (World Commission on Environment and Development, 1987). Their report, 'Our Common Future' suggested that **economic growth** should be related to **social equality** and **environmental protection**, and a sustainable level of world population as an issue to be tackled to achieve these goals. Their definition is,

> Sustainable Development is development which meets the needs of the present without compromising the ability of future generations to meet their own needs.

Seafood Processing: Technology, Quality and Safety, First Edition. Edited by Ioannis S. Boziaris.

The report suggested that overstating one of the three principles given earlier will not achieve sustainability. A balance between these competing elements is regarded as the 'Triple Bottom Line' (TBL), which has been taken up by business under the ethos of 'Corporate Social Responsibility' (CSR) as a means to legitimize their activities, although there is much debate about their sincerity and practicality. In 1992, the UN held the 'United Nations Conference on Environment and Development' in Rio de Janeiro (the 'Rio Summit') to report on progress and respond to new threats such as climate change and the tensions between environmental protection (proposed by the developed nations) and a developing nation's desire for social and economic improvement. It produced a declaration that, 'the right to development must be fulfilled so as to equitably meet developmental and environmental needs of present and future generations'.

It is not possible to address the wider sustainability debate at length here but it is appropriate to define some specific sustainability criteria as they apply to fish processing.

12.1.2 Sustainability criteria

Four common sustainability themes will be applied to the main fish processing technologies, which are:

Energy consumption: for driving machinery, producing ice, cooling and freezing, for steam generation, canning, smoking and drying/dehydration. Reduced energy consumption will conserve fossil fuels and reduce greenhouse gases (GHG) emissions, and converting wastes to energy will substitute for conventional energy sources. Energy for transportation will not be included as this forms part of the general debate about transport.

Water consumption: for transporting fish around the plant, thawing, washing and cleaning the raw material and product, cleaning plant and equipment. The substitution of seawater for fresh water, metering and controlled use, and the reuse of wastewater (where appropriate and safe) reduce the 'water footprint' of the process.

Effluent control: discharges are dilute, in large volume, containing polluting organic matter such as oils and protein in solution or as suspended solids. Effective capture of the effluent will lead to less pollution and opportunities for recovering by-products.

By-product development: the edible proportion of fish might be about 45% of the caught weight so there is scope to use the rest of the raw material as added-value by-products, for example fishmeal and oils. Oils and proteins in the effluent (or offals) have potential as biodiesel and enzymes, respectively. Whilst the importance of developing fish products for direct human consumption (DHC) must never be lost, the potential for by-products must be explored.

These four areas will be emphasized in the discussion of each of the major fish processes in this chapter.

12.1.3 Climate change

Climate change will have detrimental and beneficial impacts on fisheries. The scientific monitoring systems currently employed to address stock management are ideally suited to tracking the effect of climate change on fisheries, such as 'El Nino' events (FAO, 2009). This expertise should be adapted to deal with the uncertainty surrounding climate change and its impact on fisheries. Some of the effects of climate change could include:

- *displacement* of warm-water species towards the poles;
- *changes in habitat size and productivity* with lower latitudes (tropical/sub-tropical waters) becoming less productive and the higher latitudes more productive;
- *changes in fish physiology and seasonality* may affect reproductive cycles, migratory movement patterns, body composition and the spread of diseases and resistance to them;
- *extreme events* such as floods, droughts, storms and 'El Nino' style events will increase;
- *physical changes* such as sea-level rise, melting glaciers, ocean acidification and changes in rain patterns will affect aquatic habitats as varied as coral reefs, estuaries and lakes.

Effects will be felt mostly by fisheries dependent on specific water conditions such as coral reefs or upwellings, whilst estuarine fisheries will be subject to sea-level rise and salt invasion, and most low-lying coastal and island communities will be susceptible to extreme events. Aquaculture is concentrated in Asia where aquatic systems will be affected by water temperature rise and flooding or drought, while the promise of increased aquaculture in Africa and Latin America will be depend on adaptation of the sector to the new conditions.

Climate change will bring new opportunities such as access to new fish species thriving in the higher water temperature and salinity, and access to 'new' waters as dams are built to buffer new rainfall patterns (World Bank, 2009). Fish processing will need to address fuel, energy and processing efficiency to reduce its contribution to climate change.

12.2 Assessment tools

12.2.1 Carbon footprinting

Carbon footprinting (CF) measures GHG associated with the manufacture of products through: raw material sourcing, processing, packaging and distribution and waste treatment. Emissions are 'direct' when associated with the main process or activity and 'indirect' when associated with upstream and downstream activities. The term GHG can be defined as carbon dioxide or other gases as carbon dioxide equivalents. Sometimes it is difficult to convert an activity or process into GHG units and conversion factors are not uniform. Several definitions of CF have been proposed as described by Wiedmann and Minx (2007), who proposed the following definition,

> The carbon footprint is a measure of the exclusive total carbon dioxide emissions that is directly and indirectly caused by an activity or is accumulated over the life stages of a product.

The CF can vary depending on how many core and peripheral activities are included, e.g. the activities of suppliers to the principal manufacturer and even the GHG associated with the consumer travelling to buy the goods. This is a contentious issue as the calculation of the CF is complex and there is ample scope for misleading an uninformed public.

In the fisheries sector, emissions have been estimated for raw material supply, processing methods and transport/storage with significant differences between sectors (capture or aquaculture) and processing methods (FAO, 2009). In the capture fishery, fuel efficiency is related to the GHG emissions, fuel used and weight of catch. For aquaculture, energy consumption must include that required to produce feeds and is higher for shrimp and carnivorous fish than for omnivores, molluscs, bivalves and algae farming. Estimates for the ratio of edible protein energy output to industrial energy inputs varies from 1.4 to over 100% for these groups, respectively. The international nature of fish trading has an impact where airfreight may emit 8.5 kg of carbon dioxide per kg of fish transported, which is 3.5 times that for sea freight and 90 times that for local transportation – within 400 km of the catching point (FAO, 2009). Actions to reduce these GHG emissions can lead to new products, local markets and/or a drive for more sustainable raw material supply, processing, transport and storage options.

Carbon labelling (CL) could enable the consumer to judge the green credentials of a product from the packaging. The form of the CL and the method for determining it are open to interpretation and not yet subject to international coding or scrutiny. A system of Ecolabelling applies in the fisheries sector following FAO guidelines designed to improve fisheries practices through a market-driven approach based on consumer demand for environmentally friendly products (FAO, 2005). Ecolabels are given to products considered to have lowered environmental impact such as 'dolphin-safe tuna', fish caught from sustainable fisheries or using selective fishing gear or from organic aquaculture practices. Such labels include self-certification by producers, certification by producer trade organizations or by independent schemes. The latter include the UK-based Marine Stewardship Council (MSC) and the Marine Aquarium Council (MAC) in the United States (ICTSD, 2006). The developed economies have driven Ecolabelling as an issue and companies use it as a marketing tool for their products, although the effectiveness of Ecolabelling is not proven (ICTSD, 2006). Iles (2007) argued that seafood producers are invisible to the consumer who only recognized the well-known retailers of their products, so consumer pressure on producers was less telling than that of the production chain. The demand for ethically sourced fish products now includes the processing methods used and these should be equally ethical with respect to the environment and societal impacts (FAO, 2009). Thrane *et al.* (2009) discussed the use of ecolabelling for wild-caught seafood products and suggested that life cycle assessment (LCA) studies showed that significant environmental impacts were attributable to post-landing operations such as processing, transport, cooling and packaging. Criteria reflecting these post-landing operations should be included in an LCA, and energy/water consumption, materials and waste handling were highlighted.

12.2.2 Life cycle assessment

An LCA is the investigation and evaluation of all the environmental impacts of a given product, process or service (see Figure 12.1) underpinning CF and labelling. An early definition of LCA was,

> the process of evaluating the effects that a product has on the environment over the entire period of its life cycle ... extraction and processing; manufacture; transport and distribution; use, reuse and maintenance; recycling and final disposal
>
> (UNEP, 1996)

An LCA can be broad or narrow – a full LCA is 'cradle-to-grave', including all activity from raw material to disposal – and can be used for specific elements such as energy, water, packaging and raw materials. The LCA process based on the International Standards Organisation 14040 Series (ISO, 2006) can be divided into four activities (Figure 12.2).

Phase 1 – Goal and scope: decides the purpose of the LCA and which aspects are included. The system boundaries are set by the scope definition into four phases:

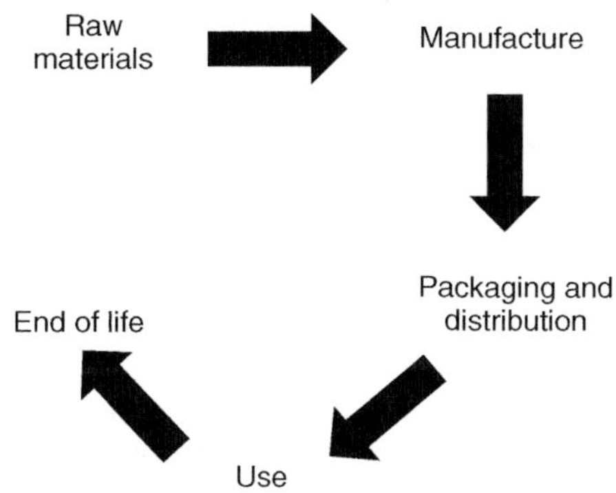

Figure 12.1 Life cycle assessment: cradle-to-grave

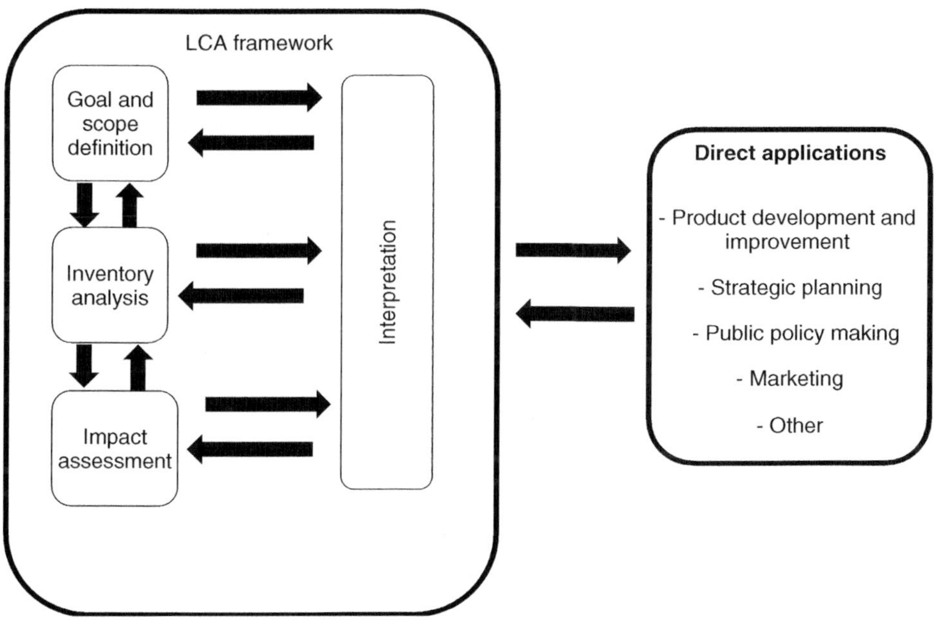

Figure 12.2 Stages in an LCA (after ISO, 2006) (reproduced from Hall, 2010, with permission from John Wiley & Sons, Ltd)

(i) pre-manufacture, (ii) manufacture, (iii) packaging and distribution and (iv) use and end of life. All phases would be, 'cradle-to-grave', while 'cradle-to- factory gate' would only include the first two. Environmental impacts are important for sustainability but so also are energy and water usage and effluent production.

If a processing plant produces more than one product there must be an allocation of impact between them by: system allocation made on the basis of the mass or economic value of the product, or (preferred by the ISO 14040 series) system expansion whereby the co-products are considered as alternatives to other products and an allowance made for this substitution in calculating the impact for the main process. The two approaches give different results.

Phase 2 – Inventory analysis: the time-consuming data collection phase including all inputs (e.g. energy, water and raw material), outputs (products) and emissions (to air, water, soil and solids) selected for inclusion. Accurate, relevant information is essential and must be available or derived from secondary data such as utility bills for water, gas and electricity. A production system is broken down into unit processes, batch or annual production, whichever best defines the system in meaningful numbers – the functional unit.

Phase 3 – Life cycle impact analysis: the inventory analysis information is processed and assigned to an environmental impact category with appropriate units which may conform to systems such as ISO 14000 Series or be process-specific. Six common environmental impact categories are:

- *global warming* mainly combustion of fossil fuels, as carbon dioxide equivalents;
- *acidification* affecting waters, forests and sometimes buildings is caused mainly by combustion for electricity, heating and transport, as sulfur dioxide equivalents;
- *eutrophication* leads to algal blooms, oxygen depletion and fish deaths caused mainly by fertilizer nitrogen run-off, as nitrate equivalents;
- *ozone depletion* caused by synthetic halocarbons, chlorofluorocarbons and hydrochlorofluorocarbons (CFCs, HCFCs);
- *land use* in the production of products, as hectares (or m^2) per year;
- *photochemical smog* from volatile organic compounds (VOCs) produced from unburnt petrol and diesel and organic solvents that cause respiratory problems and reduce agriculture yields, as ethane equivalents.

These categories are not exclusive and for certain applications the energy, water and effluent categories can be simplified or made process-specific. Once categorized, emissions should be converted to the reference units for that category using equivalence factors. For CF, inventory items are equated to GHG emissions with carbon dioxide equivalents for other gases, such as methane. A sensitivity check will determine the accuracy of the inventory data whilst a normalization process will compare the relevant data to a reference system such as an existing process. Finally, the inventory data can be weighted in terms of the most important environmental impact. The weighting criteria are, again, areas of debate and can be based on the judgement of a panel of experts, financial considerations and targets set by the company or government.

Phase 4 – Life cycle interpretation: the results of the impact analysis are compared with the original goal and scope. This analysis can take place continuously to ensure that the LCA is really achieving the goals and that the scope is correct. This iterative approach to the interpretation of data will allow incremental improvements and/or changes to the goals and scope as necessary. The final interpretations should indicate the completeness of the data, the appropriateness of the analysis and reach conclusions and lead to recommendations for process improvement. These four core elements of LCA are available in software packages providing generic categories and conversion factors, while impact assessment models and international cooperation has led to greater uniformity and consolidation of methodologies (Finkbeiner *et al.*, 2006). The importance of sustainability for the fishing sector has long been recognised leading to the early use of LCA to investigate issues (Ziegler, 2003). Recently the concept of the Social LCA (S-LCA) has developed to account for the impact of a system on people's lives – a notable extension of the original environmental LCA (E-LCA) (Benoit *et al.*, 2010).

Figure 12.3 describes a 'gate-to-finished product' approach, ignoring the raw material supply and the product outside the process plant. This is a simplified LCA (Graedel, 1998) which, although limited, does concentrate on the fish processing plant itself. The figure can be divided into three elements:

- *other systems* which precede and follow the system defined by the boundary. Individual processing operations sit within the system boundary whilst fish capture and aquaculture precede them and distribution and retailing and disposal and/or recycling follow;
- *elemental flows* represent the materials coming into and leaving the system, such as energy, water, etc. For sustainability assessment of fish processing alone these elemental flows stop at the boundaries when related to other systems, for example fuel energy for catching the fish would not be included;
- *system boundary* has been drawn to include the pre-processing operations, the process itself (e.g. canning, freezing) and production waste. Any of the processes described in other chapters could be fitted into the system boundary. Production waste can be recycled as other fish products or by-products or directed to energy generation for the process or for transport through anaerobic digestion and biodiesel.

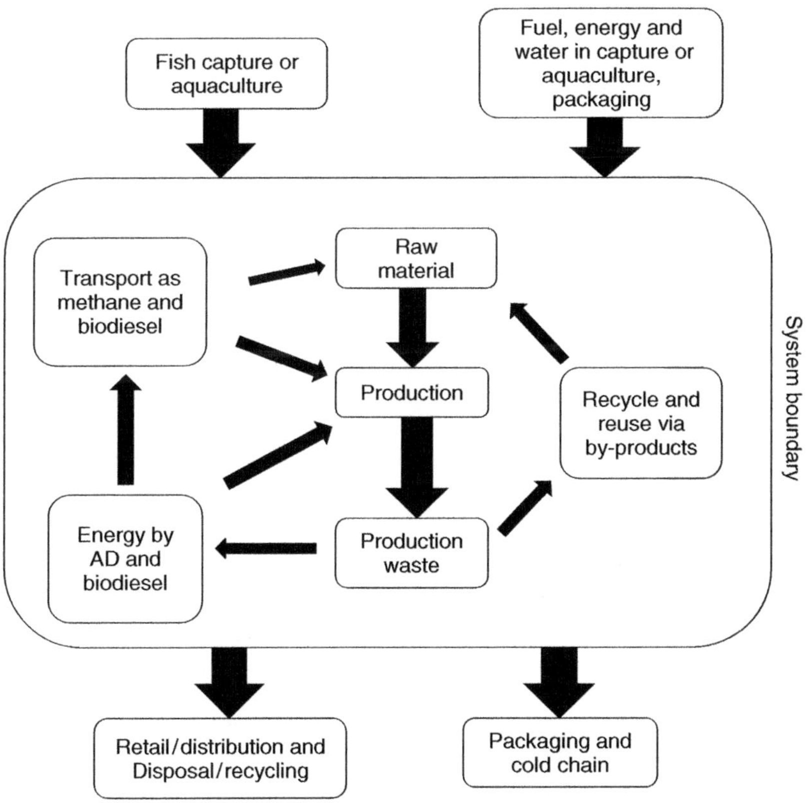

Figure 12.3 System boundaries for fish process operation (reproduced from Hall (2010) with permission from John Wiley & Sons, Ltd)

The LCA impact categories have been studied for seafood production systems (Pelletier *et al.*, 2007) that looked at the published seafood production LCAs and those from agriculture which might transfer to seafood production. The established impact categories were used frequently but impact factors specific to fisheries and seafood processing were also used. There was no treatment of socio-economic impacts and little on the fish processing itself, but energy impacts were discussed. The leaders in developing LCA have been Iceland, Norway, Sweden and Denmark based on the importance of their fishing sectors, existence of a well-developed fisheries research and government support for environmental issues. The open access database LCAFood (www.LCAFood.dk) covers a range of products and processes (Mogensen *et al.*, 2009) and quotes values based on different points along the supply chain and functional units. The application of an LCA to the UK fish industry was used to investigate the environmental impact of catching, aquaculture, and transport and processing and to make recommendations for mitigation and good practice (Cappell *et al.*, 2007), although they indicated that little fishery-related LCA was available for the UK sector. The number of databases and LCA programmes is increasing making the input and analysis of data quick and reliable. Realistic comparisons between fish (wild or farmed) and land-based livestock production may be possible with such information (Ellingsen and Aanondsen, 2006). A description of LCA for fish processing operations has been given recently (Hall, 2010).

There is little of information on fish processing in developing countries. Information seems to be extremely scarce for the smoking and salting (curing) processes, yet the benefits could be enormous from an LCA and supply chain analysis in this sector for identifying the energy inputs, possible savings and sources of post-harvest losses.

12.2.3 Supply chain

Large companies have used interventions in the supply chain to streamline operations and gain commercial benefits through increased efficiency and productivity, product development and reduced waste. The same approach can promote sustainability, but when GHG emissions, an LCA or a CF are taken into account, current operations along the supply chain might require change to have a positive environmental impact. Actions to reduce the CF, such as energy or water usage reductions, not only give overt economic benefits but can also contribute to good public relations (CSR and TBL). The supply chain approach must be applied in an all-embracing manner rather than each company in the chain looking only at the contribution of its own activities. Such a co-ordinated approach demands collaboration along the supply chain, around the central operation, with savings being identified for the product as a whole. The food industry, which processes primary products for the consumer is a prime example of the supply chain approach. The emissions associated with supplying food to the plate can be divided into direct emissions from energy consumed in the home (23%), indirect emissions from the supply chain (69%) and transport (8%) (Carbon Trust, 2006). Thus, the food industry has a large supply chain component which could have a massive sustainability impact. Fish processing with its emphasis on trade, and particularly developing-to-developed country trade, could be a prime example of beneficial supply chain interventions. Iles (2007) argued that seafood producers could be made more accountable through a supply chain approach and suggested ways to achieve this:

- identify and track companies to remove their invisibility;
- develop product chain campaigns so that companies influence each other;

- develop mechanisms to compare companies to improve industry practices;
- develop methods to track consumption, production and management changes;
- develop interactive consumer tools so that consumers get feedback on their purchasing habits which can also be fed back to the producers.

Thrane *et al.* (2009) emphasized the importance of the supply chain approach for ecolabelling, particularly the environmental impact of fish processing.

12.3 Process operations

12.3.1 Introduction

This section will describe the major fish processing operations and their sustainability issues and, as mentioned earlier, should be read in conjunction with the individual chapters for a complete appreciation of the sustainability concerns.

12.3.2 Pre-processing

The pre-processing operations include: raw material unloading, thawing; washing and grading; skinning, scaling and, beheading, gutting and filleting. These operations present significant sustainability problems as they involve the most intricate and prolonged handling of the fish, separation of parts and any pre-cooking. They involve energy for driving pumps, for machinery and heat for hot filling and for steam generation. Water usage is high for transporting materials, cleaning plant and equipment, and brine production, and becomes an effluent problem due to contamination. By-product recovery may need to be achieved, but must be done from large, dilute volumes. Recognition of the value of the by-products will lead to strategies for their recovery. The sustainability implications for these operations are applicable to the common fish processes and there is an argument that pre-processing could almost be designated as a 'process' in its own right with regard to sustainability. Table 12.1 describes the pre-processing operations, their characteristics and sustainability issues.

12.3.3 Canning

After pre-processing, canning involves two stages of filling/sealing and retorting (Figure 12.4). Filling the cans involves operations dependent on the species of fish, particularly size. Operations such as brining impart flavour, preserve texture during retorting and have sustainability implications with the addition and loss of water and pollution potential. Water usage can be reduced by calibration of the sauce/brine filler and using recovered water to wash off any spillage. A pre-cook, if done, is a considerable sustainability impact because of the energy used and the spillage of contents, which is a pollution load and a loss of valuable product. Exhausting and sealing are the final operations prior to retorting. There are three ways to achieve a good headspace vacuum: hot filling/sealing, vacuum sealing and steam sealing. The sustainability implications of these operations centre on the energy demands for heat or to create a vacuum and the pollution loads created. Careful management of these automated processes can reduce energy usage and spillage, while the usual strictures on water usage apply.

Table 12.1 Pre-processing operations in fish processing

Process operation	Characteristics	Sustainability issues
Unloading and thawing	Ship to plant	Metering water and shut-off valves
	Wet transfer with sea water	Reuse water with treatment
	Contamination with blood and slime	Use dry transfer
	Thawing uses large amounts of water	Thaw using warm moist air and improve heat transfer
Washing and grading	Contamination with blood and slime	Metering water and shut-off valves
	Water for moving fish	Use gravity for movement
Skinning and scaling	Manual or mechanically	Metering water and shut-off valves,
	Contamination with skin, scales and oil	Mechanical systems sharp and set correctly
		Water reuse
Be-heading	Manual or mechanical	Metering water and dry transfer systems
	Significant pollution load from wash water	
Filleting	Manual or mechanical	Metering water
	White and oily fish differ	Mechanical systems sharp and set correctly
	Water to remove offals contaminated	Manual filleting more accurate and flexible?

Retorting is the central operation in canning. The equipment used and the mode of operation will have profound effects on the overall energy/water usage. Retorts should reflect the nature of the fish being processed, the scale of operation and expertise to manage the plant. **Batch retorts** operated in a horizontal or vertical axis depending on batch size are inefficient of energy and water as the system must be heated and cooled for each batch. **Continuous retorts** give higher throughputs/less downtime and **hydrostatic retorts** move cans vertically up-and-down towers with high throughput, good reproducibility and efficient energy and water usage.

Steam for the sterilization process comes from a dedicated or a central plant boiler with associated pipework and valves. Some environmental losses occur through poorly insulated pipework and leaking valves. Of greater concern are losses through incomplete combustion when the air : fuel ratios are incorrect, which wastes fuel and leads to atmospheric pollution from the products of incomplete combustion. Emissions from burning fuel oil include carbon dioxide, nitrogen oxides, polyaromatic hydrocarbons and sulfur dioxide (contributing to acid rain). Selection of a low-sulfur fuel and control of combustion conditions will reduce these effects. Steam condensate should be returned to the boiler or reused elsewhere in the plant to prevent overall energy and water wastage.

Post-processing operations are cooling with chlorinated water and drying before storage and bulk packing. The major issue is the use of large volumes of water, which will be reduced by good heat transfer properties on cooling. Re-use of the water after cooling, to recover any heat or to substitute for clean water in other parts of the process, depends on the absence of contamination by leaking cans. Cooling water use can also be reduced by recycling water from other parts of the plant but is dependent on process cleanliness.

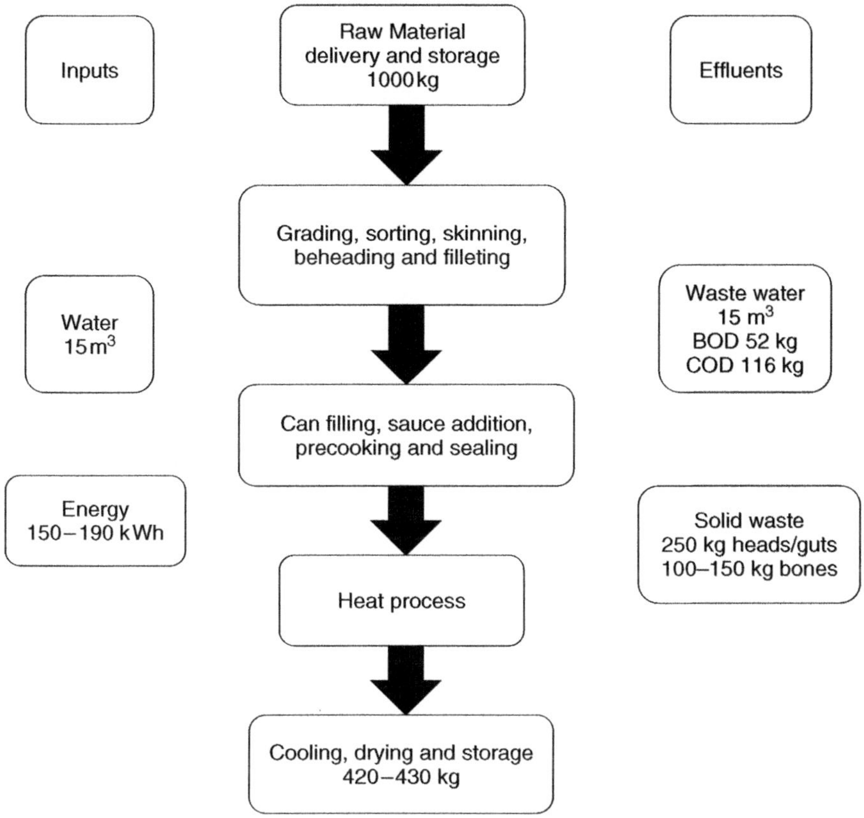

Figure 12.4 Canning operations with energy and water usage and effluents (after UNEP, 2000) (reproduced from Hall, 2010, with permission from John Wiley & Sons, Ltd)

The possibility of contamination of the can contents can be reduced by adequate chlorination of the re-used water – perhaps 85% of the water can be re-used (UNEP, 2000).

Water usage varies depending on the species being canned (especially pre-processing operations) and the heat treatment applied. A typical water usage for canned sardines would be 9 m^3 per tonne and for tuna 22 m^3 per tonne (UNIDO, 1986). The bulk of the water and biological oxygen demand (BOD) or chemical oxygen demand (COD) is generated in the pre-processing operations. Fahim *et al.* (2001) described effluent treatment possibilities for fish canning waste water from processing mackerel and sardines and the use of ferric chloride and calcium hydroxide reduced the BOD, COD and total suspended solids considerably.

The BOD and COD are indicators of water quality (Sawyer *et al.*, 2003) and high values reflect the accumulation of solubles in the wash waters during pre-processing. They indicate the potential for environmental pollution but also for the recovery of proteins and oils. Rather than recovering soluble and suspended solids from a large volume, alternative strategies are to:

- reduce the volumes of water used in total – by controlling flow rate, use of intermittent sprays and recycling;

- use physical methods of skinning, gutting and filleting – knives, scrapers and vacuum gutting;
- separate water from offals quickly to give cleaner water for recycling and less solubles/suspended solids for later recovery. This can be achieved by using gravity to collect the offal and perforated conveyors to transport it for further treatment.

Application of these ideas could reduce water usage by 50% and reduce the BOD/COD by a factor of four (UNEP, 2000); see Figure 12.4 for example figures for energy, water and effluents.

Energy usage is mainly for steam generation. Energy usage in the pre-processing operations will depend on the need for pre-cooking, hot filling and steam sealing but is still likely to be relatively small. Energy reduction can be promoted by efficient loading into cans, good insulation and the use of accurate control systems for the retort and steam boilers.

From a sustainability standpoint other options to consider would include:

- using cleaner fuels for energy generation;
- closing down controls when equipment is not in use;
- capturing low-grade heat for re-use;
- applying waste-to-energy technologies such as anaerobic digestion to produce methane to replace the mains gas supply – 'resource substitution'.

12.3.4 Smoking

Cold smoking takes place below 30 °C to allow drying and preservation by the deposition of smoke components. Examples of cold-smoked fish are kippers (herring), smoked salmon and Finnan haddock. Modern systems are horizontal (for ease of access) with fans to drive the smoke across the racks of fish, automatic humidity control, booster heaters and controlled air flows giving good process control. This is necessary as the air/smoke mixture changes as it moves over the fish, becoming cooler and less smoky due to deposition and humid as it picks up moisture from the fish.

Hot smoking takes place mostly at 70–80 °C, or even 100 °C, so the fish is cooked and can be eaten without further heating. Horizontal kilns are commonly used for hot smoking as they allow a consistent high temperature.

Modern systems: most designs are for smoking salmon incorporating features designed to give consistent product quality; energy savings; computer control backed up by temperature, smoke density and humidity management systems; fire protection; and self-cleaning in an hygienic operation. Some fish smokers in Europe and North America use traditional smoking ovens and techniques, which is a special market for premium products with strong smoked flavours. However, the bulk of fish smoking is done in the developing world where the sustainability problems faced are acute.

Traditional systems are simple designs with no control of the burning wood temperature and inefficient circulation of heat and smoke. A consistent product was only possible by a time-consuming process of unpacking and moving fish about leading to physical losses and friable products. In the high season for migratory species of fish, this low capacity and time-consuming method of smoking cannot keep pace with landings leading to spoilage and underutilization of fish. The smoking process varies with the species of fish. Wet hot smoking takes 1–2 h and the product (40–55% moisture content) has a limited shelf life (1–3 days). Dry hot smoking can take 10–18 h or even days in wet conditions, to give a dry product (less than 15% moisture content) with a long shelf life allowing transportation from the coast.

Table 12.2 Traditional African fish smoking ovens and their characteristics

Oven type	Origin	Innovation	Drawbacks
Adjetey	Ghana (1962)	Separate smoke box and smoke chamber Chimney, several trays of fish	Cost, operation and final product quality
Altona	Ghana (1970s)	Brick smoke box, metal smoke chamber, chimney	Cost, fish skewered on rods not normal practice
Ivory Coast	Ivory coast	Used oil drums for smoke box, aluminium chamber, chimney, fish on trays	Cost, uneven smoking and variable product quality
Nyegezi	Tanzania	Separate smoke box/chamber chimney	Developed for specific Tanzanian conditions so not transferable
Chorkor	Ghana (1970s)	Used traditional materials, good smoking conditions, easy handling (trays) fuel efficiency, healthier operation	

The smoking oven was traditionally circular and made of mud with a grid of wooden sticks to support the fish. These ovens have a limited life and must be rebuilt after about two years. Similar sized ovens have been made from 44-gallon (200 l) steel oil drums and are light, portable but subject to rusting, may contain toxic residues and being metal get hot during the smoking process which leads to burns for the fish and the processor. Both types leak smoke which is injurious to the eyes and lungs of the processors. Rectangular or square ovens made from mud or metal have been tried but give little advantage over the circular types. Efforts to improve the traditional smoking oven started in the 1950s onwards (Table 12.2).

A major issue for traditional fish smoking is the availability of fuelwood which is in demand for other purposes such that supplies close to the fish smoking sites are dwindling, collection becomes time consuming and the cost is increasing. The long-term sustainability of the fish smoking sector is related to forest preservation and the catch fishery, indicating the complexity of sustainability issues. Despite the importance of fuelwood supply, studies are few. Fernandes *et al.* (1995) described the conditions facing fish processors in Tanzania, Mozambique and Angola, typical of sub-Saharan Africa. Abbot and Homewood (1999) described the loss of the miombo woodlands in Malawi for domestic purposes and commercial fish smoking and suggested that current forestry management practices targeted domestic collectors but should also address wood for fish smoking. Olokor (2003) described fuelwood consumption for fish smoking in Nigeria. The average fish smoker used 16.45 kg of fuelwood daily to process 7.88 kg of fish and this equated to 396 250 kg of fuelwood to process 18.9 tonnes of fish annually.

A recent study has investigated the impact of fish smoking on the mangrove ecosystem of Cameroon (Feka and Manzano, 2008). Mangroves are a very specific, productive niche at the land-sea confluence only found in tropical and subtropical regions of the world (Barbier *et al.*, 1997). The depletion of the mangrove was primarily due to human activity and the loss has been estimated at 30% from 1980 to 2005 (UNEP, 2007). Attempts to replace raw fuelwood have included the use of sawdust in Nigeria (Akande *et al.*, 2005). Sawdust was cheap, but with transport costs, whilst fuelwood was available at the smoking location. To replace fuelwood there must be a reliable local supply, otherwise the sustainability issues surrounding energy for transport against energy from the product must be considered.

An alternative approach could be to release fuelwood for fish smoking by substitution in other areas by renewable energy technologies. Karekezi (2002) described the potential for renewable energy in Africa, of which fuelwood for fish smoking plays a part. An issue for renewable energy uptake is the role of women in energy provision for domestic cooking and fish smoking – an almost exclusively female occupation. Thus, the gender perspective of energy in the broadest sense should be recognized (Cecelski, 2000).

***Post-harvest losses*:** which affects traditionally smoked fish products is recognized as a problem in artisanal fisheries and traditional processing. Given the importance of smoked fish these losses are of great concern. The first step in preventing post-harvest losses is to improve the traditional oven so that the catch can be rapidly preserved and large seasonal catches can be dealt with. Physical losses can occur during distribution due to insect infestation, fish being eaten by animals, theft and discarding fish for which there is no market through oversupply. Other loss factors are bad weather, poor roads and unreliable transport.

***Sustainability issues*:** the antiquity of the curing processes bears testament to their simplicity and wide applicability. The introduction of modern technology has improved the quality and consistency of the products, but electricity is now relied on as an energy source. It is tempting to suggest that a return to the older processes would be a step towards sustainable fish processing. However, adherence to traditional processes is associated with poorer quality and has its own sustainability problems with the procurement of fuel. Another factor is that nowadays consumers prefer low-salt products, preservatives with a low carcinogenicity potential and lighter flavours. These factors do not favour the re-introduction of heavily salted and highly flavoured smoked products which survive in traditionally high fish-eating societies or the older generations. Nevertheless, attempts to improve product quality in current traditional processes would be extremely beneficial for increasing protein intake. In this situation the application of a full LCA and supply chain approach could pay great dividends.

12.3.5 Freezing and chilling

Freezing systems can be classified by the means of heat extraction (mechanical or cryogenic) or by the rate of freezing (movement of the ice front through the product). There is a correlation between the means of extracting heat and the rate of freezing. The freezing system will be dictated by the product although some systems are more flexible than others. The rate of freezing classification which will have energy usage consequences as follows:

- slow (sharp) freezers (0.2 cm/h)
- quick freezers (0.5–3 cm/h)
- rapid freezers (5–10 cm/h)
- ultra-rapid freezers (10–100 cm/h).

Table 12.3 lists the common freezing systems according to these two criteria.

Table 12.3 Classification of freezing systems

System	Freezing rate	Operating temp (°C)	Comments
Cooled air			
Chest	Slow	−20 to −40	whole fish, blocks, bulk storage
Blast	Quick	−30 to −40	200–1500 kg/h simple, variety of product shapes and sizes batch or continuous, multiple passes defrosting necessary
Belt (Spiral)	Quick	−30 to −40	3000 kg/h vertical tiers so space saving
Fluidised bed	Rapid	−25 to −35	10000 kg/h shrimps, prawns
Plate freezers	Quick	to −40	Horizontal for regular shapes with good contact Vertical for rapid loading and unloading on-board
Cooled liquid			
Immersion	Rapid	−6 to −15	good heat transfer at higher temperatures brine, calcium chloride, ethylene and propylene glycols
Cryogenics	Ultra-rapid	−18 to −25	45–1350 kg/h (by sublimation) expensive, liquid nitrogen or carbon dioxide, high value products prawns and shrimps

Cooled air freezers: are simple systems used for a variety of different sized/shaped products and coated products. The major criterion of performance is even air distribution over the product (Kolbe and Kramer, 2007).

Immersion freezers: because the freezing medium can envelope the product completely these systems can operate at higher freezing temperatures, for example at −17 °C compared to −40 °C for an air-blast system, but with comparable speed and improved energy efficiency (Kolbe *et al.*, 2004).

Cryogenic freezers: quality is enhanced by the exclusion of oxygen and reduced dehydration of the product but thermal shock from the very low temperatures may induce textural changes.

Liquid nitrogen (LN) systems: the sustainability issue is that storage and movement of LN in tanks and along pipes gives rise to heat leakage because of the large temperature difference between the LN and the surroundings (Kolbe and Kramer, 2007).

Liquid carbon dioxide: is typically stored in tanks of 3500–12 000 UK gallons (~16 000–54 000 l) at −18 °C and 300 psi (21 kg/cm^2) and at these higher storage temperatures the heat leakage can be effectively reduced to zero (Kolbe and Kramer, 2007).

The freezing process itself will be the major energy user in the plant although there will be demands in pre-processing operations (UNEP, 2000) because the bulk of frozen fish products are heavily processed, including: cutting, shaping, enrobing, coating and packaging, while whole fish may have been gutted. Energy savings can be made retrospectively or, better, done at the initial design stage and allied to good management practices (Gameiro, 2002). Energy consumption can be reduced by:

- reducing the freezing time;
- reducing heat leakage through poor insulation and through the doors and curtains in cold stores – good cold store management;
- reducing defrosting cycles by packaging and good plant management – balancing freezing rate against defrosting schedules;
- controlling the use of air circulation fans;
- lighting controls can be motion activated with energy-savings bulbs and lighting strips;
- recycling waste heat removed in freezing for pre-heating boiler and clean-up water.

Further energy savings can be made by matching the freezing process to the specific product type, throughput and equipment sizing. Elements for consideration include: maximize the refrigerant evaporating temperature (higher temperatures); minimize refrigerant condenser sizing; compressor selection for optimum efficiency; automatic control with accurate sensing of changing conditions; and minimize lengths of piping to reduce pressure drop (Kolbe and Kramer, 2007).

Once frozen, the fish product has to be stored and transported through a cold-chain. The relationship between time, temperature and quality in storage/transport has been developed into the time–temperature–tolerance (TTT) relationship for specific products to predict quality changes knowing the temperature history of the product. Kolbe *et al.* (2006) give data for several Pacific fish and shellfish species revealing complexities based on species, storage temperature and packaging.

There has been much concern over the use of halocarbons, commonly known as Freons, as refrigerants because of their harmful effects in ozone depletion. This has led to a controlled phasing out and replacement with chlorine-free materials (ammonia and carbon dioxide) under the Montreal Protocol of 1987 and its wide acceptance and implementation has been hailed as an exemplar of international cooperation. Implementing the Protocol threw up issues which will affect any environmental protection policy on an international scale (Rowlands, 1993) centring on the availability of substitutes, the ability of *all* user countries to introduce the substitutes and the need for exemptions.

The environmental impact of buildings is well recognized through the application of LCA (Koatji *et al.*, 2003) and the 'embodied energy' in the production of building materials has been recognised as an environmental impact factor (Sartori and Hestnes, 2007). Energy use during the practical life of the building is important, and strategies for energy conservation have been designed and building services engineers are being trained in sustainability issues (Shah *et al.*, 2007; Xiao and Wang, 2009). In the construction of freezers and cold stores concerns relate to capacity, ease of construction, temperature maintenance and prevention of moisture migration (insulation). The environmental impact of the operation of stores and their construction materials seems to have been ignored, so a contribution to developing a full LCA for these aspects is desirable.

There a few LCA and Supply Chain studies for frozen fish products. The open access database LCAFood gives a list of environmental impacts for fish products within certain system boundaries such as at the harbour, processing factory and retail store (LCAFood, 2009). For fresh product there is little added impact from the harbour to the retail sector for any of the impact categories measured. On freezing there is a substantial increase in the impact across all impact categories; an effective doubling in most cases from Ex-harbour to Ex-Factory – but a trebling in some instances. The difference between Ex-Factory and Ex-Retail impact categories for frozen products shows increases in the Global Warming category. The importance of a Supply Chain approach is exemplified here as the energy demand impact for frozen foods throughout the 'cold chain' is apparent and energy saving efforts at any point can be determined by the cumulative values for Global Warming. A Supply Chain approach would also expose the consequences of actions along the chain, whether beneficial or detrimental.

Few examples of an LCA for frozen products from the developing world exist but one recent report does give an insight into the nature of different fishing operations. Ziegler *et al.*, (2009) investigated the environmental impact of a fishery for Southern pink shrimp (*Penaeus notialis*) from Senegal with different catching styles and processing techniques. The LCA indicated where environmental impacts can be reduced but not necessarily the means to do it. The study identified the sustainability issues, their scale and the level at which they can be tackled (local, regional or national) and where a Supply Chain approach might hasten the implementation of sustainable processing.

12.3.6 Surimi production

Figure 12.5 shows the unit operations for the modern surimi process and the waste arisings associated with each operation. The crucial elements are energy consumption, water consumption, effluent control and by-product development. Morrissey *et al.* (2005) have given a full analysis of the waste and by-product issues surrounding surimi production.

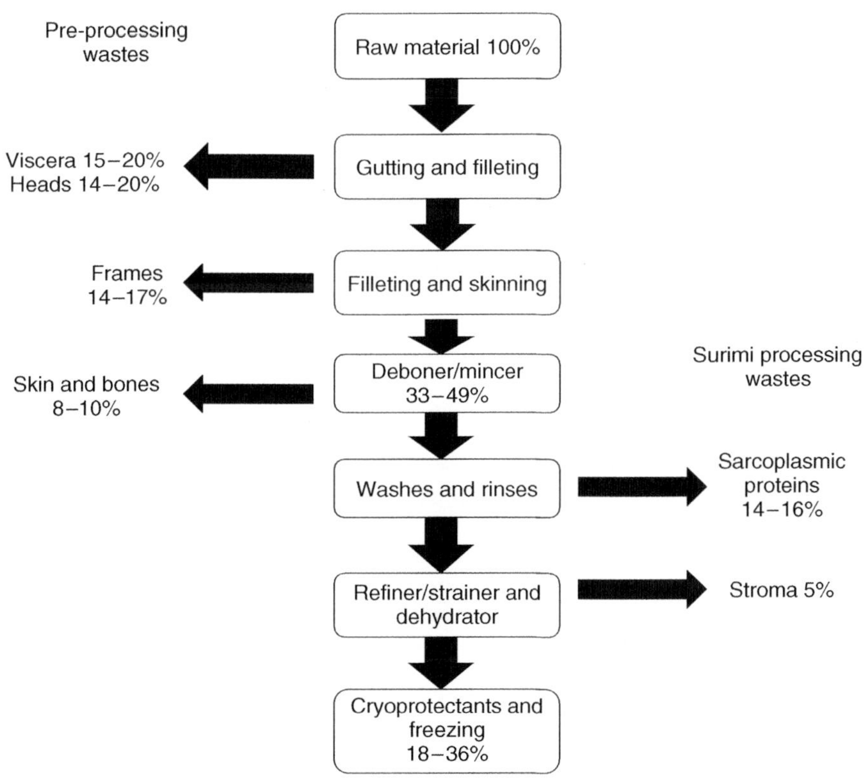

Figure 12.5 Surimi process with waste arisings and product yields (%)

Energy consumption of the surimi process may not seem high at first glance. There is the need for pre-processing and the meat-bone separator to generate fish mince. The route shown in Figure 12.5 varies via filleting/skinning, or heading/gutting, before the deboner/mincer operation. The former tends to give lower yields/better quality mince requiring fewer washes and no refining because most of the work of the deboner/mincer has been done beforehand. By the heading/gutting route more of the raw material is presented to the deboner/mincer with higher yields/lower quality through contamination. The route chosen depends on the raw material, the amount of wash water available and the final products. For sustainability there may be a preference for one or other route which must be decided by an LCA with comparisons of the energy and water usage.

Separators operate on the principle of an adjustable belt and perforated drum, rotating in opposite directions, between which the fish are squeezed, determining the yield/quality of the product. Energy usage by the separator will be determined by the speeds of rotation of the belt and the drum, the number and size of perforations on the drum and the tension applied to the belt. Energy-saving operation of the plant might lead to a poorer quality product that is not acceptable.

The heart of the surimi process is the washing cycle(s), and energy needs here are for stirring and chilling the water. The number of extraction cycles, contact time and water : mince ratio will affect energy usage and are normally optimized for protein extraction. The water temperature (5–10 °C to prevent protein denaturation) is species dependent and the proteins of tropical species may be stable at higher temperatures, an advantage in energy saving. Rinsing and refining rely on rotating drums to separate water from mince and an extra step using screw presses may be included. These steps can be removed by careful production and handling of the mince in the earlier steps. The energy demands in these latter unit operations relate to moving equipment (rotating or pressing) and if they are obviated by careful handling of the mince in earlier steps, energy saving and product quality go hand in hand.

In assessing the LCA of the surimi process, a decision must be made about which unit operations to include. Concentrating on the central process and omitting the pre-processing and freezing operations would give a smaller footprint but ignores the important role played by these processes and eliminates the opportunity to assess energy and water savings. This would be intrinsic to a supply chain approach linking all aspects of the product LCA and is particularly relevant for surimi destined for further processing into traditional products by steaming, frying and broiling and the full LCA should include these high energy processes.

Water consumption is the most conspicuous operation of the surimi process, associated with the pre-processing steps and wash and rinse cycles. The large volumes of dilute wastewater are a pollutant and inconvenient to handle. Sustainable surimi processing is based on recycling water where appropriate and the recovery of soluble/suspended components. Lin and Park (1997) approached the water usage problem by investigating the water : mince ratio combined with more washing cycles and washing contact time. In pre-processing, the wastes are easily separated from water by strainers and if collected and transported by dry methods, the water usage can be reduced. Morrissey *et al.* (2005) indicate that about one-third of the surimi process wastewater is generated in the pre-processing stage, although some is recyclable. The wash and rinse cycles generate about 25% of the wastewater and recycling can recover proteins, increasing the yield. Cleaning and sanitation account for about 44% of the wastewater and some can be recycled and good management practices reduce volumes. Soluble protein recovery is possible by flocculation, air flotation and membrane ultra-filtration. Bourtoom *et al.* (2009) reported on the recovery of proteins from surimi washwater using a pH shift and organic solvent.

Maximum protein precipitation was at pH 3.5 and increasing the temperature and the concentration of solvent affected protein recovery by increasing precipitation.

Recovery of soluble and suspended solids suggests the development of by-products (Hultin *et al.*, 2005) benefiting the process economically and environmentally. Traditional routes for the solid wastes are for fishmeal and fish protein hydrolysates. The recovery of lipids might be beneficial if oily species are used. Such species are not popular at present because of the high lipid and dark muscle content but the application of the surimi process to a wide range of species has been proposed (Guennegues and Morrissey, 2005). Conditions applied in the surimi process, chilled water and/or dilute salt solution, are the same preliminary treatments for the isolation of proteins. The same regard for the protection of surimi proteins is paramount in the isolation of bio-active proteins which suggests that by-product development from the surimi material is a distinct advantage.

12.3.7 Fish meal and fish oil

Fishmeal (and fish oil) production is a major process based around specific fishery and fish processing by-products and exemplifies several sustainability issues which will be discussed here in detail.

The standard wet-pressing method takes the components of the raw material (water, oil and fat-free solids) and separates them into fish meal and oil. The size of plant will depend on the volume of fish available, their processing characteristics and the fat content for oil recovery (not for fish with less than 3% (w/w) fat). The process operations use conventional process engineering equipment to achieve separation by heat, physical separation and concentration. A high degree of automation, computer-control and real-time adjustment of the process for optimal production and energy efficiency are possible. Figure 12.6 gives an outline of the fish meal process operations and their characteristics are described in Table 12.4. There are excellent descriptions of fish meal and oil production in the literature (Windsor and Barlow, 1981; FAO, 1986; Fosbol, 1988).

Raw material unloading is the 'pre-processing' stage and commonly carried out by pumps and gravity flumes with seawater. This water (blood water) is removed and treated before discharge. To preserve freshness, the fish can be stored under refrigeration on board ship and in ice water/refrigerated seawater on land. A feed hopper buffers the supply to the cooker and releases raw material at a controlled rate by conveyor or pump, so this stage is the first at which supply can be matched to process capacity.

The cooker conditions the raw material for the subsequent separation processes to release oil and water easily from the solids. The challenge is to obtain even heating throughout the mass of fish by heat transfer and temperature control and the cooker design is an exemplar for other operations seeking a high throughput with good process characteristics. Direct steam injection into the mass of fish is possible although not good practice as it must be removed later. The cooker is engineered to give a high surface area for heat transfer and the surfaces are coated to prevent a build-up of heat-denatured protein, which acts as a heat transfer barrier. Moderate heating regimes and good cleaning practices reinforce these design features. The mass of material in the cooker can be controlled by screw speed, levelling devices and rate of removal. Pre-straining is a simple means of separating the oil and water from the solids before pressing.

The press squeezes liquid (press liquor) from the solids (presscake), moving the material along a tapering shaft by flights where increased pressure squeezes out liquid which drains through a tight-fitting perforated sleeve. Combinations of shaft profile, flight number and pitch giving a range of compression ratios suiting different raw materials and an elevated

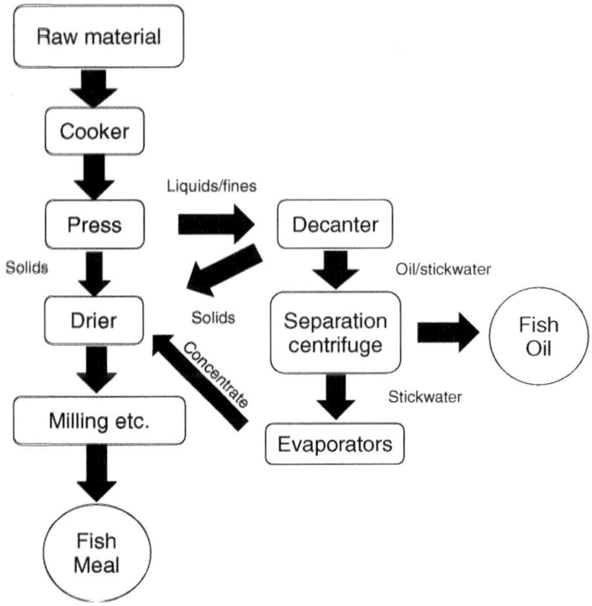

Figure 12.6 The fish meal and oil process

temperature reduces oil viscosity and aids separation. The press can be replaced by a decanting centrifuge which is a better known, quicker and simpler process operating at lower temperatures, but the solids will have a higher water content.

The decanter (continuous decanting centrifuge) separates the press liquor from the remaining solids by specific gravity differences helped by heating to about 90–95 °C. The solids (known as grax) are sent to the drier and the liquor for separation into fish oil and stickwater.

Separators and purifiers separate fish oil and stickwater. The stickwater (6–9% solids w/v) is concentrated in the evaporators and the sludge is sent to the presscake drier. The fish oil is polished by disc centrifuges operating on smaller volumes with hot water to remove impurities.

Evaporators take in the low solids stickwater, but this contains 65% (w/w) of the raw material, so recovery is essential. The large volume of water to be removed is an obvious energy burden and attention to energy savings here has been high. Heat recovery (from all sources in the plant) is used such that the heat efficiency of evaporators increases as the number of stages increases. As a rule, double effect evaporators are adequate for 30–150 t, triple effect for 200–400 t and quadruple effect for more than 500 t per 24 h. Concentrated stickwater is fed to the presscake drier at a solids concentration of 30–50% (w/v).

The drier takes the mixture of presscake, stickwater and grax and transforms them into a stable product with a moisture content of $< 12\%$ (w/w). The presscake particle size is reduced in a wet hammer mill and stickwater must be added at an appropriate point to avoid clumping. Direct fired systems give very good heat and mass transfer but may cause damage to proteins and contamination of the product by incomplete combustion of the fuel, which should be low in sulfur and nitrogen. Inlet temperatures of 500–600 °C are

Table 12.4 The fishmeal process (cleaner production hot spots **in bold**)

Process operation	Purpose	Characteristics
Unloading	Ship to processing plant	Water and pollution
Cooker	Conditioning for separation, kill microorganisms (hygiene)	90–100 °C for 20 min, indirectly steam heated screw conveyor Pre-straining possible 16–600 t/day **Energy burden**
Pressing	Separation of liquid (press liquor 70%) and solids (presscake 30%)	Single/twin screw systems. Centrifugation a possible alternative **Energy**
Decanter	Separation of remaining solids (grax) from press liquor	95–95 °C, horizontal decanting centrifuge **Energy**
Separators and purifiers	Separation of fish oil and aqueous fraction (stickwater), Polishing of the fish oil	Vertical disc centrifuge in series. 500–25 000 l/h **Energy, hot water, pollution**
Evaporators	Concentration and recovery of low solids (6%) from stickwater (20% of total yield)	Heat efficient multiple-effect evaporators (depending on throughput) reduce heat damage to proteins and vitamins. **High energy burden**
Drying	Produces a stable product from presscake, stickwater and decanter sludge (grax)	Temperatures <90 °C to prevent heat damage Direct and Indirect heating rotary driers **High energy burden**
Post-production	Cooling, milling, packaging and 'curing'	Fishmeal is susceptible to oxidation and combustion – curing over 3–4 weeks reduces the hazard Antioxidants added here **Energy**

acceptable, although the recovery of flue gas heat and air pollution abatement is difficult. Indirect steam driers move the material along a rotatory apparatus which is heated indirectly by steam with counter-current air flow. This system is less heat efficient but gives less product damage and air pollution and is used for the highest quality fish meals. Indirect hot air driers can use waste heat from the plant with no flue gas contamination.

Post-production operations after cooling to room temperature are milling to a homogenous, particular particle size and packaging.

Sustainability issues The production of fish meal and fish oil is a complicated process with many variables and various product formats and by-products. Much knowledge is derived from trial and error and an understanding of the raw material and process equipment. There is a drive for high throughputs allied to product quality, energy and water usage efficiency and pollution control – all are sustainability issues.

Energy There is a high demand for heat as most of the separation operations benefit from the material being hot. Heat is also necessary for final drying. Much design and processing effort has gone into minimizing the energy requirement in the process through high heat transfer characteristics, intimate contact between the material and heated surfaces and through the recycling of heat around the plant. Fuel oil consumption is lower if indirect heating is used as the fuel is only required to generate steam. Best available technology (BAT) suggested that fuel oil consumption could be reduced from 49 to 35 l per 1000 kg of raw material (UNEP, 2000).

Water As the process is essentially the removal of water, any additions of water are minimized and mostly derive from seawater to off-load the fish. Again, following BAT could reduce the wastewater pollution load from 30 to 22 kg COD per 1000 kg of raw material for off-loading and from 12 to 9 kg COD per 1000 kg of raw material for processing purposes (UNEP, 2000). Other uses for water are:

- as feed to boilers (fresh)
- for sludge separators (fresh)
- for cleaning equipment (fresh)
- to clean floors and fish silos (seawater)
- for condensers in concentration (seawater)
- for cooling and deodorization (seawater).

Regardless of the size of the fish meal plant, the usage of water is greatest for cooling/deodorization, then for condensation/concentration and then the other uses.

Effluents Effluents to the atmosphere are a feature of fish meal processing, principally the water vapour released in the evaporators. These vapours carry offensive odours so their removal has become standard practice. Vapors are best prevented by using fresh fish, centralizing exhausts for ease of treatment, using indirect rather than direct drying to reduce aerosol formation in the flue gases, location of the plant, and also a knowledge of prevailing weather conditions. The treatments for flue gas odour abatement include scrubbing, high temperature combustion, chemical inactivation, catalytic combustion and adsorption onto active carbon (FAO, 1986).

By-products The fish meal process uses fish deemed unsuitable for DHC and reduces them *in toto* so that by-products do not arise. Separation of fish oil is a direct aim of the process and this oil is used in aquaculture feeds and also for DHC via margarine manufacture or health products and these could be considered by-products.

12.3.8 Fermented products

Although the production of traditional fermented fish products is a long-standing industry across the world, their role in sustainable fish processing is not developed. This is not due to a lack of interest in the processes from applied microbiologists and biochemists. The intense flavours and aromas associated with the products are not appreciated in Western society and are being superseded in their place of origin by Western-style food. A future sustainable diet might demand less heat-treated products, with stronger flavours as a result, and fermented fish products would fit these circumstances. There is a dichotomy between the need for sustainable processing (less energy intense) and a healthy diet (e.g. low salt) and moving from a meat-based diet to one based on fish protein and vegetable sources. Sustainability concepts here go beyond an analysis of the processing methods to those reflecting our diet and food intake on a global scale. Fermented fish products occur across the world, reflecting the specific combination of raw material, ambient conditions, microbial flora, salt content and dietary traditions which have evolved in each region. There is no doubt that fermented products can contribute to sustainable fish processing but this will depend on a massive change in our perception of these products.

12.4 Production efficiency

12.4.1 Introduction

Mention has been made above of the main sustainability criteria for the major fish processing operations and the tools which can be used to assess their status and impact. The importance of the TBL and its application through CSR has also been emphasized. However, for sustainability to be given due prominence, management practices must be employed with fervour. Some approaches to deliver sustainability are given here.

12.4.2 Cleaner production

Sustainability can be encapsulated in the concept of Cleaner Production which aims to reduce waste, generate new products and reduce energy and water consumption in fisheries (Ayer *et al.*, 2009). These authors recognized the importance of sustainability for renewable resources like fisheries which are limited in supply and overexploited.

Cleaner Production (CP) can be defined as 'the continuous application of an integrated, preventive, environmental strategy applied to processes, products and services to increase overall efficiency and reduce risks to humans and the environment' (UNEP, 2000).

The application of CP can significantly affect environmental measures through the supply chain approach and follows a series of phases like LCA such that information gleaned from one could be used for the other. Application of CP to fish processing has been described for three different operations: filleting white fish, filleting oily fish and canning, with measurement of the inputs and outputs for each process (UNEP, 2000) considered indicative of average technology and of possible savings through the application of BAT. Obvious differences are in the wastewater pollution characteristics of each process. Canning produces a greater volume of wastewater carrying more polluting capability, while for filleting oily fish the processing generates more pollution than for white fish. The application of BAT has the potential to bring all three processes into line. The energy inputs for canning are also considerably higher than for filleting, although filleting processes both included energy inputs for ice and for freezing in the central production.

Thrane *et al.* (2009) describe the application of CP in Danish fish processing since the 1980s, using pickled herring and canned mackerel production as examples, and makes the important point that fish processing is not an exact term but encompasses a number of operations depending on the fishery. There is a difference between processing demersal fish (cod), pelagic fish (herring) and shellfish. Processing demersal fish is simpler because they are gutted at sea and presented as simple products whereas pelagic fish are converted into a variety of products after gutting onshore. Gutting at sea would remove this sustainability aspect from the LCA process if the system boundaries are defined as, 'gate-to-finished product', rather than, 'cradle-to-grave', although a supply chain approach would pick up this aspect. Table 12.4 gives a breakdown of the fish meal process and highlights 'hot spots' where CP can be applied. Overall CP opportunities in the fish meal process have been directed at energy and water savings and emissions abatement (Roeckel *et al.*, 1996; Marti *et al.*, 1994).

12.4.3 Management approaches

The barriers to the adoption of sustainable practices by industry are a combination of financial, technical and management issues. Financial considerations include the high initial capital cost of new technology, which should be offset by lower operating costs (e.g. cheaper energy) and the balance between these two elements. The application of sustainability principles will involve a minutely detailed investigation and understanding of the process operation and the full Supply Chain. These are essentially accounting issues and may require a new management vision if the new technologies are to be given a fair chance. Application of a full LCA and CF for a process which account for the full cost of carbon (and emissions) might be such a vision. Although sustainable technologies are well developed and proven, there is still uncertainty attached to them in some quarters, which could be ameliorated by the introduction of validation and certification schemes and the development of a service and maintenance sector in support of the technology. Finally, there is a need for a new generation of sustainability sector professionals to manage the technology, the regulatory issues such as planning and the contractual relationships.

In addition, the take-up of any new technology is also affected by political, macro-economic and social factors which can act as stimulants or barriers, for example Renewable Energy Certificates (ROCs) are awarded in the EU to renewable energy sold to the central electricity network, which works as a financial incentive. Conversely, low landfill taxes will encourage that practice by being cheaper than installing technology to use the waste. The EU approach is for a steadily increasing landfill tax which will one day make the alternatives attractive.

12.5 On-board processing

12.5.1 Introduction

Most processing applications operated on board are quite similar to those at land-based plants. Designing fish processing plants that will allow sustainable operation and limit their environmental effects are described in earlier sections for land-based fish processing plants. In this section, differences will be highlighted for on-board processing for the same aspects.

12.5.2 Advantages and disadvantages

On-board processing covers mainly freezing, pre-processing applications such as heading and gutting, surimi processing, egg processing, and fish meal and oil production. Köse (2010) has described both advantages and disadvantages of on-board fish processing for a sustainable approach. She outlined three types of on-board processing at sea:

(i) freezing the catch to hold before delivery to larger onshore plant for further processing such as canning or smoking; heading, gutting and filleting are done at sea;
(ii) carrying out full processing of their catch such as surimi production;
(iii) carrying out either pre-processing such as freezing, heading and gutting or full processing of the catch from other fishing boats – acting as a mother ship that collects and process fisheries products.

Three different types of vessel size are categorized for fish processing applications at sea (Johnson and Byers, 2003; Köse, 2010):

- fishing (catcher)
- processing-only (such as mother ships)
- catcher-processors.

Designing fish processing plants for sustainability depends on the type and the size of fish as well as the processing methods applied.

The sustainable issues will cover efficient energy usage, elimination of waste caused by poor product quality, and the recovery of fish by-products and liquid waste from the processing operations.

Limited space Different sizes of processing ships are reported such as 150-foot (45.7 m) LOA (Length Overall) to 636-foot (193.89 m) LOA for mother ships, and 16-foot (4.9 m) to approximately 160-foot (48.8 m) LOA for catcher vessels (Köse, 2010). The size of the ship will affect the number of required personnel or workers at the plant. Although a large ship allows a sufficient number of workers to process fish efficiently during high fishing season, the energy consumption will also be higher, particularly for catcher-processors. Moreover, the fishing season is not uniform during the operating of on-board plants, sometimes working 24 h for 1–3 weeks while at other times not having enough raw material for days or weeks. Plant owners usually sort out this situation by paying the hired workers on working hours excluding living cost.

Space in the processing plant limits the number of workers due to limited living space and operating space for both workers and processing equipment. This situation will not only cause low production yield but also longer time to finish processing of the overloaded raw material during high fishing season, resulting in quality loss. According to environmental regulations set by various organisations or countries, large fish cannot be dumped into the sea without processing into small parts. In such a situation, spoiled poor quality products are processed to be thrown into the sea leading to both environmental pollution plus economical loss. Good management in the processing plant can limit the economic loss as explained in detail in our earlier publication (Köse, 2010).

Management One of the main advantages of on-board processing is that the freshest raw material is obtained prior to processing ensuring superior product quality if managed efficiently. The benefits include the higher economic value of the end products and less environmental effects due to low discharge caused by the spoilt fish. If the product is spoilt because of management failure in raw material handling and time/temperature changes during receiving and processing, the opposite situation will occur as for space limitation.

The management of the amount and the quality of raw material during the receiving stage to suit the holding and processing capacity of the plant is important for sustainable use of the fisheries stock and for decreasing environmental effects. Additionally, efficient monitoring of the transfer of the correct amount of raw material to the processing line, hygienic aspects, and other quality and safety issues will help to achieve good quality and quantity end products with less environmental impact. As mentioned earlier, one of the main reasons for choosing on-board processing is to obtain better quality products. Good to premium quality is achieved by providing raw material for processing in the shortest time possible. This aspect is directly affected by raw material management before uploading fish to processing plants and by the speed of the processing lines.

12.5.3 Sustainability aspects

On-board processing requires additional energy use for certain operations in comparison to land-based plants, depending on the processing and type of processing ship. These differences are:

- Oil is used for the catcher-processers to obtain their raw material. In these situations, the oil cost can be covered from cheapest and freshest raw material. Some processors supply their raw material through tenders which deliver caught fish from the catcher boats. These processors are mother ships or barges. In this case, fresh raw material is not always obtained as with onshore plants. These tenders also bring fuel, drinking water and other suppliers to the fishermen (Johnson and Byers, 2003).
- Fresh water is not available to on-board plants and therefore it either must be supplied from land-based sources at high cost or fresh water has to be made from seawater. Fresh water is not only used for living activities on-board but also in high volume for some processing such as the washing steps for surimi production. Many factory ships process different species as well as utilizing their by-products as fish meal and oil, and reportedly use fish oil extracted during processing as the primary fuel to generate steam to make fresh water (over 100 000 UK gallons (~450 000 l) daily) and fish meal (Köse, 2010). Barges are sometimes lucky enough to obtain freshwater from a land source if they are situated not far offshore.

Some regulations are more flexible over certain issues relating to environmental protection. Most onshore-processing discharges have strict regulations but these are less restrictive for open waters (Köse, 2010). Morrissey and Tan (2000) reported that in the United States limited direct discharge of surimi waste water into local rivers and bays is allowable in certain areas with the permission of environmental regulatory authorities. On the other hand, surimi wastewater can be released into the open ocean. Appropriate regulations can be obtained from the environment protection departments of each government.

As mentioned earlier, on-board processors usually lack fresh water and therefore, it must be either be supplied from land sources or their own made at a high energy cost. On the other hand, on-board processing activities have the advantage of having a free seawater supply and the only cost is for pumping and chlorination. Such advantages can vary depending on the type of processing. For example, seawater can easily be used for pre-processing activities such as heading, gutting and washing fish, while fresh water is used in surimi production. Moreover, the type of equipment can have a great effect on water consumption found from the equipment specification. Water supplies can also be affected by the number of personnel in the plant. The on-board water consumption in terms of cost or environmental impact can be estimated separately for each process and type of ship by taking account of all variables that affect the water and energy consumption supply.

Park and Lin (2005) have pointed out an advantage of using fresh water generated by steam distillation. They suggested that using soft water with minimum levels of minerals such as Ca^{2+}, Mg^{2+}, Fe^{2+}, Mn^{2+} for washing during processing of surimi mince is better than usung hard water that will cause deterioration and changes of texture, and affect colour quality during frozen storage. Therefore, on-board processors such as factory trawlers that generate washwater have an advantage in getting good quality surimi with the additional advantage of fresher fish (Köse, 2010). Using recovered fish oil as an energy source to generate fresh water at sea might be a possible sustainable solution to reduce costs. However, an oily fish waste is necessary for such an alternative (Köse, 2010).

12.5.4 Plant design

Designing the handling, storage and processing facilities on board has various challenges in comparison to onshore plants. These mainly arise from (i) space limitations, (ii) safety issues including balance problems, (iii) fresh water consumption and (iv) living activities.

Plant design for different types of processing activities suitable for varying ship sizes may vary although there are necessary similarities between them. The most common processing activity at sea is freezing. Therefore, previous literature on this aspect has usually concentrated on this processing activity, and the different layouts for freezing whole or filleted fish in such factory ships have been described (McDonald, 1982; Kolbe and Kramer, 2007; Köse, 2010). Freezers and chilled rooms are usually designed according to ship size and also the type of processors, that is whether it is a mother ship, catcher-processor or a barge. Different fish species can also be important in selecting the layout of equipment on the factory ships. McDonald (1982) indicated that since small pelagic and some white fish are not normally gutted before they are frozen, they can be moved straight to the freezers once they have been sorted for size and washed. It was also pointed out that conveyors for transporting white fish are not usually suitable for small sized fish due to badly fitting gates which allow small fish through the gaps. Therefore, all the conveyors should be inspected and if necessary modified during handling of small white fish (McDonald, 1982). The type of freezers found on small boats or processing ships are air blast freezers, contact freezers and sodium chloride brine freezers. As with onshore freezing and chilling, sustainability is concerned with reducing energy usage, reducing heat loss and correct sizing of equipment (Köse, 2010).

Kolbe and Kramer (2007) discussed various aspects important to consider for freezing before designing plant layout for at sea plants. Some of these are:

(i) selecting the type of freezing technique suitable for the various species and different seasons;
(ii) the expected shelf life of the subsequently stored frozen product;
(iii) the form of the frozen product acceptable to the potential buyer;
(iv) the additional on-board handling (labour) necessary to prepare, freeze and store the product;
(v) the daily catch rate, which affects the size of the freezing system;
(vi) easy maintenance;
(vii) the related costs such as buying, installation and running costs.

Space limitations on small ships create some difficulties such as fish holds or recovery of processing waste and Kolbe and Kramer (2007) highlighted several recommendations for freezing applications. Limited space issues are particularly important during the high processing season due to handling and processing of a high volume which can also easily lead to human safety problems, loss of product quality and may create food safety issues. Good management applications can overcome some of the problems described in an earlier publication by Köse (2010). A mobile small-scale fish meal and oil plant for recovering process waste such as head, guts and filleting waste has been designed by Nicklason *et al.* (2010) for various fish species (salmon, herring, sardine, whiting, pollock, squid, flatfish). This 'Montlake Process' can be used to recover waste during on-board processing for sustainable fish processing. The processing system also allows stabilization of waste material up to 20 days using 0.8% formic acid before processing during the low fishing season as shown in Figure 12.7.

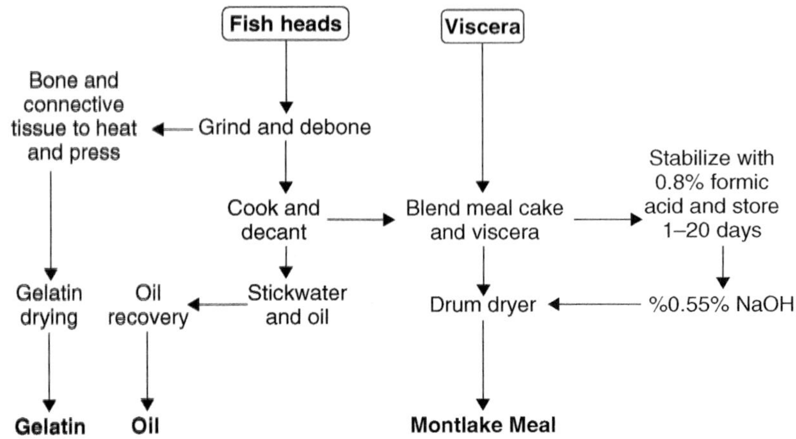

Figure 12.7 Flow diagram for salmon cannery waste processing using Montlake process (*Source*: Niklason *et al.* (2010). Reproduced with permission of University of Alaska by Sevim Köse.)

Safety regulations and requirements for ship balance, particularly for catcher-processors, may limit efficient plant layout. Ship balance must be stabilized by cooperative work between the Chief Engineer, Captain and the plant manager in terms of loading and unloading and heavy product movement in the plant such as freezers and storage rooms. Ignoring a minor imbalance is dangerous and limits future processing, loading and leads to economic and quality loss. There may be rules of either the International Maritime Organisation (IMO) or country of origin for such balance and safety issues (Köse, 2010). Due to limited space and balance problems, the mother ships often store end products on separate barges attached to them until they are collected. However, this is not possible for catcher-processors since they move regularly, therefore, they deliver their products to the ports with the additional fuel cost. Other safety issues such as electrical equipment in wet areas and fire risks are described by McDonald (1982) for freezer trawlers, which are also similar to most on-board processors. McDonald (1982) reported that most of the machinery installed aboard the modern freezer trawlers is reliable and robust enough to withstand seagoing conditions; nevertheless it must be properly maintained to avoid unnecessary and costly breakdowns.

Toyoda *et al.* (1992) suggested that the floor plan of the plant must be very carefully planned for maximum throughput, with a continuous processing line and adjustable for varying deliveries. This constraint applies to both floating processors (mother ships) and to catcher-processors, although it is more severe for the latter which cannot be as flexible as onshore plant with regard to finished product form. Furthermore, the initial capital outlay and fixed costs for an at-sea processor are much higher than for a comparable shore plant. However, the freshness of the fish used in at-sea processing virtually guarantees the superior quality and marketability of products such as surimi.

12.6 Conclusions

The goals of a sustainable fish processing industry can only be achieved through the application of appropriate measures (e.g. Cleaner Production) which can be measured and documented (LCA, Carbon Footprinting), applied across the whole of the production cycle (the Supply Chain) and as a concept must be visible to the consumer (Ecolabelling). The use of LCA will allow sustainability 'hot spots' to be identified and addressed and give transparency to the sector. Energy, waste and water reduction measures not only give savings against the traditional financial bottom line but the TBL with concomitant CSR blessings. These considerations might demand a new management style which recognizes and values the TBL as much as the financial bottom line. The universal Pre-processing Operations generate much waste (or raw material for by-product recovery) and can use lots of water, giving a large volume to treat. The importance of these operations within the overall process should be realized and Pre-processing treated as a 'process' in its own right.

The LCA, and associated tools, should also be applied to traditional fish processes such as smoking, curing and salting and across their Supply Chains as a means to reduce Post-harvest losses and providing better human nutrition. Finally, the advent of Social-Life Cycle Assessment for assessing stakeholder relationships in the fish processing Supply Chain is an exciting new dimension which can have profound impacts in poverty alleviation if allied to a true CSR ethos by all parties.

References

Abbot, J.I.O. and Homewood, K. (1999). A history of change: causes of miombo woodland decline in a protected area in Malawi. *Journal of Applied Ecology*, 36(3), 422–433.

Akande, G.R., Ajayi, A.A., Ogunweno, C. and Ash, M.T. (2005). Comparative proximate composition, physical and sensory evaluation of fish smoked with sawdust and firewood using improved Chorkor oven. In: *Issue 712 of Report and papers presented at the Seventh FAO Expert Consultation on Fish Technology in Africa, Saly-Mbour, Republic of Senegal. 10–13 December, 2001*, pp. 73–76.

Ayer, N., Cote, R.P., Tyedmers, P.H. and Martin Willson, J.H. (2009). Sustainability of seafood production and consumption: an introduction to this special issue. *Journal of Cleaner Production*, 17(3), 321–324.

Barbier, E.B., Acreman, M. and Knowler, D. (1997). *Economic Valuation of Wetlands: A Guide for Policy Makers and Planners*. RAMSAR Convention Bureau, Gland, Switzerland.

Benoit, C., Norris, G.A., Valvidia, S. *et al.* (2010). The guidelines for social life cycle assessment of products: just in time! *International Journal of Life Cycle Assessment*, 15, 156–163.

Bourtoom, T., Chinnan, M.S., Jantawat, P. and Sanguandeekul, R. (2009). Recovery and characterisation of proteins precipitated from surimi wash-water. *Food Science and Technology*, 42(2), 599–605.

Cappell, R., Wright, S. and Nimmo, F. (2007). Sustainable Production and Consumption of Fish and Shellfish: Environmental Impact Analysis. Final Report, Defra Project No 9S6182, Edinburgh.

Carbon Trust (2006). Food and Drink Processing – Introducing Energy Saving Opportunities to Business. CTV004 Sector Overview, HMSO, London.

Cecelski, E. (2000). The Role of Women in Sustainable Energy Development. NREL/SR-550–26889, National Renewable Energy Laboratory, USA.

Ellingsen, H. and Aanondsen, S.A. (2006). Environmental impacts of wild caught cod and farmed salmon. *International Journal of LCA*, 1, 60–65.

Fahim, F.A., Fleita, D.H., Ibrahim, A.M. and El-Dars, F.M.S. (2001) Evaluation of some methods for fish canning wastewater treatment. *Water, Air and Soil Pollution*, 127, 205–226.

FAO (1986). *The Production of Fish Meal and Oil*. FAO Fisheries Technical Paper T142, Food and Agriculture Organization of the United Nations, Rome, Italy.

FAO (2005). *Guidelines for the eco-labelling of fish and fishery products from marine capture fisheries*, United Nations Food and Agriculture Organisation, Rome, Italy.

FAO (2009). *The State of the World's Fisheries and Aquaculture, 2008*, Food and Agriculture Organization of the United Nations, Rome, Italy.

Feka, N.Z. and Manzano, M.G. (2008). The implications of wood exploitation for fish smoking on mangrove ecosystem conservation in the South West Province, Cameroon. *Journal of Tropical Conservation Science*, 1(3), 222–241.

Fernandes, Y., Salvador, N. and O'Keefe, P. (1995). Biomass energy efficiency in informal industry. *Energy for Sustainable Development*, 1(5), 48–50.

Finkbeiner, M., Inaba, A., Tan, R.B.H., Christianson, K. and Kluppel, H.J. (2006). The new international standards for life cycle assessment: ISO 14040 and ISO 14044. *International Journal of Life Cycle Assessment*, 11, 2, 80–85.

Fosbol, P. (1988). Production of fish meal and fish oil. *INFOFISH International*, 5/88, 28–32.

Gameiro, W. (2002). Energy costs are changing refrigeration design. Technical Paper No 6, Proceedings of the 2002 IIAR Ammonia Refrigeration Conference, Kansas City, Missouri, pp. 207–233.

Graedel, T.E. (1998). *Streamlined Life-Cycle Assessment*. Prentice-Hall.

Guennegues, P. and Morrissey, M.T. (2005). Surimi resources, in *Surimi and Surimi Seafood*, 2nd edn (ed. J.W. Park), CRC Press, Boca Raton, Florida, pp. 3–32.

Hall, G.M. (2010). Sustainability impacts of fish-processing operations, in *Fish Processing Sustainability and New Opportunities* (ed. G.M. Hall), John Wiley & Sons, Ltd, Chichester, pp. 112–137. ISBN 978-1-4051-9047-3.

Hultin, H.O., Kristinsson, H.G., Lanier, T.C. and Park, J.W. (2005). Process for recovery of functional protein by pH shifts, in *Surimi and Surimi Seafood* (ed. J.W. Park), 2nd edn. CRC Press, Boca Raton, Florida, pp. 107–140.

ICTSD (2006). *Fisheries, International Trade and Sustainable Development: Policy Discussion Paper*. ICTSD Natural Resources, International Trade and Sustainable Development Series, International Centre for Trade and Sustainable Development, Geneva.

Iles, A. (2007). Making the seafood industry more sustainable: creating production chain transparency and accountability. *Journal of Cleaner Production*, 15, 577–589.

ISO (2006). *Environmental Management – Life Cycle Assessment – Principles and Framework, ISO 14040:2006 (E) Series*, International Standards organisation, Geneva.

Johnson, T. and Byers, K. (2003). *Ocean Treasure. Commercial Fishing in Alaska*. Sea Grant, Alaska, USA. 187pp.

Karekezi, S. (2002). Renewables in Africa – meeting the energy needs of the poor. *Energy Policy*, 30, 1059–1069.

Kolbe, E., Ling, Q. and Wheeler, G. (2004). Conserving energy in blast freezers using variable frequency drives. Proceedings of the 26th National Industrial Energy Technology Conference. Houston, pp. 47–55.

Kolbe, E., Kramer, D. and Junker, J. (2006). The process of quality loss, in *Planning Seafood Cold Storage*, 3rd edn, Alaska Sea Grant, University of Alaska, Fairbanks, p. 47.

Kolbe, E. and Kramer, D. (2007). *Planning for Seafood Freezing*. Alaska Sea Grant, University of Alaska, Fairbanks.

Koatji, S., Schuurmans, A. and Edwards, S. (2003). Life-Cycle Assessment in Building and Construction: A State-of-the-art Report of SETAC Europe, SETAC Foundation.

Köse, S. (2010). Onboard fish processing, in *Fish Processing Sustainability and New Opportunities* (ed. G.M. Hall, John Wiley & Sons, Ltd, Chichester, pp. 167–206. ISBN 9781-4051-9047-3.

LCAFood (2009). *Wild fish (Vilde fisk)*. http://www.lcafood.dk/products/fish/wildfish.htm [accessed 3 June 2013].

Lin, T.M. and Park, J.W. (1997). Effective washing conditions reduce water usage for surimi processing. *Journal of Aquatic Food Product Technology*, 6(2), 65–79.

Marti, M.C., Roeckel, M., Aspe, E. and Novoa, M. (1994). Fat removal from process waters of the fish meal industry. A study of three flotation methods. *Environmental Technology*, 15, 29–39.

McDonald, I. (1982). Freezing at sea, In: Aitken A, Mackie, U M, Merritt J H, and Windsor M L, (Eds), *Fish Handling and Processing*, pp. 79–87, 2nd Edition, Ministry of Agriculture, Fisheries and Food, Torry Research Station. Edinburgh, Her Majesty's Stationary Office.

Mogensen, L., Hermansen, J.E., Halberg, N. and Dalgaard, R. (2009). Life Cycle Assessment across the food supply chain. In: Baldwin, C.J. (Ed), *Sustainability in the Food Chain*, pp. 115–144, John Wiley & Sons, Ltd.

Morrissey, M.T., Lin, J. and Ismond, A. (2005). Waste management and by-product utilisation, in *Surimi and Surimi Seafood*, 2nd edn (ed. J.W. Park), CRC Press, Boca Raton, Florida, pp. 279–323.

Morrissey, M.T. and Tan, S-M. (2000). World resources for surimi, in *Surimi and Surimi Seafood* (ed. J.W. Park), Marcel Dekker Inc., CRC Press, New York, Basel, pp. 1–22.

Nicklason, P., Stitzel, P., Barnett, H., Johnson, R. and Rust, M. (2010). Montlake Process for utilization of salmon waste in Alaska, in *A Sustainable Future: Fish Processing By-products* (eds P.J. Bechtel, and S. Smiley), Alaska Sea Grant, University of Alaska, Fairbanks, USA, pp. 221–234. DOI: 10.4027/sffpb.2010.18.

Olokor, J.O. (2003). Cost of fuelwood for fish smoking around Kainji Lake and economic prospects of the kainji solar tent fish dryer. Proceedings of the 18th Annual Conference of the Fisheries Society of Nigeria, 8–12 December, Owerri, Nigeria, pp. 105–109.

Park, J.W. and Lin, T.M.J. (2005). Surimi: manufacturing and evaluation, in *Surimi and Surimi Seafood* (ed. J.W. Park), Marcel Dekker Inc., CRC press, New York, Basel, pp. 33–106.

Pelletier, N., Ayer, N.W., Tyedmers, P.H. *et al.* (2007). Impact categories for life cycle assessment research of seafood production systems: review and prospectus. *International Journal of Life Cycle Assessment*, 12(6), 414–421.

Roeckel, M., Aspe, E. and Marti, M.C. (1996). Achieving clean technology in the fish-meal industry by addition of a new process step. *Journal of Chemical Technology and Biotechnology*, 67, 96–104.

Rowlands, I.H. (1993). The Fourth Meeting of the Parties to the Montreal Protocol: Report and reflection. *Environment*, 35(6), 25–34.

Sartori, I. and Hestnes, A.G., (2007). Energy use in the life cycle of conventional and low-energy buildings: a review article. *Energy and Buildings*, 39, 249–257.

Sawyer, C.N., McCarty, P.L. and Parkin, G.F. (2003), *Chemistry for Environmental Engineering and Science*, 5th edn, McGraw-Hill, New York

Shah, V.P., Debella, D.C. and Ries, R.J. (2007). Life cycle assessment of residential heating and cooling systems in four regions in the United States. *Energy and Buildings*, 40, 503–513.

Thrane, M., Ziegler, F. and Sonesson, U. (2009). Eco-labelling of wild-caught seafood products. *Journal of Cleaner Production*, 17, 416–423.

Toyoda, K., Kimura, I., Fujita, T., Noguchi, F.S. and Lee, C.M. (1992). The surimi manufacturing process, in *Surimi Technology* (eds T.C. Lanier and C.M. Lee), Marcel and Dekker, New York, pp. 79–112.

UNEP (1996). *Life Cycle Assessment: What It Is and How to Do It*. United Nations Environmental Programme (UNEP), New York.

UNEP (2000). *Cleaner Production Assessment in Fish Processing*, United Nations Environmental Programme (UNEP), Division of Technology, Industry and Economics, New York.

UNEP (2007). *Mangroves of Western and Central Africa*, UNEP-regional Seas Programme/UNEP-WCMC.

UNIDO (1986). *Environmental Assessment of the Fish Processing Industry*. United Nations Industrial Development Organisation (UNIDO) Sector Studies Series 28.

Wiedmann, T. and Minx, J. (2007). *ISA^{UK} Research Report 07-01, A Definition of 'Carbon Footprint'*, Integrated Sustainability Analysis, UK.

Windsor, M. and Barlow, S. (1981). *Introduction to Fishery By-products*, Fishing News Books Ltd, Farnham, UK.

World Bank (2009). *World Development Report 2010 – Development and Climate Change*, The International Bank for Reconstruction and Development/The World Bank, Washington, DC, pp. 27–30.

World Commission on Environment and Development (1987). *Our Common Future*, Oxford University Press, p. 43.

Xiao, F. and Wang, S.W. (2009). Progress and methodologies of life cycle commissioning of HVAC systems to enhance building sustainability. *Renewable and Sustainable Energy Reviews*, 13(5), 1144–1149.

Ziegler, F. (2003). Environmental impact assessment of seafood products, in (eds B. Mattsson and U. Sonesson), *Environmentally Friendly Food Processing*, Woodhead Publishing, Cambridge, pp. 70–92.

Ziegler, F., Eichelsheim, J.L., Emanuelsson, A. *et al.* (2009). Life Cycle Assessment of Southern pink shrimp products from Senegal: An environmental comparison between artisanal fisheries in the Casamance region and a trawl base in Dakar. FAO Fisheries and Aquaculture Circular No 1044, Food and Agriculture Organization of the United Nations, Rome, pp. 29.

13

Value-added Seafood

Michael Morrissey[1] and Christina DeWitt[2]

[1]Oregon State University Food Innovation Center, Portland, OR, USA
[2]Oregon State University Seafood Research and Education Center, Astoria, OR, USA

13.1 Introduction

Value-added seafood products have been important to domestic and international trade for the past half a century. While fresh, chilled and live fish and shellfish are still major players in the marketplace representing 60.2 million metric tons (MMT) or 40.5% of the world fish production and trade, processed products such as frozen, cured, canned, or other forms now represent the majority of total production at 68.1 MMT (45.9%) (Food and Agriculture Organization (FAO), 2012). The utilization of fish for human consumption has steadily increased along with overall production and there has also been a gradual shift toward increased use of small pelagic fish for direct human consumption rather than fishmeal production. There are several reasons for this, including continued demand for seafood in developing countries, recognition of the high nutritive value of seafood and technological advances in freezing, packaging and distribution that allow these products to retain high quality, have higher economic value and an extended shelf life (FAO, 2012).

While capture fisheries continue to play an important role in value-added seafood products, their resources are limited and aquaculture will be a major driver in value-added processing in the future. By 2010, aquaculture of finfish and shellfish was approximately 38% of the total fishery production and 45% of all the seafood directed to human consumption. Aquaculture is expected to exceed wild harvests by 2020 (Delgado *et al.*, 2003; FAO, 2012). While there are production issues, disease concerns and genetic improvements critical to aquaculture growth, the demand for good quality products less prone to seasonal variations will allow for the growth in value-added products from aquaculture in the future.

There are several advantages of value-added production from all sectors of the seafood industry. Value-added products can:

- provide another level of safety
- increase shelf life
- help maintain a high level of quality

Seafood Processing: Technology, Quality and Safety, First Edition. Edited by Ioannis S. Boziaris.

- open new market opportunities
- offer a solution for over-supply issues
- promote use of new technologies.

Value-added fish and shellfish products usually undergo some level of processing that will inactivate and/or kill bacteria and pathogens. The inactivation or reduction of bacteria in a food generally results in shelf-life extension and can also provide new market opportunities. The deterioration of fish and shellfish quality is attributed to the decomposition of key components of the raw material by endogenous enzymes or microbial action. Heat and cold are the two major food processing treatments used throughout the world and both have a significant effect on inactivating enzymes and destroying microorganisms. However, processing can be a two-edged sword, as food products can undergo alteration (e.g. protein denaturation) that will affect final product quality and acceptability. Freezing is a good example of a technology that has been in existence for more than 100 years but only recently has been specifically used to maintain high quality and preserve the natural attributes of fresh fish and shellfish. Thirty years ago conventional wisdom was to freeze fish when it was nearing the end of its fresh shelf life and to hope that consumers would accept this product at a discounted rate compared to the fresh product. Over the past two decades, however, the seafood industry has been taking a closer look at freezing as a means of maintaining the high quality attributes of seafood. Freezing will also contribute to an increased shelf life and, consequently, boost distribution and market opportunities. Rapid freezing at sea or at the harvest ponds has become analogous to preserving the intrinsic quality of fish/shellfish and provides the fishermen/ harvesters with new opportunities to develop novel, high-end markets. Sloan (2011), in her review of the top ten trends in foods, lists frozen seafood, including seafood entrees, in the top category of consumer packaged goods.

Innovative technologies such as high-pressure processing, e-beam radiation, radio frequency heating and others also present an additional layer of safety for numerous seafood products. Interestingly, these technologies, which originated as a means to improve safety, have initiated new product development and created a unique suite of products not found in other protein commodities.

The general premise of value-added has changed over the past 10 years. While most of us are familiar with traditional processing of seafood (canned, smoked, cured), there are other concepts that need to be addressed when discussing valued-added seafood products. It is important to understand the driving forces for new products and how new demands can be met with new technologies and new philosophies. The consumer is striving to understand the whole food system rather than just purchasing the end product at retail and that has significantly changed how the industry does business and presents products to the consumer. Therefore, this chapter will discuss several different aspects of value-added seafood products. It will begin with a short overview of some of the value-added approach systems that exist in the food industry that can be also applied to seafood. It will also look at different drivers such as new resources, seafood 'values', nutritional aspects and new technologies. It will end with a general discussion of value-added seafood and possible directions for the future.

13.2 Value-added product development

One of the primary objectives of value-added product development (VAPD) is to create new markets for industries that may have shown stagnant growth in the marketplace

(Fuller, 2011; Cooper, 2001). Current consumer trends may provide opportunities that did not previously exist and allow VAPD to capture new and expanding markets at both domestic and external levels (Cagan and Vogel, 2002). For example, the poultry industry has done extremely well in satisfying consumer demand and it is astounding to visit a retail operation and witness the variety of poultry products (from just a few species) that are offered to the consumer. Although consumer needs continue to evolve and become more sophisticated, there are basic five principles of (i) quality, (ii) safety, (iii) convenience, (iv) taste and (v) affordability for any new product that are often key to a successful launch. While it is often not possible to capture all five attributes in a single value-added product, the most successful product ventures generally provide the majority of these principles. There are several systems that have been developed to assist in VAPD such as Stage-Gate and Culinology; they have had varying degrees of success for seafood businesses (Grobe *et al.*, 2002; Cooper and Edgett, 2005). While the variety of species is considered a benefit from a culinary standpoint, it can be problematic to the industry as a whole when attempting to create a systematic approach to VAPD. For example, the main safety and quality concerns of oyster processing are quite different from those of tilapia processing. Furthermore, the consumer may also have different priorities in these two sectors. While safety may be paramount for the consumer in an oyster product, convenience and cost may be the important purchasing criteria for tilapia. Industry needs must also be considered in VAPD. Understanding supply issues, competition, regulations and industry proficiency is vital to the expansion or development of new markets. These might be reflected in sector concerns or single company needs as new markets are developed. Successful entry into these markets is often determined by the ability of the industry to understand these needs and know how to address them. The example of the pin-bone-out (pbo) salmon fillet from Chile, introduced into the US marketplace in the mid-1990s, is a good example of how quickly a product can be successful, especially when consumer and industry needs converge. Consumer demands have changed over the past decade and the need for assuring the consumer that the fish or shellfish they are eating has come from a sustainable fishery is often as important as the quality of the seafood itself. In addition, the general recognition that seafood is not only a food of sustenance but also one highly valued for its nutritional quality is becoming more important to the consumer as well. One needs to also ask what resources will best meet the demands for increasing populations as well as provide a source of value-added seafood products. Innovative technologies can provide interesting opportunities for high-quality, safe seafood products that are appealing to the consumer. Non-thermal processing technologies are becoming more mainstream and provide wonderful opportunities for seafood products with an additional margin of safety for several items. Finally, new regulations have had a major impact on the importance of VAPD over the past decade. Seafood safety has become a key factor at both the wholesale and retail level as consumers demand safe, wholesome products for themselves and their families. These new technologies have also provided novel opportunities for VAPD, which have been successfully introduced in the marketplace. The following discussion will demonstrate how some of these drivers (market demand, values, nutrition, resources and new technologies) have changed the dynamics of value-added seafood.

13.3 Market-driven

When the needs of both the consumer and the industry converge, we see great opportunities for success in the arena of seafood value-added products. For example, this occurred in the mid-1990s in the farmed salmon industry. During this period, farmed salmon production was increasing by more than 10% per year in both Norway and Chile, while Alaska was experiencing record harvests of wild-caught salmon. The oversupply caused dramatic declines in the wholesale market price of both wild-caught and farmed salmon, and the industries were under considerable pressure to create changes in the traditional ways that they had marketed salmon to the public. A few decades ago, salmon was considered a high-end product more often known to consumers through white-tablecloth restaurants rather than home consumption. Consumers shied away from cooking the product at home, as it was not considered very convenient to cook, there were several small bones that often had to be removed for meal preparation, it would results in strong odours in the kitchen, and of course the consumer often had few ideas how to prepare and cook salmon in their own home. These sentiments changed dramatically when pin-bone-out (pbo) skinless salmon fillets from Chile were introduced into the US marketplace. Seemingly overnight, salmon was easy to prepare and affordable. Furthermore, the fillets were of good quality and, more importantly, each fillet was consistent in taste and texture from month to month. With the boneless salmon fillet, the consumer began to view seafood in a new light. The US consumer also found that salmon suited the favourite American pastime of grilling together with steaks and burgers. The wholesale price of salmon fillet continued to drop from \$4/lb in 1994 to less than \$3/lb in 1997 so that the retail price to the consumer at larger grocery chains was often \$4.99/lb or less (Johnson, 2004). This was a considerable bargain to seafood lovers who five years previously would have struggled to find salmon fillets (with the bones) at <\$10/lb. The workmanship and consistency of the farmed salmon fillets was also far superior to what was usually experienced by the consumer. The mystery of cooking fish seemed to disappear with simple instructions of grilling fish on the barbecue. It was easy to prepare, kept odours outside, impressed the guests and, quite simply, was delicious. It was a culinary hit. The public responded appropriately and the number of salmon consumers in the United States increased rapidly. Salmon consumption doubled from 0.7 lb/yr in 1990 to 1.4 lb/yr by 1997 and continued to rise to 2.2 lb/yr by 2002. Salmon products are now offered in practically all major grocery store outlets and in a range of restaurants throughout the United States. Although the wild-caught salmon industry initially experienced financial issues due to the reduced prices of farmed salmon, they too eventually benefited from this rise in salmon consumption and they also began to recognize the importance of responding to consumer needs. The pbo farmed-salmon fillet in the mid-1990s is one of the true stars of seafood value-added products. Although simple in concept, it demonstrates the power of understanding consumer needs and having an industry respond to those needs.

More recent trends in food consumption patterns include understanding market demands that focus on different demographics and the needs within that age group. This could include the adventurous palates of the 'Gen Y'ers' (age 17–34) who have a more exploratory appetite and are quick to accept new ideas or even new species such as octopus or periwinkles and who keep an open mind about what they eat. Other demographic sectors such as the 'baby boomers' (age 65–84) eat more traditional food, are more health conscious and view seafood products as a means of providing a healthy and varied diet. There are an estimated 100 million Americans over the age of 50 and their sheer numbers will force food companies to develop products specifically for this age group (Food Channel, 2012). This age group also has more disposable income and is willing to spend more on high-end consumer products including seafood. Convenience is always an important driver for new seafood markets and the recent food trend toward small plates, such as crab cakes or sliders, fish tacos and other snacks provide new openings for market driven seafood products.

New market demands have also opened up opportunities for traditional fisheries. One can also argue that the increased trade in live fish and shellfish on a domestic and global basis is also a form of value-added seafood products. The ability to fill this market has come about through technological advances, improved logistics as well as a willingness to pay high prices for the final product. An example of this trend is the increased export of Dungeness crab from the west coast of the United States to China. The value of this product has increased fourfold due to increased demand in the Chinese markets and the ability to meet that demand using technologies to distribute and sell live crab (Tobias, 2011).

13.4 Values-driven

Values-added agriculture and fisheries have become an increasingly important concept for sustainability, marketing and sales over the past decade. 'Values-added' can be defined as foods that are produced and marketed using attributes other than their inherent nutritional or food characteristics. These can include environmental impacts during harvesting, production and processing; social aspects of the workforce; management practices including rules and regulations; or other attributes such as location of origin (Allen, 2006). Sustainability is often a critical factor in the 'values-added' arena and can be an important driver in the marketing of seafood throughout the world. There are several programmes that highlight sustainability including the Marine Stewardship Council (MSC), The United Nations Food and Agriculture Organization (FAO), Responsible Fisheries Management and the Alaska Seafood Marketing Institute (ASMI) certification programme to name a few. These programmes focus on the sustainability of fisheries stocks and have a biological emphasis, as well as stressing low environmental impact. Many private companies have their own sustainability programme that covers more of the value chain from capture to processing and distribution and often have third party certification. The MSC certified fisheries programme (www.msc.org) is one of the oldest sustainability programmes and is well accepted (Constance and Bonnano, 2000). The programme has developed standards for sustainable fishing and seafood traceability and works closely with major fisheries organizations to certify fisheries and their management practices. MSC began in the 1990s through a partnership with the World Wildlife Fund and Unilever Corp. and has grown to over 180 fisheries with another 104 fisheries under assessment to date. The number of fisheries products under this certification programme rose to 2366 by 2009 and represented a growth of 66% year-on-year. Currently, 23% of consumers recognize the MSC label. More importantly, large retail operations require that their fishery resource is sustainable and certified. This remarkable transition to requiring sustainable products is also crucial for the aquaculture sector and there are several programmes underway to certify proper management of aquaculture as well. These include third party certification programmes such as Best Aquaculture Practices (www.aquaculturecertification.org), and the Aquaculture Stewardship Council (www.asc-aqua.org), which is similar to the MSC programme.

While fisheries management issues are the major focus of sustainability, there are other 'values-added' concepts that have recently become important in the marketplace. These concepts often begin in the agricultural sector and over time influence how the consumer views seafood and its markets. One of the more recent concepts and 'values-added' concepts is 'buying local' and direct marketing. The steady increase of farmers markets has created opportunities for direct marketing of fish, as well as an emphasis on buying local seafood for retail, restaurants and specialty stores. Although the seasonality of some fisheries will directly impact availability and supply in the marketplace, seafood has made progress in developing alternative market opportunities. Recent surveys have shown that there is increased consumer interest for buying local to support the regional fishing industry and fishing families. Local buying also provides perceptions of higher quality and 'greener' distribution, as can be seen from various websites and promotional campaigns. Another effort to promote local seafood marketing is through the promotion of Community Supported Fisheries (CSF) operations that are based in successful community supported agriculture (CSA). These programmes support local small farms in agriculture in which community members buy an annual share of the harvest for a set fee and are assured a weekly delivery of crops from local farms. There is a willingness to support these operations for values that go beyond food and for the feeling of a sense of community and caring about regional food systems and their continued success (Brown, 2012). The concept of 'value-chains' that has emerged in agriculture is starting to make inroads in the fisheries/seafood industry as well. This is an effort by small and mid-size industries to de-commodify seafood products based on environmental and social attributes. A value-chain is an alliance of enterprises collaborating vertically to achieve a more rewarding position on the marketplace. This concept is critical to 'values-added' seafood as distribution is often the major bottleneck in the successful marketing of these products.

Seafood was one of the first industry sectors to embrace the 'traceability' concept. It was the first commodity in the United States to be regulated under the Country of Origin Labeling (COOL) Act (www.surefish.com/COOL.htm). Since this requires a degree of traceability, many seafood companies began their own traceability programmes (Thompson *et al.*, 2005). While traceability is looked upon as a regulatory programme to prevent consumer fraud and ensure the wholesomeness of the seafood supply, several companies have added elements of marketing to the traceability component to promote seafood and highlight 'values-added' attributes to their products. An example of this is the Gulf Wild programme that allows the customer to verify fish species, see the harvesting vessel's name and captain's background, know what fish house and city where the seafood was landed, down to a 10-mile square location in the Gulf where the fish was caught (http://mygulfwild.com/track-your-fish). This type of information can be a powerful marketing tool to assure the consumer that their fish or seafood product is harvested in a sustainable manner as well as giving support to local fisheries. There are several other traceability programmes that exist and are being developed. A good overview of seafood traceability is described in Seafood Traceability in Canada (Magera and Beaton, 2009).

13.5 Health-driven

Seafood has been long viewed as a healthy food choice. Initially, it was touted as a low-fat protein source, but not for its health-promoting lipids (Anderson and Wiener, 1995). However, studies with the Greenland Inuit (Bang and Dyerberg, 1980) suggested an association between fish consumption and reductions in death from coronary heart disease (CHD). Kromhout *et al.* (1985) reported a 44% reduction in CHD death for men consuming 1–2 servings (100–200 g) of fish per week compared to men not eating fish. Research over the next few decades supported the association between fish consumption and reductions in CHD incidence and mortality and identified two long-chain omega-3s fatty acids (LC-omega-3) that are primarily found in marine species and have a major role in these beneficial effects: eicosapentaenoic acid (EPA) and docosahexaenoic acid

(DHA) (Simopoulos, 1991; Mozaffarian, 2008). EPA and DHA cannot be synthesized in substantial amounts by the human body and must be obtained through the diet, with the main source being seafood, especially fatty fish. Other studies have shown the reduced risk of other cardiovascular disease outcomes such as sudden death (Mozaffarian and Rimm, 2006) and stroke, increased duration of gestation, and improved visual and cognitive development (FAO/WHO, 2010). Seafood also contains a number of vitamins (for example, A, B-complex and D) and minerals (such as selenium, iodine, iron and zinc) that have been linked to various health benefits (Nesheim and Yaktine, 2007).

Based on the identification and characterization of the health benefits of seafood and EPA/DHA intake, a number of health-based guidance values and nutritional recommendations have been developed by various health organizations. Scientific committees from organizations such as the Institute of Medicine, the FAO and World Health Organization have all affirmed the importance of seafood in the diet. Recently the US Department of Agriculture's Dietary Guidelines Advisory Committee and the US Department of Health recommended consumption of at least two servings per week of a variety of seafood (USDA, 2012). While it is recognized that there are some risks involved with seafood consumption, it is widely accepted that the benefits greatly outweigh these and that the general populace needs to increase its consumption of fish and shellfish (Hellberg *et al.*, 2012).

The seafood industry has recognized the marketing advantage of LC-omega-3 that is uniquely found in seafood. Many of the traditional and new value-added products developed over the past 10 years have labels that highlight the nutritional benefits such as high omega-3s content or that have the heart-healthy symbol connected with the American Heart Association. Consumers in both developed and developing countries are becoming increasingly aware of the role of health and food (Kearney, 2010). Sloan (2011) reports that there is a trend toward health-based foods and that more consumers are searching for natural foods that have a nutritional focus. Seafood and value-added products serve this market well; there will be opportunities to develop new products that focus on the nutritional quality, as well as the eating quality of seafood products. Sometimes LC-omega-3 is added back to the seafood product as in the case of fortified surimi seafood. The company Shining Ocean is currently producing a Crab-Smart Natural product (http://www.kanimi.com/prod_crabsmartnatural.php) in which 400 mg of LC-omega-3 is added to the surimi-based end product. Other traditional surimi companies are making product lines based on the health aspects of seafood. TransOcean produces a surimi product from Alaska pollock (http://www.trans-ocean.com/omega3.html) which is enriched with fish oil to boost the LC-omega-3 content. The main concern has been that these fatty acids are highly unsaturated and that lipid oxidation will occur and off-flavours develop. Research has shown that seafood products can be fortified with fish oils with minimal deterioration of the sensory characteristics of the product (Pérez-Mateos *et al.*, 2004; Pietrowski *et al.*, 2011).

While all seafood has some level of LC-omega-3, there is a wide range of lipid levels depending on the species. Certain species of salmon, such as Chinook, can have over 1000 mg of combined EPA and DHA per g of fish while freshwater species such as tilapia can have less than 100 mg/g. Some of the underutilized species such as small pelagic fish (e.g. sardines, anchovies) also have high amounts of EPA and DHA but are often directed toward non-food uses. Although these species are canned by traditional methods and represent a low-priced and highly nutritious product for the general consumer, new opportunities can be generated for entrepreneurs and chefs to add-value to new dishes and create new products at the high end as well.

13.6 Resource-driven

There is general agreement that capture fisheries are at their limit and there is little opportunity for growth in this sector. Capture fisheries peaked in 1996 at 86.4 MMT and then decreased slightly to stabilize at about 80 MMT. The FAO (2012) has stated that 59% of the capture fisheries are fully exploited, while under or moderately exploited stocks represent 3 and 20% of the fish stocks. Therefore, there is little room for expansion of resources in the wild-capture sector and most of the growth in fisheries and satisfying increased world demand will come from aquaculture. World aquaculture production continues to increase at about 5% per year while wild capture has levelled off or slightly decreased. Marine aquaculture has grown faster (7.2%) than inland aquaculture (2.1%) and should continue to hold this level of growth over the next decade as new technologies and new species facilitate growth in this sector. Consequently, the role of aquaculture in most developing countries often has two purposes: satisfying domestic markets and consumer demands, and a mechanism for foreign investment and economic benefits through international markets and revenue generation (Delgado *et al.*, 2003). The dramatic increase in aquaculture production is a two-edged sword as there is increased concern about environmental damage and other impacts on developing countries where growth is highest. For the United States markets, most aquaculture products are imported with shrimp, salmon and tilapia leading the way, while there is some domestic production of species such as catfish. Note that the production levels of these species are minor compared to the aquaculture of carp and shellfish which are used for domestic consumption in China and Southeast Asia. Although farmed salmon and shrimp production is significantly less than carp, these species are major commodities for international trade. Tilapia production has significantly increased recently and is having an impact on the sales of white fish in the United States and Europe. Tilapia production output doubled from 0.8 to 1.6 MMT in 1999 and increased to 3.5 MMT in 2008, with the most production coming from Asia (Josupeit, 2012). Tilapia now represents the fifth most consumed species of seafood in the United Stated and it is present in many forms, from fillets to value-added products. The inherent characteristics of tilapia, moderately firm, white flesh, mild taste and relatively low cost have made it an ideal choice for value-added products, especially frozen main courses in the United States and other marketplaces. While individually quick-frozen (IQF) fillets are the most popular, new products that combine tilapia with different sauces as the central dish are becoming more common. Tilapia production will continue to grow and could double over the next decade. Other potential resources from aquaculture include the barramundi which is gaining acceptance as a sustainable fishery resource and different forms of catfish and *Pangasius* species. Although marine aquaculture is expanding, it will take several years before production levels have a major impact in the value-added arena. In wild-caught fisheries, small pelagics represent one opportunity for increased utilization as value-added products for human consumption.

13.7 Technology-driven

VAPD can be driven by new technologies that allow for processing breakthroughs in product safety, quality or efficiency of production. Emerging technologies, such as high-pressure processing, radio-frequency heating, flexible retort packaging, etc., have spurred increased research activity and the potential for improving the safety and quality of seafood products. Novel packaging technologies have allowed for innovative developments in VAPD and new market opportunities for many traditional products. One aquaculture sector that has embraced some of these technologies has been the oyster industry, which is surprising as there have been few changes in the processing of oysters over the past century. The implementation of new technologies in oyster processing has focused on a unique combination of safety and quality parameters that have revolutionized the industry and created several new commercial products. The need for reducing illnesses brought on by the natural pathogens of *Vibrio vulnificus* and *Vibrio parahaemolyticus* caused a dramatic increase in sponsored research activities in this field. Researchers and their industry counterparts not only increased the safety of the product but also developed several new unique products for the consumer. The industry implemented these technologies relatively quickly as they responded to the safety concerns of federal agencies and consumer groups. The first new method used alternating hot and cold temperatures that appeared to 'shock' *V. vulnificus* and kill off a number of the cells. This became known as the 'Ameripure' process and uses a heating bath at 50 °C followed by a cold water treatment to reduce the pathogen to a safe level (Andrews *et al.*, 2003). Another technology that was brought to a new level was rapid freezing of oysters. Research at the University of Florida, USA, has shown that rapidly freezing oysters significantly reduced *V. vulnificus* levels and also preserved many of the intrinsic characteristics of the oyster for the consumer (Seminario *et al.*, 2011). The use of liquid nitrogen in the freezing process preserved much of the flavour and texture characteristics as well as greatly improving the shelf life of the product. Several companies are now freezing shellfish, thereby creating a portfolio of safe products for the consumer to choose from.

High pressure processing (HPP) is an emerging technology that has been implemented by some oyster growing operations in the United States. This technology was a response to Food and Drug Administration concerns over possible hazards associated with raw oysters. Originally implemented by Motivatit Co. in Louisiana, HPP was shown by several researchers to inactivate *Vibrio* pathogens (Calik *et al.*, 2002). The process relies on a conveyor system, which moves shell stock 2½ bushels at a time immersed in water, into a vessel. The intake end of the vessel is then sealed shut and 310 MPa of pressure are applied to the contents. This amount of pressure severs the adductor muscle from the shell, which results essentially in a perfectly shucked oyster (He *et al.*, 2002). Pressure is reduced using the mechanized control panel and the oysters are placed onto another conveyor where the meats and shell are separated. Research has demonstrated that the optimum HPP pressure needed for the shucking process is also effective for the reduction of *V. parahaemolyticus* (Shiu, 1999; Calik *et al.*, 2002). The industry has used also HPP to develop a new, popular convenience product for retail (Novak, 2003). This product is a whole oyster in a shrink-wrapped band that is internally shucked and ready for consumption. The consumer needs only to cut the band to have oysters on the half-shell. The HPP project represents a major opportunity to preserve and enhance the economic impacts of the oyster industry. For the first time in over 100 years, a new technology has been introduced to replace the traditional method of processing oysters. This technology is also currently being investigated as a pathogen intervention step for uncooked frozen meat products (Fernández *et al.*, 2007; Realini *et al.*, 2011; Vaudagna *et al.*, 2012). The advantage of using pressure on frozen products may be the improvement of quality. It is well known that HHP of uncooked refrigerated products browns the product (conversion of oxymyoglobin to metmyoglobin) and reduces water holding capacity of the product due to protein denaturation. HHP of frozen uncooked product does not cause product discoloration and there is evidence that microbial loads are also reduced (Fernández *et al.*, 2007).

The pH-shift process is another technology that has been under development since the late 1990s (Undeland *et al.*, 2002). This process is based on the solubility of myofibrillar proteins at extreme pH and low ionic strength. Conventionally, myofibrillar proteins have been classified as salt-soluble rather than water-soluble proteins. However, Hultin and Kelleher (1999, 2000) demonstrated that with a sufficient reduction of ionic strength (dilution), myofibrillar proteins can be solubilized in water. They then expanded this concept by studying the solubility of myofibrillar proteins in low ionic (dilute, 1 : 10 w/v) conditions at varying pH. Under these conditions, myofibrillar proteins can be completely solubilized below pH 4 and above pH 10. They also demonstrated that when solubilized in this manner, these proteins can stay suspended even when centrifugal forces as high as 10 000 *g* are applied. This was a particularly important finding because phospholipids, which are highly susceptible to oxidation, are sedimented at this level of centrifugal force. It was this finding that resulted in several patents for both acid and alkaline solubilization of muscle proteins (Hultin and Kelleher, 1999, 2000; Kelleher, 2006; Hultin and Riley, 2011). Other important constituents that can be sedimented away from the solubilized myofibrillar proteins include collagen and bones. Once unwanted materials are removed, the myofibrillar protein can then be simply concentrated through isoelectric precipitation and collected either through filtration or through centrifugation at about 3000 *g*. The pH-shift process was originally envisioned as a means to recover protein from small, difficult to process fish species or as a means to recover valuable myofibrillar protein remaining on the frames, skin and bones of fish after processing. However, subsequent research has demonstrated that the proteins recovered from the pH-shift process have the unique ability to form a gel without the addition of salt. The current theory is that acid or alkaline solubilization of proteins acts similarly to salt solubilization by uncoiling the protein and exposing potential binding positions for gelation (James and Mireles DeWitt, 2004; Kim and Park, 2008; Raghavan and Kristinsson, 2008; Ingadottir and Kristinsson, 2010; Davenport and Kristinsson, 2011). As a result, additional potential applications for the pH-shift process include production of low-salt restructured fish or surimi and incorporation of solubilized but not precipitated proteins through direct injection into muscle food products. In summary, the pH-shift process is currently being used to either obtain protein from difficult to process, low-value fish or to functionally alter proteins for injection back into seafood products. The latter application is used to firm the texture of seafood and to improve water retention during processing, storage and cooking of fish.

Injection technology is also seeing a renaissance in seafood processing. The brining of seafood is a time-honoured tradition for smoking and drying. It has also been used to ease removal of the exoskeleton from shrimp and to slow water loss in scallops. The development of automated brine injectors has allowed processors to improve processing efficiency through quick incorporation of brine (salt) into the product. In the traditional processes of wet (soaking) and dry brining the processor needs to consider many factors including: appropriate brine to product mass ratios, brine composition and temperature, product fat content, thickness of product, fresh versus previously frozen, and, for high-salt brines, dilution due to water osmosis from the product, which can prevent sufficient salt uptake for food safety purposes. All of these factors can impact the final salt content trying to be achieved in a seafood product. Injection of brine directly into products removes many of these variables from the processing equation. As a result, brine injection has for many years been used for incorporation of salt into products intended for canning and for both hot-smoked and cold-smoked products. However, a new application for brine injection is to maintain moisture as opposed to adding flavour or for purposes of food safety. This involves adding salt and/or phosphates to alter protein structure (through myofibrillar solubilization) sufficiently to enhance the water holding capacity of the product (Bendall, 1954; Hamm, 1961, 1970; Offer and Trinick, 1983). The aim is to prevent moisture losses that occur during handling and storage of the product.

In addition, this form of brine injection processing can be used to improve flavour, add nutritive constituents (e.g. omega-3 fatty acid), antioxidants and antimicrobials to improve the nutritional composition, protect valuable nutrients from degradation and decrease product perishability (Vogel *et al.*, 2006; Mireles DeWitt CA *et al.*, 2007; Vann *et al.*, 2007; Cerruto-Noya *et al.*, 2009; Parsons *et al.*, 2011). For all the beneficial applications of low salt brine injection technology, there is one serious drawback. This technology can also be used for purposes of economic fraud. Since the salt content is not sufficiently high enough to be easily detected from a flavour standpoint as in a marinated product, processors can be tempted to use the process to add water to the product. When brines are used to add water, the processor's label should inform the end user of the ingredients in the brine and the percentage of water added.

Other notable technologies currently used and/or under development are modified atmosphere packaging (MAP), flexible retort packaging for canning, filtered smoke application and natural/organic curing. MAP has been proposed for seafood for many years. Numerous studies have demonstrated its control of both spoilage and pathogenic microorganisms. An excellent review was recently provided by Sivertsvik *et al.* (2002). This type of packaging typically uses carbon dioxide, nitrogen and oxygen at varying ratios. The pathogens of major concern when packaging seafood under anaerobic conditions is *Clostridium botulinum* type E and non-proteolytic type B and when temperatures are below 10 °C, *Listeria monocytogenes* can grow under anaerobic, chilled conditions. The 10 °C limit was used because Sivertsvik *et al.* (2002) later reported that the Advisory Committee on the Microbiological Safety of Food in the UK concluded that at temperatures less than 10 °C and 10 days storage, risks for botulism are low. If the shelf life is longer, however, they suggest that combinations of preservative factors such as pH (≤ 5), water activity (≤ 0.97), and salt ($>3.5\%$ in the aqueous phase) should be established to prevent growth of psychrotrophic *C. botulinum*. The incorporation of CO_2 into packaging is usually accomplished by evacuating air and then flushing the gas mixture into the package. This method, however, creates a large headspace which increases packaging, transportation and storage costs. Sivertsik *et al.* (2002) also describe other means of creating the modified atmosphere and including the sublimation of dry ice in a vacuum package, reaction of exudate with sodium carbonate and citric acid in the drip pad, or dissolving CO_2 in the product prior to packaging under elevated pressure and low temperature.

Flexible retort packages are also gaining popularity as an alternative to cans (Khurana *et al.*, 2004). The material cost to the processor of retort package can be up to five times cheaper than cans and their thin profile can reduce shipping space almost sevenfold (personal communication, 2012). In addition, increased heat penetration efficiency can significantly reduce cooking time and thereby increase product quality. Finally, the large flat surface area of these packages allow increased space for printing. Innovations in this packaging technology include packages that can stand on their end, with easy to tear tops that can be removed for direct package microwaving and direct consumption from the package (www.kapak.com). In addition, a zipper retortable pouch has also been developed by Pyramid Flexible Packaging (La Habra, CA). These packages can withstand temperatures up to 135 °C.

The addition of sodium nitrite for curing smoked tuna fish, salmon, sablefish, shad and hena serves to not only fix colour, but also to enhance the inhibitory effectiveness of salt against *C. botulinum* type E spores (Flick and Kuhn, 2012). The past 10 years, however, has seen advances in natural/organic curing processes (Sebranek *et al.*, 2012). Many of these products are labelled 'uncured' or 'no nitrates or nitrites added'; however, this does not mean they do not contain nitrates or nitrites. 'Natural curing', a term not defined and/or recognized by USDA, started with the use of natural sources of nitrate in combination with a nitrate-reducing starter culture to form nitrite. This process mimics the ancient meat curing processes that used natural nitrate as saltpetre with natural nitrate-reducing bacteria. However, instead of saltpetre, a concentrated vegetable extract from celery which contains about 3% (w/w) nitrate and a purified strain of efficient nitrate-reducing bacteria are used. The process requires an incubation step at 38–40 °C prior to smoking and cooking to give the culture enough time to reduce the nitrate. This incubation time, however, can be shortened through a 'pre-conversion' process that involves incubating the celery concentrate with the culture prior to adding the mixture to the product. Recently, spray-dried Swiss chard powder has been used an allergen-free alternative to celery powder. These innovations in 'natural curing' have been applied to processed meats over the past 10 years. It is unclear whether the seafood industry is yet taking advantage of these commercially available 'natural' sources of nitrate or nitrite. There is currently no scientific literature available on the use of natural curing agents in smoked seafood.

Preservation and colour fixation of fresh tuna, mahi mahi, marlin, swordfish and tilapia with filtered natural wood smoke is another fairly recent innovation in the US seafood industry. The smoke is filtered to remove taste, odour and particulates leaving carbon monoxide (CO) to bind the haeme molecule and convert the unstable red oxymyoglobin molecule to a stable red carboxymyoglobin pigment. Treatment of products with carbon monoxide not only imparts colour stability, but also improves the product's oxidative stability. One advantage the use of filtered smoke may have over pure carbon monoxide is that it carries other compounds (typically phenolics) that act to reduce microbial growth (Pivarnik *et al.*, 2011).

13.8 Conclusions

The concept of value-added processing has evolved over the past decade. Attention to consumer demands as well as industry needs will be key to the introduction of value-added products and their success in the marketplace. While traditional products such as canned, smoked and frozen foods will continue to occupy an important place in seafood markets, new consumer demands for 'value' attributes such as sustainability, local support and traceability are changing the interaction between buyer and seller. The positive health aspects of seafood will continue to be a driver for the development of new products and the use of resources that highlight these characteristics. Increased worldwide demand for seafood will be met primarily through increases in aquaculture which continues to grow, mainly in developing countries. There continues to be a strong interaction between value-added products and seafood safety that is facilitated by new technologies that maintain high quality and provide an additional layer of safety for new seafood products. The production of value-added seafood products affords new opportunities for entrepreneurs and established companies to meet future market challenges at the local, regional and international level.

References

Allen, J.H. (2006). Assessing the market dynamics of value-added agriculture and food businesses in Oregon: Challenges and opportunities. Center for Sustainable Processes and Practices, Portland State University.

Anderson, P.D. and Wiener, J.B. (1995). Eating fish. *Risk versus Risk: Tradeoffs in Protecting Health and the Environment* (eds J.D. Graham and J.B. Wiener). Cambridge, MA: Harvard University Press, pp. 104, 123.

Andrews, L.S., DeBlanc, S., Veal, C.D. and Park, D.L. (2003). Response of *Vibrio parahaemolyticus* 03: K6 to a hot water/cold shock pasteurization process. *Food Additives and Contaminants*, 20(4), 331–334.

Bang, H.O. and Dyerberg, J. (1980). Lipid metabolism and ischemic heart disease in Greenland Eskimos. *Advances in Lipid Research* 3, 1–22.

Bendall, J.R. (1954). The swelling effect of polyphosphates on lean meat. *Journal of the Science of Food and Agriculture*, 5(10), 468–475.

Brown, P.L. (2012). For Local Fisheries a Line of Hope. NewYork Times (1 October). http://www.nytimes.com/2012/10/03/dining/a-growing-movement-for-community-supported-fisheries.html?pagewanted=all&_r=1& [accessed 3 June 2013].

Calik, H., Morrissey, M.T., Reno, P.W. and An, H. (2002). Effect of high-pressure processing on *Vibrio parahaemolyticus* strains in pure culture and pacific oysters. *Journal of Food Science*, 67(4), 1506–1510.

Cagan, J. and Vogel, C.M. (2002). *Creating Breakthrough Products*, Prentice Hall, Upper Saddle River, NJ.

Cerruto-Noya, C.A., VanOverbeke, D.L. and DeWitt, C.A. (2009). Evaluation of 0.1% ammonium hydroxide to replace sodium tripolyphosphate in fresh meat injection brines. *Journal of Food Science*, 74(7), C519-C525.

Constance, D.H. and Bonanno, A. (2000). Regulating the global fisheries: the World Wildlife Fund, Unilever, and the Marine Stewardship Council. *Agriculture and Human Values*, 17(2), 125–139.

Cooper, R.G. and Edgett, S.J. (2005). *Lean, Rapid, and Profitable New Product Development*, Product Development Institute, Ancaster, Ontario.

Cooper, R.G. (2001). *Winning at New Products: Accelerating the Process from Idea to Launch*, Basic Books.

Cousminer, J. (1999). *Practicing Culinology*, Food Product Design.

Davenport, M.P. and Kristinsson, H.G. (2011). Channel catfish (*Ictalurus punctatus*) muscle protein isolate performance processed under different acid and alkali pH values. *Journal of Food Science*, 76(3), E240-E247.

Delgado, C.L., Wada, N., Rosegrant, M.W., Meijer, S. and Ahmed, M. (2003). *Fish to 2020*, World Fish Centre and International Food Policy Research Institute, Penang, Malaysia.

FAO (2012). *The State of the World Fisheries and Aquaculture*, Food and Agriculture Organization, Rome, Italy. http://www.fao.org/docrep/016/i2727e/i2727e00.htm [accessed 3 June 21013].

FAO/WHO (2010). Joint FAO/WHO Expert Consultation on the Risks and Benefits of Seafood Consumption. FAO Fisheries and Aquaculture Report No. 978, Rome. http://www.fao.org/docrep/014/ba0136e/ba0136e00.pdf

Fernández, P.P., Sanz, P.D., Molina-García, A.D. *et al.* (2007). Conventional freezing plus high pressure–low temperature treatment: physical properties, microbial quality and storage stability of beef meat. *Meat Science*, 77(4), 616–625.

Flick Jr,, G.J. and Kuhn, D.D. (2012). Smoked, cured, and dried fish. *The Seafood Industry*, 404–426.

Food Channel (2012). Top 10 trends in seafood. www.foodchannel.com/articles/article/top-10-trends-in-seafood [accessed 3 June 2013].

Fuller, G.W. (2011). *New Food Product Development: from Concept to Marketplace.* CRC Press.

Grobe, D., Sylvia, G. and Morrissey, M.T. (2002). Designing a culinology-based research and development framework for seafood products. *Journal of Aquatic Food Product Technology*, 11(2), 61–71.

Hamm, R. (1961). Biochemistry of meat hydration. *Advances in Food Research*, 10, 355–463.

Hamm, R. (1970). Interactions between phosphates and meat proteins. Symposium: Phosphates in Food Processing, Avi Publishing, Westport, CT, p. 65.

He, H., Adams, R.M., Farkas, D.F. and Morrissey, M.T. (2002). Use of high-pressure processing for oyster shucking and shelf-life extension. *Journal of Food Science*, 67(2), 640–645.

Hellberg, R.S., DeWitt, C.A.M. and Morrissey, M.T. (2012). Risk–benefit analysis of seafood consumption: a review. *Comprehensive Reviews in Food Science and Food Safety*, 11(5), 490–517.

Hultin, H.O. and Kelleher, S.D. (1999). US Patent No. 6,005,073. US Patent and Trademark Office, Washington, DC.

Hultin, H. and Kelleher, S.. (2000). US Patent No. 6,136,959. US Patent and Trademark Office, Washington, DC.

Hultin, H.O. and Riley, C. (2011). US Patent No. 8,021,709. US Patent and Trademark Office, Washington, DC.

Ingadottir, B. and Kristinsson, H.G. (2010). Gelation of protein isolates extracted from tilapia light muscle by pH shift processing. *Food Chemistry*, 118(3), 789–798.

James, J.M. and DeWitt, C.A. (2004). Gel attributes of beef heart when treated by acid solubilization isoelectric precipitation. *Journal of Food Science*, 69(6), C473–C479.

Johnson, H. (2004). 2005 Annual Report on the United States Seafood industry.

Josupeit, H. (2012). World supply and demand of tilapia. http://www.globefish.org/world-supply-and-demand-of-tilapia.html [accessed 3 June 2013].

Kearney, J. and Kearney, J. (2010). Food consumption trends and drivers. *Philosophical Transactions of the Royal Society B: Biological Sciences*, 365(1554), 2793–2807.

Kelleher, S.D. (2006). *US Patent No. 7,033,636.* Washington, DC: US Patent and Trademark Office.

Khurana, A., Awuah, G. and Charbonneau, J. (2004). Thermal processing considerations for retortable pouches: draft literature review. National Food Processors Association, Washington, DC.

Kim, Y.S. and Park, J.W. (2008). Negative roles of salt in gelation properties of fish protein isolate. *Journal of food science*, 73(8), C585-C588.

Kromhout, D., Bosschieter, E.B. and De Lezenne, C.C. (1985). The inverse relation between fish consumption and 20-year mortality from coronary heart disease. *The New England Journal of Medicine*, 312(19), 1205.

Magera, A. and Beaton, S. (2009). Seafood Traceability in Canada. http://www.seachoice.org/wp-content/uploads/2011/09/Seafood_Traceability_in_Canada.pdf [accessed 3 June 2013].

Mireles DeWitt, C.A., Nabors, R.L. and Kleinholz, C.W. (2007). Pilot plant scale production of protein from catfish treated by acid solubilization/isoelectric precipitation. *Journal of food science*, 72(6), E351–E355.

Mozaffarian, D. (2008). Fish and *n*-3 fatty acids for the prevention of fatal coronary heart disease and sudden cardiac death. *The American Journal of Clinical Nutrition*, 87(6), 1991S–1996S.

Mozaffarian, D. and Rimm, E.B. (2006). Fish intake, contaminants, and human health. *JAMA: Journal of the American Medical Association*, 296(15), 1885–1899.

Nesheim, M.C. and Yaktine, A.L. (2007). *Seafood Choices: Balancing Benefits and Risks*. National Academy Press.

Novak, T. (2003). Pressed for success. *Oregon's Agricultural Progress Magazine*, 58(1): 30–35.

Offer, G. and Trinick, J. (1983). On the mechanism of water holding in meat: the swelling and shrinking of myofibrils. *Meat Science*, 8(4), 245–281.

Parsons, A.N., VanOverbeke, D.L., Goad, C.L. and Mireles DeWitt, C.A. (2011). Retail display evaluation of steaks from select beef strip loins injected with a brine containing 1% ammonium hydroxide. Part 1: Fluid loss, oxidation, color, and microbial plate counts. *Journal of Food Science*, 76(1), S63–S71.

Pérez-Mateos, M., Boyd, L. and Lanier, T. (2004). Stability of omega-3 fatty acids in fortified surimi seafoods during chilled storage. *Journal of agricultural and food chemistry*, 52(26), 7944–7949.

Pietrowski, B.N., Tahergorabi, R., Matak, K.E., Tou, J.C. and Jaczynski, J. (2011). Chemical properties of surimi seafood nutrified with ω-3 rich oils. *Food Chemistry*, 129(3), 912–919.

Pivarnik, L.F., Faustman, C., Rossi, S. *et al.* (2011). Quality assessment of filtered smoked yellowfin tuna (*Thunnus albacares*) steaks. *Journal of food science*, 76(6), S369–S379.

Raghavan, S. and Kristinsson, H.G. (2008). Conformational and rheological changes in catfish myosin during alkali-induced unfolding and refolding. *Food Chemistry*, 107(1), 385–398.

Realini, C.E., Guàrdia, M.D., Garriga, M., Pérez-Juan, M. and Arnau, J. (2011). High pressure and freezing temperature effect on quality and microbial inactivation of cured pork carpaccio. *Meat Science*, 88(3), 542–547.

Sebranek, J.G., Jackson-Davis, A.L., Myers, K.L. and Lavieri, N.A. (2012). Beyond celery and starter culture: advances in natural/organic curing processes in the United States. *Meat Science*, 92(3), 267.

Seminario, D.M., Balaban, M.O. and Rodrick, G. (2011). Inactivation kinetics of *Vibrio vulnificus* in phosphate-buffered saline at different freezing and storage temperatures and times. *Journal of Food Science*, 76(2), E232–E239.

Shiu, S.E. (1999). Effect of high hydrostatic pressure (HHP) on the bacterial count and quality of shucked oysters. MS Thesis, Corvallis, OR.

Simopoulos, A.P. (1991). Omega-3 fatty acids in health and disease and in growth and development. *The American journal of clinical nutrition*, 54(3), 438–463.

Sivertsvik, M., Jeksrud, W.K. and Rosnes, J.T. (2002). A review of modified atmosphere packaging of fish and fishery products–significance of microbial growth, activities and safety. *International Journal of Food Science and Technology*, 37(2), 107–127.

Sloan, E. (2011). Top ten food trends. *Food Technology*, 65(4): 24–41.

Thompson, M., Sylvia, G. and Morrissey, M.T. (2005). Seafood traceability in the United States: current trends, system design, and potential applications. *Comprehensive Reviews in Food Science and Food Safety*, 4(1), 1–7.

Tobias, L. (2011). Oregon's Dungeness crab learn to fly to China for a whole new market. http://www.oregonlive.com/business/index.ssf/2011/03/oregons_dungeness_crab_learn_t .html [accessed 3 June 2013].

Undeland, I., Kelleher, S.D. and Hultin, H.O. (2002). Recovery of functional proteins from herring (*Clupea harengus*) light muscle by an acid or alkaline solubilization process. *Journal of agricultural and food chemistry*, 50(25), 7371–7379.

USDA (2012). *Dietary Guidelines for Americans.* http://www.cnpp.usda.gov/dietaryguidelines.htm [accessed 3 June 2013].

Vann, D.G. and Mireles DeWitt, C.A. (2007). Evaluation of solubilized proteins as an alternative to phosphates for meat enhancement. *Journal of Food Science*, 72(1), C072–C077.

Vaudagna, S.R., Gonzalez, C.B., Guignon, B., Aparicio, C., Otero, L. and Sanz, P.D. (2012). The effects of high hydrostatic pressure at subzero temperature on the quality of ready-to-eat cured beef *Carpaccio*. *Meat Science*, 92, 575–581.

Vogel, B.F., Yin Ng, Y., Hyldig, G., Mohr, M. and Gram, L. (2006). Potassium lactate combined with sodium diacetate can inhibit growth of *Listeria monocytogenes* in vacuum-packed cold-smoked salmon and has no adverse sensory effects. *Journal of Food Protection*, 69(9), 2134–2142.

Part II

Quality and Safety Issues

14
Seafood Quality Assessment

Jörg Oehlenschläger
Seafood Consultant, Buchholz, Germany

14.1 Why is quality assessment of aquatic animals multifarious and complex?

The quality assessment of fish and other seafood is hampered by the immense variation between species and by the considerable variability between specimens of each species (Love, 1988). Weight and length vary with age and parallel with other factors such as proximate composition, which is subject to change based on the state of maturity, fishing area, season and nutritional status. This implies that each fish specimen is unique and therefore before analysing fish a careful consideration has to be made if this variation is essential for the results and the conclusions drawn from them or if a more general statement is sufficient (pooled samples).

A drastic example of this variation is found in the Atlantic mackerel (*Scomber scombrus*), a pelagic swarm fish. Its fillets can contain a fat content of 35% pre-spawning and only 5% post-spawning. When captured during the spawning season, specimens with between 5 and 35% fat can be found together. This causes severe problems because in parallel with fat content other parameters such as water content and the concentrations of lipophilic organic pollutants may also vary.

In the fish body itself, a certain degree of variability is found in basic components like water, fat and protein, which are not evenly distributed in the edible part but vary from head section to tail section or dorsal part to ventral part.

The challenge when assessing aquatic animals is that immediately after capture or harvest an alteration in intrinsic properties is initiated firstly by metabolic (autolytic) and secondly by microbiological processes which last until the final state of spoilage is reached. Several analytical methods (chemical, physical, microbiological and sensory) can be used to determine spoilage depending on its state. For fish, the status of spoilage is usually referred to as days in ice.

Seafood Processing: Technology, Quality and Safety, First Edition. Edited by Ioannis S. Boziaris.

In addition, various special problems can arise. Aquatic animals may contain parasites (e.g. nematodes, cestodes, trematodes) that affect human health (e.g. anisakidosis) when swallowed and may lead to allergic reactions (Buendia, 1997). Long-lived predatory fish species at the top of the marine food web may accumulate quantities of mercury during their lifespan that can exceed legal limits. High amounts of cadmium can accumulate in the digestive glands (hepatopancreas) of molluscs such as cephalopods and mussels (Oehlenschläger, 2002). Toxins from dinoflagellates can concentrate in bivalve molluscs causing several diseases such as diarrhoeic shellfish poisoning (DSP), paralytic shellfish poisoning (PSP), neurotoxic shellfish poisoning (NSP), and amnesic shellfish poisoning (ASP) and, while in fish the toxins can cause ciguatera poisoning.

Seafood products which have not undergone a sufficient thermal treatment and which are consumed as ready-to-eat products (e.g. cold smoked products, gravlax products) pose an inherent microbial risk.

Fish and other aquatic animals from areas which are polluted by anthropogenic activities (e.g. rivers, inshore waters, estuaries, seas with no or limited water exchange with world oceans such as Baltic Sea, Mediterranean Sea, Caspian Sea or Black Sea) can carry a high burden of persistent organic pollutants (e.g. dioxins, PCBs).

Live aquatic animals from certain areas of the world can carry viruses and microorganisms (e.g. *Vibrio* sp.) which are harmful to human health and must be destroyed or removed prior to marketing the products made from them.

A study of the attitudes of the European fish sector towards monitoring quality conducted in 2001 with 453 participants from 12 European countries concluded that the fish industry needs rapid methods for measuring the quality of fish and fish products. One result of the study was that instrumental techniques for measuring individual attributes such as colour, texture or odour were not considered important. Freshness and quality were considered to be complex concepts that cannot be quantified by a single attribute measurement. The key role of sensory tests was commonly recognized and the evaluation of the various sensory attributes was considered very important for quality control. The need for information and implementation of the tools already available, both sensory and instrumental, is apparent in the fish sector. Further, the study highlighted the need of a small, handheld instrument for fast and reliable quality determination (Jørgensen *et al.*, 2003).

Only a few books and reviews about seafood quality assessment have been published so far (e.g. Connell, 1990; Sørensen, 1992; Huss *et al.*, 1992; Botta, 1995; Luten *et al.*, 1997, 2003; Olafsdottir *et al.*, 1997a, 1997b, 2004; Rehbein and Oehlenschläger, 2009; Ozogul, 2010; Nollet and Toldra, 2010).

14.2 Fish composition

14.2.1 Introduction

There are more than 30 000 known species (FishBase 2012: 32 400 species, 299 200 names) of fish, the biggest group in the animal kingdom. However, only some 700 of the fish species and about 200 crustacean and mollusc species are commercially fished and/or farmed and used for food production. Actually approximately 40% of the world seafood production of 140 million tonnes is farmed. The amount of seafood captured and farmed worldwide is biannually published by the FAO (Food and Agriculture Organization of the United Nations).

14.2.2 Categories of fish species

Fish species can be divided into categories, for example according to their habitat – marine or freshwater species (Oehlenschläger and Rehbein, 2009). Some species such as the European eel and most salmon species can live in both environments. Typical representatives of freshwater fish are: carp (*Cyprinus carpio*), pike (*Esox lucius*), perch (*Perca fluviatilis*), pikeperch (*Stizostedion lucioperca)*, tench (*Tinca tinca*), and of marine fish: cod (*Gadus morhua*), Alaska pollack (*Theragra chalcogramma*), redfish (*Sebastes* sp.), mackerel (*Scomber scombrus*) and herring (*Clupea harengus*). Their anatomic shape is also used for classification: round fish (circular in cross section) such as saithe (*Pollachius virens*), cod and hake (*Merluccius merluccius*); fish with a more elliptical cross section (tuna species, herring, salmon (*Salmo* sp. or *Oncorhynchus* sp.); flat fish species such as plaice (*Pleuronectes platessa*), halibut (*Hippoglossus hippoglossus*) and flounder (*Platychthys flesus*); and snake-shaped fish such as eel (*Anguilla* sp.), lamprey (*Lampetra* sp.) or moray (*Muraena* sp.).

The categories ground fish and pelagic fish are frequently used in the fish industry reflecting their main habitat use. Ground fish species live mostly close to the bottom of the sea (cod, flat fish) and pelagic fish swim in the open sea in big swarms, for example the small fatty species herring, sprat (*Sprattus sprattus*) and sardine (*Sardina pilchardus*).

Most commercial species belong to the skeletal (or bony) fish (Osteichthyes), which means that a fully developed bony skeleton is present. Some fish species contain real bones which mean that these bones are not attached to the backbone but formed by hardened connective tissue free in the muscle. Sharks and rays are representatives of the group of cartilaginous fish (Chondrichthyes) that do not have bones but have cartilaginous tissue instead.

Fish species can also be divided by their nutritional properties, for example their fat content in the edible part (muscle, fillet) makes up four categories: lean fish species, species with low fat content, medium fatty fish species, and fatty (oily) fish species, which in some species can vary with their maturity and nutritional status from 1% to 30%. Examples of some common fish species are given in Table 14.1. In contrast to fish, all species of crustacean and mollusc shellfish contain only small amounts of lipids (1–3%).

Table 14.1 Total fat content in muscle tissue of different categories of marine and freshwater fish species

Fat category	Total fat content [%], on a wet weight basis	Trade name or common name
Lean fish species	<1	Marine: cod, saithe, haddock, whiting, ling, blue ling, cusk, grenadier, blue whiting, angler, lemon sole, Alaska pollack Freshwater: pike, perch, pikeperch
Species with low fat content	>1 to 5	Marine: white halibut, wolffish, plaice, hake, dab, grey mullet, red mullet, redfish, whitch, sole, turbot, brill Freshwater: trout, tench, carp
Medium fatty fish species	>5 to 10	Marine: Redfish, sardine, swordfish, seabream, catfish, albacore, dogfish, tuna, conger, salmon Freshwater: bream
Fatty fish species	>10	Marine: Sprat, black halibut, mackerel, herring, eel Freshwater: eel

14.2.3 Fish muscle

The body musculature of fish commonly referred to as fillet reaches from head to tail. This muscle forms the major part of the edible portion of fish. The proportion of fish flesh to total body weight varies between 25 and 60% depending of the filleting method (manual or mechanised), species, shape, age and the physiological/nutritional status of the fish.

Fish flesh consists of light and dark musculature. The light musculature is used for rapid, sudden movements and gets its energy mainly from anaerobic glycolysis; the dark musculature is well supplied with blood and rich in myoglobin, is used for continuous swimming and has an aerobic metabolism. The dark muscle is therefore well developed in fast moving pelagic species (herring, mackerel and tuna).

14.2.4 Nutritional composition

Fish muscle protein is rich in essential amino acids, has a high biological value and can be digested easily due to the low content of connective tissue, which is as low as 1–2% compared to warm blooded animals with 10–13%.

The content of non-protein nitrogen (NPN) components in fish flesh is high. Main components are creatine (200–700 mg/100 g), trimethylamine oxide (TMAO) (100–1000 mg/100 g), adenosine nucleotides (200–400 mg/100 g), free amino acids and dipeptides. Cartilaginous species additionally contain high amounts of urea. NPN components amount to >400 mg/100 g fish tissue and contribute to 15% of their nitrogen content.

The lipid content of fish exhibits huge variations in quantity and fatty acid composition, but the protein content is almost constant. In fish species with a low or moderate fat content in the muscle tissue, the lipids are located mainly in the liver for energy storage; in fatty species, lipids are deposited in the muscle tissue, as a subcutaneous layer or in the intestinal tract.

Marine fish lipids differ from the lipids of terrestrial animals and of vegetable oil mainly by their high content of long chain highly unsaturated fatty acids of the $n-3$ series, namely eicosapentaenoic acid, 20:5 ($n-3$) and docosahexenoic acid, 22:6 ($n-3$), often referred to as polyunsaturated fatty acids (PUFA) which are of high nutritional impact. The content of PUFAs in fatty fish species can be high: herring, salmon and tuna species >2 g/100 g and mackerel >4 g/100 g. The cholesterol content of fish muscle is generally low (35 mg/100 g) (Souci *et al.*, 2008). Cholesterol burden is high in shrimps and prawns (~200 mg/100 g), in cephalopods (>200 mg/100 g) and in fish gonads (>500 mg/100 g) and their products (e.g. caviar) (Oehlenschläger, 2006).

Vitamins The vitamin content in captured fish can vary considerably within one species with age, size, sex, season, diet, state of health and geographic location. In farmed fish, vitamin content reflects the composition of the corresponding components in the fish feed.

Fish liver is a rich source of fat-soluble vitamins (A, D, E and K). Dark fish muscle contains more fat-soluble vitamins than white muscle. Fish and fish products are commonly regarded as the most important natural food sources of vitamin D: the higher the fat content of the fish flesh the higher the vitamin D content. Fish flesh is only a low to modest source of vitamin E.

The dark muscle of fish contains higher levels of the water-soluble vitamins than the white muscle. Fish flesh is a modest source of riboflavin (vitamin B_2). Only the dark muscle, roe and liver of several fish species contain higher concentrations of riboflavin.

Lean fish contain lower levels of niacin than fatty fish like mackerel, salmon and tuna whereas liver, roe and sperm can contain even higher amounts of niacin.

Fish and shellfish are very good sources of vitamin B_6. Mackerel, herring, tuna, sardine and salmon are especially rich in this vitamin and fish liver contains particularly high concentrations. Seafood is an important food source of vitamin B_{12}, with liver and heart being particularly very rich sources. Fish with a high proportion of dark muscle, like herring and mackerel, contain more vitamin B_{12} than fish like cod and flat fish species (Ostermeyer, 1999; Ostermeyer and Schmidt, 2006; Kim, 2010).

Minerals The two most important elements found predominantly in marine fish species are the essential elements selenium and iodine (Oehlenschläger, 2010a, 2010b). Fish is the only major natural source for these elements. Regular dietary intake of these elements is of eminent importance as there is an insufficient supply in many European populations. Other trace elements such as zinc, chromium, vanadium, molybdenum, and manganese are also present. Oysters and octopus are the richest source of zinc (up to 100 mg/kg) (Carvalho *et al.*, 2005).

14.3 Fish freshness

14.3.1 What is fish freshness and how can it be defined?

The term 'freshness' in relation to fish has been discussed incessantly (Los Angeles Times, 1990). Books (Botta, 1995; Olafsdottir, 2007b) and numerous papers on the subject have been published. A look in the dictionary Merriam-Webster (2012) shows that freshness is defined as 'the quality or appeal of being new', the synonyms are newness, and originality, some near antonyms are out datedness, and staleness.

As a result of three years concerted action, Oehlenschläger and Sörensen (1997) stated that freshness of fish means that a fish in its entire properties is not far away from those properties it had in the living state, or that only a short period of time has passed since the fish has been caught or harvested. The fish is *per definitionem* new, novel and not used.

The flavour and odour compounds that characterize fish immediately after catch disappear in the first few days during storage in ice, and the fish flesh becomes almost flavourless and odourless for a while. The time from catch/slaughter until the fish has lost the initial freshness characteristics (such as sweet flavour) has been defined as the 'freshness period'. 'Shelf life' is defined as the time a food product is fit for human consumption. The terms shelf life and freshness period should be used with some precaution due to the uncertainty in the estimation. Various factors can affect the shelf life and the freshness period, such as the fish species and handling of the fish (Matís, 2012).

For Bremner (1997), freshness is more a concept, but not an object which can be directly measured. Therefore the use of this concept and the use of the word itself need to be carefully defined for the context in which they are used. The definitions can be broad or narrow and can be carefully constructed to suit the particular circumstances.

Later, Bremner and Sakaguchi (2000) published two alternative approaches to define the complex phenomenon freshness in stating that there are fundamentally two philosophically different ways of considering 'freshness':

1. 'freshness' is considered as some *ineffable property that the fish possesses*, and
2. 'freshness' is a result of a *variety of definite properties* that the fish possesses.

The first is a valid approach but it can only lead in one direction, which is, that 'freshness' must be assessed psychometrically. Sensory methods are thus mandatory and a panel can be used to directly assess freshness. The second approach, in which it is considered that *the fish possesses a variety of properties*, which can be assessed by various means, is more practical and does not lead to similar difficulties. These properties are not 'freshness' but, for practical purposes, could be designated as 'freshness' indicators (measurable parameters). Although this is a technological solution, given the above considerations, it is a very practical one if the definitions are clear and the terminology is consistent. 'When *freshness* is used, it must be carefully defined so that its scope takes into consideration the peri-mortal circumstances. This means that, for given circumstances, freshness must be defined by the users or researchers.'

14.3.2 Freshness and quality relationship

Freshness makes a major contribution to the quality of fish and of fishery products. For all kinds of products, freshness of the raw material used for processing is essential for the overall quality of the final product. It is impossible to process a high-quality product from raw material with poor freshness characteristics. It may be argued that for some matured/ripened fishery products, fish should preferably be not completely fresh to obtain the required properties. A closer look at these products, however, reveals that the starting material used for a controlled deterioration (ripening) process must be in a state of good freshness. A well-defined planned ageing or controlled decomposition (ripening) of fish for processing certain products is strongly dependent on the perfect freshness of the initial raw material to keep the entire process under control.

However, 'freshness' does not mandatorily mean quality. Also perfect fresh fish (even live fish) may offer some properties which exclude or limit this fish as a raw material for direct human consumption or for a high-quality product: for example parasites, poor nutritional properties, water content too high, tainting, pollutants, toxins, heavy metals, and mechanical damage. Therefore, whereas freshness is essential for quality, it is singularly not necessary a quality factor *per se*.

Freshness and all other contributing quality attributes must be well balanced to form a good and acceptable product with a satisfying *overall* quality.

Besides freshness, many other parameters also contribute to the complex quality of a seafood-based product such as the sensory properties (taste, odour, texture, appearance); the availability/disposability over a long time period; the safety; desired convenience characteristics; appealing packaging, colours and pictures; an attractive price; consistency, uniformity and conformity with standards and legislation; nutritional properties; ethical (stunning and slaughter); and ethnic (halal, kosher) constraints.

14.3.3 Some indicators for the freshness determination of fish

Based on the second definition of Bremner and Sakaguchi (2000), a set of freshness indicators can be defined and formulated. The properties which wet fish in ice (fresh, unprocessed, not frozen) should have when regarded as perfect fresh can be defined for the following indicators (Oehlenschläger and Sörensen, 1997; Olafsdottir, 2007a):

Rigor mortis. Fish regarded as fresh should be pre-rigor, in rigor or should just have passed rigor; it must be 'new', meaning that the time passed between harvesting or catch and marketing (auction) and consumption or processing into final products (including freezing) should be as short as possible.

Sensory. If graded according to the European Union (EU) quality grading scheme, it should be preferably of extra 'E' or upper 'A' quality. The sensory assessment of a cooked sample of the edible part should give high scores for all attributes – appearance, flavour, odour, taste, colour and texture. Each fish species has its own characteristic sensory attributes of flavour, appearance, odour and texture; attributes change with time and temperature after harvest; choice of sensory methods depends on the purpose of the sensory evaluation and whether it is used in incoming control, product development, quality control, consumer studies or research.

Volatiles. The concentrations of basic volatiles like trimethylamine (TMA), dimethylamine (DMA) or total volatile basic nitrogen (TVB-N) are low, and that of trimethylamine oxide (TMAO) is high. Fresh odours (of plants, cucumber and mushroom character) are high in concentration whereas spoilage odours (amines) are insignificant; typical fishy odour is lacking.

Physical methods. The pH is clearly below 7.0 and the impedance measured is high, the microstructure is intact; mechanical properties of fresh fish are commonly measured post rigor; elasticity decreases, flesh is softening; appearance: flesh is translucent, at later stages opaque.

Microorganisms. Only a few microorganisms are present on skin and gills, the muscle is sterile; the proportion and activity of specific spoilage bacteria in microbiological flora determines remaining shelf life (freshness).

ATP and ATP breakdown products. *k*-values should be low; *k*-values increase with time post mortem – this parameter, however, must be checked for every species and condition; noticeable signs of decomposition or spoilage (autolytic or microbial) are lacking.

Proteins. Low proteolysis rate in proteins.

Lipids. High ratio of antioxidants to pro-oxidants in lipids, the muscle is in a status where lipid oxidation is inhibited.

Sensory and measurable noticeable signs of decomposition or spoilage (autolytic or microbial) are lacking; proteins, lipids and non-protein nitrogen (NPN)-compounds are intact; degradation products are present in traces only.

14.4 Sensory methods

Sensory evaluation of quality is applied in seafood research, for the determination of shelf life, for quality control in fish industry, for product development, for regulatory purpose and for consumer studies. For seafood and seafood products, all sensory tests used for food can be used. The objective sensory tests can be divided into two groups: discriminative tests and descriptive tests. Discriminative tests are used to determine if a difference exists between samples (e.g. triangle test, ranking test). Descriptive tests are used to determine the kind and intensity of the differences (e.g. profiling and quality tests). The subjective test is an affective test based on a measure of preference or liking of consumers (hedonic tests). Some tests, however, such as the EU quality grading scheme, the Torry scheme and the Quality Index Method (QIM) are specifically developed for fish (Martinsdottir *et al.*, 2009a, 2009b):

14.4.1 EU quality grading scheme

Quality grading of fish by outer inspection of the whole fish means that different quality raw materials can be selected for different purposes and products. A correlation with measurable properties, however, is difficult or impossible. One example of a grading scheme for fresh fish is the EU quality grading scheme which is laid down in an EU regulation 'Council Regulation (EC) No 2406/96 of 26 November 1996 (Anonymous, 1996), laying down common marketing standards for certain fishery products'. This regulation has to be applied at the first point of sale (e.g. auctions). Fish are graded into three freshness categories (E, A and B); fish not fulfilling the criteria is not admitted.

The freshness ratings in this regulation apply to the following products or groups of products, by reference to appraisal criteria specific to each of them:

- **whitefish** (haddock, cod, saithe, pollack, redfish, whiting, ling, hake, Ray's bream, anglerfish, pouting and poor cod, bogue, picarel, conger, gurnard, mullet, plaice, megrim, sole, dab, lemon sole, flounder, scabbard fish);
- **bluefish** (albacore or long finned tuna, bluefin tuna, bigeye tuna, blue whiting, herring, sardines, mackerel, horse mackerel, anchovy);
- **selachii** (dogfish, skate);
- **cephalopods** (cuttlefish); and
- **crustaceans** (shrimps, Norway lobster).

In the grading scheme for whitefish, the skin, skin mucus, eyes, gills, peritoneum (in gutted fish), smell of gills and the texture of the flesh are assessed. A grade is given based on the result of this complex assessment.

The main criticism is that the EU scheme does not consider the specific spoilage pattern of individual species and that no training for inspectors is obligatory. Further, the three freshness grades are too rough and cannot be correlated with days in ice. In 1992, a useful multilingual glossary in 12 European languages for the EU grading scheme was published (Howgate *et al.*, 1992). Despite criticism, the scheme is still in use. An excerpt of the scheme for whitefish given in Table 14.2 gives an impression how the EU scheme looks.

14.4.2 The Torry scheme for cooked fillets

For the sensory assessment of fish fillets, it is common to cook the fillets without addition of salt and spices and to evaluate their odour, flavour and texture. The first detailed scheme developed for evaluating the freshness of cod is the Torry scale which is the most used industry scale for evaluating the freshness of cooked fish. The Torry-scale is a descriptive 10-point scale and has been developed for lean, medium fat and fat fish species. Scores are given from 10 (very fresh in taste and odour) down to 3 (spoiled). Descriptions below 3 are redundant since the fish is no longer fit for human consumption. An average score of 5.5, when the sensory panel detects evident spoilage characteristics, has been used as the limit for edibility.

14.4.3 Quality Index Method

The QIM is a freshness grading system for seafood and is based on a scheme originally developed by the Tasmanian Food Research Unit (Bremner, 1985). QIM is widely accepted

Table 14.2 Excerpt from the EU quality grading scheme for whitefish (covering 24 species)

	Criteria			
	Freshness category			Not admitted
	Extra	A	B	
Skin	Bright, iridescent pigment (except for redfish) or opalescent; no discolouration	Pigmentation bright but not lustrous	Pigmentation in the process of becoming discoloured and dull	Dull pigmentation
Skin mucus	Aqueous, transparent	Slightly cloudy	Milky	Yellowish green, opaque mucus
Eye	Convex (bulging); black, bright pupil; transparent cornea	Convex and slightly sunken; black dull pupil; slightly opalescent cornea	Flat; opalescent cornea; opaque pupil	Concave in the centre; grey pupil; milky cornea
Gills	Bright colour; no mucus	Less coloured; transparent mucus	Brown/grey becoming discoloured; thick, opaque mucus	Yellowish; milky mucus

as a reference method in research; however, a substitution of the EU grading scheme by this method is missing and implementation in the fish sector is poor.

QIM has the advantages of being rapid, cheap to use, non-destructive and objective compared to other sensory methods. QIM is easy to work with as the method includes instructions and guidance, it is a convenient method for teaching inexperienced people to evaluate fish, and to train and monitor the performance of panellists. As the Quality Index (QI) is designed to increase linearly with storage time (days in ice), the information may be used in production management (Hyldig *et al.*, 2010b).

QIM is based on the characteristic changes that occur in seafood when stored in ice. A score from 0 to 3 points is given for changes in outer appearance (e.g. eyes, skin and gills) and changes that occur in odour and texture. The colour of blood and fillets (cut surface) is additionally evaluated in gutted fish. For some ungutted fish species, disintegration of the intestines can be evaluated. The descriptions for each score for each parameter to be evaluated are described in the QIM scheme with concise, simple and brief keywords. Each parameter within the QIM scheme is evaluated independently and individually by each panellist. The description for score zero should be characteristic for the parameter at the day of catch/harvest, while higher scores given for later stages of storage should reflect the changes occurring for that parameter.

As an example, the parameters 'form of eye' for a fish species is described by three scores which reflect the changes of the form of the eyes during storage time in ice. At the very beginning of storage time the eye is convex (score = 0), then it becomes flat (score = 1), and at last sunken or concave (score = 2).

Typically, the number of parameters in a QIM scheme is around 10, each with two to four score categories. Otherwise the assessor might get confused and the evaluation might require too much time. When all parameters in the scheme have been given a score, the scores for all characteristics are summarized to give an overall sensory score, which is called the QI. The scientific development of QIM schemes for various species aims at having the QI increase linearly with storage time in ice.

So far QIM schemes have mainly been developed for whole fresh unprocessed fish, but several schemes have been developed for other types of seafood and products, such as raw, frozen/thawed fillets and cooked fillets. Currently the literature offers publications on cod (*Gadus morhua*), herring (*Clupea harengus*), Atlantic mackerel (*Scomber scombrus*), horse mackerel (*Trachurus trachurus*) and European sardine (*Sardina pilchardus*), thawed whole cod, thawed cod fillets, thawed cooked cod fillets (*Gadus morhua*), redfish (*Sebastes mentella/marinus*), deep water shrimp, fjord shrimp and peeled shrimp (*Pandalus borealis*), plaice (*Pleuronectes platessa*), brill (*Rhombus laevis*), dab (*Limanda limanda*), haddock (*Melanogrammus aeglefinus*), pollock (*Pollachius virens*), sole (*Solea vulgaris*), and turbot (*Scophthalmus maximus*), gilthead seabream (*Sparus aurata*), farmed Atlantic salmon (*Salmo salar*), frozen hake (*M. capensis and M. Paradoxus*) Mediterranean Hake (*Merluccius merluccius*), octopus (*Octopus vulgaris*) cod fillets (*Gadus morhua*), flounder (*Paralichthys patagonicus*), air and MA-packed maatje herring (*Clupea harengus*), Mediterranean anchovies (*Engraulis encrasicholus*), cuttlefish (*Sepia officinalis*) and broadtail shortfin squid (*Illex coindetii*), farmed Atlantic halibut (*Hippoglossus hippoglossus L.*). Many QIM schemes and relevant literature about existing schemes and references for the development of QIM schemes can be found on the homepage of QIM Eurofish (qim-eurofish, 2012).

Two detailed and elaborate papers about the QIM and about the sensory quality of fish have been published recently by Hyldig *et al.* (2012a) and by Hyldig (2012b).

14.5 Chemical methods

14.5.1 Traditional methods as TVB-N, TMAO, TMA, DMA

TVB-N Volatile amines are the characteristic substances responsible for the fishy odour and flavour encountered in marine fish after having passed the initial phase of freshness. Most consumers consider this aminic smell to be typical for fish in general and the consumer's decision of 'like' or dislike strongly depends on the concentration of these amines in fishery products. Amines, however, are present in freshly caught marine fish at a very low level, except ammonia and the non-volatile TMAO, and develop only in later stages of storage depending on species, temperature, time, hygienic regime, microbial action and other factors. The most relevant amines are ammonia, DMA, TMA, TMAO as the main source for DMA and TMA, and TVB-N or TVN, which is a sum parameter consisting mostly of differing proportions of ammonia-N, DMA-N and TMA-N. Considerable literature about the formation and detection of volatile amines including TMAO exists (Howgate, 2009, 2010a, 2010b; Oehlenschläger 1997a, 1997b).

Formation of amines (especially TMA) increases when spoilage microorganisms (bacteria) have entered the muscle tissue and microbial degradation of low-molecular substances and proteins starts. Depending on the fish species and on the storage temperature, it takes about 10 days before microbial action on muscle tissue leads to an increase of amines. From this point on, amine formation follows an exponential function and can be used as a measure for spoilage. Because the amine concentration remains almost constant throughout the first 10 days of storage in ice, amines (with the exception of TMAO, which is also degraded by autolytic processes) are not freshness indicators. Determinations of amines have proven useful in numerous species, products and for legislative purposes.

TVB-N is still the most frequently used method for quality assessment in the fish industry. It is often used as a method for a rough quality check for incoming material. During iced storage experiments with roundfish, the TVB-N contents stay typically at the level found in freshly caught fish during the first week of iced storage, or decreases even somewhat decreases because one of its constituents, ammonia, is leached out by melting ice. After this period, TVB-N increases and reaches values of 30–50 mg/100 g when the borderline of fitness for human consumption is reached. TVB-N is only of limited use in the small fatty pelagic species. In these species it sometimes seems that no TVB-N is formed (obviously all compounds formed are leached out) as it can be measured only to a small extent. For TVB-N, the same statement as for TMA-N can be made: it is a good spoilage indicator (but somewhat inferior to TMA-N), but it is of no use as a freshness indicator. When used as a spoilage indicator, conclusions drawn should always be checked by sensory assessment.

In the EU (Anonymous, 2005) a regulation is in force in which limit values for TVB-N for certain categories of fishery products and the analytical methods to be used are laid down. It is regulated that unprocessed fishery products belonging to certain species categories (category a: *Sebastes* spp., *Helicolenus dactylopterus, Sebastichthys capensis*; category b: species belonging to the Pleuronectidae family (with the exception of halibut: *Hippoglossus* spp.); category c: *Salmo salar*, species belonging to the Merlucciidae family, species belonging to the Gadidae family) shall be regarded as unfit for human consumption where organoleptic assessment has raised doubts as to their freshness and chemical checks reveal that the following TVB-N limits are exceeded: for category a: 25 mg of nitrogen/100 g of flesh, for category b: 30 mg of nitrogen/100 g of flesh, and for category c: 35 mg of nitrogen/100 g of flesh.

The routine methods which may be used to check the TVB-N limit are a microdiffusion method according to Conway and Byrne (1933), a direct distillation method described by Antonacopoulos and Vyncke (1989), and distillation of an extract deproteinized by trichloracetic acid. The reference method to be used for determining the concentration on TVB-N in fish and fishery products involves distilling an extract deproteinized by perchloric acid. The procedure is applicable at TVB-N concentrations of 5 mg/100 g to at least 100 mg/100 g. The distillation must be performed using an apparatus which complies with a diagram in the regulation. Vyncke (1995) used the determination of TVB-N in eye fluid instead of fish muscle and found this method to be feasible with cod (*Gadus morhua*).

Ammonia The ammonia content in iced, stored marine fish initially remains at the same level as for freshly caught fish (10 mg/100 g). In small fish, there is a decrease of ammonia content in this phase, based on leaching effects. At days 7–12 (depending of species), an increase in ammonia can be observed. This increase continues throughout the whole remaining storage time but there is a large scattering of values. Ammonia is only a poor indicator of freshness and cannot be used as an effective spoilage indicator.

DMA-N The development of DMA is different in fish which contain the enzyme trimethylamine oxide demethylase (TMAOase) from fish which lack this enzyme. In fish containing TMAOase (gadoid species like cod, haddock, whiting), a steady increase of DMA-N can be observed in the first week of iced storage from 0.1 mg up to values of 4–7 mg DMA-N/100 g due to degradation of TMAO into DMA and formaldehyde. Often a sudden stop of development can then be observed, the later phases of storage show DMA-N content scattering at values around 6 mg/100 g. The reason for the initial development of DMA-N and the break at approximately day 7 is that the enzyme TMAOase responsible for DMA formation is inhibited by oxygen. During the first stage of storage, the fish muscle remains intact, but in later phases the fish muscle becomes porous due to autolytic processes and the onset of microbial action, making it possible for oyxgen to diffuse into the muscle tissue.

In these species DMA-N can be used as an excellent freshness indicator for the first phase of storage. In fish species which lack the enzyme, no DMA-N is formed during iced storage.

TMA-N Similarly to ammonia, TMA-N remains at its initial level until the onset of microbial action at around day 10 when TMA-N is mainly formed from breakdown of TMAO. From day 10 of storage of wet fish there is a continuous increase in TMA-N until the fish is completely spoiled. Because of its universal production in all marine fish species, TMA is an excellent indicator for the onset of spoilage and for the different stages of spoilage. TMA-N cannot be used as a freshness indicator: between day 0 and day 10 in ice it gives no information about the days in ice. Recently the development of a microconductometric biosniffer for detection of TMA which is a 'good parameter for estimating fish freshness' has been published. The biosensor consists of two parts: a microconductometric transducer and a structured enzymatic membrane using a flavin-containing monooxygenase 3 (FMO3), which is known for its sensitivity and selectivity for TMA (Fillit *et al.*, 2008).

TMAO-N TMAO-N is the only amine which decreases in concentration during iced storage. In iced storage experiments it was found that the decrease is almost linear. It was also observed that the TMA-N production continues after the TMAO-N concentration was zero. Obviously other sources than TMAO for TMA-N are present in fish. The TMAO-N content can be used as a freshness and a spoilage indicator in some marine species. Unfortunately it is, however, essential to know the concentration of TMAO-N in the freshly caught fish. Therefore this indicator is limited to some well-known, well-described and biochemically characterized species.

The amines present in marine fish can be determined by a number of different methods which are all well documented in the literature. As indicators for freshness, only DMA-N can be used in the first phase of storage if the fish species contains TMAOase. Ammonia is only of limited usefulness except for elasmobranch fishes, in which it is formed during storage by the cleavage of urea. TMA-N and TVB-N are both good, sometimes excellent, indicators of spoilage, if no sensory assessment is possible. As freshness indicators they are of limited value only because they remain unchanged throughout the first 8–14 days of iced storage, depending on the fish species. Besides the single amines, some mathematically transformed figures (TMA-index) or the quotient between TMA-N and TVB-N have also been proposed as powerful quality criteria. A closer look at these alternatives, however, did not prove any advantage of these over traditional methods.

14.5.2 Biogenic amines

Biogenic amines can be formed post mortem by the decarboxylation of specific amino acids in fish and shellfish tissue. Biogenic amines produced in fish are mainly produced by the action of exogenous enzymes released by the various microorganisms related with seafood; however, endogenous decarboxylase enzymes naturally occurring in fish or shellfish tissue may also contribute. Histamine, putrescine, cadaverine, tyramine, tryptamine, β-phenylethylamine, spermine and spermidine are the most relevant biogenic amines in seafood.

The feasibility of quality assessment using biogenic amines content such as histamine, cadaverine, putrescine and agmatine was studied in sardine (*Sardina pilchardus*), Atlantic horse mackerel (*Trachurus trachurus*), chub mackerel (*Scomber japonicus*) and Atlantic mackerel (*Scomber scombrus*) during storage at room temperature (20–23 °C) and in ice (2–3 °C). Levels of histamine, cadaverine and putrescine increased gradually in all species as decomposition progressed, regardless of storage temperatures and attained maximum limits for human consumption after 24 h of storage at room temperature. In contrast, amine production in iced fish was considerably reduced and histamine concentration increased slowly until day 7, after which a considerable increase was observed, nevertheless generally below 100 mg/kg. The evaluation of the changes in fish quality showed that no correlation existed between the concentration of biogenic amines and the degree of fish decomposition. Thus, the use of histamine or other amines as a freshness index of the studied fish species was not considered appropriate. The proposal that visible signs of decomposition protect consumers from the action of hazardous biogenic amines therefore seems doubtful (Mendes, 2009; Ruiz-Capillas and Jimenez-Colmenero, 2010). Freezing fish can significantly reduce bacterial counts; however, it will not inhibit the activity of decarboxylase enzymes that were produced before freezing. The current analytical methods for the determination of biogenic amines in food including seafood have been reviewed by Önal (2007).

14.5.3 K-value

The extent of breakdown of ATP (adenosintriphosphate) and related products was expressed as K-value, defined as a percentage of the amount of inosin and hypoxanthin to the total amount of ATP-related compounds (Saito *et al.*, 1959):

$$\text{K-value (\%)} = \frac{(\text{Ino} + \text{Hx})}{(\text{ATP} + \text{ADP} + \text{AMP} + \text{Ino} + \text{Hx})} \times 100$$

K-value is an index that has a linear function with time during fish storage at chill temperatures and is used extensively as a commercial index in Japan for estimating fish freshness. Since ATP is rapidly degraded to IMP after death in the majority of fish species and remains very low during storage, other indices which do not require determination of ATP, ADP or AMP have been proposed to evaluate fish freshness. The usefulness of these freshness indices depends on the fish species being examined.

A K-value of 20% has been defined by Japanese researchers as the limit for raw fish ('sashimi' grade) consumption. The time in ice reported for several species to reach this value ranges between one day for Pacific cod (*Gadus macrocephalus*) and 14 days for sharp-toothed eel (*Muraenesox cinereus*) and flying fish (*Cypselurus opisthopus hiraii*). In farmed fish, the period ranged between 3 and 4 days for seabass, 7–8 days for sea bream and 8–10 days for Senegalese sole depending on the batch studied.

The measurement K-value or other related values based on ATP degradation are considered the best techniques for evaluating freshness in fish stored at temperatures above freezing. The increase of K-value is detected very early, the maximum values and the rate of increase remain quite stable for each fish species and has been found to be closely related to sensory evaluation in many fish species (Tejada, 2009).

14.6 Physical methods

14.6.1 pH

Measuring pH in an aqueous muscle homogenate or using an injection electrode directly in the muscle tissue is the simplest method for quality assessment of seafood. Immediately after catch, the pH in fish muscle is between 5.8 and 6.5 depending on numerous factors such as species, stress and nutritional status. During iced storage, pH gradually increases, reaching the point of spoilage values at 7.5 to 8. A general rule is that when a pH of 7.0 in fish muscle is reached or exceeded, the borderline of edibility is reached. The pH value is not a good indicator of freshness and/or spoilage. pH values vary too much between specimens so that no correlation can be calculated from single values. However, recently a review paper on the correlation between freshness and pH during cold storage was published which claimed that pH measurement is a reliable tool for the indication of freshness, rather than a sensory evaluation method which inherently has many uncertainties (Abbas *et al.*, 2008).

14.6.2 Eye fluid refractive index

Traditionally, the visual appearance of the fish eye has been widely used by consumers and trade to identify the freshness of fish because it changes dramatically shortly after catch. Loss of water and an increase in membrane permeability reduces the intraocular pressure of the eyes causing blur and/or light transmission properties. The alteration of the eye affects the refraction properties of the light due to an increase in the concentration of soluble components. Therefore it was proposed to use the refractive index of the fish eye for freshness determination. Only a few applications of this method have been published currently. Recent applications have been reported for sardine (Gokoglu and Yerlikaya, 2004) and for anchovy (Yapara and Yetimb, 1998).

14.7 Instrumental methods and automation

The preferred instrumental method for the determination of quality of seafood should ideally be cheap, easily handled and operated by non-scientifically educated personnel, non-invasive and non-destructive, able to measure freshness and spoilage throughout the whole span from catch/harvest to spoilage, used for all species and all types of seafood – and is therefore non-existent. All existing instrumentation is more or less an approximation or an approach towards the ideal instrument. However, combining the results obtained by different instruments measuring different principles and calibrating them with sensory scores of QIM for attributes like appearance, smell and texture, gives an Artificial Quality Index (AQI) that can be as accurate and precise as the QIM sensory score. The outcome provides a basis for the construction and industrial exploitation of multisensor devices for defining the quality of fish (Olafsdottir *et al.*, 2004). A short overview about some rapid methodologies (near infrared spectroscopy, imaging spectroscopy, NMR spectroscopy and X-ray imaging) used for the determination of basic composition of seafood has been published by Nilsen *et al.* (2010).

14.7.1 Fischtester and Torrymeter

The basis for electrical measurements on fish, using instruments like the Intellectron Fischtester VI or the Torrymeter (Distell Fish Freshness Meter), is the fact that the resistance of fresh fish measured immediately after catch is about 2000 Ω while spoiled fish has only 50 Ω of resistance left. The conductivity of fresh fish is approximately 500 μS and that of spoiled fish around 20 000 μS (Oehlenschläger and Nesvadba, 1997; Oehlenschläger, 2003, 2005; Kent and Oehlenschläger, 2009).

Cell membranes in fish muscle tissue are progressively disrupted by autolytic enzymatic degradation and later by microbial action. Cell fluid leaking into the intercellular space leads to a decrease of electrical resistance and capacity. This phenomenon can be measured by instruments which record a combination of capacity and resistance using electrodes applied either to one side of the fish or through the fish.

The readings of the instruments decrease almost with a linear function throughout the iced storage of marine fish. Therefore electrical measurements are – after an appropriate calibration for the species – an ideal tool for monitoring freshness and/or spoilage.

The Intellectron Fischtester VI was thoroughly investigated and found to be suitable for fresh fish grading, for the determination of days in ice and the prediction of days in ice left for lean fish species, for medium fatty and fatty fish species, and for aquacultured fish species. The instruments have also been used to differentiate between fresh and frozen/thawed fish and to examine the influence of catching methods on fish quality.

The decrease of Fischtester readings during the storage of wet fish in melting ice is linear throughout the whole storage period. In a series of experiments, results obtained with Fischtester VI were compared with those obtained with QIM. The correlation coefficients of one experiment between days in ice and QIM scores and Fischtester readings were 0.9873 and 0.9901, respectively. The correlation between Fischtester VI and QIM scores was −0.9654.

As Fischtester and Torrymeter measurements are based on the existence of intact cell membranes, they fail when the cells are disrupted or broken (damage, bruising, frozen/thawed fish).

14.7.2 VIS/NIR spectroscopy

The operating principle of the VIS/NIR spectroscopy measurement technique is based on sending light onto the sample and then measuring the light coming from the sample at different wavelengths. The wavelength region ranges from 400 to 2500 nm. This covers the visible (400–700 nm) and near infrared (700–2500 nm) regions.

VIS/NIR spectroscopy refers to experimental set-ups where the sample is illuminated with broad banded VIS/NIR light and the emitted VIS/NIR light spectrum from the sample is recorded. When the light is sent through the sample, it is named 'transmission' or 'transmittance' measurement. When the illumination and detector unit is located on the same side of the sample the measurement is referred to as 'reflection' or 'reflectance' measurement. A 'transflection/transflectance' measurement, also called 'diffuse reflectance', refers to the situation when the sample is illuminated and measured on the same side, but not in the same location.

In traditional VIS/NIR spectroscopy, the sample measurement results in a spectrum in the visible and/or near infrared region. In the past decade a new technique referred to as Imaging Spectroscopy has been developed. In addition to the spectral information this technique also gives spatial information of the sample. In other words, at each location at the sample a full spectrum is recorded. Imaging Spectroscopy can be used for transmission, reflection as well as transflection measurements (Nielsen and Heia, 2009).

Spectroscopy applied to fish or fish products is not a direct technique as further analysis of the recorded spectrum is needed. A common methodology to use is multivariate analysis, also referred to as chemometrics. These techniques model the relationship between the recorded spectral information and the information searched for. The complex nature of the VIS/NIR spectral region seldom permits the use of single wavelengths for qualitative or quantitative purposes. Typically, a model based on a higher number of wavelengths is required to extract useful information from the spectroscopic data.

A recent application of VIS/NIR spectroscopy demonstrated that visible spectroscopy can be used to predict the QIM score of cod caught both by long line and gillnet, and the QIM score for both catches may be predicted from VIS spectroscopy using only one model. It is shown that a relatively narrow band in the mid-visible spectral range is sufficient when measuring freshness. This is considered interesting in view of its simplicity for development of commercial instrumentation (Nilsen and Essaiassen, 2005).

The assessment of chemical constituents has been the main activity within VIS/NIR spectroscopy and fish quality determination. The fact that the sample material can be further processed and used for other purposes after VIS/NIR measurements makes the technique a viable tool for on-line analysis. Until now there have been few applications for safety and authentication issues.

VIS/NIR spectroscopy has recently been applied in the identification of authenticity of fresh and frozen/thawed swordfish (*Xiphias gladius* L) (Fasolato *et al.*, 2012) and Sivertsen *et al.* (2011) have described the application of VIS/NIR spectroscopy for automatic freshness assessment of cod (*Gadus morhua*) fillets.

14.7.3 Electronic nose

The development, the current status and future achievable developments of the electronic nose as a device to mimic the discrimination of the mammalian olfactory system for smells has been published in an extensive review by Röck *et al.* (2008). The analysis of odours, namely the investigation of gaseous samples containing volatile compounds, is a typical subject of analytical chemistry where several methods are available to separate mixtures in individual compounds (e.g. gas chromatography).

In order to analyse a gaseous mixture without separation, a set of selective sensors can be applied where ideally each sensor senses only one of the many molecular species. However, the poor selectivity of chemical sensors was considered as one of the main problems limiting their distribution for practical applications.

More than 10 000 odorous compounds are known to exist in nature but only a few of these are likely to be important in solving any discrimination task by an electronic nose. Since the early 1990s, a number of commercial instruments were launched on the market for monitoring freshness or quality of various foods, but there were some drawbacks. The main reason was that the instruments were not fully developed; improvements of the selectivity, sensitivity and reproducibility of the gas sensors have been the key issues. Electronic noses have been suggested for various applications related to quality evaluation of different foods like monitoring freshness and the onset of spoilage or bioprocesses in food. Many of these applications detecting volatile compounds produced because of the growth of fungi, moulds or microbes or changes occurring in food because of oxidation. In recent years attempts to use electronic nose technology to track the spoilage processes occurring in fish have been reported in numerous papers. Most of these are feasibility studies, showing the ability of the electronic nose to discriminate between different spoilage levels or the storage time of samples. Instruments with different sensor technologies have been used, for example metal-oxide chemoresistor sensors, MOSFET sensors, amperometric sensors, conducting polymer sensors and quartz microbalance sensors.

The currently available electronic noses are still based on an oversimplified olfaction model taking into consideration very little of the complexity of the natural olfaction. A next generation of electronic noses has been devised including more biological similitudes such as sample separation in an environment resembling the function of nasal mucus or introducing novel data analysis methods based on biological paradigms. Furthermore, the introduction of transducers based on consumer electronics may greatly help the process of capillary diffusion of the electronic nose simply embedding the sensors into current technology, such as computers or cellular phones (Di Natale and Olafsdottir, 2009; Di Natale, 2010). A recent result of the research about electronic noses is a method for building a simple and reproducible electronic nose using commercially available metal oxide sensors (MOS) to monitor the freshness of hairtail fish at 15, 10, and 5 °C (Tian *et al.*, 2012). The correlation coefficient with TVB-N was 0.97, although the accuracy was low.

14.7.4 Colour measurement

Colour is the visual perception that results from the detection of light after it has interacted with an object. The perceived colour of an object is affected by three entities: the physical and chemical composition of the object, the spectral composition of the light source illuminating the object, and the spectral sensitivity of the eye. Since human beings are sensitive to the colour of foods, appetite is stimulated or dampened in almost direct relation to the observer's reaction to colour. The consumer often assesses the initial quality of the food product by its colour and appearance: colour and/or lightness such as that of canned tuna are important for identity and quality grading. However, the colour of food is not stable, as it changes with increasing spoilage. The appearance and the brilliant colours of a freshly caught fish make it irresistible as an item of food. Just a few hours later, parallel with colours, the attractiveness fades away.

Instruments measuring colour photoelectrically can be divided into two classes, trichromatic colorimeters and spectrophotometers. Colorimeters are tristimulus devices that make use of red, green and blue filters that emulate the response of the human eye to light and colour. The more modern tristimulus instruments are linked to computers with automatic compensation and the provision of a number of colour spaces. A colorimeter uses a light source to light the specimen being measured. The light reflected from the object's surface than passes through the glass filters to simulate the standard observers' functions for a particular illumination. A photodetector beyond each filter then detects the amount of light passing through the filters. These signals are then displayed as X, Y and Z values.

Colour measurements have been commonly and successfully applied in aquaculture, minced fish products, surimi and surimi-based products, when assessing the processing effect on the colour of fish and fishery products, in refrigerated and frozen storage, as well as after thermal processing (heating and smoking) and high-pressure processing.

One of advantage of colour measurement is that it is not expensive and can therefore be applied widely. Two extensive overviews on colour measurement of fish and fishery products including surimi, surimi-based products and processing effects are given by Schubring (2009, 2010).

14.7.5 Texture measurement

Instrumental methods used for texture measurement are the Kramer test, the Warner–Bratzler shear test, puncture, tensile and compression tests, the texture profile analysis, and viscoelastic methods such as stress-relaxation, creep and oscillatory measurements.

The instrumental measurement methods have benefits and disadvantages, most disadvantages originating from the complex structure of fish muscle and the changes that occur during cooking which prevent the correct measurement of some properties for processed products. For a valid measurement it may be necessary to combine some methods. The development of quick tests with portable instruments is a challenge for fish quality assessment. Combining measurements of texture and other aspects of fish quality such as colour and odour, as demonstrated in the AQI concept should be pursued (Olafsdottir *et al.*, 2004).

Texture has been defined as the 'sensory and functional manifestation of the structural and mechanical properties of foods detected through the senses of vision, hearing, touch, and kinaesthetic'. The sensory quality of fish depends mainly on the texture characteristics. From the moment of catch/harvest or slaughter to consumption of the fish, several biochemical changes associated with the onset and resolution of rigor mortis occur, influencing the texture of the fish. Before the onset of rigor mortis the muscle is soft and elastic. In the state of rigor mortis, it becomes hard, due to the contraction of the muscle fibre forming the actomyosin complex, and with its resolution the muscle becomes soft but less elastic. The extent of these changes and their effect on muscle texture depend on factors including species, pre-slaughter conditions or catching methods, post-mortem handling and treatment, and storage time and temperature.

Freezing and frozen storage are among the most used technologies for long-term storage. During frozen storage, fish muscle may undergo a series of changes such as loss of water holding capacity, and unwanted changes in texture characteristics which result in a hard and dry product. Many factors affect texture changes during frozen storage including *ante mortem* factors, pre-processing, degree of comminution, or time and temperature of storage (Careche and Barroso, 2009).

Other technologies applied to fish may affect the textural properties in different ways, either causing biochemical, structural or microbiological changes, or preventing some unfavourable changes that otherwise could occur. The complexity of textural parameters in foods, the perception of texture, and the consumer attitudes towards this property have been reviewed by Szczesniak (2002), as well as the most complete sensory method, the Sensory Texture Profile (Szczesniak, 1998). Hyldig and Nielsen (2001) list some texture properties of fish evaluated in the literature.

14.7.6 NMR (Nuclear Magnetic Resonance)

The most common magnetic nuclei for the analysis of foods including seafood are ^{1}H, ^{13}C and ^{31}P. Nuclei in different chemical environments have slightly different resonance frequencies. NMR in chemistry has traditionally been limited to the analysis of liquid samples or solutions.

NMR-based methods have several inherent advantages. The NMR instrument interacts with the investigated object by means of electromagnetic waves in the radiofrequency range. This makes most NMR techniques non-invasive and non-destructive for the sample, rapid, harmless for the operator and non-polluting for the environment. The applications of NMR methods in food research can be divided into three main types in accordance to the type of equipment used as they can provide versatile information about the chemical composition and structure of biological systems at various levels. The types are: magnetic resonance imaging (MRI), low field (LF) NMR, and high resolution NMR (HR NMR). MRI is a technique that offers a unique opportunity to produce cross-section images of intact whole fish.

As MRI does not destroy the sample, so it is also a powerful tool to follow various dynamic processes in time. Thus, MRI can be used in fish processing as a research tool for the optimization of various unit operations such as drying, freezing, thawing, rehydration and salting. In aquaculture, MRI can be useful for studying various issues related to feed composition and flesh quality.

^{1}H NMR imaging of fish has been used to determine the spatial distribution and quantification of fat and water, and for assessment of fish quality changes after freezing and thawing, ^{23}Na imaging of fish for salt distribution and quantification challenges. Direct imaging of ^{23}Na nuclei using MRI is an efficient way to visualize salt distribution in cured seafood.

LF NMR (often also called time-domain NMR) is a fast and powerful method that can be used for non-invasive measurements of the mobility of water and fat in different tissues or compartments in muscle-based foods. The main advantages of the LF NMR method compared with the other NMR techniques are low investment costs, small size, no maintenance costs (permanent magnet), high stability, high degree of automation and easy operation. LF NMR can be implemented at-line under comparatively harsh industrial conditions.

The structure of muscles can be studied indirectly using LF ^{1}H NMR. Different tissue compartments or muscle structures, as affected for instance by fish processing, can be studied since protons in different environments exhibit different T_1 and T_2 relaxation properties. The relaxation data may be interpreted in terms of structural changes occurring after fish processing operations.

High resolution NMR has been shown to be particularly well suited in the profiling of biological material, and it has emerged as a popular technique for the analysis of foodstuffs including fish and fish products (Aursand *et al.*, 2009). HR NMR gives a 'fingerprint' of the composition of the sample analysed and allows identification/quantification of metabolites with relevance for the quality and nutritional value.

During the past decade, HR NMR has become an important tool in quality assessment of fish and fish products in research. The techniques have been found valuable particularly for studies of marine lipids since the pure extracts can be analysed non-destructively and non-invasively. Lipid extracts have previously been analysed by NMR to determine lipid class composition, i.e. triacylglycerols, phospholipids (including individual phospholipids such as phosphatidylcholine, phosphatidylethanolamine), cholesterol and free fatty acids. Also specific esterified fatty acids and their individual acyl-positions of the fatty acids in lipid molecules are identified (Falch *et al.*, 2006). The DHA and $n-3$ fatty acid content can also be quantified directly from intact material (e.g. fish tissues) using the MAS (by magic angle spinning the broad lines of the spectrum become narrower, increasing the resolution) technique (Aursand *et al.*, 2006). In a survey of nutrients and a quality assessment of crab paste by 1H NMR spectroscopy and multivariate data analysis, it was recently demonstrated that NMR-based metabolomics technology could be used as a powerful tool to analyse the complex aqueous composition of crab paste, which is largely dominated by amino acids, organic acids, nucleotides, amines and sugars (Ye *et al.*, 2012). Another application of NMR spectroscopy in combination with multivariate data analysis was used to investigate the molecular components of the aqueous extract of samples of bottarga, a salted and dried mullet (*Mugil cephalus*) roe, manufactured in Sardinia (Italy) from mullet of known and unknown geographical provenance (Locci *et al.*, 2011).

14.8 Imaging technologies and machine vision

Kroeger (2009) has published a concise document on the background and the principles of image analysis and concludes that a visual inspection system is capable of investigating particular aspects of quality assessment for fishery products. The trends in the application of imaging technologies to inspect fish and fish products was later reviewed by Mathiassen *et al.* (2011), in particular reviewing the applications of visible/near-infrared (VIS/NIR) imaging, VIS/NIR imaging spectroscopy, planar and computed tomography (CT), X-ray imaging and magnetic resonance imaging (MRI). Another review on this topic entitled 'Machine Vision Applications to Aquatic Foods' by Gümüs *et al.* (2011) describes the determination of composition, measurement and evaluation of size and volume, measurement of shape parameters, quantification of the outside or meat colour of aquatic foods, and as well as detection of defects during quality evaluation. A third recent review (Dowlati *et al.*, 2012) underlines the importance of this emerging technology. The authors state: 'Machine vision can evaluate the quality of fish and fish products, providing fast identification and measurement of selected objects and classifying the samples into categories based on shape, size, color and other visual attributes. The attractiveness of machine vision for the fish industry is maintained via inspection of diverse raw and processed aquatic foods in a flexible and non-destructive manner.'

14.9 Conclusion

Despite the development of numerous allegedly specially tailored instrumental analytical methods in seafood research laboratories worldwide during the past 20 years that are able to measure different well-defined properties of fish with high accuracy and repeatability, there is no single robust method available which reliably and satisfactorily measures the overall quality of seafood. Progress has been made in identifying and measuring spoilage parameters and certain quality related criteria, but the newly developed methods have been used so far almost exclusively in research laboratories or to some extent in cooperation projects with industry; none of them has widespread use in the industry as a routine method. The methods used in daily practise in the SMEs which form the majority of seafood enterprises worldwide are those which have been used for decades (TVB-N, pH, sensory inspection). What has changed is the equipment – from manual to automatic operation.

By combining the results obtained by different methods and calibrating them with sensory scores of the QIM for attributes like appearance, smell and texture, a so called AQI could be calculated that can be as accurate and precise as the QIM sensory score (Olafsdottir *et al.*, 2004). However, the only single method which currently gives a reliable impression of the complex phenomenon of quality or 'freshness' of fish is the sensory assessment of fish using a trained panel. A chapter in a book written by Connell and Shewan (1980) more than 30 years ago about sensory and non-sensory assessment of fish reflects to a great extent what is still common practice in many establishments today.

Tom Gill (1992) ended a short presentation about biochemical and chemical indices of seafood quality during a conference in Copenhagen in 1991 by saying 'it remains to be seen which of these [new] technologies will obtain the greatest commercial success but it is perhaps inevitable that objective quality testing will become routine as fish prices continue to increase.' Now, over 20 years later the answer is: none of the new technologies has become a routine method in daily practice despite fish being more expensive than ever.

References

Abbas, K.A, Mohamed, A., Jamilah, B. and Ebrahimian, M. (2008). A review on correlations between fish freshness and pH during cold storage. *American Journal of Biochemistry and Biotechnology*, 4, 416–421.

Antonacopoulos, N. and Vyncke, W. (1989). Determination of volatile basic nitrogen in fish: A third collaborative study by the West European Fish Technologists' Association (WEFTA). *Zeitschrift für Lebensmittel-Untersuchung und Forschung*, 189(4), 309–316.

Anonymous (1996). Council regulation (EC) No. 2406/96 of 26 November laying down common marketing standards for certain fishery products. *Official Journal of European Communities* No. L334/1–14, 23 December 1996.

Anonymous (2005). Commission Regulation (EC) No 2074/2005 of 5 December 2005 laying down implementing measures for certain products under Regulation (EC) No 853/2004 of the European Parliament and of the Council and for the organisation of official controls under Regulation (EC) No 854/2004 of the European Parliament and of the Council and Regulation (EC) No 882/2004 of the European Parliament and of the Council, derogating from Regulation (EC) No 852/2004 of the European Parliament and of the Council and amending Regulations (EC) No 853/2004 and (EC) No 854/2004. *Official Journal of the European Union* L 338/27–59.

Aursand, M., Gribbestad, I. and Martinez, I. (2006). Omega-3 fatty acid content of intact muscle of farmed Atlantic salmon (*Salmo salar*) examined by ^{1}H MAS NMR spectroscopy, in *Handbook of Modern Magnetic Resonance* (ed. G. Webb), Kluwer/Plenum Publishers, London, UK.

Aursand, M., Veliyulin, E., Standal, I.B. *et al.* (2009). Nuclear magnetic resonance, in *Fishery Products – Quality, Safety and Authenticity* (eds H. Rehbein and J. Oehlenschläger), Wiley-Blackwell, Oxford, pp. 252–272.

Botta, J.R. (1995). *Evaluation of Seafood Freshness Quality*. VCH, New York, Weinheim, Cambridge.

Bremner, H.A. (1985). A convenient, easy-to-use system for estimating the quality of chilled seafood. *Fish Processing Bulletin*, 7, 59–70.

Bremner, A. (1997). If freshness is lost, where does it go? in *Methods to Determine the Freshness of Fish in Research and Industry* (eds G. Olafsdottir *et al.*), International Institute of Refrigeration, Paris, pp. 36–51.

Bremner, H.A. and Sakaguchi, M. (2000). A Critical Look at Whether 'Freshness' Can Be Determined, *Journal of Aquatic Food Product Technology*, 9, 5–25.

Buendia, E. (1997). Anisakis, anisakidosis, and allergy to Anisakis. *Allergy*, 52, 481–482.

Careche, M. and Barroso, M. (2009). Instrumental texture measurement, in *Fishery Products – Quality, Safety and Authenticity* (eds H. Rehbein and J. Oehlenschläger), Wiley-Blackwell, Oxford, pp. 214–239.

Carvalho, M.L., Santiago, S. and Nunes, M.L. (2005). Assessment of the essential element and heavy metal content in edible fish muscle. *Analytical and Bioanalytical Chemistry*, 328, 426.

Connell, J.J. (1990). *Control of Fish Quality*, 3rd edition, Fishing News Books, Oxford.

Connell, J.J. and Shewan, J.M. (1980). Sensory and non-sensory assessment of fish. In: Connell, J.J. (Ed.), *Advances in fish science and technology*, pp. 56–64, Fishing News Books, Farnham, Surrey.

Conway, E.J. and Byrne, A. (1933). An absorption apparatus for the micro-determinationof certain volatile substances. I. The micro-determination of ammonia. *Biochemical Journal*, 27, 419–429.

Di Natale, C. and Olafsdottir, G. (2009). Electronic nose and electronic tongue, in *Fishery Products – Quality, Safety and Authenticity* (eds H. Rehbein and J. Oehlenschläger), Wiley-Blackwell, Oxford, pp. 105–126.

Di Natale, C. (2010). Chemical sensors, in *Handbook of Seafood and Seafood Products Analysis* (eds L.M.L. Nollet and F. Toldra), CRC Press, Boca Raton, pp. 3–10.

Dowlati, M., Mohtasebi, S.S. and de la Guardia, M. (2012). Application of machine-vision techniques to fish-quality assessment. *Trends in Analytical Chemistry*, 40, 168–180.

Falch, E., Størseth, T.R. and Aursand, M. (2006). Multi-component analysis of marine lipids in fish gonads with emphasis on phospholipids using high resolution NMR spectroscopy. *Chemistry and Physics of Lipids*, 144(1), 4–16.

Fasolato, L., Balzan S., Riovanto R. *et al.* (2012). Comparison of visible and near-infrared reflectance spectroscopy to authenticate fresh and frozen-thawed swordfish (*Xiphias gladius* L). *Journal of Aquatic Food Product Technology*, 21, 493–507.

Fillit, C., Jaffrezic Renault, N., Bessueille, F. *et al.* (2008). Development of a microconductometric biosniffer for detection of trimethylamine. *Materials Science and Engineering* C, 28, 781–786.

FishBase (2012). http://www.fishbase.org/search.php [17 June 2013].

Gill, T.A. (1992). Biochemical and chemical indices of seafood quality, in *Quality Assurance in the Fish Industry* (eds H.H. Huss *et al.*), Elsevier, Amsterdam, pp. 377–387.

Gokoglu, N. and Yerlikaya, P. (2004). Use of eye fluid refractive index in sardine (*Sardina pilchardus*) as a freshness indicator. *European Food Research and Technology*, 218, 295–297.

Gümüş, B. (2011). Machine vision applications to aquatic foods: a review. *Turkish Journal of Fisheries and Aquatic Sciences*, 11, 171–181.

Howgate, P. (2009). Traditional methods, in *Fishery Products – Quality, Safety and Authenticity* (eds H. Rehbein and J. Oehlenschläger), Wiley-Blackwell, Oxford, pp. 19–41.

Howgate, P. (2010a). A critical review of total volatile bases and trimethylamine as indices of freshness of fish. Part 1. Determination. *Electronic Journal of Environmental, Agricultural and Food Chemistry*, 9, 29–57.

Howgate, P. (2010b). A critical review of total volatile bases and trimethylamine as indices of freshness of fish. Part 2. Formation of the bases, and application in quality assurance. *Electronic Journal of Environmental, Agricultural and Food Chemistry*, 9, 58–88.

Howgate, P., Johnston, A. and Whittle, K.J. (1992). *Multilingual guide to EC freshness grades for fishery products*. Torry Research Station, Food Safety Directorate, Ministry of Agriculture, Fisheries, and Food, Aberdeen, Scotland.

Huss, H.H., Jakobsen, M. and Liston, J. (eds) (1992). *Quality Assurance in the Fish Industry*, Elsevier, Amsterdam.

Hyldig, G. and Nielsen, D. (2001). A review of sensory and instrumental methods used to evaluate the texture of fish muscle. *Journal of Texture Studies*, 32, 219–242.

Hyldig, G. (2010a). Sensory descriptors, in *Handbook of Seafood and Seafood Products Analysis* (eds L.M.L. Nollet and F. Toldra), CRC Press, Boca Raton, pp. 481–498.

Hyldig, G., Martinsdottir, E., Sveinsdottir, K., Schelvis, R. and Bremner, A. (2010). Quality index methods, in *Handbook of Seafood and Seafood Products Analysis* (eds L.M.L. Nollet and F. Toldra), CRC Press, Boca Raton, pp. 463–480.

Hyldig, G., Bremner, A., Martinsdottir, E. and Schelvis, R. (2012a). Quality Index Methods. In: Nollet, L.M.L. (Ed.), *Handbook of Meat, Poultry and Seafood Quality*. 2nd ed., pp. 437–458, WILEY-BLACKWELL, Oxford.

Hyldig, G. (2012b). Sensory quality of fish, in: *Handbook of Meat, Poultry and Seafood Quality*, 2nd edn (ed. L.M.L. Nollet), Wiley-Blackwell, Oxford, pp. 459–478.

Jørgensen, B.M., Oehlenschläger, J., Olafsdottir, G. *et al.* (2003). A study of the attitudes of the European fish sector towards quality monitoring and labeling, in *Quality of Fish from Catch to Consumer – Labelling, Monitoring and Traceability* (eds J.B. Luten *et al.*), Wageningen Academic Publishers, pp. 57–74.

Kent, M. and Oehlenschläger, J. (2009). Measuring electrical properties, in *Fishery Products – Quality, Safety and Authenticity* (eds H. Rehbein and J. Oehlenschläger), Wiley-Blackwell, Oxford, pp. 286–300.

Kim Y.-N. (2010). Vitamins, in *HANDBOOK of Seafood and Seafood Product Analysis* (eds L.M.L. Nollet and F. Toldra), CRC Press, Boca Raton, pp. 327–350.

Kroeger, M. (2009a). Quality and safety assessment by image processing, in *Fishery Products – Quality, Safety and Authenticity* (eds H. Rehbein and J. Oehlenschläger), Wiley-Blackwell, Oxford, pp. 240–251.

Locci, E, Piras, C., Mereu, S., , Cesare Marincola, F. and Scano, P. (2011). ^{1}H NMR metabolite fingerprint and pattern recognition of mullet (*Mugil cephalus*) Bottarga. *Journal of Agriculture and Food Chemistry* 59, 9497–9505.

Love, R.M. (1988). *The Food Fishes their Intrinsic Variation and Practical Implications*, Farrand Press, London.

Los Angeles Times (1990). Sea of Confusion Over 'Fresh Fish' Definition: Food: Congress May Intervene with Bills to Protect the Seafood Buyer. http://articles.latimes.com/1990-05-09/business/fi-308_1_fresh-seafood [4 June 2013].

Luten, J.B., Børresen, T. and Oehlenschläger, J. (eds) (1997). *Seafood from Producer to Consumer, Integrated Approach to Quality*, Elsevier, Amsterdam.

Luten, J.B., Oehlenschläger, J. and Olafsdottir, G. (Eds.) (2003). *Quality of Fish from Catch to Consumer – Labelling, Monitoring and Traceability*. Wageningen Academic Publishers.

Martinsdottir, E., Schelvis, R., Hyldig, G. and Sveinsdottir, K. (2009a). Sensory evaluation of seafood: general principles and guidelines, in *Fishery Products – Quality, Safety and Authenticity* (eds H. Rehbein and J. Oehlenschläger), Wiley-Blackwell, Oxford, pp. 411–424.

Martinsdottir, E., Schelvis, R., Hyldig, G. and Sveinsdottir, K. (2009b). Sensory evaluation of seafood: methods, in *Fishery Products – Quality, Safety and Authenticity* (eds H. Rehbein and J. Oehlenschläger), Wiley-Blackwell, Oxford, pp. 425–443.

Matís (2012). Shelf life and freshness. http://www.kaeligatt.is/english/shelf-life-and-freshness/ [4 June 2013].

Mathiassen, J.R, Misimi, E., Bondø, M., Veliyulin, E. and Ove Østvik, S. (2011). Trends in application of imaging technologies to inspection of fish and fish products. *Trends in Food Science and Technology*, 22, 257–275.

Mendes, R. (2009). Biogenic amines, in *Fishery Products – Quality, Safety and Authenticity* (eds H. Rehbein and J. Oehlenschläger), Wiley-Blackwell, Oxford, pp. 42–67.

Mirriam Webster (2012). *Collegiate Dictionary*. Available at: http://www.merriam-webster .com/thesaurus/freshness [4 June 2013].

Nilsen, H. and Esaiassen, M. (2005). Predicting sensory score of cod (*Gadus morhua*) from visible spectroscopy. *Lebensmittel-Wissenschaft und-Technologie*, 38, 95–99.

Nielsen, H.A. and Heia, K. (2009). VIS/NIR spectroscopy, in *Fishery Products – Quality, Safety and Authenticity* (eds H. Rehbein and J. Oehlenschläger), Wiley-Blackwell, Oxford, pp. 89–104.

Nilsen, H., Heia, K. and Esaiassen, M. (2010). Basic composition: rapid methodologies, in *Handbook of Seafood and Seafood Products Analysis* (eds L.M.L. Nollet and F. Toldra), CRC Press, Boca Raton, pp. 121–138.

Nollet, L.M.L. and Toldra, F. (2010). *Handbook of Seafood and Seafood Products Analysis*. CRC Press, Boca Raton.

Oehlenschläger, J. and Nesvadba, P. (1997). Methods for freshness measurement based on electrical properties of fish tissue, in *Methods to Determine the Freshness of Fish in Research and Industry* (eds G. Olafsdottir *et al.*), International Institute of Refrigeration, Paris, pp. 363–368.

Oehlenschläger, J. and Rehbein, H. (2009). Basic facts and figures, in *Fishery Products – Quality, Safety and Authenticity* (eds H. Rehbein and J. Oehlenschläger), Wiley-Blackwell, Oxford, pp 1–18.

Oehlenschläger, J. and Sörensen, N.-K, (1997). Criteria of seafood freshness and quality aspects, in *Methods to Determine the Freshness of Fish in Research and Industry* (eds G. Olafsdottir *et al.*), International Institute of Refrigeration, Paris, pp. 30–35.

Oehlenschläger, J. (1997a). Suitability of ammonia-N, dimethylamine-N, trimethylamine-N, trimethylamine oxygen-N and total volatile basic nitrogen as freshness indicators in seafood, in *Methods to Determine the Freshness of Fish in Research and Industry* (eds G. Olafsdottir *et al.*), International Institute of Refrigeration, Paris, pp. 92–99.

Oehlenschläger, J. (1997b). Volatile amines as freshness/spoilage indicators. A literature review, in *Seafood from Producer to Consumer, Integrated Approach to Quality* (eds J.B. Luten *et al.*), Elsevier, Amsterdam, pp. 571–588.

Oehlenschläger, J. (2002). Identifying heavy metals in fish, in Bremner, H.A. (Ed.), *Safety and Quality Issues in Fish Processing*. pp. 95–113, Woodhead Publishing Ltd, Cambridge.

Oehlenschläger, J. (2003). Measurement of freshness of fish based on electrical properties, in *Quality of Fish from Catch to Consumer: Labelling, Monitoring and Traceability* (eds J.B. Luten, J. Oehlenschläger and G. Olafsdottir), Wageningen Academic Publishers, pp. 237–249.

Oehlenschläger, J. (2005). The Intellectron Fischtester VI – an almost forgotten but powerful tool for freshness and spoilage determination of fish at the inspection level, in *Fifth World Fish Inspection and Quality Control Congress* (eds J. Ryder and L. Ababouch), FAO, Rome, pp. 116–122.

Oehlenschläger, J. (2006). Cholesterol content in seafood. Data from the last decade: a review, in *Seafood Research from Fish to Dish* (eds J.B. Luten *et al.*), Wageningen Academic Publishers, pp. 41–58.

Oehlenschläger, J. (2010a). Introduction – Importance of analysis in seafood and seafood products, variability and basic concepts, in *Handbook of Seafood and Seafood Products Analysis* (eds L.M.L. Nollet and F. Toldra), CRC Press, Boca Raton, pp. 3–10.

Oehlenschläger, J. (2010b). Minerals and trace elements, in *Handbook of Seafood and Seafood Products Analysis* (eds L.M.L. Nollet and F. Toldra), CRC Press, Boca Raton, pp. 352–376.

Önal, A. (2007). A review: Current analytical methods for the determination of biogenic amines in foods. *Food Chemistry* 103, 1475–1486.

Olafsdottir, G, Martinsdottir, E., Oehlenschläger, J. *et al.* (1997a). Methods to evaluate freshness in research and industry. *Trends in Food Science and Technolology*, 8, 258–265.

Olafsdottir, G., Luten, J., Dalgaard, P. *et al.* (eds) (1997b). *Methods to Determine the Freshness of Fish in Research and Industry*, International Institute of Refrigeration, Paris.

Olafsdottir, G., Nesvadba, P., Di Natale, C. *et al.* (2004). Multisensor for fish quality determination, *Trends in Food Science and Technology*, 15, 86–93.

Ostermeyer, U. and Schmidt, T. (2006). Vitamin D and provitamin D in fish – Determination by HPLC with electrochemical detection. *European Food Research Technology*, 222, 403–413.

Ostermeyer, U. (1999). Vitamine in Fischen. *Inf Fischwirtsch Fischereiforsch*, 46, 42–50.

Ozogul, Y. (2010). Methods for freshness quality and deterioration, in *Handbook of Seafood and Seafood Products Analysis* (eds L.M.L. Nollet and F. Toldra), CRC Press, Boca Raton, pp. 189–214.

QIM Eurofish (2012). http://www.qim-eurofish.com/default.asp?ZNT=S0T1O13 [4 June 2013].

Rehbein, H. and Oehlenschläger, J. (2009b). *Fishery Products – Quality, Safety and Authenticity*, Wiley-Blackwell, Oxford.

Röck, F., Barsan, N, and Weimar, U. (2008). Electronic nose: current status and future trends. *Chemical Reviews*, 108, 705–725.

Ruiz-Capillas, C. and Jimenez-Colmenero, F. (2010). Biogenic amines in seafood products. in *Handbook of Seafood and Seafood Products Analysis* (eds L.M.L. Nollet and F. Toldra), CRC Press, Boca Raton, pp. 833–850.

Saito, T., Arai, K. and Matsuyoshi (1959). A new method for estimating the freshness of fish. *Bulletin of the Japanese Society of Scientific Fisheries*, 24, 749–750.

Schubring, R. (2009). Colour measurement, in *Fishery Products – Quality, Safety and Authenticity* (eds H. Rehbein and J. Oehlenschläger), Wiley-Blackwell, Oxford, pp. 127–172.

Schubring, R. (2010). Quality assessment of fish and fishery products by color measurement, in *Handbook of Seafood and Seafood Products Analysis* (eds L.M.L. Nollet and F. Toldra), CRC Press, Boca Raton, pp. 395–424.

Sivertsen, A.H., Kimiya T. and Heia K. (2011). Automatic freshness assessment of cod (*Gadus morhua*) fillets by Vis/NIR Spectroscopy. *Journal of Food Engineering*, 103, 317–323.

Sørensen, N.K. (1992). Physical and instrumental methods for assessing seafood quality, in *Quality Assurance in the Fish Industry* (eds H.H. Huss, M. Jakobsen, and J. Liston), Elsevier, Amsterdam, pp. 321–332.

Souci, S.W., Fachmann, W. and Kraut, H. (eds) (2008). *Food Composition and Nutrition Tables*, 7th edn, Medpharm Scientific Publishers, Stuttgart, pp. 475–616.

Szczesniak, A.S. (1998). Sensory texture profiling – historical and scientific perspectives. *Food Technology*, 52, 54–57.

Szczesniak, A.S. (2002). Texture is a sensory property. *Food Quality and Preference*, 13, 215–225.

Tejada, M. (2009), ATP-derived products and K-value determination, in *Fishery Products – Quality, Safety and Authenticity* (eds H. Rehbein and J. Oehlenschläger), Wiley-Blackwell, Oxford, pp. 68–88.

Tian, X.-Y., Cai Q. and Zhang Y.-M. (2012). Rapid classification of hairtail fish and pork freshness using an electronic nose based on the PCA method. *Sensors*, 12, 260–277.

Vyncke, W. (1995). The determination of the total volatile bases in eye fluid as a non-destructive spoilage assessment test for fish. *Archiv fuer Lebensmittelhygiene*, 46, 96–98.

Yapara, A. and Yetimb, H. (1998). Determination of anchovy freshness by refractive index of eye fluid. *Food Research International*, 31, 693–695.

Ye, Y.F. (2012). Survey of nutrients and quality assessment of crab paste by ^1H NMR spectroscopy and multivariate data analysis. *Chinese Science Bulletin*, 57, 3353–3362.

15

Microbiological Examination of Seafood

Ioannis S. Boziaris and Foteini F. Parlapani

Department of Ichthyology and Aquatic Environment, School of Agricultural Sciences, University of Thessaly, Volos, Greece

15.1 Introduction

Microbiological quality assessment is necessary to ensure the safety and assess the quality and freshness of seafood throughout the production chain, from harvesting, processing and distribution to the consumer's table. Microorganisms are the main cause of spoilage in fresh and lightly processed seafood (Ashie *et al.*, 1996; Gram and Huss, 1996; Gram and Dalgaard, 2002). Additionally, food-borne pathogenic bacteria can occur on seafood causing food-borne infection and intoxications (Feldhusen, 2000; Huss *et al.*, 2000). The detection of these food-borne pathogens and the identification and enumeration of microbial spoilage populations is very important for the food industry and related food authorities.

Conventional microbiological methods are still considered as standards methods and rely on the use of appropriate culture media for enrichment, isolation, identification and enumeration of microorganisms. These methods are laborious and time-consuming, so alternative rapid methods of determination are urgently needed by the seafood industry. Indirect methods, such as measuring microbial metabolism rather than population counts, for example impedimetry, or methods that rely on immunological or genetic characteristics, may be used for rapid microbiological examinations.

In this chapter, the microbiological parameters of seafood and the methodologies usually used for their determination in quality control and research are presented. Conventional culture techniques, indirect methods and other rapid methods and molecular techniques applied in seafood microbiological examination are presented and discussed.

Seafood Processing: Technology, Quality and Safety, First Edition. Edited by Ioannis S. Boziaris.

15.2 Seafood microbiology

15.2.1 Indigenous microbiota

The indigenous microbiota of aquatic products consists mainly of bacteria that can cause spoilage while some can also be human pathogens. Natural bacterial populations can be found in skin, gills and the digestive tract (Shewan, 1962; Liston, 1980; Gennari *et al.*, 1999). Gram-negative bacteria, usually psychrotrophs, predominate especially in fish from cold (North Atlantic, North Sea) or more warm (Mediterranean Sea) temperate waters, while higher amounts of Gram-positive bacteria can be found in fish from tropical waters (Gram *et al.*, 1990; Gram and Huss, 1996). *Pseudomonas, Shewanella, Psychrobacter, Moraxella, Acinetobacter, Flavobacterium, Vibrio, Photobacterium* and *Aeromonas*, are the most common Gram-negative genera that make up initial microbiota, while the most common Gram-positive genera are *Micrococcus, Corynebacterium, Bacillus* and *Clostridium* (Gram *et al.*, 1990; Gram and Huss, 1996). Enterobacteriaceae can also be found but thought to be non-indigenous microbiota originating from human activity.

15.2.2 Contamination (exogenous) microbiota

Contamination with non-indigenous microorganisms may occur during processing. Microorganisms from the environment, food contact surfaces and workers can be transferred to food (Jay *et al.*, 2005). It is known that humans carry many bacteria of which some may be food-borne pathogens such as *Staphylococcus aureus*. Enterobacteriaceae are commonly found in any environment, while contamination with enteric pathogens can also occur in poor hygienic conditions. *Listeria monocytogenes* can attach on food contact surfaces forming biofilms and contaminating seafood (Truelstrup and Vogel, 2011). The conditions imposed by processing and those that occur during storage and distribution greatly affect the growth and consequently the dominant microbial diversity of seafood. For fresh or lightly preserved seafood, extrinsic factors like temperature and atmosphere (e.g. CO_2 presence) have the most affect on microbiological growth, while for preserved products the alteration of intrinsic properties (pH, a_w, etc.) also contribute to the selection of the dominant microbiota (Gram, 2009).

15.2.3 Spoilage microbiota

In fresh or lightly preserved seafood, the main spoilage mechanism is the growth and metabolic activity of microorganisms, which results in the production of various metabolites that directly affect the organoleptic properties of the product (Ashie *et al.*, 1996; Dainty, 1996; Gram and Huss, 1996). The main microorganisms responsible for the production of off-flavours/odours (chemical spoilage indices), which lead to seafood rejection, are specific spoilage organisms (SSOs). Thorough reviews of the SSO concept can be found in the scientific literature (Gram and Huss, 1996; Huis in't Veld, 1996; Dalgaard, 2000). Accumulation of microbial metabolites at levels high enough to cause organoleptic rejection usually occurs when the total microbial population or population of SSOs reach the level of 10^7–10^9 cfu/g. The time required for microorganisms to reach the minimum spoilage level determines shelf life. Shelf life usually correlates well with the population of total or spoilage microorganisms (Dalgaard, 1995a; Dalgaard *et al.*, 1997; Koutsoumanis and Nychas, 1999, 2000; Koutsoumanis, 2001; Boziaris *et al.*, 2011). Hence, identification and enumeration of spoilage microorganisms is important to allow action to be taken in time, to avoid negative impacts and also to evaluate the quality status of seafood.

Various microorganisms have been characterized as SSOs. For a given set of primary processing and storage conditions, the SSOs of fresh seafood are determined, mainly by geographic origination (temperate or tropical) and the waters (marine or fresh waters) where those aquatic organisms live (Liston, 1980; Ashie, *et al.*, 1996; Gram and Huss, 1996). *Pseudomonas* and *Shewanella* species are the spoilers of marine fish and crustaceans stored aerobically at low temperatures (Gram and Huss, 1996; Dalgaard, 2000; Koutsoumanis and Nychas, 1999, 2000; Lopez-Caballero *et al.*, 2006; Boziaris *et al.*, 2011). Freshwater chilled fish is usually spoiled by the action of *Pseudomonas* (Gram *et al.*, 1990; Gram and Huss, 1996). Under elevated CO_2 concentrations (modified atmosphere packaging, MAP), *Photobacterium phosphoreum* and lactic acid bacteria usually predominate (Dalgaard, 2000; Koutsoumanis *et al.*, 2000). Seafood processed by curing usually spoils due to microorganisms that grow and predominate under reduced pH and a_w, such as lactic acid bacteria and yeasts (Gram, 2009; Boziaris *et al.*, 2013). However, in such processed products other spoilage mechanisms like autolysis and chemical oxidation can also contribute to deterioration of product quality (Ashie *et al.*, 1996).

15.2.4 Pathogenic microorganisms

Several bacterial pathogens have been implicated in seafood-borne infections or intoxications. These bacterial pathogens can originate from indigenous microbiota, or contaminate fish when they are still in the water and/or during processing. Pathogenic bacteria naturally present in the aquatic environment are *Vibrio cholerae, V. parahaemolyticus, V. vulnificus, Aeromonas hydrophilla* and *Clostridium botulinum* (Huss *et al.*, 2000). Enteric bacteria, like *Salmonella* spp., *Shigella* spp., pathogenic *Escherichia coli*, etc., can be also present due to the faecal contamination of water. Other microorganisms such *L. monocytogenes, S. aureus, Bacillus cereus, C. perfrigens* together with bacterial pathogens of faecal origin can contaminate the product during processing (Feldhusen, 2000).

Enteric viruses such as hepatitis A virus (HAV), Norwalk-like viruses and Astrovirus are important human pathogens and they are sometimes associated with seafood, especially with shellfish. Shellfish such as oysters, cockles and mussels are filter feeding organisms. Therefore, they can concentrate a number of viruses which may cause outbreaks of hepatitis or gastroenteritis if the shellfish are eaten raw or are insufficiently cooked (Koopmans *et al.*, 2002; Casas *et al.*, 2007).

15.3 Microbiological parameters of seafood analysis

Microbiological analysis is essential to estimate quality and potential shelf life and ensure the microbiological safety of seafood. Table 15.1 summarizes important seafood microbiological parameters for seafood that are set out as requirements in various guidelines and legislation.

The total amount of bacteria serves as a general microbiological parameter which indicates the hygienic quality, freshness and the potential shelf life of the product. Freshness and shelf life can be evaluated by the total aerobic plate counts (APC) or psychrotrophic plate count (PPC) or alternatively by the population of SSOs. Guidelines

Table 15.1 Microbiological parameters of seafood

Seafood	Microbiological parameter	Sampling plan		Limit		Reference
		n	c	m	M	
Cooked crustaceans, molluscan shellfish	*Salmonella*[a]	5	0	Absence in 25 g		Commission Regulation (EC) (2007)
Live bivalve molluscs and live echinoderms, tunicates, gastropods	*Salmonella*[a]	5	0	Absence in 25 g		
	E. coli[a]	1	0	230 MPN/100 g		
Shelled and shucked products of cooked crustaceans and molluscan shellfish[b]	*E. coli*[b]	5	2	1 cfu /g	10 cfu /g	
	Coagulase-positive staphylococci[b]	5	2	100 cfu/g	1000 cfu/g	
Ready-to-eat foods able to support the growth of *L. monoc.*	*L. monocytogenes*[a]	5	0	100 cfu/g		
	L. monocytogenes[c]	5	0	Absence in 25 g		
Ready-to-eat foods unable to support the growth of *L. monoc.*	*L. monocytogenes*[a]	5	0	100 cfu/g		
Fresh and frozen fish; cold smoked fish	APC	5	3	5×10^5 cfu/g	10^7 cfu/g	ICMSF (1986)
	E. coli	5	3	11 cfu/g	500 cfu/g	
Precooked breaded fish	APC	5	2	5×10^5 cfu/g	10^7 cfu/g	
	E. coli	5	2	11 cfu/g	500 cfu/g	
Frozen raw crustaceans	APC	5	3	10^6 cfu/g	10^7 cfu/g	
	E. coli	5	3	11 cfu/g	500 cfu/g	
Frozen cooked crustaceans	APC	5	2	5×10^5 cfu/g	10^7 cfu/g	
	E. coli	5	2	11 cfu/g	500 cfu/g	
	S. aureus	5	0	10^c cfu/g	-	
Cooked, chilled, and frozen crabmeat	APC	5	2	10^5 cfu/g	10^6 cfu/g	
	E. coli	5	1	11 cfu/g	500 cfu/g	
	S. aureus	5	0	10^c cfu/g	-	
Fresh and frozen bivalve molluscs	APC	5	0	5×10^5 cfu/g	-	
	E. coli	5	0	16 cfu/g	-	
Cooked or ready- to-eat	*E. coli*	5	1	4 cfu/g	40 cfu/g	Canadian Food Inspection Agency (2011)
Raw molluscan shellfish	*E. coli*	5	1	230 MPN/100 g	330 MPN/100g	
All other types of products	*E. coli*	5	2	4 cfu/g	40 cfu/g	

All types of products	Coagulase-positive staphylococci	5	1	10[c] cfu/g	10[d] cfu/g	
	Salmonella	5	Absent in each 25 g sample or in pooled samples of 125 g			
Cooked or ready- to-eat products	*V. cholerae*	5	Absent in each 25 g sample or in pooled samples of 125 g			
Raw crustaceans	*E. coli*	5	1	10[b] cfu/g	10[c] cfu/g	ANZFA (2001)
	V. cholerae	5	0	0 cfu/g		
	Coagulase-positive staphylococci	5	2	10[b] cfu/g	10[c] cfu/g	
	Salmonella	5	0	Absence in 25 g		
	APC	5	0	5×10^5 cfu/g	5×10^6 cfu/g	
Cooked crustaceans	*E. coli*	5	1	10 cfu/g	10[b] cfu/g	
	V. cholerae	5	0	0 cfu/g		
	V. parahaemolyticus	5	2	10[b] cfu/g	10[c] cfu/g	
	Coagulase-positive staphylococci	5	2	10[b] cfu/g	10[c] cfu/g	
	Salmonella	5	0	Absence in 25 g		
	APC	5	0	10^5 cfu/g	10^6 cfu/g	
Ready-to-eat processed finfish, other than fully retorted finfish	*L. monocytogenes*	5	1	0 cfu/g	10[b] cfu/g	
Bivalve molluscs, other than scallops	*E. coli*	5	1	2.3 MPN/g	7.0 MPN/g	
Bivalve molluscs[d]	*L. monocytogenes*	5	0	Absence in 25 g		
Cooked crustaceans and molluscan shellfish	*V. parahaemolyticus*	5	2	10 cfu/g	100 cfu/g	
Shelled and shucked products of cooked crustaceans and molluscan shellfish	*E. coli*	5	2	1 cfu/g	10 cfu/g	

[a]Products placed on the market during their shelf-life
[b]End of the manufacturing process
[c]Before the food has left the immediate control of the food processor
[d]They that have undergone processing other than depuration

and microbiological specifications indicate that APC of fresh fish suitable for human consumption should not exceed 10^6–10^7 cfu/g (ICMSF, 1986). However, fresh fish is not considered spoiled and the spoilage cannot become detectable by organoleptic evaluation if APC or SSO counts are not as high as 10^8–10^9 cfu/g (Gram and Huss, 1996). On the other hand, lower bacterial populations (10^5–10^6 cfu/g) have been recorded at the end of the shelf life of crustaceans such as Norway lobster (Boziaris *et al.*, 2011) and crabs (Robson *et al.*, 2007).

Determination of Enterobacteriaceae serves as a general hygienic indicator (Jay *et al.*, 2005). More specifically, determination of *Escherichia coli* is usually required. *E. coli* can serve as a hygienic indicator of the final seafood product and/or for the control of the manufacturing process (Table 15.1). Determination of pathogens such as *L. monocytogenes*, *Salmonella* spp. and pathogenic *Vibrios* are also required for a variety of seafood (Table 15.1).

15.4 Microbiological analysis using conventional culture techniques

Microbiological techniques based on culture media are still considered as standard methods. The most usual technique for the enumeration of microorganisms is the colony count method on agar media. Bacterial population is determined by inoculating dilutions of microbial suspension on the surface of solid dry agar medium (spread plates) or mixing the test suspension with liquefied agar medium (pour plates). Other culture techniques used are the Most Probable Number (MPN), mostly used for the determination of Enterobacteriaceae, coliform and *E. coli*, especially when low numbers are suspected in the product. The conventional culture method for detection of pathogens includes mixing and homogenization of the product with general purpose pre-enrichment medium to help potentially stressed or injured cells to recover, subsequently selective enrichment into liquid medium to increase the population of the target microorganism, and finally plating onto selective agar media and confirmation of isolated colonies of the target microorganisms by phenotypic tests.

15.4.1 Enumeration of total viable counts

Estimation of the total culturable bacteria is one of the most common microbiological parameters used to indicate the microbiological quality of foods. According to the International Organization for Standardization (ISO, 2003) and the standard ISO 4833:2003, total microbial population in foods is determined by counting colonies on pour plates of Plate count agar (PCA), after aerobic incubation at 30 °C for 72 h (Aerobic plate counts, APC). However, for determination of APC in seafood, lower incubation temperatures (20–25 °C) and various non-selective media are usually used. Table 15.2 presents various agar media and incubation temperatures used for the enumeration of microorganisms in seafood, taking data from a few out of the plethora of publications found in the scientific literature. Incubation at 20 °C for 72 h generates about 10-fold higher counts than at 35 °C (Levin, 2010). Levin (2010) also reported that maximum counts can be found after incubation at 3 °C for 6 days, since psychrotrophic bacteria are allowed to grow. Various researches have also used temperatures ranging from 4 to 10 °C for enumeration of psychrotrophs (Table 15.2). For the enumeration of psychrotrophic

Table 15.2 Media and incubation conditions used for the quantification of spoilage microorganisms in various seafood

Microorganisms	Media	Seafood product	Reference
Aerobic plate count (APC)	PCA (20 °C, 4 d)	Pearl spot	Lalitha *et al.* (2005)
	PCA (25 °C, 3–4 d)	Sea salmon	Hozbor *et al.* (2006)
	PCA (30 °C, 2–3 d)	Smoked rainbow trout	Cakli *et al.* (2006)
		Mussels	Caglak *et al.* (2008)
		Gutted sea bass	Torrieri *et al.* (2006)
	PCA (32 °C, 2 d)	Cod fillet	Corbo *et al.* (2005)
	IA (20 °C, 3 d)	Norway lobster	Boziaris *et al.* (2011)
		Sea bream	Lougovois *et al.* (2003)
	IA (22 °C, 3 d)	Salmon	Sveinsdottir *et al.* (2002)
	IA with 1% NaCl (15 °C, 3 d)	Tiger prawns	Lopez-Caballero *et al.* (2006)
	IA (20 °C, 3 d)	Atlantic salmon fillet	Sivertsvik *et al.* (2003)
		Ready- to- cook spotted wolf fish	Rosnes *et al.* (2006)
	LH with 1% NaCl (15 °C, 7 d)	Various fish	Dalgaard *et al.* (1997)
		Chilled cooked peeled shrimps	Mejlholm *et al.* (2005)
		Fresh and thawed salmon	Emborg *et al.* (2002)
Psychrotrophic plate count (PPC)	IA (5 °C, 14 d)	Nile perch	Gram *et al.* (1990)
	LH (10 °C, 7 d)	Mediterranean boque	Koutsoumanis and Nychas (1999)
		Sea bass	Koutsoumanis *et al.* (2002)
	LH (10 °C, 7 d)	Mullet	Pournis *et al.* (2005)
	PCA (7 °C, 10 d)	Smoked rainbow trout	Cakli *et al.* (2006)
		Mussels	Caglak *et al.* (2008)
	PCA (10 °C, 7 d)	Yellowfin tuna	Guizani *et al.* (2005)
		Rainbow trout fillet	Gimenez *et al.* (2002)
	PCA (5 °C, 7 d)	Cod fillet	Corbo *et al.* (2005)
	PCA (5 °C,14 d)	Sea salmon	Hozbor *et al.* (2006)
	PCA 1% NaCl (8 °C, 5–7 d)	Atlantic salmon fillet	Sivertsvik *et al.* (2003)
	PCA 1% NaCl (8 °C, 7–10 d)	Spotted wolf fish	Rosnes *et al.* (2006)
Photobacterium phosphoreum	IA (10 °C, 7 d)	Cod	Dalgaard (1995b)
	IA with 1% NaCl, (15 °C, 5 d)	Hake	López-Caballero *et al.* (2002)
	LH with 1% NaCl (15 °C, 5 d)	Cold-smoked spoiled salmon	Stohr *et al.* (2001)
Pseudomonas sp.	CFC (20 °C, 2 d)	Norway lobster	Boziaris *et al.* (2011)
		Mediterranean boque	Koutsoumanis and Nychas (1999)
		Sea bass	Koutsoumanis *et al.* (2002)

(*continued overleaf*)

Table 15.2 (*continued*)

Microorganisms	Media	Seafood product	Reference
	CFC (25 °C, 2 d)	Red mullet	Pournis *et al.* (2005)
		Tiger prawns	Lopez-Caballero *et al.* (2006)
	GSP agar acc. to Kielwein (1969; 1971) (25 °C, 3–4 d)	Sea salmon	Hozbor *et al.* (2006)
H_2S-producing bacteria	IA (20 °C, 4 d)	Mediterranean Boque	Koutsoumanis and Nychas (1999)
		Sea bass	Koutsoumanis *et al.* (2002)
	IA (20 °C, 2–3 d)	Red mullet	Pournis *et al.* (2005)
	IA (20 °C, 5 d)	Pearl spot	Lalitha *et al.* (2005)
	IA (20 °C, 3 d h)	Sea bream	Lougovois *et al.* (2003)
		Norway lobster	Boziaris *et al.* (2011)
		Atlantic salmon fillet	Sivertsvik *et al.* (2003)
		Spotted Wolf fish	Rosnes *et al.* (2006)
	IA (22 °C, 3 d)	*Salmo salar*	Sveinsdottir *et al.* (2002)
	IA with 1% NaCl (15 °C, 3 d)	Tiger prawns	Lopez-Caballero *et al.* (2006)
	IA (25 °C, 3 d)	Fresh and thawed salmon	Emborg *et al.* (2002)
		Chilled cooked peeled shrimps	Mejlholm *et al.* (2005)
Yeasts	RBC (29 °C, 5 d)	Pearl spot	Lalitha *et al.* (2005)
	RBC (25 °C, 3 d)	Cured seafood	Boziaris *et al.* (2012)
	Oxytetracycline yeast extract agar (30 °C, 3–5 d)	Smoked rainbow trout	Cakli *et al.* (2006)
	Yeast extract agar (30 °C 3–5 d)	Mussels	Caglak *et al.* (2008)
Lactic acid bacteria	MRS (30 °C, 3–5 d)	Smoked rainbow trout	Cakli *et al.* (2006)
		Mussels	Caglak *et al.* (2008)
		Tiger prawns	Lopez-Caballero *et al.* (2006)
	MRS (25 °C, 5 d)	Cured seafood	Boziaris *et al.* (2012)
		Mullet	Pournis *et al.* (2005)
	MRS (30 °C, 4 d under anaerobiosis)	Cod fillet	Corbo *et al.* (2005)
	Nitrate actidion polymyxin agar (NAP) (25 °C, 3 d)	Fresh and thawed salmon	Emborg *et al.* (2002)
		Chilled cooked peeled shrimps	Mejlholm *et al.* (2005)
	NAP (25 °C, 5 d)	Cold-smoked fish	Tomé *et al.* (2006)
	MRS agar with 1% sorbic acid (25 °C, 5 d, under anaerobiosis)	Pearl spot	Lalitha *et al.* (2005)

plate counts (PPC) in food including fish, ISO (2001a) recommends PCA without addition of extra salt or minerals (ISO 17410:2001) and aerobic incubation at 6.5 °C for 10 days.

PCA, Iron agar (IA) and Long and Hammer agar (LH) (Van Spreekens, 1974) with or without added NaCl have been used as media to determine APC (Table 15.2). According to Gram (1992), IA gives significantly higher counts than PCA after 3–4 days incubation at 25 °C. In many studies, IA with 1% w/v NaCl has been used (Gram *et al.*, 1987; Lopez-Caballero *et al.*, 2006). Several comparative studies between various non-selective media with or without added NaCl have been carried out. Valle *et al.* (1998) found that addition of 1.38% (w/v) NaCl in Tryptone soy agar (TSA) or PCA is required for optimal enumeration of APC. The total bacterial population of chilled stored sea bream counted on TSA spread plates was higher compared to PCA or IA pour plates, with or without added NaCl (Kakasis *et al.*, 2011). Many marine bacteria cannot grow on media unless they contain certain salts and minerals that PCA lacks (Atlas, 1995). Khan *et al.* (2005) suggested the use of Marine agar (MA) to evaluate the general microbial quality of bivalves instead of PCA or PCA with 1% NaCl, while Gornik *et al.* (2011) used MA to determine APC on Norway lobster tail meat.

Other studies have reported that PPC on PCA is underestimated by 1 log cfu/g compared to growth media containing sodium chloride (NaCl) such as LH. For this reason, the Nordic Committee of Food Analysis (2006) recommends the NMKL 184 method, which uses LH with 1% NaCl incubated aerobically at 15.0 ± 1.0 °C for 5–7 days for the enumeration of psychrotrophic microorganisms in seafood.

Broekaert *et al.* (2011) using molecular identification found that LH and MA gave the best quantitative and qualitative APC results for fish quality analysis, compared to PCA and IA, but some species, such as *Pseudomonas fragi* and *Acinetobacter* sp. were unable to grow on those media. So far, there does not seem to be an agar medium capable of supporting the growth of all cultivable bacteria from fish.

15.4.2 Determination of spoilage microorganisms

Identification of spoilage microorganisms Identification of fish spoilage bacteria using the classical microbiological approach was first carried out in the 1960s (Shewan *et al.*, 1960a, 1960b). According to phenotypic examination such as Gram-reaction, oxidase test, Hugh and Leifson reaction (Hugh and Leifson, 1953), production of acid and/or gas from glucose, motility, sensitivity to pteridine compound (0/129), etc., the isolated colony could be characterized as *Pseudomonas, Vibrio, Aeromonas, Achromobacter, Flavobacterium*, Coryneforms, *Micrococcus*, etc. Gram *et al.* (1987) identified most of the 'black' colonies with the ability to reduce TMAO to TMA and produce H_2S, isolated on IA from spoiled fish at 0 °C, as *Alteromonas putrefaciens*. Jorgensen and Huss (1989) studied H_2S-producing organisms isolated on IA from spoiling cod and were identified as *S. putrefaciens*, the same as described by Gram *et al.* (1987). Dalgaard *et al.* (1993) and Dalgaard (1995b) isolated *P. phosphoreum* and *S. putrefaciens* from spoiled packed cod using spread and pour IA plates, respectively. The large cells from glass-like colonies were identified as *P. phosphoreum* and more than 99% of strains isolated as black colonies were tentatively identified as *S. putrefaciens*. Currently, mesophilic, halotolerant strains are recognized as *S. algae* and the psychrotrophic strains as *S. putrefaciens* or *S. baltica* (Gram and Dalgaard, 2002). Other researchers using the classical approach or a combination of culture and non-culture methods have studied the spoilage microbiota of iced fish caught from Mediterranean Sea and found that the predominant spoilage microbiota is mainly *Pseudomonas* and *Shewanella* (Gennari *et al.*, 1999; Tryfinopoulou *et al.*, 2002, 2007).

Enumeration of spoilage microorganisms Various selective media are used for the enumeration of SSOs (Table 15.2). *Pseudomonas* spp. are enumerated using cetrimide-fucidin-cephaloridine agar (CFC agar) after 24–48 h incubation at 20–25 °C. The selectivity of CFC was examined by Trifinopoulou *et al.* (2001) who found that apart from *Aeromonas*, species of Enterobacteriaceae and *S. putrefaciens* were able to grow on CFC medium. However, reliable estimation of *Pseudomonas* population can be achieved in samples where its population level is higher than 1–2 log cfu/g compared to the other microorganisms able to grow on CFC. GSP Agar (Pseudomonas-Aeromonas Selective Agar) according to Kielwein (1969, 1971) can be used for simultaneous determination of *Pseudomonas* and *Aeromonas*. The medium contains penicillin as a selective inhibitor and starch which is degraded by *Aeromonas* with acid production, but not by *Pseudomonas*, causing phenol red to change to yellow. Hozbor *et al.* (2006) found that *Aeromonas* was significant part of bacterial spoilage association of sea-salmon stored in ice.

Lactic acid bacteria in seafood are usually enumerated on Man–Rogosa–Sharp agar (MRS) (De Man *et al.*, 1960) or nitrite actidione polymyxin agar (NAP) (Davidson and Cronin, 1973) (Table 15.2). MRS-medium with a lower pH (e.g. 5.7) and the addition of yeast/mould inhibitor (e.g. sorbic acid) has also been used as selective medium for lactic acid bacteria (Reuter, 1985) (Table 15.2).

Brochothrix thermosphacta populations in seafood are enumerated on streptomycin sulfate thallous acetate agar (STAA) spread plates after incubation at 20 °C for 3–4 days (Table 15.2). For *B. thermosphacta*, STAA is the only medium so far available.

Yeast and moulds, mostly found in cured seafood, can be enumerated using chloroamphenicol-based media, like rose bengal chloroamphenicol (RBC) and tryptose-glucose-yeast extract containing chloroamphenicol, or oxytetracycline-glucose-yeast extract agar. Yeast and mould populations can be enumerated on spread plates after incubation at 25–30 °C for 3–5 days (Table 15.2).

Although various media are available for enumeration of the previous SSOs, so far no single selective medium can satisfactorily recover and differentiate *Shewanella* and *Photobacterium* from other bacteria. IA, which is specifically used for the enumeration of H_2S-producing microorganisms (black colonies on IA), is also often used for the enumeration of the total microbial population (Table 15.2). Dalgaard (1995b) found that that 99% of the black colonies on IA were *S. putrefaciens*, but other H_2S-producing bacteria that do not belong to *Shewanella* genus, such as Enterobacteriaceae, *Aeromonas* and *Vibrio* can also grow on IA (Tryfinopoulou *et al.*, 2007). *P. phosphoreum* can grow on spread plates of IA or LH with 1% NaCl after incubation at 5 °C for 14 days, or 15 °C for 5 days (Dalgaard *et al.*, 1993; Dalgaard, 1995b; Dalgaard *et al.*, 1996; Lopez-Caballero *et al.*, 2002). *P. phosphoreum* appears on those media as large coccobacilli in small glass-like colonies (Dalgaard *et al.*, 1993; Dalgaard, 1995b), but biochemical and/or other tests have to be performed to confirm *P. phosphoreum* numbers involved in seafood spoilage (Dalgaard, 1995b; Lopez-Caballero *et al.*, 2002). *P. phosphoreum* is a psychrophilic bacterium that is sensitive to elevated temperatures (25–35 °C), so the temperature has to remain low (below 10 °C) to grow the maximum population. This analysis has to be performed with diluents, pre-poured agar plates, pipette tips, etc., that remain chilled throughout the analytical procedure (Dalgaard *et al.*, 1993; Levin, 2010).

15.4.3 Hygienic indicators

The presence of Enterobacteriaceae is a widely accepted indicator of failures in the manufacturing process or of post-process contamination. However, the presence of *E. coli* is the parameter that is usually taken into account when assessing the hygienic quality of seafood, especially shellfish (Table 15.2).

For Enterobacteriaceae, ISO proposes two methods, ISO 21528–1:2004 and ISO 21528–2:2004, employing MPN and the colony count method, respectively (ISO, 2004a, 2004b). MPN method is proposed when the Enterobacteriaceae number is expected to be as low as 1–100 per ml or per g of test sample. The MPN method uses buffered peptone water (BPW) and buffered brilliant green bile glucose broth (EE broth) for pre-enrichment and selective enrichment, respectively, and the use of violet red bile glucose agar (VRBGA) for confirmation. The enumeration of Enterobacteriaceae, according to the colony count method is performed on VRBGA. Characteristic colonies are pink to red or purple with or without haloes of the same colour. Sub-culture of presumptive colonies on a non-selective medium and confirmation by glucose fermentation and presence of oxidase tests are required.

E. coli is also determined by the colony count technique or the MPN method according to ISO 16649–2:2001 or ISO 16649–3:2005, respectively (ISO, 2001b, 2005). The colony count technique determines the number of colony-forming units (cfu) of β-glucuronidase-positive *E. coli* cells, per ml or per g of sample after incubation at 44 °C for 18–24 h using tryptone-bile-glucuronide agar medium (TBX), which contains 5-bromo-4-chloro-3-indolyl-D-glucuronide (BCIG). Only the colonies that are blue or blue-green are counted. With the MPN method for the enumeration of β-glucuronidase-positive *E. coli*, enrichment in selective broth (mineral-modified glutamate) at 37 °C for 24 h takes place and subsequently samples from the tubes showing acid production (yellow colour) are sub-cultured on TBX for confirmation. With both methods, strains of *E. coli* which cannot grow at 44 °C and especially those that are β-glucuronidase-negative, such as *E. coli* O157, will not be detected. The specific detection of strains of this serotype is described in ISO 16654:2001 (ISO, 2001c).

Staphylococcus aureus is also included in microbiological criteria (Table 15.1). The natural source of *S. aureus* is human skin, hair and superficial mucous membranes Enumeration, according to ISO 6888-1:1999, can be carried out by spread plating on Baird-Parkers egg yolk agar and incubation for 24 h at 37 °C, while positive cultures need to be confirmed by testing for coagulase activity (ISO, 1999). Presence of high numbers indicates poor sanitary or production practice and/or the possible presence of enterotoxin.

15.4.4 Pathogen detection

Conventional culture methods are still the most reliable and accurate techniques for pathogen detection. Detection protocols for food-borne pathogens usually use one or two enrichment steps, followed by plating on selective agar media, and any presumptive pathogen colony is confirmed by morphological, biochemical, immunological or molecular tests (Figure 15.1). These protocols are laborious and take quite long time to give results. It usually takes 2–3 days to obtain the presumptive results and up to 5 or more days for confirmation. The confirmation step can be shorter and sometimes more accurate and reliable by employing PCR (Figure 15.1).

To substantially reduce analysis time, new immune-based approaches such as enzyme-linked immunosorbent assay (ELISA) or PCR-based techniques are employed (Figure 15.1). It should be mentioned that ELISA tests need about 10^5 cells to be reactive and therefore before performing the tests, enrichment is necessary by incubating the food sample overnight so that the target organism has reach a detectable level. Similarly, for detection based on PCR, an enrichment step has to be included when it is necessary

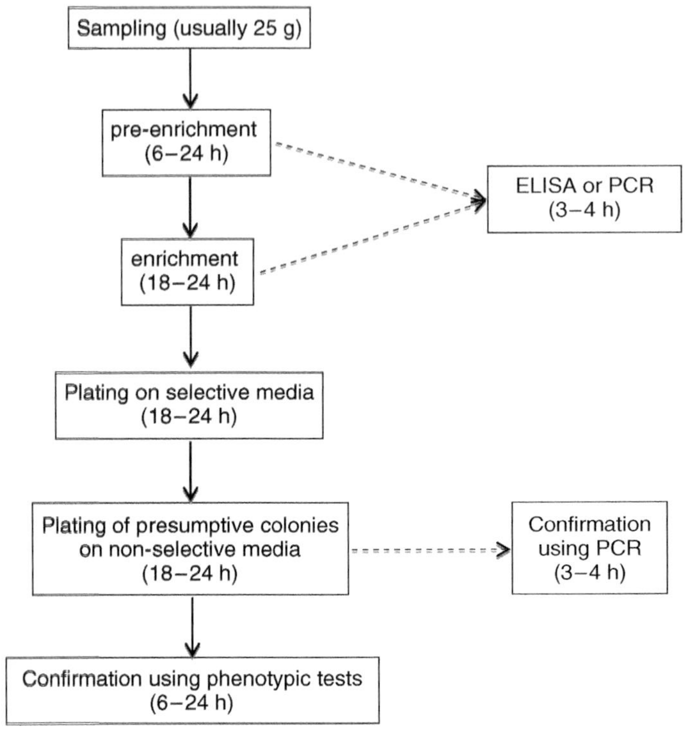

Figure 15.1 Generalized flow diagram of pathogen detection procedures

to detect low numbers of pathogenic microorganisms (1–100 cells). The total time to detect a pathogen by these approaches should include the enrichment time of the target pathogens. ELISA and PCR methods are discussed in sections 15.7 and 15.8, respectively.

Determination of pathogens such as *L. monocytogenes, Salmonella* spp. and pathogenic *Vibrios* are required for a variety of seafood (Table 15.1). The standard procedures for the detection of these pathogens are presented.

Pathogenic *Vibrio* species The microbiological analysis for pathogenic *Vibrio* concerns *V. parahaemolyticus, V. cholerae* and *V. vulnificus*. These pathogens are associated with the consumption of undercooked or raw seafood such as shellfish. ISO/TS 21872-1:2007 and ISO/TS 21872-2:2007 describe the standard protocols for the detection of *V. parahaemolyticus* and *V. cholerae* and species other than *V. parahaemolyticus* and *V. cholerae* (e.g. *V. vulnificus*), respectively (ISO, 2007a, 2007b). According to the standard protocols, pre-enrichment at 41.5 °C for 6 h and second enrichment at 41.5 °C for 18 h in alkaline saline peptone water (ASPW) is required. For frozen products, pre-enrichment at 37 °C for 6 h is recommended. To obtain single colonies, inoculation on Thiosulfate Citrate Bile and Sucrose (TCBS) agar and another optional selective medium after incubation at 37 °C for 24 h is required. The optional selective medium can be soya peptone triphenyl tetrazolium chloride agar (TSAT) or sodium dodecyl sulfate polymyxin sucrose (SDSPS) agar according to ISO/TS 21872–1:2007 (ISO, 2007a) and colistin polymixin β-cellobiose agar (CPC), sodium dodecyl sulfate polymixin B sucrose agar (SDS) or modified colistin polymixin cellobiose agar (mCPC) according to ISO/TS 21872-2:2007 (ISO, 2007b). The typical presumptive pathogenic *Vibrio* spp. are sub-cultured onto saline nutrient agar (SNA) followed by biochemical confirmation.

***Salmonella* spp.** The standard culture media protocol for *Salmonella* spp. includes pre-enrichment and selective enrichment, plating on selective culture media and confirmation of presumptive colonies according to ISO 6579:2002 (ISO, 2002). Pre-enrichment takes place in BPW at 37 °C for 18 h. Then selective enrichment is performed using two different media, Rappaport–Vassiliadis with soya broth (RVS) and Muller–Kauffmann tetrathionate novobiocin (MKTTn) broth at 37 °C for 24 h. Two different media are used for plating, xylose lysine deoxycholate (XLD) agar and another optional selective medium (e.g. brillian green agar). The presumptive *Salmonella* spp. colonies have to undergo biochemical and serological confirmation after sub-culturing on non-selective medium (e.g. NA).

Listeria monocytogenes Two reference methods have been developed by ISO for detection and enumeration of *L. monocytogenes*, ISO 11290-1:1996/Amd 1:2004 and ISO 11290-2:1998/Amd 1:2004, respectively (ISO, 1996, 1998, 2004c, 2004d). According to the detection protocol, primary and secondary enrichment occurs in Half Fraser and Fraser broth at 30 °C for 24 h and at 37 °C for 48 h, respectively. Subsequently plates on Agar Listeria according to Ottaviani and Agosti (ALOA) and on another optional selective medium (e.g. PALCAM or Oxford) are prepared. The presumptive colonies are confirmed by biochemical tests after sub-culturing on non-selective medium.

15.5 Microbiological examination using indirect rapid methods

The term 'rapid methods' refers to improved, alternative methods that can enable early, accurate and faster detection, isolation, characterization and/or enumeration of microorganisms. Indirect methods determine microbial populations by monitoring their metabolic activity rather than their populations.

15.5.1 Determination of bacterial ATP

One of the first attempts to employ indirect techniques in seafood analysis was the determination of adenosine triphosphate (ATP) to estimate the total number of viable cells in fish (Ward *et al.*, 1986). Bacterial ATP is extracted and measured by the luciferin/luciferase bioluminescent reaction. The intensity of light emitted is directly proportional to the quantity of ATP and therefore the number of viable cells in the medium. The method can detect populations as low as 10^2–10^3 bacterial cells and is very rapid since the results are obtained in less than 1 h. However, the results are not consistent due to the different amount of ATP per cell for different microorganisms or different physiological states. Additionally, background ATP from food material, biological fluids, etc., interferes with microbial ATP (Fung, 2002). This method is now widely used for the monitoring the cleaning and disinfection of food processing surfaces.

15.5.2 Electrical methods

Another indirect method measures the impedance, conductance or capacitance changes of liquid growth medium during the growth of microorganisms in it. Production of metabolic products changes these electric properties due to the production of charged molecules (e.g. lactic acid) from the catabolism of molecules that do not have an electrical charge (e.g. sugars). The time required to detect (T_d) a change of these properties, is inversely proportional to the logarithm of the initial microbial population and occurs when microbial population levels reach 10^6 cfu/ml (Fung, 2002). It is obvious that determination of the microbial population in the sample requires a calibration curve of T_d against plate counts.

The development of a sensitive and selective conductance method for quantitative determination of *P. phosphoreum* in fresh fish was presented by Dalgaard *et al.* (1996). Less than 50 cells per g of fish could be detected in less than 45 h. Russell (1998) used capacitance to quantify *P. flourescens* from rainbow trout, Atlantic salmon, red grouper and tilapia fillets using Brain Heart Infusion (BHI) as a growth medium with *P. fluorescens* selective additive (US patent 1998). Koutsoumanis *et al.* (2002) used metronidazole, carbenicilline, cetrimide, cycloheximide, diamide (MCCCD) as the conductance medium, first described by Salvat *et al.* (1997), to quantitatively determine the *Pseudomonas* spp. population in European sea bass (*Dichentrachus labrax*). Metcalfe and Marshall (2004) used capacitance to quantify the total microbial population and evaluate microbiological quality of raw shrimps. The application of capacitance microbiology to detect and enumerate *V. cholerae* and *V. vulnificus* was attempted by Marshall *et al.* (1999) with encouraging results, while Dupont *et al.* (1996) succeeded in quantifying *E. coli* in bivalve shellfish in 96% of cases, and only 0.7% of *E. coli* cultures failed to give a conductance signal within 20 h.

15.5.3 Other indirect methods

Quantification of H_2S-producing bacteria in cod, salmon, wolf fish and coal fish was carried out using the formation rate of iron sulfide during growth of the bacteria in a liquid growth medium containing cysteine, sodium thiosulfate and ferric citrate as specific substrates for iron sulfide formation (Skjerdal *et al.*, 2004). The iron sulfide turns the medium grey and masks the background fluorescence of the medium when the H_2S-producing bacteria population in the assay is approximately 10^9 cfu/ml. A linear correlation between the H_2S-producing bacteria counts and T_d was obtained over the entire range from 1 to 10^9 cfu/g, corresponding to detection times of 17 and 1 h, respectively.

A more recent indirect method based on measurement of the respiratory activity of aerobic microorganisms growing in a liquid medium was used to quantify APC in salmon, cod, plaice, mackerel and whiting (Hempel *et al.*, 2011). The method employed fluorescence-based oxygen sensing probes and fluorescent reader detection to monitor growth of aerobic cells via respiration. The samples produced characteristic profiles, with a sharp increase in fluorescence above the baseline level at a certain threshold time, which could be correlated with initial microbial population. The method can detect populations from a range of 10^4–10^7 cfu/g, with accuracy of 1 log cfu/g in 2–12 h depending on the level of initial population.

15.6 Microscopy based rapid methods

15.6.1 Direct Epiflouresence Filter Technique (DEFT)

DEFT is a technique that relies on direct microscopy using the stain acridine orange which binds onto nucleic acid and fluoresces orange. After sample pre-treatment with enzymes and detergent, the microbial cells are concentrated on a polycarbonate membrane filter before staining with the dye (Jasson *et al.*, 2010). This technique has not been used extensively in seafood microbiological analysis, despite the early study of Pettipher and Rodrigues (1982), who found a high correlation between DEFT and plate counts of fish samples. Although it is a rapid method, it lacks sensitivity (10^4–10^5 cells/ml).

15.6.2 Fluorescent *In Situ* Hybridization (FISH)

Although FISH is a molecular technique, it is also a microscopy-based method, hence it is presented in this section. In this method, oligonucleotide targeted probes labelled with a fluorescent dye, hybridize the target RNA of microbial cells that are chemically fixed on a glass slide. After washing to remove any unbound probes, the stained cells are detected using epifluorescence microscopy. The detection limit is about 10^4 cells/ml (Jasson *et al.*, 2010). The practical application of FISH technology is more difficult in food, compared to water samples, as food particles interfere with detection of the bacterial cells by epifluorescent microscopy. For detection of low bacterial numbers, enrichment is necessary. Ootsubo *et al.* (2003) developed a FISH filter cultivation (FISHFC) method, which includes a brief cultivation step on a membrane filter before *in situ* hybridization, for rapid (7 h) enumeration of Enterobacteriaceae in food samples at population levels as low as 10^2 cells/g. With the FISHFC technique, Fuchizawa *et al.* (2008) detected and enumerated *Listeria* spp. in smoked salmon in less than 14 h.

15.6.3 Flow cytometry

In flow cytometry, bacterial cells are forced to pass through a capillary in front of a focused light beam or laser. The instrument measures the forward and side scatter and fluorescence. Scattering is related to surface properties, while fluorescence is emitted by cellular components. Fluorescent dyes or other staining compounds or probes that can reveal the viability or metabolic state of the cell can be used (Jasson *et al.*, 2010). Flow cytometry can be used for automated cell counting and microbial cell detection. Despite the ability to measure individual cells, the handling of small volumes makes flow cytometry unable to detect very low bacterial concentration. For microbial detection, an enrichment step is necessary (Veal *et al.*, 2000). Endo *et al.* (2002) applied flow cytometry to enumerate the total number of microbial cells on yellow-fin horse mackerel. The total measurement time was about 30 min including preparation of the microbial cells for flow cytometry. Good correlation was obtained with the standard colony count methods in the range 10^5–10^9 cells/g.

15.7 Immuno-based techniques

Methods based on highly specific antigen–antibody binding have become widespread during recent decades for detection and/or identification of food-borne pathogens. Specialized monoclonal or polyclonal antibodies bind on specific sites on cell surfaces allowing visualization of the reaction through precipitation or other means, such as colour or fluorescence development.

The most widely used immunological technique is the ELISA. In general, the target antigen of the sample binds to the antibodies that are fixed on the surface of microtitre plate wells. After washing to remove any unbound antigen, enzyme-labelled antibodies are added and target the antigens in the wells. The visualization is possible, after washing to remove any unbound antibodies, by adding a substrate that the antibody-linked enzyme converts to a visual signal. The method is rapid, but the detection limit is about 10^5-10^6 cells/ml, which corresponds to 10^6-10^7 cells/g of food. This is a reason that immuno-based techniques have not been used for determination of spoilage microorganisms despite the development of antibodies for *S. putrefaciens* (Fonnesbech *et al.*, 1993). However, it can be used for pathogen detection in seafood (Kumar *et al.*, 2008a, 2011). An enrichment step is required, which increases analysis time, but detection of food-borne pathogens can be substantially decreased down to 1 or 2 days, compared to 5 days or more required with conventional culture techniques. Many ELISA tests are automated in order to reduce the hands-on time and increase reproducibility.

15.8 Molecular methods for microbial determination

It is known that less than 1% of the total microbiota present in environmental samples grows on plates (Ward *et al.*, 1990). Comparison of direct microscopy and plate counts have yielded similar total numbers of cells (Dalgaard *et al.*, 1993) and no examples have been reported where 'unculturable' bacteria have shown any importance in seafood spoilage (Rudi *et al.*, 2004; Hovda *et al.*, 2007a, 2007b; Olofsson *et al.*, 2007; Parlapani *et al.*, 2013). However, culture-dependent methods are time-consuming, phenotypic identification lacks the discriminatory power of molecular techniques and culture media are not able to support growth of all cultivable bacteria (Broekaert *et al.*, 2011; Svanevik and Lunestad, 2011; Parlapani *et al.*, 2013). Additionally, stressed or sublethally injured cells are not able to recover and grow on selective media, while other populations are inhibited by microorganisms present in higher numbers (Hugenholtz *et al.*, 1998). Comparative studies between conventional culture-based and molecular methods have shown that complementary results are obtained and in some cases the PCR-based methods revealed microorganisms that escape the classical approach.

15.8.1 Exploration of fish and seafood microbiota

The use of 16S rRNA gene is currently the most common approach for community analysis due to the phylogenetic properties and the large amount of both variable and conserved regions available. The sequence of the 16S rRNA gene has been widely used to determine phylogenetic relationships among bacteria as well as to identify unknown bacteria to the genus or species level (Sacchi *et al.*, 2002).

Microorganisms that grow on selective or non-selective agar media can be easily isolated and identified by 16S rRNA sequence analysis using only a small amount of DNA. However, various important microorganisms in seafood do not grow on some general growth media (Broekaert *et al.*, 2011). An alternative molecular approach for exploration of microbiota in any sample (e.g. fish flesh) is the direct extraction of DNA. This makes it possible to reveal microorganisms that escape standard plating approaches. The first challenge is the extraction of DNA from the seafood samples followed by PCR amplification of the nucleotide sequence of the 16S rRNA gene. At the final step, 16S rRNA gene fragments from the different bacteria present in the mixed amplification products are analysed using various approaches.

The microbial composition can be mainly evaluated through fingerprinting (pattern analysis) of the PCR product and/or separation of amplified 16S rRNA gene products using various techniques and sequencing (Table 15.3). Fingerprinting is usually carried out by techniques such as terminal restriction fragment length polymorphism (T-RFLP) and denaturing gradient gel electrophoresis (DGGE) or thermal gradient gel electrophoresis (TGGE). In T-RFLP, the PCR products are digested using restricted enzymes, followed by electrophoresis to obtain the banding patterns. In DGGE or TGGE, chemical denaturants or temperature, respectively, are used to separate the amplified PCR products (Justé *et al.*, 2008). To identify differential signals obtained from the above gel-based approaches, many researchers used products of analysis for sequencing to characterize the predominant microbial flora during fish spoilage (Rudi *et al.*, 2004, Hovda *et al.*, 2007a, 2007b; Jaffrès *et al.*, 2009; Svanevik and Lunestad, 2011; Macé *et al.*, 2012). Another approach is to separate the amplified 16S rRNA gene products using cloning (Rudi *et al.*, 2004; Olofsson *et al.*, 2007; Jaffrès *et al.*, 2009; Parlapani *et al.*, 2013).

Rudi *et al.* (2004) determined the microbial composition of salmon and coalfish under MAP by using a combination of molecular techniques, including T-RFLP analysis and 16S rRNA gene sequencing from isolates or DNA extracted directly from the fish flesh. It was found that *P. phosphoreum* was associated with coalfish, while *Brochothrix* spp. and *Carnobacterium* spp. were associated with salmon. Hovda *et al.* (2007a, 2007b) used DGGE with further sequencing of the PCR products to identify the predominant microbial community. *P. phosphoreum, Pseudomonas* spp. and *B. thermosphacta* were found to dominate in halibut, while *Photobacterium* sp., *S. putrefaciens* and *Pseudomonas* spp. dominated in cod. Olofsson *et al.* (2007) studied the spoilage microbiota of freshly chilled cold-smoked salmon under vacuum packaging using 16S rRNA sequencing of both isolated bacteria and bacterial DNA from fish tissue. *Lactobacillus* spp. and *Photobacterium* spp. (not only *P. phosphoreum*, but others that are yet unclassified) were the predominant bacteria. Of the bacterial microbiota, 10% were unknown species. Parlapani *et al.* (2012) studied the microbiota of spoiled sea bream fillets isolated on TSA, under various storage conditions using a combination of traditional and molecular methodologies. Although *Pseudomonas* was the predominant microorganism, the molecular approach revealed other microorganisms, such as *Psychrobacter* and *Carnobacterium*, that escaped the traditional approach. Extracting DNA directly from the fish flesh, Parlapani *et al.* (2013) investigated the initial and spoilage microbial diversity of iced stored sea bream using 16S rRNA gene amplification, cloning and sequencing. It was found that *Aeromonas salmonicida*, which escaped the standard procedures, co-dominated together with *Pseudomonas* and *Shewanella*. Scientists will certainly continue to employ and improve molecular techniques to study the microbial communities and expand the studies to the microbiological analysis of food/seafood.

Recent advances in sequencing technology have changed how microbial communities are explored. Pyrosequencing analysis with primers targeting the V1 to V3 region of

Table 15.3 PCR based approaches and methodologies for the identification, detection and enumeration of microorganisms in seafood

Purpose	Seafood	Source of extracted DNA	Methodologies	Target gene	Reference
Identification of microbiota	Sliced VP smoked salmon	Fish tissue	Cloning, Amplification Ribosomal DNA Restriction Analysis (ARDRA), sequencing	16s rRNA	Cambon-Bonavita *et al.* (2001)
	Salmon and Coalfish under MAP	Isolates on various agar media and fish tissue	T-RFLP cloning sequencing	16S rRNA	Rudi *et al.* (2004)
	Cold-smoked VP salmon	Isolates on TGE agar and fish tissue	cloning sequencing	16S rRNA	Olofsson *et al.* (2007)
	Farmed halibut and Atlantic cod in MAP	Isolates on PCA and Fish tissue	DGGE sequencing	16S rRNA	Hovda *et al.* (2007a, 2007b)
	Cooked and peeled shrimps	Isolates on LH, IA and fish tissue	RFLP TTGE cloning sequencing	16S rRNA and 16S-23S Intergenic Spacer Regions (ISRs)	Jaffrès *et al.* (2009)
	Various fermented seafood	Fish tissue	tag pyrosequencing and DGGE sequencing	16s rRNA	Roh *et al.* (2010)
	Korean salted and fermented seafood (jeotgal)	Isolates on various agar media	sequencing	16S rRNA	Guan *et al.* (2011)
	Atlantic mackerel	Isolates on IA and fish tissue	DGGE sequencing	16s rRNA	Svanevik and Lunestad (2011)
	Various fish	Isolates on various general media	DGGE sequencing	16S rRNA, gyrB	Broekaert *et al.* (2011)
	Raw salmon steaks under VP or MAP	Isolates on various agar media and fish tissue	TTGE cloning sequencing	16s rRNA	Macé *et al.* (2012)
	Seabream	Fish tissue	cloning sequencing	16s rRNA	Parlapani *et al.* (2013)
Exploration of *Shewanella* diversity	Sea Bream under air and MAP	Isolates on IA	sequencing	16s rRNA	Tryfinopoulou *et al.* (2007)

Detection of *V. parahaemolyticus*	Oyster	Tissue	qPCR	*tdh*	Blackstone *et al.* (2003)
	Mussels	Tissue	PCR	*toxR*, *tlh*, *tdh*, *trh*	Di Pinto *et al.* (2008)
	Mussels	Tissue	PCR	*tdh trh*	Ottaviani *et al.* (2005)
	Mussels	Tissue	PCR-ELISA	*tl*	Di Pinto *et al.* (2012)
	Octopus Manta ray Pangasious Boiled frozen mussels	Tissue	Multiplex qPCR	*tdh* *trh*	Garrido *et al.* (2012)
Detection of *E. coli*, *S. typhimurium*, *V. vulnificus*, *V. cholerae*, *V. parahaemolyticus*	Oysters	Tissue	Multiplex PCR	*uidA*, *cth*, *invA*, *ctx*, *tl*	Brasher *et al.* (1998)
Detection of *V. vulnificus*	Oysters	Tissue	Multiplex PCR	*vvhA vcg* 16S rRNA CPS	Han and Ge (2010)
Detection of *V. parahaemolyticus* *V. cholerae*, *V. vulnificus*	Oysters	Tissue	Multiplex PCR, DNA Microarrays	*vvh*,*viuB* *ompU*, *toxR*, *tcpI*, *hlyA*, *tlh*, *tdh*, *trh*	Panicker *et al.* (2004)
Detection of *V. parahaemolyticus*, *V. cholera*, *V. vulnificus*	Raw and cooked seafood and fish samples	Tissue	Multiplex qPCR	*vvha*, *toxR* *sodB*, *tdh*, *trh1*, *trh2*, *ctx*, *toxR*	Messelhäusser *et al.* (2010)
Detection of *V. cholerae*, *V. parahaemolyticus*, *V. vulnificus* *V. alginolyticus*, *V. mimicus*	Bivalve	Tissue	Microarray, PCR-Ligase Detection Reaction-Universal Array (PCR-LDR-UA)	*tdh*, *trh* and other specific genes	Cariani *et al.* (2012)
Detection of *V. cholerae* 01, 0139, Non-01, and Non-0139	Oysters	Tissue	TaqMan PCR	*hlyA*	Lyon (2001)

(*continued overleaf*)

Table 15.3 (*continued*)

Purpose	Seafood	Source of extracted DNA	Methodologies	Target gene	Reference
Detection of *C. botulinum*	Fish shellfish	Tissue	PCR-ELISA	*bont*	Fach *et al.* (2002)
Detection of *Salmonella*	Various seafood	Tissue	PCR	*trh*	Kumar *et al.* (2008a)
	Various seafood	Tissue	PCR	*invA*	Kumar *et al.* (2008b)
Detection of *Salmonella spp. Shigella spp.*	Mussels	Tissue	Multiplex PCR	*invA* *virA*	Vantarakis *et al.* (2000)
Detection and Identification of histamine-forming bacteria	Tuna fillets	Fish tissue	sequencing	histidine decarboxylase (*hdc*), 16S-23S ISRs, 16s rRNA	Ferrario *et al.* (2012)
Quantification of *C. botulinum* type E	Various seafood under MAP	Tissue	qPCR	BoNT/E	Kimura *et al.* (2001)
Quantification of Total viable bacteria	Cod fillets	Fish tissue	qPCR – Ethidium bromide Mono Azide (EMA)	16S rRNA	Lee and Levin (2007)
Quantification of TMA producing bacteria	Whiting and plaice	Fish tissue	qPCR	*torA*	Duflos *et al.* (2010)
Quantification of *P. phosphoreum*	Salmon packed under MAP	Fish tissue	qPCR	*gyrB*, 16S rRNA	Macé *et al.* (2013)
Quantification of *Pseudomonas* spp.	Cod	Fish tissue	qPCR	carbamoyl phosphate synthase gene (*carA*)	Reynisson *et al.* (2008)
Quantification of *B. thermosphacta*	Cooked shrimp and salmon	Fish tissue	qPCR	16S rRNA	Mamlouk *et al.* (2012)

the bacterial 16S rRNA gene has been used to better understand microbial diversity in various habitats. Roh *et al.* (2010) examined bacterial diversity in fermented seafood using the DGGE method and tag pyrosequencing. The second approach determined hundreds of sequences, revealing ones that escape classical molecular procedures such as DGGE. As a consequence, tag pyrosequencing can be a powerful tool for microbiota characterization in fermented seafood compared with classical molecular approaches.

15.8.2 Detection and quantification of microorganisms

Another alternative approach to classical detection of pathogens in seafood is the application of PCR-based methods. Real-time PCR (hereafter named qPCR), a combination of PCR and ELISA (PCR-ELISA) and microarray techniques are successfully used to detect food-borne pathogens in seafood (Table 15.3).

Various authors have reported the application of qPCR for detection and quantification of pathogens, such as *Vibrio* spp. and *C. botulinum* type E (Kimura *et al.*, 2001; Robert-Pillot *et al.*, 2010). The qPCR assays, especially those using TaqMan probes, have been reported for detection of pathogenic *Vibrio* species in seafood (Lyon, 2001; Cai *et al.*, 2006; Canigral *et al.*, 2010). For *V. parahaemolyticus*, thermostable direct haemolysin (*tdh*) and related haemolysin (*trh*) genes are the only genes accepted as pathogenicity markers by ISO (ISO, 2007a) and FDA (Kaysner and DePaola, 2004). Blackstone *et al.* (2003) developed and evaluated a qPCR assay in order to detect the presence of *tdh*, a marker of pathogenicity in *V. parahaemolyticus*, in pure cultures and in oyster enrichments.

The development and evaluation of PCR and ELISA into a single analytical technique (PCR-ELISA) have been also used for the detection of *V. parahaemolyticus* as well as *C. botulinum* in seafood (Fach *et al.*, 2002; Di Pinto *et al.*, 2012). This technique combines an immunological method (ELISA) to quantify the PCR products (Landgraf *et al.*, 1991).

In addition, another technique for detecting pathogens is the multiplex PCR using multiple biomarkers, for example cytolysin/haemolysin gene (*vvhA*), the virulence-correlated gene (*vcg*), 16S rRNA and capsular polysaccharide operon (CPS), in characterized *V. parahaemolyticus* (Garrido *et al.*, 2012) and *V. vulnificus* (Han and Ge, 2010) strains in seafood. Garrido *et al.* (2013) also developed and applied to food and seafood samples a complete protocol for the detection of *Salmonella* spp. and *L. monocytogenes* by using one single enrichment broth, a DNA-extraction method and a multiplex qPCR detection system. The qPCR method was able to detect three cells of *Salmonella* and *L. monocytogenes* in 25 g of samples.

Recently, the DNA microarray hybridization approach, mainly coupled with multiplex PCR, has been effective for the detection of amplicons generated by PCR from multiple pathogenic strains that are present in seafood (Panicker *et al.*, 2004; Gonzalez *et al.*, 2004). The DNA array involves the immobilization of numerous oligonucleotide DNA probes at high densities on a solid matrix to which fluorescence-labelled, PCR-amplified DNAs are hybridized.

PCR based methods are also used for the detection of food-borne viruses. Traditionally, virus detection takes place in cell culture, but low sensitivity and high cost are important problems. Moreover, many viruses such as HAV are difficult or impossible to cultivate. Therefore, molecular techniques such as reverse transcription-polymerase chain reaction (RT-PCR) have been successfully used for the detection of HAV in shellfish (Green and Lewis, 1998; Ribao *et al.*, 2004; Sincero *et al.*, 2006; Casas *et al.*, 2007). Currently, scientists continue to employ and improve molecular techniques to detect viruses in food/seafood.

Total microbial population and SSOs such as *Pseudomonas* spp. and *Brochotrix thermosphacta* have been quantified in seafood. Mamlouk *et al.* (2012) developed a qPCR method using a propidium monoazide (PMA) sample step for monitoring the *B. thermosphacta* population in cooked shrimp and salmon in 3–4 h with the 16S rRNA gene sequence as the target. The study showed a high correlation between the *B. thermosphacta* counts obtained by qPCR and plate counts on shrimp and salmon. Reynisson *et al.* (2008) developed a qPCR to quantify *Pseudomonas* spp. in cod in 5 h using the carbamoyl phosphate synthase gene (*car*A) as targets with SYBR green. A high degree of correlation with plate counts was observed. Lee and Levin (2007) developed a qPCR assay combined to the fluorescent dye ethidium bromide monoazide (EMA) for distinguishing between the total number of viable and dead bacterial cells on cod fillets. However, detection of viable *E. coli* cells was achieved by Sheridan *et al.* (1998) using mRNA reverse transcription-PCR. Although qPCR is recognized as a power tool for detection and quantification of pathogens and other microorganisms, the literature currently lacks studies for quantification of SSOs in seafood.

15.9 Conclusions

Microbiological analysis for quality control and research purposes can be conducted using the classical approach based on culture media or other alternative techniques. The standard methods for detection of microbial pathogens still rely on culture media; however, many alternative methods are in use or will be employed as standard methods in the future. Due to the very low detection limit that is usually required, it seems to be difficult to exclude the enrichment stage. The most promising methods for microbiological analysis are the molecular methods, especially PCR-based methods. These methods have been increasingly developed and improved in the past few years and they are currently established as useful and reliable, offering rapid detection, identification and quantification of microorganisms in seafood.

References

Ashie, I.N.A., Smith, J.P. and Simpson, B.K. (1996). Spoilage and shelf-life extension of fresh fish and shellfish. *Critical Reviews in Food Science and Nutrition*, 36, 87–121.

ANZFA (2001). User guide to Standard 1.6.1 – Microbiological Limits for Food with additional guideline criteria. Australia New Zealand Food Authority. http://www.foodstandards.gov.au/code/userguide/pages/microbiologicallimit1410.aspx [accessed on 18 June 2013].

Atlas, R.M. (1995). *Handbook of Microbiological Media for the Examination of Food*, CRC Press, Boca Raton, FL, USA, p.10.

Blackstone, G.M., Nordstrom, J.L., Vickery, M.C.L. *et al.* (2003). Detection of pathogenic *Vibrio parahaemolyticus* in oyster enrichments by real time PCR. *Journal of Microbiological Methods*, 53, 149–155.

Boziaris, I.S., Kordila, A. and Neofitou, C. (2011). Microbial spoilage analysis and its effect on chemical changes and shelf-life of Norway lobster (*Nephrops norvegicus*) stored in air at various temperatures. *International Journal of Food Science and Technology*, 46, 887–895.

Boziaris, I.S., Stamatiou, A.P. and Nychas, G.-J.E. (2013). Microbiological aspects and shelf life of processed seafood products. *Journal of the Science of Food and Agriculture*, 93, 1184–1190.

Brasher, C.W., DePaola, A., Jones, D.D. and Bej, A.K. (1998). Detection of microbial pathogens in shellfish with Multiplex PCR. *Current Microbiology*, 37(2), 101–107.

Broekaert, K., Heyndrickx, M., Herman, L., Devlieghere, F. and Vlaemynck, G. (2011). Seafood quality analysis: Molecular identification of dominant microbiota after ice storage on several general growth media. *Food Microbiology*, 28, 1162–1169.

Caglak, E., Cakli, S. and Kilinc, B. (2008). Microbiological, chemical and sensory assessment of mussels (*Mytilus galloprovincialis*) stored under modified atmosphere packaging. *European Food Research and Technology*, 226, 1293–1299.

Cai, T., Jiang, L., Yang, C. and Huang, K. (2006). Application of real-time PCR for quantitative detection of *Vibrio parahaemolyticus* from seafood in eastern China. *FEMS Immunology and Medical Microbiology*, 46, 180–186.

Cakli, S., Kilinc, B., Dincer, T. and Talasa, S. (2006). Comparison of the shelf lives of map and vacuum packaged hot smoked rainbow trout (*Onchoryncus mykiss*). *European Food Research and Technology*, 224, 19–26.

Cambon-Bonavita, M.-A., Lesongeur, F., Menoux, S., Lebourg, A. and Barbier, G. (2001). Microbial diversity in smoked salmon examined by a culture-independent molecular approach – a preliminary study. *International Journal of Food Microbiology*, 70, 179–187.

Canadian Food Inspection Agency (2011). Bacteriological guidelines for fish and fish products. Fish products standards and methods manual. http://www.inspection.gc.ca/food/fish-and-seafood/manuals/standards-and-methods/eng/1348608971859/1348609209602?chap=0 s17c7 [accessed on 18 June 2013].

Canigral, I., Moreno, Y., Alonso, J.L., Gonzalez, A. and Ferrus, M.A. (2010). Detection of *Vibrio vulnificus* in seafood, seawater and waste water samples from a Mediterranean coastal area. *Microbiological Research*, 165, 657–664.

Cariani, A., Piano, A., Consolandi, C. *et al.* (2012). Detection and characterization of pathogenic vibrios in shellfish by a Ligation Detection Reaction-Universal Array approach. *International Journal of Food Microbiology*, 153, 474–482.

Casas, N., Amarita F. and Martínez de Marañón, I. (2007). Evaluation of an extracting method for the detection of Hepatitis A virus in shellfish by SYBR-Green real-time RT-PCR. *International Journal of Food Microbiology*, 120, 179–185.

Commission Regulation (EC) (2007). No 1441/2007 of 5 December 2007 amending Regulation (EC) No 2073/2005 on microbiological criteria for foodstuffs. *Official Journal of the European Communities* L 322/12, 7.12.2007, pp. 12–29.

Corbo, M.R., Altieri, C., Bevilacqua, A. *et al.* (2005). Estimating packaging atmosphere–temperature effects on the shelf life of cod fillets. *European Food Research and Technology*, 220, 509–513.

Dainty, R.H. (1996). Chemical/biochemical detection of spoilage. *International Journal of Food Microbiology*, 33, 19–33.

Dalgaard, P., Gram, L. and Huss, H.H. (1993). Spoilage and shelf life of cod fillets packed in vacuum or modified atmospheres. *International Journal of Food Microbiology*, 19, 283–294.

Dalgaard, P. (1995a). Modelling of microbial activity and prediction of shelf life of packed fresh fish. *International Journal of Food Microbiology*, 19, 305–318.

Dalgaard, P. (1995b). Qualitative and quantitative characterization of spoilage bacteria from packed fish. *International Journal of Food Microbiology*, 26, 319–333.

Dalgaard, P., Mejlholm, O. and Huss, H.H. (1996). Conductance method for quantitative determination of *Photobacterium phosphoreum* in fish products. *Journal of Applied Bacteriology*, 81, 57–64.

Dalgaard, P., Mejlholm, O., Christiansen, T.J. and Huss, H.H. (1997). Importance of *Photobacterium Phosphoreum* in relation to spoilage of modified atmosphere packed fish products. *Letters in Applied Microbiology*, 24, 373–378.

Dalgaard, P. (2000). Fresh and lightly preserved seafood, in *Shelf-Life Evaluation of Foods* (eds C.M.D. Man and A.A. Jones), Aspen Publishers, London, pp 110–139.

Davidson, C.M. and Cronin, F. (1973). Medium for the selective enumeration of lactic acid bacteria from foods. *Applied Microbiology*, 26, 439–440.

De Man, J.C., Rogosa, M. and Sharpe, M.E. (1960). A medium for the cultivation of lactobacilli. *Journal of Applied Bacteriology*, 23, 130–135.

Di Pinto, A., Ciccarese, G., De Corato, R., Novello, L. and Terio, V. (2008). Detection of pathogenic *Vibrio parahaemolyticus* in southern Italian shellfish. *Food Control*, 19, 1037–1041.

Di Pinto, A., Terio, V., Di Pinto, P., Colao, V. and Tantillo, G. (2012). Detection of *Vibrio parahaemolyticus* in shellfish using polymerase chain reaction–enzyme-linked immunosorbent assay. *Letters in Applied Microbiology*, 54, 494–498.

Duflos, G., Theraulaz, L., Giordano, G., Mejean, V. and Malle P. (2010). Quantitative PCR Method for evaluating freshness of whiting (*Merlangius merlangus*) and plaice (*Pleuronectes platessa*). *Journal of Food Protection*, 73(7), 1344–1347.

Dupont, J., Menard, D., Herve, C. *et al.* (1996). Rapid estimation of *Escherichia coli* in live marine bivalve shellfish using automated conductance measurement. *Journal of Applied Bacteriology*, 80, 81–90.

Emborg, J., Laursen, B.G., Rathjen, T. and Dalgaard, P. (2002). Microbial spoilage and formation of biogenic amines in fresh and thawed modified atmosphere-packed salmon (*Salmo salar*) at 2 °C. *Journal of Applied Microbiology*, 92, 790–799.

Endo, H., Nakamura, J., Ren, H. and Hayashi, T. (2002). Flow cytometry for rapid determination of number of microbial cells grown on fish. *Food Science and Technology Research*, 8, 342–346.

Fach, P., Perelle, S., Dilasser, F. *et al.* (2002). Detection by PCR–enzyme-linked immunosorbent assay of *Clostridium botulinum* in fish and environmental samples from a coastal area in northern France. *Applied and Environmental Microbiology*, 68(12), 5870–5876.

Feldhusen, F. (2000). The role of seafood in bacterial foodborne diseases. *Microbes and Infections*, 2(13), 1651–1660.

Ferrario, C., Pegollo, C., Ricci, G., Borgo, F. and Fortina, M.G. (2012). PCR Detection and Identification of Histamine-Forming Bacteria in Filleted Tuna Fish Samples. *Journal of Food Science*, 77(2), 115–120.

Fonnesbech, B., Frockjaer, H., Gram, L. and Jespersen, C.M. (1993). Production and specificities of poly- and monoclonal antibodies against *Shewanella putrefaciens*. *Journal of Applied Bacteriology*, 74, 444–451.

Fuchizawa, I., Shimizu, S., Kawai Y. and Yamazaki, K. (2008). Specific detection and quantitative enumeration of *Listeria* spp. using fluorescent in situ hybridization in combination with filter cultivation (FISHFC). *Journal of Applied Microbiology*, 105, 502–509.

Fung, D. (2002). Rapid methods and automation in microbiology. *Comprehensive Reviews in Food Science and Food Safety*, 1, 3–22.

Garrido, A., Chapela, M.-J., Ferreira, M. *et al.* (2012). Development of a multiplex real-time PCR method for pathogenic *Vibrio parahaemolyticus* detection (*tdh* and *trh*). *Food Control*, 24, 128–135.

Garrido, A., Chapela, M.-J., Román, B. *et al.* (2013). A new multiplex real-time PCR developed method for *Salmonella* spp. and *Listeria monocytogenes* detection in food and environmental samples. *Food Control*, 30, 76–85.

Gennari, M., Tomaselli, S. and Cotrona, V. (1999). The microflora of fresh and spoiled sardines (*Sardina pilchardus*) caught in Adriatic (Mediterranean) Sea and stored in ice. *Food Microbiology*, 16, 15–28.

Gimenez, B., Roncales, P. and Beltran, J.A. (2002). Modified atmosphere packaging of filleted rainbow trout. *Journal of the Science of Food and Agriculture*, 82, 1154–1159.

Gonzalez, S.F., Krug, M.J., Nielsen, M.E., Santos, Y. and Call, D.R. (2004). Simultaneous detection of marine fish pathogens by using multiplex PCR and a DNA microarray. *Journal of Clinical Microbiology*, 42(4), 1414–1419.

Gornik, S.G., Albalat, A., Macpherson, H., Birkbeck, H. and Neil, D.M. (2011). The effect of temperature on the bacterial load and microbial composition in Norway lobster (*Nephrops norvegicus*) tail meat during storage. *Journal of Applied Microbiology*, 111, 582–592.

Gram, L., Trolle, G. and Huss, H.H. (1987). Detection of specific spoilage bacteria on fish stored at high (20 °C) and low (0 °C) temperatures. *International Journal of Food Microbiology*, 4, 65–72.

Gram, L., Wedell-Neergaard, C. and Huss, H.H. (1990). The bacteriology of fresh and spoiling Lake Victorian Nile perch (*Lates niloticus*). *International Journal of Food Microbiology*, 10, 303–316.

Gram, L. (1992). Evaluation of the bacteriological quality of seafood. *International Journal of Food Microbiology*, 16, 25–39.

Gram, L. and Huss, H.H. (1996). Microbiological spoilage of fish and fish products. *International Journal of Food Microbiology*, 33, 121–137.

Gram, L. and Dalgaard, P. (2002). Fish spoilage bacteria – problems and solutions. *Current Opinion in Biotechnology*, 13, 262–266.

Gram, L. (2009). Microbiological spoilage of fish and seafood products, in *Compendium of the Microbiological Spoilage of Food and Beverages* (eds W.H. Spenser and M.P. Doyle), Springer Science, pp. 87–119.

Green, D.H. and Lewis, G.D. (1999). Comparative detection of enteric viruses in wastewaters, sediments and oysters by reverse transcription-PCR and cell culture. *Water Research*, 33(5), 1195–1200.

Guan, L., Cho, K.H. and Lee, J.-H. (2011). Analysis of the cultivable bacterial community in *jeotgal*, a Korean salted and fermented seafood, and identification of its dominant bacteria. *Food Microbiology*, 28(1), 101–113.

Guizani, N., Al-Busaidy, M.A., Al-Belushi, I.M., Mothershaw, A. and Rahman, M.S. (2005). The effect of storage temperature on histamine production and the freshness of yellowfin tuna (*Thunnus albacares*). *Food Research International*, 38, 215–222.

Han, F. and Ge, B. (2010). Multiplex PCR assays for simultaneous detection and characterization of *Vibrio vulnificus* strains. *Letters in Applied Microbiology*, 51, 234–240.

Hempel, A., Borchert, N., Walsh, H. *et al.* (2011). Analysis of total aerobic viable counts in raw fish by high-throughput optical oxygen respirometry. *Journal of Food Protection*, 74, 776–782.

Hovda, M.B., Lunestad, B.T., Sivertsvik, M. and Rosnes, J.T. (2007a). Characterisation of the bacterial flora of modified atmosphere packaged farmed Atlantic cod (*Gadus morhua*) by PCR-DGGE of conserved 16S rRNA gene regions. *International Journal of Food Microbiology*, 117, 68–75.

Hovda, M.B., Sivertsvik, M., Lunestad, B.T., Lorentzen, G. and Rosnes, J.T. (2007b). Characterisation of the dominant bacterial population in modified atmosphere packaged farmed halibut (*Hippoglossus hippoglossus*) based on 16S rDNA-DGGE. *Food Microbiology*, 24, 362–37.

Hozbor, M.C., Saiz, A.I., Yeannes, M.I. and Fritz, R. (2006). Microbiological changes and its correlation with quality indices during aerobic iced storage of sea salmon (*Pseudopercis semifasciata*). *LWT – Food Science and Technology*, 39, 99–104.

Hugenholtz, P, Goebel, B.M. and Pace, N.R. (1998). Impact of culture independent studies on the emerging phylogenetic view of bacterial diversity. *Journal of Bacteriology*, 180, 4765–4774.

Hugh, R. and Leifson, E. (1953). The taxonomic significance of fermentative versus oxidative Gram-negative bacteria. *Journal of Bacteriology*, 66, 24–26.

Huis in't Veld, J.H.J. (1996). Microbial and biochemical spoilage of foods: an overview. *International Journal of Food Microbiology*, 33, 1–18.

Huss, H.H, Reilly, A. and Karim Ben Embarek, P. (2000). Prevention and control of hazards in seafood. *Food Control*, 11, 149–156.

ICMSF (1986). Sampling plan and recommended microbiological limits for Seafood. International Commission on Microbiological Specifications for Foods. www.fao.org/docrep/X5624E/x5624e08.htm [accessed on 18 June 2013].

ISO (1996). ISO 11290-1. Microbiology of Food and Animal Feeding Stuffs. Horizontal Method for the Detection and Enumeration of *Listeria monocytogenes*. Part 1: Detection Method. International Organization for Standardization, Geneva, Switzerland.

ISO (1998). ISO 11290-2. Microbiology of Food and Animal Feeding Stuffs. Horizontal Method for the Detection and Enumeration of *Listeria monocytogenes*. Part 2: Enumeration Method. International Organization for Standardization, Geneva, Switzerland.

ISO (1999). ISO 6888-1:1999. Microbiology of food and animal feeding stuffs. Horizontal method for the enumeration of coagulase-positive staphylococci (*Staphylococcus aureus* and other species). Part 1: Technique using Baird-Parker agar medium. International Organization for Standardization, Geneva, Switzerland.

ISO (2001a). ISO 17410:2001. Microbiology of food and animal feeding stuffs. Horizontal method for the enumeration of psychrotrophic microorganisms. International Organization for Standardization, Geneva, Switzerland.

ISO (2001b). ISO 16649-2:2001. Microbiology of food and animal feeding stuffs. Horizontal method for the enumeration of β-glucuronidase positive *Escherichia coli*. Part 2: Colony-count technique at 44 °C using 5-bromo-4-chloro-3-indolyl-D-glucuronide. International Organization for Standardization, Geneva, Switzerland.

ISO (2001c). ISO 16654:2001. Microbiology of food and animal feeding stuffs. Horizontal method for the detection of Escherichia coli O157. International Organization for Standardization, Geneva, Switzerland.

ISO (2002). ISO 6579:2002. Microbiology of food and animal feeding stuffs. Horizontal method for the detection of *Salmonella* spp. International Organization for Standardization, Geneva, Switzerland.

ISO (2003). ISO 4833. Microbiology of food and animal feeding stuffs. Horizontal method for the enumeration of micro organisms – Colony-count technique at 30 °C. International Organization for Standardization, Geneva, Switzerland.

ISO (2004a). ISO 21528-1:2004. Microbiology of food and animal feeding stuffs. Horizontal methods for the detection and enumeration of Enterobacteriaceae. Part 1: Detection and enumeration by MPN technique with pre-enrichment. International Organization for Standardization, Geneva, Switzerland.

ISO (2004b). ISO 21528-2:2004. Microbiology of food and animal feeding stuffs. Horizontal methods for the detection and enumeration of Enterobacteriaceae. Part 2: Colony-count method. International Organization for Standardization, Geneva, Switzerland.

ISO (2004c). ISO 11290-1. Microbiology of Food and Animal Feeding Stuffs. Horizontal Method for the Detection and Enumeration of *Listeria monocytogenes*. Part 1: Detection Method. AMENDMENT 1: Modification of the Isolation Media and the Haemolysis Test, and Inclusion of Precision Data. International Organization for Standardization, Geneva, Switzerland.

ISO (2004d). ISO 11290-2. Microbiology of Food and Animal Feeding Stuffs. Horizontal Method for the Detection and Enumeration of *Listeria monocytogenes*. Part 2: Enumeration Method, AMENDMENT 1: Modification of the Enumeration Medium. International Organization for Standardization, Geneva, Switzerland.

ISO (2005). ISO/TS 16649-3:2005. Microbiology of food and animal feeding stuffs. Horizontal method for the enumeration of β-glucuronidase positive *Escherichia coli*. Part 3: Most Probable Number technique using 5-bromo-4-chloro-3-indolyl-D-glucuronide. International Organization for Standardization, Geneva, Switzerland.

ISO (2007a). ISO/TS 21872-1:2007. Microbiology of food and animal feeding stuffs – Horizontal method for the detection of potentially enteropathogenic Vibrio spp. Part 1: Detection of *Vibrio parahaemolyticus* and *Vibrio cholerae*. International Organization for Standardization, Geneva, Switzerland.

ISO (2007b). ISO/TS 21872-2:2007 Microbiology of food and animal feeding stuffs. Horizontal method for the detection of potentially enteropathogenic Vibrio spp. Part 2: Detection of species other than *Vibrio parahaemolyticus* and *Vibrio cholerae*. International Organization for Standardization, Geneva, Switzerland.

Jaffrès, E., Sohier, D., Leroi, F. *et al.* (2009). Study of the bacterial ecosystem in tropical cooked and peeled shrimps using a polyphasic approach. *International Journal of Food Microbiology*, 131, 20–29.

Jasson, V., Jacxsens, L., Luning, P., Rajkovic, A. and Uyttendaele, M. (2010). Alternative microbial methods: an overview and selection criteria. *Food Microbiology* 27, 710–730.

Jay, J.M., Loessner, M.J. and Golden, D.A. (2005). *Modern Food Microbiology*, 7th edn, Springer Science+Business Media, Inc., New York.

Jørgensen, B.R. and Huss, H.H. (1989). Growth and activity of *Shewanella putrefaciens* isolated from spoiling fish. *International Journal of Food Microbiology*, 9, 51–62.

Justé, A., Thomma, B.P.H.J. and Lievens, B. (2008). Recent advances in molecular techniques to study microbial communities in food-associated matrices and processes. *Food Microbiology*, 25, 745–761.

Kakasis S., Parlapani F.F. and Boziaris, I.S. (2011). Performance and selectivity of media used for the enumeration of bacterial populations in seafood. 33 ° Scientific Conference of Hellenic Association for Biological Sciences, 19–21 May 2011, Edessa, pp. 100–101.

Kaysner, C.A. and DePaola, A. (2004). *Bacteriological Analytical Manual Online. Vibrio*. US Food and Drug Administration. Center for Food Safety and Applied Nutrition. http://www.fda.gov/Food/FoodScienceResearch/LaboratoryMethods/ucm070830.htm [accessed on 18 June 2013].

Khan, M.A., Parrish, C.C. and Shahidi, F. (2005). Enumeration of total heterotrophic and psychrotrophic bacteria using different types of agar to evaluate the microbial quality of blue mussels (*Mytilus edulis*) and sea scallops (*Placopecten magellanicus*). *Research International*, 38, 751–758.

Kielwein, G. (1969). Ein Nährboden zur selektiven Züchtung von Pseudomonaden und Aeromonaden. *Archiv fur Lebensmittelhygiene*, 20, 131–133.

Kielwein, G. (1971). Pseudomonaden und Aromonaden in Trinkmilch: Ihr Nachweis und ihre Bewertung. *Archiv fur Lebensmittelhygiene*, 22, 15–19.

Kimura, B., Kawasaki, S., Nakano, H. and Fujii, T. (2001). Rapid, Quantitative PCR monitoring of growth of *Clostridium botulinum* type E in modified-atmosphere-packaged fish. *Applied and Environmental Microbiology*, 67, 206–216.

Koopmans, M., Von Bondsdorff, C.H., Vinjé, J., De Medici, D. and Monroe, S. (2002). Foodborne viruses. *FEMS Microbiology Reviews*, 26, 187–205.

Koutsoumanis, K. and Nychas, G.-J.E. (1999). Chemical and sensory changes associated with microbial flora of Mediterranean Boque (*Boops boops*) stored aerobically at 0, 3, 7 and 10 °C. *Applied and Environmental Microbiology*, 65, 698–706.

Koutsoumanis, K. and Nychas, G.-J.E. (2000). Application of a systematic experimental procedure to develop a microbial model for rapid fish shelf-life predictions. *International Journal of Food Microbiology*, 60, 171–184.

Koutsoumanis, K., Taoukis, P.S., Drosinos, E.H. & Nychas, G.-J.E. (2000). Applicability of an Arrhenious model for the combined effect of temperature and CO_2 packaging on the spoilage microflora of fish. *Applied and Environmental Microbiology*, 66, 3528–3534.

Koutsoumanis, K. (2001). Predictive modeling of the shelf life of fish under nonisothermal conditions. *Applied and Environmental Microbiology*, 67, 1821–1829.

Koutsoumanis, K., Giannakourou, M.C., Taoukis, P.S. and Nychas, G.-J.E. (2002). Application of shelf life decision system (SLDS) to marine cultured fish quality. *International Journal of Food Microbiology*, 73, 375–382.

Kumar, R., Surendran, P.K. and Thampuran, N. (2008a). Evaluation of culture, ELISA and PCR assays for the detection of *Salmonella* in seafood. *Letters in Applied Microbiology*, 46, 221–226.

Kumar, R., Surendran, P.K. and Thampuran, N. (2008b). An eight-hour PCR-based technique for detection of *Salmonella* serovars in seafood. *World Journal of Microbiology and Biotechnology*, 24, 627–631.

Kumar, B.K., Raghunath, P., Devegowda, D. *et al.* (2011). Development of monoclonal antibody based sandwich ELISA for the rapid detection of pathogenic *Vibrio parahaemolyticus* in seafood. *International Journal of Food Microbiology*, 145, 244–249.

Lalitha, K.V., Sonaji, E.R., Manju, S. *et al.* (2005). Microbiological and biochemical changes in pearl spot (*Etroplus suratensis* Bloch) stored under modified atmospheres. *Journal of Applied Microbiology*, 99, 1222–1228.

Landgraf, A., Reckmann, B. and Pingoud, A. (1991). Direct analysis of polymerase chain reaction products using enzyme-linked immunosorbent assay techniques. *Analytical Biochemistry*, 198, 86–91.

Lee, J.-L. and Levin, R.E. (2007). Quantification of total viable bacteria on fish fillets by using ethidium bromide monoazide real-time polymerase chain reaction. *International Journal of Food Microbiology*, 118, 312–317.

Levin, R.E. (2010). Assessement of seafood spoilage and the microorganisms involved, in *Handbook of seafood and seafood products analysis* (eds L.L. Leo and F. Toldra), CRC Press, Taylor and Francis Group, Boca Raton, FL, USA, pp. 515–535.

Liston, J. (1980). Microbiology in fishery science, in *Advances in Fish Science and Technology* (eds J.J. Connell and staff of Torry Research Station), Fishing News Books, Farnham, UK, pp. 138–157.

López-Caballero, M.E., Huidobro, A., Pastor, A. and Tejada, M. (2002). Microflora of gilthead seabream (*Sparus aurata*) stored in ice. Effect of washing. *European Food Research and Technology*, 215, 396–400.

Lopez-Caballero, M.E., Martinez-Alvarez, O., Gomez-Guillen, M.C. and Montero, P (2006). Effect of natural compounds alternative to commercial antimelanosisc on polyphenol oxidase activity and microbial growth in cultured prawns (*Marsupenaeous tiger*) during chill storage. *European Food Research Technology*, 223, 7–15.

Lougovois, V.P., Kyranas, E.R. and Kyrana, V.R. (2003). Comparison of selected methods of assessing freshness quality and remaining storage life of iced gilthead sea bream (*Sparus aurata*). *Food Research International*, 36, 551–560.

Lyon, W.J. (2001). TaqMan PCR for Detection of *Vibrio cholerae* O1, O139, Non-O1, and Non-O139 in Pure Cultures, Raw Oysters, and Synthetic Seawater. *Applied and Environmental Microbiology*, 67(10), 4685–4693.

Macé, S., Cornet, J., Chevalier, F. *et al.* (2012). Characterisation of the spoilage microbiota in raw salmon (*Salmo salar*) steaks stored under vacuum or modified atmosphere packaging combining conventional methods and PCR-TTGE. *Food Microbiology*, 30, 164–172.

Macé, S., Joffraud, J.-J., Cardinal, M. *et al.* (2013). Evaluation of the spoilage potential of bacteria isolated from spoiled raw salmon (*Salmo salar*) fillets stored under modified atmosphere packaging. *International Journal of Food Microbiology*, 160, 227–238.

Mamlouk, K., Macé, S., Guilbaud, M. *et al.* (2012). Quantification of viable *Brochothrix thermosphacta* in cooked shrimp and salmon by real-time PCR. *Food Microbiology*, 30, 173–179.

Marshall, D.L., Domma, D.N. and Grodner, R.M. (1999). A potential capacitance detection and enumeration method for *Vibrio cholerae* and *Vibrio vulnificus*. *Journal of Aquatic Food Product Technology*, 8, 5–19.

Mejlholm, O., Bøknæs, N. and Dalgaard, P. (2005). Shelf life and safety aspects of chilled cooked and peeled shrimps (*Pandalus borealis*) in modified atmosphere packaging. *Journal of Applied Microbiology*, 99, 66–76.

Messelhäusser, U., Colditz, J., Thärigen, D. *et al.* (2010). Detection and differentiation of *Vibrio* spp. in seafood and fish samples with cultural and molecular methods. *International Journal of Food Microbiology*, 142, 360–364.

Metcalfe, A.M. and Marshall, D.L. (2004) Capacitance method to determine the microbiological quality of raw shrimp (*Penaeus setiferus*). *Food Microbiology*, 21, 361–364.

Nordic Committee of Food Analysis (2006). NMKL 184. Aerobic count and specific spoilage organisms in fish and fish products. *NMK Newsletter* No. 61, 1. http://www.nmkl.org/Engelsk/Newsletter/61-eng.pdf [accessed 18 June 2013].

Olofsson, T.C., Ahrne, S. and Molin, G. (2007). The bacterial flora of vacuum-packed cold-smoked salmon stored at 7 °C, identified by direct 16S rRNA gene analysis and pure culture technique. *Journal of Applied Microbiology*, 103, 1364–5072.

Ootsubo, M., Shimizu, T., Tanaka, R. *et al.* (2003). Seven-hour fluorescence in situ hybridization technique for enumeration of Enterobacteriaceae in food and environmental water sample. *Journal of Applied Microbiology*, 95, 1182–1190.

Ottaviani, D., Santarelli, S., Bacchiocchi, S. *et al.* (2005). Presence of pathogenic *Vibrio parahaemolyticus* strains in mussels from the Adriatic Sea, Italy. *Food Microbiology*, 22, 585–590.

Panicker, G., Call, D.R., Krug, M.J. and Bej, A.K. (2004). Detection of pathogenic *Vibrio* spp. in shellfish by using multiplex PCR and DNA microarrays. *Applied and Environmental Microbiology*, 70(12), 7436–7444.

Parlapani F.F., Kormas, K.Ar. and Boziaris, I.S. (2012). Use of 16S rRNA gene analysis for the identification of dominant microbiota in sea-bream fillets stored at various conditions. FOOD MICRO Conference, 3–7 September 2012, Istanbul, p. 610.

Parlapani F.F., Meziti A., Kormas, K.Ar. and Boziaris I.S. (2013). Indigenous and spoilage microbiota of farmed sea bream stored in ice identified by phenotypic and 16S rRNA gene analysis. *Food Microbiology*, 33, 85–89.

Pettipher, G.L. and Rodrigues, U.M. (1982). Rapid enumeration of microorganisms in foods by the direct epifluorescent filter technique. *Applied and Environmental Microbiology*, 44, 809–813.

Pournis, N., Papavergou, A., Badeka, A., Kontominas, M.G. and Savvaidis, I.N. (2005). Shelf-life extension of refrigerated mediterranean mullet (*mullus surmuletus*). using modified atmosphere packaging. *Journal of Food Protection*, 68, 2201–2207.

Reuter, G. (1985). Elective and selective media for lactic acid bacteria. *International Journal of Food Microbiology*, 2, 55–68.

Reynisson, E., Lauzon, H.L., Magnusson, H., Hreggvidsson, G.O. and Marteinsson, V.T. (2008). Rapid quantitative monitoring method for the fish spoilage bacteria *Pseudomonas*. *Journal of Environmental Monitoring*, 10, 1357–1362.

Ribao, C., Torrado, I., Vilariño, M.L. and Romalde J.L. (2004). Assessment of different commercial RNA-extraction and RT-PCR kits for detection of hepatitis A virus in mussel tissues. *Journal of Virological Methods*, 115, 177–182.

Robert-Pillot, A., Copin, S., Gay, M., Malle, P. and Quilici, M.L. (2010). Total and pathogenic *Vibrio parahaemolyticus* in shrimp: fast and reliable quantification by real-time PCR. *International Journal of Food Microbiology*, 143, 190–197.

Robson, A.A., Kelly, M.S. and Latchford, J.W. (2007). Effect of temperature on the spoilage rate of whole, unprocessed crabs: *Carcinus maenas, Necora puber* and *Cancer pagurus*. *Food Microbiology*, 24, 419–424.

Roh, S.W., Kim, K.-H., Nam, Y.-D. *et al.* (2010). Investigation of archaeal and bacterial diversity in fermented seafood using barcoded pyrosequencing. *Multidisciplinary Journal of Microbial Ecology*, 4, 1–16.

Rosnes, J.T., Kleiberg, G.H., Sivertsvik, M., Lunestad, B.T. and Lorentzen, G. (2006). Effect of modified atmosphere packaging and superchilled storage on the shelf-life of farmed ready-to-cook spotted wolf-fish (*Anarhichas minor*). *Packaging Technology and Science*, 19, 325–333.

Rudi, K., Maugesten, T., Hannevik, S.E. and Nissen, H. (2004). Explorative multivariate analyses of 16S rRNA gene data from microbial communities in modified-atmosphere-packed salmon and coalfish. *Applied and Environmental Microbiology*, 70(8), 5010–5018.

Russell, S.M. (1998). Capacitance microbiology as a means of determining the quantity of spoilage bacteria on fish fillets. *Journal of Food Protection*, 61, 844–848.

Sacchi, C.T., Whitney, A.M., Mayer, L.W. *et al.* (2002). Sequencing of 16S rRNA gene: a rapid tool for identification of *Bacillus anthracis*. *Emerging Infectious Diseases*, 8, 1117–1123.

Salvat, G., Rudelle, S., Humbert, F., Colin, P. and Lahellec, C. (1997). A selective medium for the rapid detection by an impedance technique of *Pseudomonas* spp., associate with poultry meat. *Journal of Applied Microbiology*, 83, 456–463.

Sheridan, G.E., Masters, C.I., Shallcross, J.A. and MacKey, B.M. (1998). Detection of mRNA by reverse transcription-PCR as an indicator of viability in *Escherichia coli* cells. *Applied and Environmental Microbiology*, 64, 1313–1318.

Shewan, J., Hobbs, G. and Hodgkiss, W. (1960a). A determinative scheme for the identification of certain genera of Gram-negative bacteria, with special reference to the Pseudomonadaceae. *Journal of Applied Bacteriology*, 23, 379–390.

Shewan, J., Hobbs, G. and Hodgkiss, W. (1960b). The *Pseudomonas* and *Achromobacter* groups of bacteria in the spoilage of marine white fish. *Journal of Applied Bacteriology*, 23, 463–468.

Shewan, J.M. (1962). The bacteriology of fresh and spoiling fish and some related chemical changes. *Recent Advances in Food Science*, 1, 167–193.

Sincero, T.C.M., Levin, D.B., Simões, C.M.O. and Barardi, C.R.M. (2006). Detection of hepatitis A virus (HAV) in oysters (*Crassostrea gigas*). *Water Research*, 40, 895–902.

Sivertsvik, M., Rosnes, J.T. and Kleiberg, G.H. (2003). Effect of modified atmosphere packaging and superchilled storage on the microbial and sensory quality of atlantic salmon (*Salmo salar*) fillets. *Journal of Food Science*, 68(4), 1467–1472.

Skjerdal, O.T., Lorentzen, G., Tryland, I. and Berg, J.D. (2004). New method for rapid and sensitive quantification of sulphide-producing bacteria in fish from arctic and temperate waters. *International Journal of Food Microbiology*, 93, 325–333.

Stohr, V, Joffraud, J.J., Cardinal, M. and Leroi, F. (2001). Spoilage potential and ensory profile associated with bacteria isolated from cold-smoked salmon. *Food Research International*, 34, 797–806.

Svanevik, C.M. and Lunestad, B.T. (2011). Characterisation of the microbiota of Atlantic mackerel (*Scomber scombrus*). *International Journal of Food Microbiology*, 151, 164–170.

Sveinsdottir, K., Martinsdottir, E., Hyldig, G., Jorgensen, B. and Kristbergsson, K. (2002). Application of Quality Index Method (QIM) scheme in shelf-life study of farmed Atlantic salmon (*Salmo salar*). *Journal of Food Science*, 67(4), 1570–1579.

Tomé, E., Teixeira, P. and Gibbs, P.A. (2006). Anti-listerial inhibitory lactic acid bacteria isolated from commercial cold smoked salmon. *Food Microbiology*, 23, 399–405.

Torrieri, E., Cavella, S., Villani, F. and Masi, P. (2006). Influence of modified atmosphere packaging on the chilled shelf life of gutted farmed bass (*Dicentrarchus labrax*). *Journal of Food Engineering*, 77, 1078–1086.

Truelstrup, H.L. and Vogel, B.F. (2011). Desiccation of adhering and biofilm *Listeria monocytogenes* on stainless steel: survival and transfer to salmon products. *International Journal of Food Microbiology*, 146, 88–93.

Tryfinopoulou, P., Drosinos, E.H. and Nychas, G.-J.E. (2001). Performance of Pseudomonas CFC-selective medium in the fish storage ecosystems. *Journal of Microbiological Methods*, 47, 243–247.

Tryfinopoulou, P., Tsakalidou, E. and Nychas, G.-J.E. (2002). Characterization of *Pseudomonas* spp. associated with spoilage of gilt-head sea bream stored under various conditions. *Applied and Environmental Microbiology*, 68, 65–72.

Tryfinopoulou, P., Tsakalidou, E., Vancanneyt, M. *et al.* (2007). Diversity of *Shewanella* population in fish *Sparus aurata* harvested in the Aegean Sea. *Journal of Applied Microbiology*, 103, 711–721.

US Patent (1998). Selective additive and media for *Pseudomonas fluorescens*. Inventor S.M. Russell. Patent number 5,741,663.

Valle, M., Eb, P., Tailliez, R. and Malle, P. (1998). Optimization of the enumeration of total aerobic bacterial flora in the flesh of seafish. *Journal of Rapid Methods and Automation in Microbiology*, 6, 29–42.

Van Spreekens, K.J.A. (1974). The suitability of a modification of Long and Hammer's medium for the enumeration of more fastidious bacteria from fresh fishery products. *Archiv für Lebensmittelhygiene*, 25, 213–219.

Vantarakis, A., Komninou, G., Venieri, D. and Papapetropoulou, M. (2000). Development of a multiplex PCR detection of *Salmonella* spp. and *Shigella* spp. in mussels. *Letters in Applied Microbiology*, 31, 105–109.

Veal, D.A., Deere, D., Ferrari, B., Piper, J. and Attfield, P.V. (2000). Fluorescence staining and flow cytometry for monitoring microbial cells. *Journal of Immunological Methods*, 243, 191–210.

Ward, D.R., LaRocco, K.A. and Hopn, D.J. (1986). Adenosine triphosphate bioluminescent assay to enumerate bacterial numbers on fresh fish. *Journal of Food Protection*, 49, 647–650.

Ward, D.M., Weller, R. and Bateson, M.M. (1990). 16S rRNA sequences reveal numerous uncultured microorganisms in a natural community. *Nature (London)*, 345, 63–65.

16

Fish and Seafood Authenticity – Species Identification

Fátima C. Lago, Mercedes Alonso, Juan M. Vieites and Montserrat Espiñeira

ANFACO-CECOPESCA, Vigo, Pontevedra, Spain

16.1 Molecular techniques applied to seafood authentication

In recent years, species identification has become increasingly important as a target goal for molecular techniques within the food sector. This is because of the growing interest of consumers in quality assurance for products available in the markets, and particularly fisheries and aquaculture products. Parallel to this increase, the control of quality and alimentary security has improved along with production and commercialization.

To protect consumers and to prevent fraud in the marketing of fishery and aquaculture products, the European Union has established European Commission Regulation (EC) 104/2000 (EC, 2000). The measures for correct labelling included in this regulation were developed subsequently in European Commission Regulation (EC) 2065/2001 (EC, 2001). These regulations detail rules about the information required for the consumer on product labels, such as the commercial designation and scientific name of species, the method of production and the area in which it was caught. For cultivated species, the name of the country in which the product will undergo the final developmental stage is required. Compliance with these regulations aids consumer choice and provides full and accurate information about the product.

'Seafood authentication' indicates that the correct labelling of seafood products, ensuring adequate species identification, geographic origin and method of production (wild or cultured), has been checked. The main aspect of seafood authentication is species authentication, which preserves food quality and gives assurance on market transparency, thus avoiding fraud. Species authentication confirms that the commercial and scientific name provided on the label is the one that belongs to the species included in the product.

Seafood Processing: Technology, Quality and Safety, First Edition. Edited by Ioannis S. Boziaris.

Diverse techniques have been developed for species authentication; some of these apply to geographic origin identification from isotopes by nuclear magnetic resonance spectroscopy, from otoliths by isotope ratio mass spectrometry, and by analysis of fatty acid profiles (Aursand *et al.*, 1999; Joensen *et al.*, 2000; Secor *et al.*, 2002). Specialized techniques have been adapted to discriminate between wild and farmed fish, such as analysis of fatty acid profiles and metal or isotope analysis (Villarreal, Rosenblum and Fries, 1994; Aursand, Mabon and Martin, 2000; Foran *et al.*, 2004). Most species authentication techniques have been developed mainly to focus on species identification.

For years, the identification of fish species has been strongly linked to their morphological characteristics, using existing taxonomic keys to recognize and differentiate one species from another. This is the only suitable method for the identification of specimens by fishermen onboard fishing vessels and by retailers ashore in the harbour. This method depends on the skill and qualification of the technician, does not require expensive equipment and does not require alteration of the fish, so it is kept intact and in good condition for sale and consumption. However, during the different stages of processing the external morphology of the fish can completely change, making identification difficult or impossible.

To combat this problem, various molecular techniques based on the intrinsic components of tissues and independent of morphological characters have been developed. Molecular techniques that allow fish authentication can be divided into two main groups, depending on the molecular marker target: protein-based techniques and DNA-based techniques.

16.1.1 Molecular markers

Molecular markers are biomolecules distributed throughout the genome that are inherited, and whose analysis allows the assignment of a given sample to a particular species, population or individual. A monomorphic molecular marker is invariable in all studied organisms, but when there are differences in the molecular weight, enzyme activity, structure or restriction sites, it is said to be polymorphic. Sometimes the degree of variation is such that they are called hypervariables. These molecular markers may be DNA, or proteins with similar attributes, although proteins are not distributed in the genome.

Techniques using protein analysis focus on various physicochemical differences such as size, net charge and amino acid composition. The first protein markers were developed in the late 1970s using the identification of proteins and isozymes by electrophoresis in starch or polyacrylamide gels according to their solubility in water. These markers opened up the knowledge of the structure and genetic heterogeneity among different species, varieties and populations with different geographical origins. However, this technique had a very important limitation; it was not capable of detecting sufficient polymorphism between related varieties or species because proteins are the result of gene expression, which may differ between tissues, stage of development, environment, and season.

In response to these limitations, molecular markers using DNA were developed using the advances made in DNA technology, obtaining stability in species identification. Although DNA analysis is more complex than protein analysis, the great stability of this molecule and the large amount of information that gives, turns it into a powerful molecular marker for the identification of species.

16.1.2 Reference Material (RM) and Tissue Banks (TBs)

For successful development of the methodology it is necessary to collect specimens of all species of interest located throughout the world use as Reference Material (RM), even though this may be tedious and difficult task.

RM means any substance whose properties are sufficiently well established to be used in different measuring functions. RM provides a benchmark for a measurement. It is therefore used in instrument calibration, evaluation methods, quality control of products and assigning values to other materials.

The RM should be homogeneous, so that the values that are determined in a sample of a batch can be applied to any other sample, within the limits of uncertainty indicated. Moreover, it must be highly stable throughout its shelf life so the conditions of storage and use should be well defined.

Certified Reference Material (CRM) is RM that has had its properties certified by a technically valid procedure, and that is accompanied by a traceability certificate or other documentation issued by a certifying body.

In the absence of CRM, and given the scarcity of commercial RM that fits our needs and allows proper quality control procedures for species identification, working RMs need to be prepared, characterized and identified to ensure the reliability and accuracy of the determinations developed. These working RMs are exhaustively characterized to convert them into traceable CRM. Also, they can then be deposited in Tissue Banks (TBs) to store the biological material, so that useful molecules such as DNA, RNA, proteins or enzymes for research and analysis can be extracted by molecular techniques. The TBs are formed from sets of specimens collected for research purposes and which serve as RM for the determination of other specimens. The physical evidence of the organisms is deposited in the TB and scientific data and information is associated with each one. The material can yield different sequences of molecular markers, which are used to create and expand our own databases of sequences and also serve as comparison and reference for taxonomic studies. So, different living organisms can be identified and classified, even new species. Moreover, not only can species that are currently widely marketed be correctly identified, but other species that may be a future resource can also be correctly identified.

TBs are a natural historic archive with many uses where specimens and their associated information are preserved and form the base of population genetics, taxonomy, systematics, and ecology, as well as phylogenetic, biogeographic and conservation studies. All of them are a crucial part of the knowledge of biological diversity and of the advancement of biological sciences.

16.1.3 Databases (DBs)

A database (DB) is a collection of information where records within the same context are stored and organized so that a computer program can quickly select the relevant information whenever it is needed.

The most relevant genetic DBs are of nucleic acids and proteins. These collections contain records of nucleotide sequences, proteins, protein structures, genomes, gene expressions, references, taxonomy, metabolism, transcription factors, etc., and its main role is to provide free access to the records created from publications by the scientific community. DB information includes the name of the author, the source of publication, the organism and other important accompanying characteristics.

Three free public databases of nucleic acids of major importance are located in the United States, Europe and Japan, and are respectively:

- The National Center for Biotechnology Information (NCBI): GenBank (http://www.ncbi.nlm.nih.gov/);
- The European Bioinformatics Institute (EBI): European Molecular Biology Laboratory Bank (EMBL-BANK) (http://www.ebi.ac.uk/);
- The DNA Data Bank of Japan (DDBJ) (http://ddbj.sakura.ne.jp/).

These three DBs are closely coordinated, despite being maintained by different agencies in different countries, so a sequence sent to any of these DBs will be quickly available at the other two, allowing access to almost all the DNA sequence information from any of the three sites.

All have free online access. NCBI has a very useful web page of multiple biological databases (sequences, proteins, genomes, taxonomic, publications, etc.) interconnected so that a single search and retrieval data system called ENTREZ can be used. This search system allows users to interact with these DBs by providing access and allowing retrieval of information from many DBs at once. The EBI uses the Sequence Retrieval System (SRS) as a management package for DBs, but unlike ENTREZ, it is not distributable, so SRS is obtained and installed locally.

Another particularly useful tool offered by the NCBI is the software BLAST (Basical Logical Alignment Search Tool). Among other applications, this software makes it possibile to compare the sequence obtained in the laboratory with the sequences present in the DB automatically by the alignment of algorithms, allowing us to identify unknown species.

BLAST searches similar sequences to a sequence 'query' in sequence DB. For this, it uses different search strategies, but this command requires at least three arguments for carrying out this search: –i a sequence 'query' (i = input); –d a sequence DB (d = database); and –p, which represents the kind of search (p = program: blastp, blastn, blastx, . . .). So, BLAST performs homology tests between at least two sequences: a query sequence or unknown sequence, and sequences belonging to a DB. First, BLAST provides a quantitative measure of the similarity of the query sequence with each one of the DB sequences with which it shows high similarity by the local alignment by pairs. For this, the unknown sequence is entered in the search window of this tool. A list of all stored sequences which resemble the introduced sequence is obtained, ordered from highest to lowest degree of similarity. Although it cannot guarantee that it has found the right solution, it is able to calculate the significance of its results.

Normally, the sequences generated in different studies can be published and loaded in these public DBs. One problem is that sequence data are often loaded without having completed a full or correct study, or even before the publication is accepted. This can lead to erroneous sequences being present in the DB. To avoid these mistakes, databases need to be carefully generated.

Furthermore, although the DNA DBs are constantly growing for certain organisms, data are not available for all sequences of interest. This is another reason to create a wide and complete DB, composed from sequences obtained from all reference or pattern sequences.

16.2 Molecular techniques based on protein analysis

Protein analysis has been used for both research and species identification as well as various other purposes. For the application of protein analysis to species identification a muscular structure, which has variations that allow specific identification, is required. The proteins that shape this muscular tissue are very similar in all animal species and can be divided into myofibril proteins, proteins for connective tissue or from the stroma, and sarcoplasmic proteins (Westermeier, 2001; Roca, 2003).

Myofibril proteins make up between 65 and 75% of muscular proteins, and include myosin, actin, tropomyosin and troponin. They are extracted using medium or high ionic strength saline solutions and because the amino-acid sequences are very well preserved between species, they can be important for species identification when the product has been subjected to heat denaturation (Westermeier, 2001; Roca, 2003).

The proteins from the connective tissue or the stroma represent 3–10% of the protein muscular content; among them are elastin and collagen, the latter being the most abundant. They are insoluble in saline solutions so acidic or basic solutions are required for their dissolution (Westermeier, 2001; Roca, 2003).

The sarcoplasmic proteins make up between 30 and 35% of the muscular proteins, and are easily extracted with water or dilute saline solutions. Sarcoplasmic proteins include the majority of enzymes involved in the intermediary metabolism of the muscular cells, which makes them very important for species identification (Westermeier, 2001; Roca, 2003).

Diverse methodologies can be used for fish species identification by protein analysis: electrophoretic techniques, high-performance liquid chromatography (HPLC) and immunological techniques.

16.2.1 Electrophoretic techniques

Electrophoresis is an analytical technique of separation based on the migration of dissolved macromolecules in a particular environment, through a matrix, when an electrical field is applied between two electrodes. The movement of the molecules is given by their electrophoretic mobility, which depends on their size and net charge in the matrix used for the analysis. The higher the charge/size ratio, the faster an ion migrates through the electric field. Molecules with a higher net charge will tend to move faster than those with a lower net charge. For the same net charge, the molecules with a lower molecular weight (MW) will move faster. There is a linear correlation between the distance travelled by the molecules and the decimal of the logarithm value of the MW, which makes it possible to determine the MW of these molecules (Weber, *et al.*, 1972; Westermeier, 2001; Montowska and Pospiech, 2007).

The species identification is made by comparing the electrophoretic profile obtained from the muscular proteins of the tested samples, with band patterns obtained from reference samples under the same conditions. The comparison can be visual, although this increases the subjectivity of the results, or made using a densitometer or an image analyser, where the density of the optical bands allows band detection by appropriate software. To compare the band patterns it is necessary for reference samples to undergo the same treatments and be analysed under the same conditions and even in the same gel if possible, because slight experimental variations can alter the protein profiles obtained (AOAC, 1990). The reproducibility of the experimental conditions can be affected by multiple factors that are difficult to control, an example is the internal temperature that the electrophoresis system reaches, which despite the use of refrigeration systems can vary from one migration to another (Westermeier, 2001).

To obtain electrophoretic profiles for authentication, the treatment or processing to which the foodstuff is subjected is very crucial.

Heat treatments denature and precipitate the proteins. Species identification by electropherograms of the sarcoplasmic proteins is difficult, limiting its application to fresh, slightly cooked or frozen food analysis. The use of electrophoretic techniques in cooked products is made by extraction of the precipitated proteins with urea, which allows protein extraction in denaturing conditions. However, the temperature and pressure applied to canned products denature the proteins so much that it isn't possible to obtain replicable electrophoretic patterns (Mackie, 1968, 1969; Harano *et al.*, 2008).

Preservation by freezing also affects protein conformation because the water available in frozen products decreases, so the proteins aggregate and precipitate, making protein extraction and reproducibility of the methodology difficult (Matsumoto, 1980; Torrejón, 1996).

The **presence of additives** in the processed product limits the reproducibility of the results. Many of the additives can modify the pH of the product, which subsequently causes the proteins to aggregate. The presence of additives makes the development of the electrophoretic patterns difficult because the control samples have to undergo the same treatment as the tested sample to ensure that the electrophoretic profiles will be comparable. Furthermore, processed foods that are in contact with sauces or spices need additional washing steps to minimize the effect of these substances. All these preparations make the methodology more complicated and more expensive (Matsumoto, 1980; AOAC, 1990).

Various researchers have developed electrophoretic methods to identify fish species in samples having different treatments, without however resolving the problems and limitations of thermally processed products (Etienne *et al.*, 2000; Mackie *et al.*, 2000). There are various electrophoretic techniques available and which one is used and depends on the degree of resolution desired and the type of sample for analysis (e.g. fresh, frozen, heat treated, etc.). The techniques used most often for species identification are (i) isoelectrofocusing (IEF), (ii) urea isoelectrofocusing (urea-IEF), (iii) polyacrylamide gel electrophoresis with SDS (SDS-PAGE), (iv) bidimensional electrophoresis (2DE) and (v) capillary electrophoresis (CE) (Hall, 1997).

Isoelectric Focusing (IEF) Isoelectric focusing is an electrophoretic technique in a pH gradient that allows the separation of components that differ by only 0.001 pH units. Each protein has a pH where the molecule has no net charge because it contains the same number of positive and negative charges – this is the isoelectric point (pI). When an electric field is applied in a pH gradient, the proteins migrate towards the different electrodes according to their electric charge. When the protein reaches its isoelectric point (pI, net charge is zero), its migration stops and it precipitates (Figure 16.1). Therefore, in the isoelectric focusing technique, the separation depends on the isoelectric point of the protein and not on the charge or size (Westermeier, 2001).

The IEF technique has been used to identify several fish species. This methodology uses the sarcoplasmic fraction of the proteins mainly for identifying species in products that have not been heat treated. The sarcoplasmic fraction includes the enzymes that take part in metabolic processes and other proteins such as parvalbumin, myoalbumin or globulin. IEF obtains characteristic band patterns to identify some of the fish species that are most likely to be substituted in markets, for example, blowfish species differentiation made by

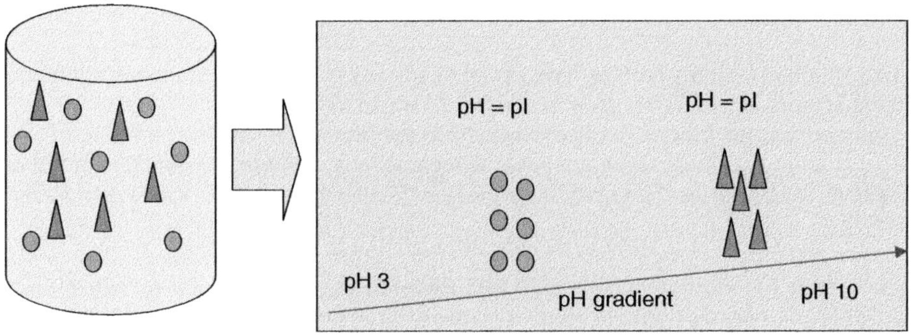

Figure 16.1 Schematic representation of the IEF

Chen *et al.* (2003). IEF has also been used to identify the most common meat species in fresh samples and in mixtures (Slattery and Sinclair, 1983; Skarpeid *et al.*, 1998).

The myofibrillar fraction of the muscle and the heat-resistant sarcoplasmic fraction (parvalbumins) are used in this technique to identify species in processed products. Its effectiveness has been demonstrated by Esteve-Romero *et al.* (1996) where parvalbumins were used to differentiate various salmonid species and flatfishes, and in mixtures of species.

Urea Isoelectric Focusing (Urea-IEF) This technique is a modification of IEF, but with some differences in the sample preparation process. Proteins denaturated by heat can be dissolved in 8 M urea solutions. The urea increases the solubility of the proteins and improves their differentiation. Low molecular weight proteins such as myosins (14–23 kDa), troponins (19–30 kDa) and parvalbumins (12 kDa) are the proteins usually used for distinguishing species using urea-IEF. A comparison of the pI of the parvalbumins in native and denaturalized conditions, in the presence of urea, showed an increase in the pI value. This can be explained by the change of conformation and dissociation of acid and basic groups induced by the denaturation of the protein effected by the presence of urea (Etienne *et al.*, 2000).

Various researchers have used the technique for thermally processed products. Etienne *et al.* (1999) developed a standardized method to identify fresh and heat-processed fish using urea-IEF (Etienne *et al.*, 1999, 2000; Mackie *et al.*, 2000).

In comparison with other electrophoretic techniques, IEF and urea-IEF have many advantages:

- During electrophoresis, the proteins are located in narrow zones, which improves resolution and sensitivity.
- At the end of electrophoresis, the system is balanced and therefore the variation in the experimental parameters has less influence on the protein pattern obtained, so photographs can be used to compare the bands of the sample with the bands of the reference patterns.
- The use of commercially prepared polyacrylamide gels, coupled to semi-automated equipment increases the reproducibility of the results and decreases the time required for analysis.

However, these IEF techniques show important inconveniences:

- The obtained protein profiles have many bands and their interpretation is sometimes complicated.
- This laborious technique requires specialized operators and suitable equipment.
- The sampling process is complex and the technique is expensive, which makes it more difficult to use in food analysis laboratories (Westermeier, 2001; Rodríguez Ramos, 2004).

Polyacrilamide Gel Electrophoresis with SDS (SDS-PAGE) SDS-PAGE is a conventional electrophoresis where sample proteins are dissolved in sodium dodecyl sulphate anionic detergent (SDS), usually at a concentration of 2% (w/v). The proteins lose their individual charges and acquire a negative net charge because a SDS protein–anion complex is formed (Weber and Osborn, 1969). The amount of detergent incorporated per mass unit is the same for all proteins and consequently the mobility into the electrophoretic matrix depends exclusively on the mass. The separation of the proteins, once dissolved, is made in polyacrylamide gels, in which SDS is also added (Figure 16.2).

SDS-PAGE is used to identify species in samples subjected to heat treatment because SDS detergent allows the accurate extraction of the denatured proteins present in these samples. This technique is also simpler than the IEF gels with urea (Scobbie and Mackie, 1988). Etienne *et al.* (2000) applied SDS-PAGE and urea-IEF to identify fish species in cooked products, validating these methodologies in a collaborative study. Complementing previous work, these methodologies have been applied in species identification and also in smoked products (Mackie *et al.*, 2000).

Electrophoresis in SDS-PAGE has seldom been used to identify product species not subjected to thermal treatment. As discussed in the previous section, IEF has been the technique of choice for this type of product because of its higher resolving power.

Amongst the advantages of the SDS-PAGE technique are:

- the possibility of using commercial gels which allow faster analysis times;
- species identification in products subjected to heat treatment.

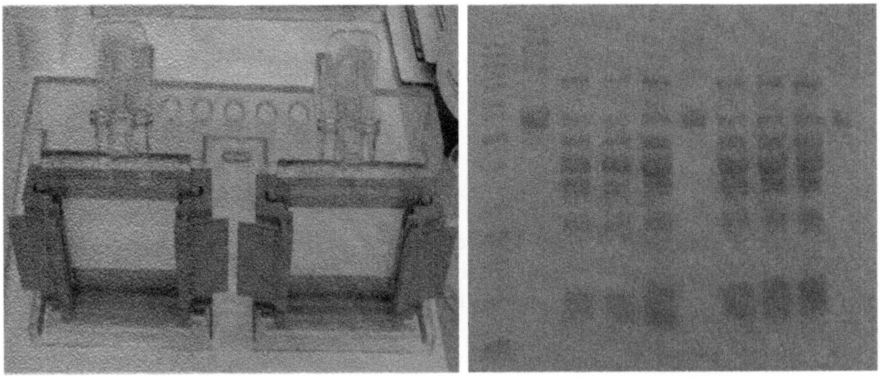

Figure 16.2 Images SDS-PAGE: left, gels; right, electrophoretic profile

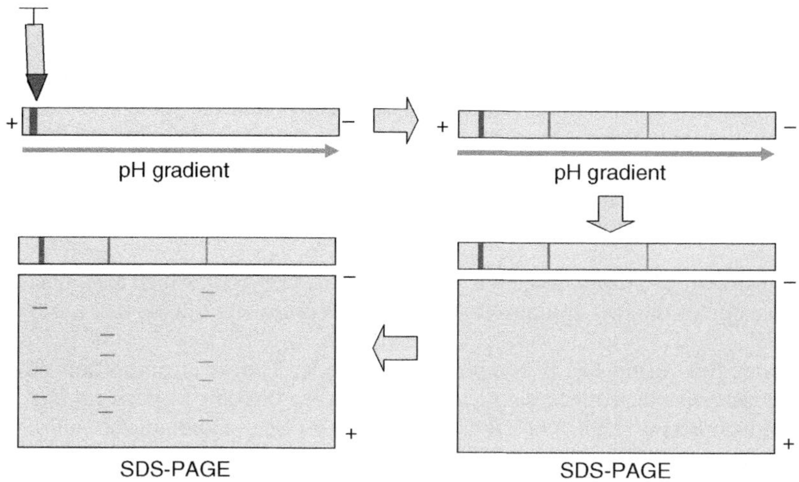

Figure 16.3 Overview of the development of 2D electrophoresis

The drawbacks of the SDS-PAGE technique are the same as those mentioned in the previous section for IEF (Westermeier, 2001):

- the complexity in the protein profiles obtained;
- sample processing is difficult and the methodology cost is high;
- the requirement for specialized workers and instruments.

Bidimensional Electrophoresis (2DE) This technique can be considered as the consecutive union of isoelectric focusing and SDS-PAGE electrophoresis. The proteins of the primary sample are separated by their pI and secondarily by their size. Joining both methodologies gives a better resolution than that obtained using separate techniques (Westermeier, 2001; Rodríguez, 2004) (Figure 16.3).

2DE, besides the complexity of the obtained profiles and the need of specialized staff, is a very labour-intensive and time-consuming technique. Hence, it has not been used for the identification of many species.

2DE, like the techniques that compose it, has been used to identify species in both fresh and heat-treated seafood (Valenzuela *et al.*, 1999; Chen *et al.*, 2004; Berrini *et al.*, 2006).

Capillary Electrophoresis (CE) Capillary electrophoresis has also been used for species differentiation. Although its basis was established in 1937, its use has not been widespread until the recent commercialization of proper instrumental equipment (Cancalon, 1995a, 1995b). This technique separates molecules with the same charge/mass ratio and different masses by subjecting them to an electrical field inside a silica tube of diameter between 50 and 150 μm.

The advantages of this method, in contrast to other analytical techniques are:

- It allows the detection and quantification of different molecules simultaneously because the equipment is fitted with a system that deletes the filling buffer of the column and

replaces it with another automatically, allowing the different components of a sample to be analysed without the need of taking over.
- Analysis can be performed in less than 10 min (only 3 min are required for the cleaning and rebalance of the column between two consecutive analyses).
- Only small sample volumes are required.

Although this technique initially lacked sensitivity, improvements in the detection systems have solved this problem. Its main limitation is, however:

- the adjustment of proper detection systems for each reagent, which also must be very sensitive due to the tiny volumes that are used (Westermeier, 2001; Roca, 2003).

Protein analysis using CE is completely automatic. This is an important advantage over other current electrophoretic methods, which are difficult to perform and require specialized technicians (Hall, 1997; Recio, Ramos and Lopez-Fandino, 2001).

Different types of CE exist, but the most common is capillary zone electrophoresis (ECZ), which uses reagents with wide pH intervals (from acids to bases) for separating the different constituents of a sample. ECZ uses analysis of the sarcoplasmic fraction of the muscle proteins to identify different species in both fresh and frozen products (LeBlanc *et al.*, 1994; Gallardo *et al.*, 1995; Cota-Rivas and Vallejo-Cordoba, 1997; Vallejo-Cordoba and Cota-Rivas, 1998). Gallardo *et al.* (1995) have differentiated fish species, while Cota-Rivas and Vallejo-Córdoba (1997) and Vallejo-Cordoba B. and Cota-Rivas M. (1998) have differentiated meat species.

16.2.2 High-Performance Liquid Chromatography (HPLC)

High-performance liquid chromatography is an analytical method that separates molecules using their polarity. In all chromatographic separation, the sample is moving within a mobile phase that can be gas or liquid. The mobile phase passes through a stationary phase that is immiscible with it and is fixed to a column or a solid surface. The two phases have to be selected so that the constituents of the sample are distributed differently between the mobile phase and the stationary phase. Those reagents that are strongly retained will move through the stationary phase slowly with the flux of the mobile phase; in contrast, constituents that are weakly joined to the stationary phase will move quicker. Because of their different mobilities, the sample constituents are divided into bands or discrete zones that can be qualitatively and quantitatively analysed (Figure 16.4). These results are collected as graphics called chromatograms (Westermeier, 2001; Roca, 2003).

The reverse-phase HPLC technique (RP-HPLC), which is the most used for animal species identification, separates the proteins according to their distribution between a polar mobile phase and an organic phase that is fixed to a matrix (Ashoor *et al.*, 1988). Chromatographic profiles belonging to each species are obtained, which can be compared to reference chromatograms for species identification.

The RP-HPLC shows important advantages over electrophoretic techniques:

- it is fast and simple;
- it has high resolution power;

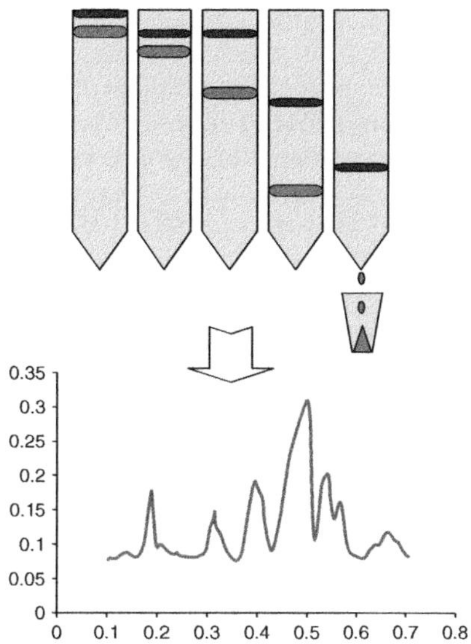

Figure 16.4 Scheme of separation by liquid chromatography

- no toxic reagents are used;
- once the chromatograms are obtained, the combined analysis of the reference samples is not necessary due to the high reproducibility of the results (McCormick, *et al.*, 1988);
- this technique could be used for quantification, since the detection systems could be used to estimate the amount of protein of a species present in a mixture (Toorop *et al.*, 1997).

The biggest disadvantage of this technique is that many samples get damaged during preparation. This is because this process generates degradation or clumping of sarcoplasmic proteins, which cause changes in the resulting chromatograms.

HPLC technique is also useful for the identification of fish species in fresh or frozen products (Osman *et al.*, 1987; Armstrong *et al.*, 1992), for example the identification of gadoids and the detection of significant differences between species of salmonids, cyprinids and perciformes (Knuutinen and Harjula, 1998).

16.2.3 Immunological techniques

Immunological techniques are analytical procedures based on the specific reaction between an antigen (which is the substance to be analysed) and its corresponding antibody.

Using immunological techniques to identify species has important advantages over the use of electrophoretic techniques and HPLC, such as: simplicity, the ability to analyse a high number of samples simultaneously, only small amounts of sample are required, reduced time and cost, the use of less complex material, and the chance of semi-automation and application in field tests, and miniaturized kits. On the negative side, a disadvantage is the possibility of a crossed responsiveness between the proteins of closely related species if antibodies that are not specific for a particular species are used (Wolf *et al.*, 2000). These methods are less used for the evaluation of a huge range of species, because it would be necessary to prepare a high number of specific antibodies (Roca, 2003; Rodríguez, 2004).

Immunological tests used to identify species include various techniques, such as immuno-precipitation, immuno-diffusion, immuno-electrophoresis and various immuno-assays. These techniques have been widely used in the analysis of meat products (Martin *et al.*, 1988a, 1988b, 1988c; Hsieh *et al.*, 1998; Chen and Hsieh, 2002), and have also been used to identify fish species in cooked and dried products, including haddock, hake and eel species (Dominguez *et al.*, 1997; Ochiai and Watabe, 2003). These techniques can also be used to identify species in fresh meat and cooked samples (Martin *et al.*, 1988c; Hsieh *et al.*, 1998).

16.2.4 Limitations of fish species identification techniques based on analysis of proteins

There is a huge variety of techniques for protein analysis that can be used for species identification. Technique selection depends on the available equipment, the degree of resolution required and the type of sample being tested. However, when assessing the methodologies, the disadvantages need to be taken into account. There are major limitations of the methodologies based on protein analysis:

- limited applicability – they cannot be used for analysis of highly processed products because the proteins are denatured during the treatments;
- the reproducibility of the results is impeded by both processing treatment and the incorporation of additives into the product, since they limit the ability to obtain comparable patterns and hamper the protein extraction process;
- complex chromatographic profiles make the results difficult to interpret;
- limited resolution capacity – prevents the distinction between closely related species;
- in general, the techniques are difficult to perform, are time-consuming and require specialized technicians.

16.3 Molecular techniques based on DNA analysis

DNA analysis is constantly being developed and has become one of the most accurate molecular tools. Its use has spread into many diverse areas such as control and law enforcement concerning food and seafood authenticity in general.

DNA has many advantages compared to proteins. Although DNA may be altered using various food processes, it is a far more resistant and heat-stable than a protein, so it is possible to amplify small regions of DNA which are sufficient to enable identification even in DNA fragmentation. Besides, given the degeneration of the genetic code and the presence of non-coding regions, this molecule provides more information than proteins.

There is a huge variety of DNA-based methods available to identify fish species. The main differences between them, besides the development of different methodologies, are their application range, complexity and cost, which are important factors to consider when selecting a concrete methodology.

16.3.1 PCR (Polymerase Chain Reaction)

PCR is an analytical method based on the detection and amplification of specific DNA fragments by means of the primers used. The primers are a pair of small nucleotide sequences which limit the region to be amplified.

The basic reaction of the PCR is a succession of cycles each with a DNA denaturation step to separate the two chains, a primer alignment with template DNA step, and a polymerization step to synthesize a new DNA between the two primers. From there it goes back to the denaturation step to start a new cycle. Specific primers are used to generate fragments that, after being separated and visualized by agarose gel electrophoresis, can be used to identify species. This specificity arises because the primers are constructed from highly conserved regions of DNA that have a very few polymorphisms between many species. Therefore, prior knowledge of the gene to be amplified is required.

PCR technology is being applied in many research fields. Fish species identification by PCR can be performed in two ways:

- By the absence or presence of the fragment to amplify. Thus, if there is specific fragment amplification for a particular species, it is possible to conclude what this species is.
- Depending on the size of the amplified fragment. If amplification is observed, the size of the amplified product needs to be checked to perform the identification (as specificity depends on size).

This methodology is conditional on the size differences of the amplified fragments being detectable by electrophoresis, as some species that are closely related phylogenetically can amplify fragments of similar size and so prevent identification at species level.

For appropriate use of a PCR method, suitable controls need to be included to avoid the possibility of false positives or false negatives. This is because the lack of amplification of fragments may not be due to the absence of DNA, but to technical problems. Table 16.1 gives the main studies on species identification with PCR listed according to different taxonomic groups.

16.3.2 Polymerase Chain Reaction – Restriction Fragment Length Polymorphism (PCR–RFLP)

This technique is used to detect DNA fragments of different molecular weight, caused by fragmentation with restriction enzymes that recognize specific sites of a previously amplified region, and then separating them by electrophoresis. This methodology requires prior knowledge of the fragment to be analysed as a restriction enzyme needs to be selected that allows us to detect polymorphisms. Different size fragments are obtained that cause a band pattern specific to a particular species (see Figure 16.5).

Although it is a simple, robust, easy to perform and less expensive technique than other PCR-based techniques, is also less effective for species identification from the sequence because the patterns obtained are difficult to interpret and less reproducible (Ram *et al.*, 1996; Cespedes *et al.*, 1998). Sometimes there may be an incomplete digestion that could impact on the elimination of intraspecific variation or create additional restriction sites (Lockley and Bardsley, 2000). The main studies on species identification using PCR-RFLP listed according to different taxonomic groups, can be seen in Table 16.2.

Table 16.1 Main published research about species identification by means of PCR classified according to different taxonomic groups

Taxonomic group	Publications
Scombroids	Lin and Hwang, 2008
Dolphinfish	Rocha-Olivares and Pablo Chavez-Gonzalez, 2008
Grouper	Han *et al.*, 2011
Salonids	Wasko and Galetti, 2003
Sardines	Durand *et al.*, 2010
Horse mackerels	Iguchi *et al.*, 2012
Eels	Sezaki *et al.*, 2005
Swordfish	Hsieh *et al.*, 2004
Gadoids	von der Heyden *et al.*, 2007
Mussels	Santaclara *et al.*, 2006
	Daguin *et al.*, 2001
	Inoue *et al.*, 1995
Other bivalves	Fernandez-Tajes *et al.*, 2012
	Freire *et al.*, 2011
	Marshall *et al.*, 2007
	Larsen *et al.*, 2005
Sea cucumbers	Wen *et al.*, 2012
Meats	Rojas *et al.*, 2011
	Nau *et al.*, 2009
	Martin *et al.*, 2007a, 2007b, 2007c
	Chen *et al.*, 2005

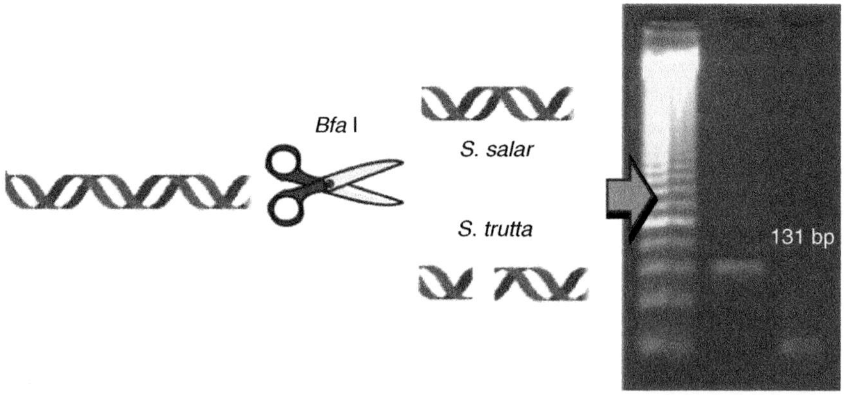

Figure 16.5 Action of B*fa* I restriction enzyme over *Salmo salar* and *Salmo trutta* sequences

16.3.3 Real-Time PCR (RT-PCR)

Among the methods based on DNA analysis for species identification, the use of the RT-PCR has increased considerably. This is because RT-PCR not only allows the identification of species in products where there are more than one species, but also enables their quantification.

Table 16.2 Main published research about species identification by means of PCR-RFLP classified according to different taxonomic groups

Taxonomic group	Publications
Scombroids	Ram *et al.*, 1996
	Lin and Hwang, 2007
Anchovies	Santaclara *et al.*, 2006
	Rea *et al.*, 2009
	Sebastio *et al.*, 2001
Anglerfish	Armani *et al.*, 2012
	Espiñeira *et al.*, 2008
	Sanjuan *et al.*, 2002
Salmonids	Rasmussen *et al.*, 2010
	Espiñeira *et al.*, 2009
	Russell *et al.*, 2000
	Carrera *et al.*, 1999a, 1999b
Horse mackerels	Aranishi *et al.*, 2005
	Turan *et al.*, 2009
	Karaiskou *et al.*, 2007
	Takashima *et al.*, 2006
Flatfish	Céspedes *et al.*, 1998
	Comesaña *et al.*, 2003
	Sanjuan *et al.*, 2002
Mullets	Trape *et al.*, 2009
Swordfish	Chow *et al.*, 1997
	Hsieh *et al.*, 2005; Hsieh *et al.*, 2007
Sharks	Mendonca *et al.*, 2009
Gadoids	Perez *et al.*, 2005; Perez and Presa, 2008
	Di Finizico *et al.*, 2007
	Akasaki *et al.*, 2006
	Aranishi *et al.*, 2005
	Quinteiro *et al.*, 2001
Skates	Alvarado *et al.*, 2005
Gobies	Larmuseau *et al.*, 2008
Barbs	Aksakal and Erdogan, 2007
Snappers	Zhang *et al.*, 2007
	Chow *et al.*, 1993
Scabbardfish	Quinta *et al.*, 2004
Mussels	Santaclara *et al.*, 2006
	Toro, 1998
Other bivalves	Fernández *et al.*, 2000; Fernández *et al.*, 2001; Fernández and Mendez, 2007
	Elliott *et al.*, 2002
Cephalopods	Santaclara *et al.*, 2007
	Colombo *et al.*, 2002
Sea cucumbers	Wen *et al.*, 2010
Various fish species	Nebola *et al.*, 2010
	Hisar *et al.*, 2006
	Klossa-Kilia *et al.*, 2002
	Asensio *et al.*, 2001a
	Wolf *et al.*, 2000
	Cocolin *et al.*, 2000
	Heindel *et al.*, 1998
Meats	Sun and Lin, 2003

RT-PCR is a highly sensitive technique that can amplify and quantify a specific DNA sequence, detecting the PCR product in real time. The amount of DNA, complementary DNA (cDNA) or RNA can be calculated by determining the cycle in which the PCR product is detected.

In this technique the presence of amplification products is transformed into fluorescence, and this fluorescence is detected and monitored while the amplification takes place. The linear relationship between the PCR product and the fluorescence intensity is used to calculate the amount of target DNA present at the beginning of the reaction. More specifically, RT-PCR results are expressed in terms of 'threshold cycle' (Ct), which is the cycle in which the fluorescence becomes detectable by the detector. The Ct is inversely proportional to the initial number of copies of the specific target DNA: the larger the initial number of target DNA copies, the earlier the fluorescence threshold will be reached. The use of fluorescent markers, which are components that emit fluorescence when bound to the target DNA, allows the amplification and detection phases to be combined. However, the chemical markers can distinguish between non-specific and specific detection.

Non-specific detection uses fluorophores known as intercalating agents that bind to double-stranded DNA sequences in the PCR reaction. This union emits a fluorescent signal that is processed in real time. Therefore, an increase in PCR product leads to an increase of the fluorescence detected in each cycle, enabling the quantification (Valasek and Repa, 2005). In this chemistry, the design and adjustment of the assays is direct, and material costs are low. Despite this, they have low specificity because they link indistinctly to both non-specifically generated products and dimers of primer, very common in the PCR reactions. Therefore, a denaturation curve ('melting curve') has to be worked out at the end of the reaction (see Figure 16.6) to find out if the samples with a positive amplification curve correspond to specific or non-specific products.

Specific detection requires the design and synthesis of one or more probes, each one specific for every assay. A probe is defined as a DNA fragment complementary to a DNA intermediate part that we want to amplify. This probe is affixed to a fluorescent molecule called a fluorophore or reporter, and a 'quencher' which is another molecule that inhibits the fluorescence. Only when the probe is displaced from its site by the action of the DNA polymerase, is it released from the action of the 'quencher' and can emit fluorescence. The wavelengths and amount of light absorbed and emitted are characteristic for each fluorophore, and permit their specific detection by combining thermal cycling during the assay.

These detection systems are able to distinguish between the sequence of interest and the dimers of primer or non-specific amplifications because, as opposed to non-specific

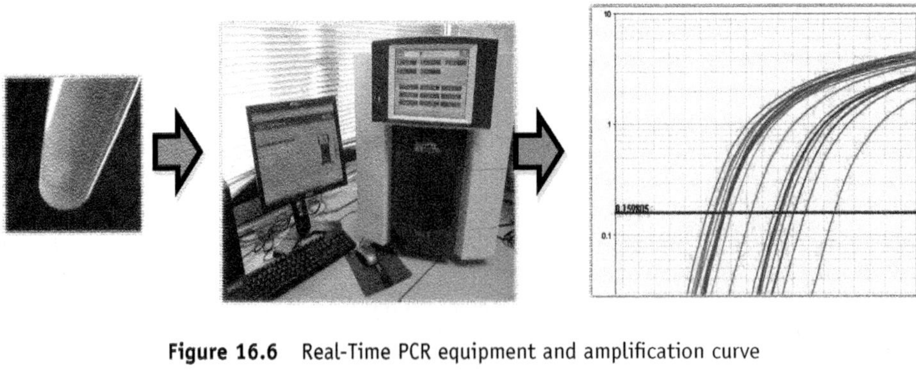

Figure 16.6 Real-Time PCR equipment and amplification curve

chemicals, the specificity resides not only in the primers, but also on the probe, so that the fluorescence signal is only generated if the probe hybridizes specifically with the complementary target sequence. In the absence of target DNA, the fluorescence is inhibited or is very small. Furthermore, as these probes can be labelled with different fluorophores, which differ in their excitation and emission spectra, this chemistry allows detection of amplification products of various different sequences in a single PCR reaction (multiplex PCR).

Despite the speed of this technique, several different aspects need to be considered when making a species assignment for a given sample. First, because there are species which are closely related phylogenetically that have few nucleotide differences in their sequences, as much nucleotide information as possible needs to be obtained to assign a sample to a given species. RT-PCR only evaluates the hybridation positions of the probe and of the primers, but there are techniques such as sequencing, which obtain a more mapping information, thus offering more reliable results.

It is also necessary for the selected molecular marker to have a high intraspecific conservation, otherwise, some haplotype variation for which the primers/probe set is not specific may not be considered. This might produce false negative results if this haplotype is not detected, despite its presence in the sample, or false positive results if a non-target species containing a haplotype variation that permits linking of the set of primers/probe is detected.

Finally, note that the mechanical and chemical thermal processes that a product is subjected to in its elaboration, and the presence of certain sauces, acids, fats and other additives, which in themselves affect the quality and quantity of DNA, may also affect the value of Ct, increasing it or even giving false negatives. Table 16.3 shows the main studies on species identification using RT-PCR listed according to different taxonomic groups.

16.3.4 Forensically Informative Nucleotide Sequencing (FINS)

FINS technique was described by Bartlett and Davidson in 1992, who proposed the genetic identification of species through phylogenetic analysis of DNA sequences (Bartlett and Davidson, 1992).

Bartlett and Davidson were pioneers in the development of this method for the direct sequencing of DNA fragments amplified by PCR. The obtained sequences are aligned and compared with known reference sequences.

In this sense, as is mentioned in Sections 16.1.2 and 16.1.3, the existence of tissue banks from where genetic information from different species of interest can be obtained is very important. From the genetic analysis of these reference patterns, reliable information on DNA sequences could be obtained to enrich our own databases and confirm that the information available in international databases is correct. Thus, assuring the accuracy of these data can guarantee the robustness of the molecular methodologies, as for the FINS technique which compares the sequence of a test sample with pattern sequences.

The basis of this technique is the comparison of sequences of unknown samples against the pattern of species sequences. Therefore, to apply this method, sequence patterns for the studied species need to be obtained. From the set of an unknown species sequences and reference sequences belonging to pattern species, a matrix of distances is generated, and the results of analysis of this matrix are shown in a phylogenetic tree. This tree is built using different biostatistical methods based on distances or on characters, and it groups

Table 16.3 Main published research about species identification by means of Real Time PCR classified according to different taxonomic groups

Taxonomic group	Publications
Scombroids	Lopez and Pardo, 2005
	Dalmasso *et al.*, 2007
Anglerfish	Herrero *et al.*, 2011
Salmonids	Herrero *et al.*, 2011
Sardines	Herrero *et al.*, 2011
	Armani *et al.*, 2012
Eels	Watanabe *et al.*, 2004
Flatfish	Herrero *et al.*, 2012
Swordfish	Herrero *et al.*, 2011
Croaker	Bayha *et al.*, 2008
Grouper	Trotta *et al.*, 2005
	Chen *et al.*, 2012
Gadoids	Hird *et al.*, 2012
	Herrero *et al.*, 2010
	Taylor *et al.*, 2002
	Bertoja *et al.*, 2009
	Sanchez *et al.*, 2009
	Hird *et al.*, 2005
	Fox *et al.*, 2005
Cephalopods	Herrero *et al.*, 2012
	Espiñeira and Vieites, 2012
Meats	Rojas *et al.*, 2011

sequences of the same species into the same node. A visual identification of the unknown species can be made according to the node where is positioned (see Figure 16.7). Emphasis is on the use of the neighbour-joining method (Saitou and Nei, 1987), based on genetic distances, for the construction of phylogenetic trees as it provides statistical support for different groups obtained in the phylogenetic tree by application of the bootstrap test. This test verifies the reliability of the assignment of a species to a particular node, so that bootstrap values greater or equal to 70 usually correspond with a probability greater or equal to 95% that the cluster or node might be real (Hillis and Bull, 1993).

Nowadays, two decades after development of the FINS technique, the great technological breakthroughs and the extension of this technology to many fields of knowledge have led to the reduction in the cost of the equipment and materials used, and the simplification of the tedious work protocols that existed initially. Also, the study of different taxonomic groups, increasingly larger and broader, makes it possible to use this technology in any laboratory, resulting in very accurate identification results. These advances made possible its arrival to quality control laboratories, and thus its application to many fields.

Among the molecular techniques based on DNA work, and even though it is a young methodology, the FINS technique has been one of the most widely used and praised genetic methods for species identification. It is a very robust and powerful technique

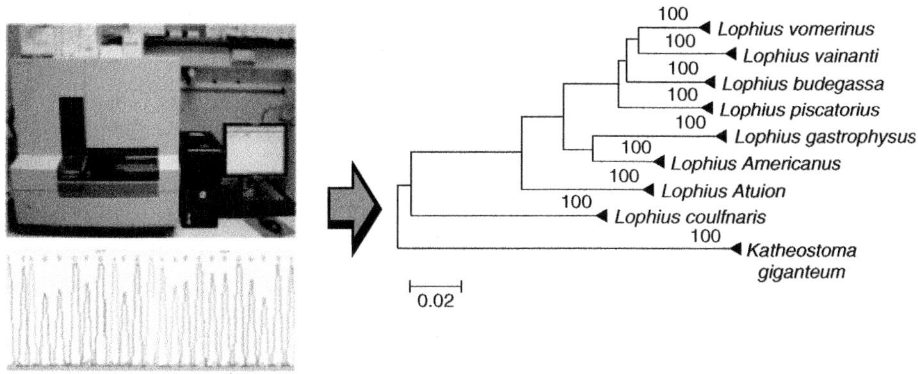

Figure 16.7 Phylogenetic analysis using FINS technique

for determining interspecific and intraspecific variability, suitable for the identification of closely related species, even in highly processed products. FINS is an accurate and reproducible method based on widely studied procedures of proven success, such as PCR and sequencing.

Among its main applications are: authentication of animal species in food with specific genetic material identification of each species, advice on labelling to the processing sector, traceability of food from its origin, conservation of genetic resources and control of protected species.

Compared with other molecular techniques, FINS methodology has many advantages. It is highly specific, sensitive, with high precision and is very reproducible. Also, it provides a wealth of information and is not affected by intraspecific variability, i.e this method overcomes the drawback of intraspecific variability and allows detection of new species that have not previously been studied.

However, the equipment is very expensive, the process is not automated and highly qualified personnel are needed to carry out this very difficult technique. Table 16.4 shows the main studies on species identification by means of FINS listed according to different taxonomic groups.

16.3.5 Other methodologies for fish species identification

There are other genetic techniques, apart from those presented above, which have been used with great success to identify different fish species.

Among these there are microsatellites (Edwards *et al.*, 1998; Barroso *et al.*, 2003; Beacham *et al.*, 2004; Hassanien and Gilbey, 2005; Rezaei *et al.*, 2011). This methodology consists of tandem repeats of two, three or four nucleotides in the nuclear DNA with a large mutation capacity in comparison with other regions of the DNA. Because the microsatellites are highly polymorphic, this technique is widely used in the determination of the individual genetic profile of a single species (Ryan *et al.*, 2005; Charrier *et al.*,

Table 16.4 Main published research about species identification by means of FINS classified according to different taxonomic groups

Taxonomic group	Publications
Scombroids	Espiñeira *et al.*, 2009
	Infante *et al.*, 2007
	Manchado *et al.*, 2004
	Ram *et al.*, 1996
	Barlett and Davidson, 1991
Anchovies	Santaclara *et al.*, 2006
Anglerfish	Espiñeira *et al.*, 2008
	Sanjuan *et al.*, 2002
Salmonids	Espiñeira *et al.*, 2009
Sardines	Lago *et al.*, 2011
	Jerome *et al.*, 2003
Horse mackerels	Lago *et al.*, 2011
	Karaiskou *et al.*, 2003 and 2007
Flatfish	Espiñeira *et al.*, 2008
Swordfish	Herrero *et al.*, 2011
Gadoids	Perez and Presa, 2008
Mussels	Santaclara *et al.*, 2006
Other bivalves	Espiñeira *et al.*, 2009
Cephalopods	Espiñeira *et al.*, 2010
	Santaclara *et al.*, 2007
Skates	Lago *et al.*, 2011
	Spies *et al.*, 2006
Eels	Lago *et al.*, 2012
Meats	Lago *et al.*, 2011

2006; Pampoulie *et al.*, 2006) or in population genetics (D'Amato *et al.*, 1999; Seibert and Ruzzante, 2006). Random Amplification of Polymorphic DNA (RAPD) uses a collection of 10-nucleotides with an arbitrary sequence to amplify, through PCR, specific areas distributed by the genome at random. These fragments can be separated by electrophoresis in agarose gel to give specific banding patterns for each species (Callejas and Ochando, 1998, 2001; Ruzainah *et al.*, 2003; Das *et al.*, 2005; Lakra *et al.*, 2007).

Another method used to identify different fish species is Single-Strand Conformation Polymorphism amplified by PCR (PCR-SSCP). It is based on the mobility of the single strands of DNA, relative to their folded conformations, so two strands of DNA that differ by a single base will migrate differently giving unique band patterns for each species (Rehbein *et al.*, 1997; Asensio *et al.*, 2001b; Weder *et al.*, 2001; Colombo *et al.*, 2005; Rehbein, 2005; Garcia-Vazquez *et al.*, 2006; Sriphairoj *et al.*, 2010; Aranceta-Garza *et al.*, 2011).

Another method that may also be used to identify fish species is Amplified Fragment Length Polymorphism by PCR (PCR-AFLP) (Chong *et al.*, 2000; Han and Ely, 2002; Zhang and Cai, 2006; Hsu *et al.*, 2010; Sriphairoj *et al.*, 2010; Chiu *et al.*, 2012). This technique combines the use of restriction enzymes and oligonucleotides for PCR, to obtain very specific molecular markers without prior sequence knowledge. The DNA is cut with two restriction enzymes, one frequent cutting and the other rare cutting. Terminal oligonucleotides compatible with the enzymes used are linked to the fragments generated, and are amplified by PCR using primers complementary to these regions. The number of amplified bands can decrease or increase depending on whether there is more or less complementarity between the oligonucleotide and the restriction site. Furthermore, this technique has been widely used in population genetics to determine slight differences within the same population (Zhang *et al.*, 2004, 2007; Maldini *et al.*, 2006).

Finally, the Electrophoresis Gels in Denaturing/Thermic Gradient (D/TGGE) is a methodology that separates amplicons of the same size but different sequence by electrophoresis in gradient gel based on differences in their denaturation profile, as this is specific to each sequence, and so to each species. It has been widely used in microbial ecology (Ercolini, 2004); however, it has rarely been used in fish species identification (Comi *et al.*, 2005; Deagle *et al.*, 2005; Dooley *et al.*, 2005a, 2005b; Zhang *et al.*, 2007).

Although all these techniques have been used for fish species identification, probably because they are relatively easy to develop and are economic, all have numerous disadvantages, which make other methodologies, such as FINS, or other sequencing techniques more preferable for this task. RAPD patterns are difficult to replicate, which makes them less suitable for the construction of reference databases. Also, like AFLP, it is not suitable for the analysis of material that may be degraded because it is highly susceptible to slight changes in the quality and quantity of the target DNA (Teletchea, 2009). PCR-SSCP is not very reproducible and intraspecific variation coexists with the possibility of displaying more than two bands that differ in intensity because the single-stranded DNA can have multiple conformational states (depending on the electrophoretic conditions), may lead to misidentifications. Moreover, the main disadvantage of the methodology D/TGGE is low discriminatory power and technical difficulties in maintaining stable experimental conditions.

16.3.6 Accredited assays as a quality seal

Accreditations arise with the aim to show society, authorities, as well as businesses and consumers, that the available products and services comply with certain requirements, usually associated with its quality and safety.

Some of these requirements are established by law and therefore have a regulatory nature, but others may be specified in standards, specifications or other voluntary documents, so these controls can be optional. Although many of them are not mandatory, models have been created and applied by many countries to give additional value to their products and services and to ensure their effectiveness.

According to the European Commission,

> ... accreditation is essential for the proper functioning of a transparent market and oriented to the quality in Europe. It is essential for the industry to be fully competitive. It is fundamental for public authorities, both national and European, to achieve a sufficient level of confidence in the certificates issued anywhere in Europe, and thus facilitate the free movement of products throughout the European Economic Area (EEA). It is essential for the conformity assessment bodies, so that they can independently demonstrate their technical competence and ensure transparent competition and quality oriented between them.
>
> (Royal Decree, 1995).

In Spain, the National Accreditation Entity (ENAC), the organization appointed by the Administration to establish and maintain the system of national accreditation, always follows the policies and recommendations of the European Union. Internationally, their qualifications are recognized in over 60 countries as ENAC is signer of the Mutual Recognition Agreements (MLA) established worldwide. Like ENAC, other countries operate other entities such as A2LA (American Association for Laboratory Accreditation) in America, RvA (Dutch Accreditation Council) in the Netherlands, and UKAS (United Kingdom Accreditation Service) in the United Kingdom.

Within ENAC accreditations, the List of Assays Under Accreditation (LEBA) in each category is particularly important (ENAC, 2004), referring to fish species identification. It is to be noted that the Specialized Molecular Biology Unit of ANFACO-CECOPESCA is the first and only Spanish laboratory with this accreditation under the category of species identification assay by DNA sequencing and phylogenetic analysis (FINS) (ANFACO-CECOPESCA.) (www.anfaco.es). Also, note that LEBA is an open and growing list that in addition to fish species, includes other taxonomic groups such as cephalopods, mussels and meat, and that adds value and international validity to any report issued for identification of these species.

References

Alvarado, J.R., Frisk, M.G., Miller, T.J. *et al.* (2005). Genetic identification of cryptic juveniles of little skate and winter skate. *Journal of Fish Biology*, 66(4), 1177–1182.

Aksakal, E. and Erdogan, O. (2007). Use of PCR-RFLP analysis of mtDNA cytochrome-b gene to determine genetic differences in *Capoeta* spp. *Israeli Journal of Aquaculture-Bamidgeh*, 59(4), 206–211.

Akasaki, T., Yanagimoto, T., Yamakami, K., Tomonaga, H. and Sato, S. (2006). Species identification and PCR-RFLP analysis of cytochrome b gene in cod fish (order Gadiformes) products. *Journal of Food Science*, 71(3), C190–C195.

AOAC (1990). *Official Method of Analysis*, 15th edn (Association of Official Analytical Chemists, Washington, DC, pp. 883–889.

Aranceta-Garza, F., Perez-Enriquez, R. and Cruz, P. (2011). PCR-SSCP method for genetic differentiation of canned abalone and commercial gastropods in the Mexican retail market. *Food Control*, 22(7), 1015–1020.

Aranishi, F. (2005). PCR-RFLP analysis of nuclear nontranscribed spacer for mackerel species identification. *Journal of Agricultural and Food Chemistry*, 53(3), 508–511.

Aranishi, F., Okimoto, T. and Izumi, S. (2005a). Identification of gadoid species (Pisces, Gadidae) by PCR-RFLP analysis. *Journal of applied genetics*, 46(1), 69–73.

Aranishi, F., Okimoto, T., Ohkubo, M. and Izumi, S. (2005b). Molecular identification of commercial spicy pollack roe products by PCR-RFLP analysis. *Journal of Food Science*, 70(4), C235–C238.

Armani, A., Castigliego, L., Tinacci, L. *et al.* (2012). A rapid PCR-RFLP method for the identification of Lophius species. *European Food Research and Technology*, 235(2), 253–263.

Armstrong, S.G., Leach, D.N. and Wyllie, S.G. (1992). The use of HPLC protein profiles in fish species identification. *Food Chemistry*, 44(2), 147–155.

Asensio, L., Gonzalez, I., Fernandez, A. *et al.* (2001a). Genetic differentiation between wreck fish (*Polyprion americanus*) and Nile perch (*Lates niloticus*) by PCR-RFLP analysis of an alpha-actin gene fragment. *Archiv Fur Lebensmittelhygiene*, 52(6), 131–134.

Asensio, L., Gonzalez, I., Fernandez, A. *et al.* (2001b). PCR-SSCP: a simple method for the authentication of grouper (*Epinephelus guaza*), wreck fish (*Polyprion americanus*), and Nile perch (*Lates niloticus*) fillets. *Journal of Agricultural and Food Chemistry*, 49(4), 1720–1723.

Ashoor, S.H., Monte, W.C. and Stiles, P.G. (1988). Liquid-chromatographic identification of meats. *Journal of the Association of Official Analytical Chemists*, 71(2), 397–403.

Aursand, M., Mabon, F., Axelson, D. and Martin, G.J. (1999). Origin testing of fish and fish oil products by using 2H and 13C NMR spectroscopy. 29th WEFTA Conference, Thessaloniki, Greece.

Aursand, M., Mabon, F. and Martin, G.J. (2000). Characterization of farmed and wild salmon (*Salmo salar*) by a combined use of compositional and isotopic analyses. *Journal of the American Oil Chemists Society*, 77(6), 659–666.

Barroso, R.M., Hilsdorf, A.W.S., Moreira, L.M. *et al.* (2003). Identification and characterization of microsatellites loci in *Brycon opalinus* (Cuvier, 1819) (Characiforme, Characidae, Bryconiae). *Molecular Ecology Notes*, 3(2), 297–298.

Bartlett, S.E. and Davidson, W.S. (1991). Identification of *Thunnus* tuna species by the polymerase chain reaction and direct sequence analysis of their mitochondrial cytochrome b genes. *Canadian Journal of Fisheries and Aquatic Sciences*, 48(2), 309–317.

Bartlett, S.E. and Davidson, W.S. (1992). FINS (forensically informative nucleotide sequencing): a procedure for identifying the animal origin of biological specimens. *Biotechniques*, 12(3), 408–411.

Bayha, K.M., Graham, W.M. and Hernandez, F.J. (2008). Multiplex assay to identify eggs of three fish species from the northern Gulf of Mexico, using locked nucleic acid Taqman real-time PCR probes. *Aquatic Biology*, 4(1), 65–73.

Beacham, T.D., Lapointe, M., Candy, J.R. *et al.* (2004). Stock identification of Fraser River sockeye salmon using microsatellites and major histocompatibility complex variation. *Transactions of the American Fisheries Society*, 133(5), 1117–1137.

Berrini, A., Tepedino, V., Borromeo, V. and Secchi, C. (2006). Identification of freshwater fish commercially labeled 'perch' by isoelectric focusing and two-dimensional electrophoresis. *Food Chemistry*, 96(1), 163–168.

Bertoja, G., Giaccone, V., Carraro, L., Mininni, A.N. and Cardazzo, B. (2009). A rapid and high-throughput real-time PCR assay for species identification: application to stockfish sold in Italy. *European Food Research and Technology*, 229(2), 191–195.

Callejas, C. and Ochando, M.D. (1998). Identification of Spanish barbel species using the RAPD technique. *Journal of Fish Biology*, 53(1), 208–215.

Callejas, C. and Ochando, M.D. (2001). Molecular identification (RAPD) of the eight species of the genus *Barbus* (Cyprinidae) in the Iberian Peninsula. *Journal of Fish Biology*, 59(6), 1589–1599.

Cancalon, P.F. (1995a). Capillary electrophoresis – a new tool in food analysis. *Journal of AOAC International*, 78(1), 12–15.

Cancalon, P.F. (1995b). Capillary electrophoresis – a useful technique for food analysis. *Food Technology*, 49(6), 52–58.

Carrera, E., Garcia, T., Cespedes, A. *et al.* (1999a). PCR-RFLP of the mitochondrial cytochrome oxidase gene: a simple method for discrimination between Atlantic salmon (*Salmo salar*) and rainbow trout (*Oncorhynchus mykiss*). *Journal of the Science of Food and Agriculture*, 79(12), 1654–1658.

Carrera, E., Garcia, T., Cespedes, A. *et al.* (1999b). Salmon and trout analysis by PCR-RFLP for identity authentication. *Journal of Food Science*, 64(3), 410–413.

Cespedes, A., Garcia, T., Carrera E. *et al.* (1998). Polymerase chain reaction restriction fragment length polymorphism analysis of a short fragment of the cytochrome b gene for identification of flatfish species. *Journal of Food Protection*, 61(12), 1684–1685.

Charrier, G., Durand, J.D., Quiniou, L. and Laroche, J. (2006). An investigation of the population genetic structure of pollack (*Pollachius pollachius*) based on microsatellite markers. *ICES Journal of Marine Science*, 63(9), 1705–1709.

Chen, F.C. and Hsieh, Y.H.P. (2002). Porcine troponin I: a thermostable species marker protein. *Meat Science*, 61(1), 55–60.

Chen, S., Zhang, J., Chen, W. *et al.* (2012). Quick method for grouper species identification using real-time PCR. *Food Control*, 27(1), 108–112.

Chen, T.Y., Shiau, C.Y., Noguchi, T., Wei, CI. and Hwang, DF. (2003). Identification of puffer fish species by native isoelectric focusing technique. *Food Chemistry*, 83(3), 475–479.

Chen, T.Y., Shiau, C.Y., Wei, C.I. and Hwang, D.F. (2004). Preliminary study on puffer fish proteome – species identification of puffer fish by two-dimensional electrophoresis. *Journal of Agricultural and Food Chemistry*, 52(8), 2236–2241.

Chen, Y., Wu, Y.J., Xu, B.L., Wan, J. and Qian, Z.M. (2005). Species-specific polymerase chain reaction amplification of camel (*Camelus*) DNA extracts. *Journal of AOAC International*, 88(5), 1394–1398.

Chiu, T.H., Su, Y.C., Paia, J.Y. and Changa, H.C. (2012). Molecular markers for detection and diagnosis of the giant grouper (*Epinephelus lanceolatus*). *Food Control*, 24(1–2), 29–37.

Chong, L.K., Tan, S.G., Yusoff, K. and Siraj, S.S. (2000). Identification and characterization of Malaysian river catfish, *Mystus nemurus* (CandV): RAPD and AFLP analysis. *Biochemical Genetics*, 38(3–4), 63–76.

Chow, S., Okamoto, H., Uozumi, Y., Takeuchi, Y. and Takeyama, H. (1997). Genetic stock structure of the swordfish (*Xiphias gladius*) inferred by PCR-RFLP analysis of the mitochondrial DNA control region. *Marine Biology*, 127(3), 359–367.

Chow, S.N., Clarke, M.E. and Walsh, P.J. (1993). PCR-RFLP analysis on 13 Western Atlantic snappers (Subfamily Lutjaninae – a simple method for species and stock identification. *Fishery Bulletin*, 91(4), 619–627.

Cocolin, L., D'Agaro, E., Manzano, M., Lanari, D. and Comi, G. (2000). Rapid PCR-RFLP method for the identification of marine fish fillets (seabass, seabream, umbrine, and dentex). *Journal of Food Science*, 65(8), 1315–1317.

Colombo, F., Cerioli, M., Colombo, M.M. *et al.* (2002). A simple polymerase chain reaction-restriction fragment length polymorphism (PCR-RFLP) method for the differentiation of cephalopod mollusc families Loliginidae from Ommastrephidae, to avoid substitutions in fishery field. *Food Control*, 13(3), 185–190.

Colombo, F., Manglagalli, G. and Renon, P. (2005). Identification of tuna species by computer-assisted and cluster analysis of PCR-SSCP electrophoretic patterns. *Food Control*, 16(1), 51–53.

Comesana, A.S., Abella, P. and Sanjuan, A. (2003). Molecular identification of five commercial flatfish species by PCR-RFLP analysis of a 12S rRNA gene fragment. *Journal of the Science of Food and Agriculture*, 83(8), 752–759.

Comi, G., Iacumin, L., Rantsiou, K., Cantoni, C. and Cocolin, L. (2005). Molecular methods for the differentiation of species used in production of cod-fish can detect commercial frauds. *Food Control*, 16(1), 37–42.

Cota-Rivas, M. and Vallejo-Cordoba, B. (1997). Capillary electrophoresis for meat species differentiation. *Journal of Capillary Electrophoresis*, 4(4), 195–199.

Dalmasso, A., Fontanella, E., Piatti, P. *et al.* (2007). Identification of four tuna species by means of real-time PCR and melting curve analysis. *Veterinary Research Communications*, 31, 355–357.

Daguin, C., Bonhomme, F. and Borsa, P. (2001). The zone of sympatry and hybridization of *Mytilus edulis* and *M-galloprovincialis*, as described by intron length polymorphism at locus *mac*-1. *Heredity*, 86, 342–354.

D'Amato, M.E., Lunt, D.H. and Carvalho, G.R. (1999). Microsatellite markers for the hake *Macruronus magellanicus* amplify other gadoid fish. *Molecular Ecology*, 8(6), 1086–1088.

Das, P., Prasad, H., Meher, P.K., Barat, A. and Jana, R.K. (2005). Evaluation of genetic relationship among six *Labeo* species using randomly amplified polymorphic DNA (RAPD). *Aquaculture Research*, 36(6), 564–569.

Deagle, B.E., Tollit, D.J., Jarman, S.N. *et al.* (2005). Molecular scatology as a tool to study diet: analysis of prey DNA in scats from captive Steller sea lions. *Molecular Ecology*, 14(6), 1831–1842.

Di Finizio, A., Guerriero, G., Russo, G.L. and Ciarcia, G. (2007). Identification of gadoid species (Pisces, Gadidae) by sequencing and PCR-RFLP analysis of mitochondrial 12S and 16S rRNA gene fragments. *European Food Research and Technology*, 225(3–4), 337–344.

Dominguez, E., Perez, M.D., Puyol, P. and Calvo, M. (1997). Use of immunological techniques for detecting species substitution in raw and smoked fish. *Zeitschrift Fur Lebensmittel-Untersuchung Und-Forschung a-Food Research and Technology*, 204(4), 279–281.

Dooley, J.J., Sage, H.D., Brown, H.M. and Garrett, S.D. (2005a). Improved fish species identification by use of lab-on-a-chip technology. *Food Control*, 16(7), 601–607.

Dooley, J.J., Sage, H.D., Clarke, M.A.L., Brown, H.M. and Garrett, S.D. (2005b). Fish species identification using PCR-RFLP analysis and lab-on-a-chip capillary electrophoresis: Application to detect white fish species in food products and an interlaboratory study. *Journal of Agricultural and Food Chemistry*, 53(9), 3348–3357.

Durand, J.D., Diatta, M.A., Diop, K. and Trape, S. (2010). Multiplex 16S rRNA haplotype-specific PCR, a rapid and convenient method for fish species identification: an application to West African clupeiform larvae. *Molecular Ecology Resources*, 10(3), 568–572.

EC (2000). Council Regulation (EC) No. 104/2000, of 17 December 2000 on the common organisation of the markets in fishery and aquaculture products. *Official Journal of the European Communities*, L 17/22. http://eur-lex.europa.eu/LexUriServ/LexUriServ.do?uri=OJ:L:2000:017:0022:0052:EN:PDF [accessed 18 June 2013].

EC (2001). Commission Regulation (EC) No. 2065/2001, of 22 October 2001 laying down detailed rules for the application of the Council Regulation (EC) No. 104/2000 as regards informing consumers about fishery and aquaculture products. *Official Journal of the European Communities*, L 278/6. http://eur-lex.europa.eu/LexUriServ/LexUriServ.do?uri=OJ:L:2001:278:0006:0008:EN:PDF.

Edwards, Y.J.K., Elgar, G., Clark, M.S. and Bishop, M.J. (1998). The identification and characterization of microsatellites in the compact genome of the Japanese pufferfish, *Fugu rubripes*: perspectives in functional and comparative genomic analyses. *Journal of Molecular Biology*, 278(4), 843–854.

Elliott, N.G., Bartlett, J., Evans, B. and Sweijd, N.A. (2002). Identification of southern hemisphere abalone (*Haliotis*) species by PCR-RFLP analysis of mitochondrial DNA. *Journal of Shellfish Research*, 21(1), 219–226.

ENAC (2004). Testing Laboratories: Accreditation for Testing Categories. NT-18 Rev. 1.

Ercolini, D. (2004). PCR-DGGE fingerprinting: novel strategies for detection of microbes in food. *Journal of Microbiological Methods*, 56(3), 297–314.

Espiñeira, M., Gonzalez-Lavin, N., Vieites, J.M. and Santaclara, F.J. (2008). Authentication of anglerfish species (*Lophius* spp) by means of polymerase chain reaction-restriction fragment length polymorphism (PCR-RFLP) and forensically informative nucleotide sequencing (FINS) methodologies. *Journal of Agricultural and Food Chemistry*, 56(22), 10594–10599.

Espiñeira, M. and Vieites, J.M. (2012). Rapid method for controlling the correct labeling of products containing common octopus (*Octopus vulgaris*) and main substitute species (*Eledone cirrhosa* and *Dosidicus gigas*) by fast real-time PCR. *Food Chemistry*, 135(4), 2439–2444.

Espiñeira, M., Vieites, J.M. and Santaclara, F.J. (2009). Development of a genetic method for the identification of salmon, trout, and bream in seafood products by means of PCR-RFLP and FINS methodologies. *European Food Research and Technology*, 229(5), 785–793.

Espiñeira, M., Vieites, J.M. and Santaclara, F.J. (2010). Species authentication of octopus, cuttlefish, bobtail and bottle squids (families Octopodidae, Sepiidae and Sepiolidae) by FINS methodology in seafoods. *Food Chemistry*, 121(2), 527–532.

Esteve-Romero, J.S., Yman, I.M., Bossi, A. and Righetti, P.G. (1996). Fish species identification by isoelectric focusing of parvalbumins in immobilized pH gradients. *Electrophoresis*, 17, 1380–1385.

Etienne, M., Jerome, M., Fleurence, J. *et al.* (1999). A standardized method of identification of raw and heat-processed fish by urea isoelectric focusing: A collaborative study. *Electrophoresis*, 20(10), 1923–1933.

Etienne, M., Jerome, M., Fleurence, J. *et al.* (2000). Identification of fish species after cooking by SDS-PAGE and urea IEF: A collaborative study. *Journal of Agricultural and Food Chemistry*, 48(7), 2653–2658.

Fernandez, A., Garcia, T., Asensio, L. *et al.* (2000). Identification of the clam species *Ruditapes decussatus* (grooved carpet shell), *Venerupis pullastra* (pullet carpet shell), and *Ruditapes philippinarum* (Japanese carpet shell) by PCR-RFLP. *Journal of Agricultural and Food Chemistry*, 48(8), 3336–3341.

Fernandez, A., Garcia, T., Asensio, L. *et al.* (2001). PCR-RFLP analysis of the internal transcribed spacer (ITS) region for identification of 3 clam species. *Journal of Food Science*, 66(5), 657–661.

Fernandez-Tajes, J., Arias-Perez, A., Gaspar, M.B. and Mendez, J. (2012). Identification of *Ensis siliqua* samples and establishment of the catch area using a species-specific microsatellite marker. *Journal of AOAC International*, 95(3), 820–823.

Fernandez-Tajes, J. and Mendez, J. (2007). Identification of the razor clam species *Ensis arcuatus, E-siliqua, E-directus, E-macha*, and *Solen marginatus* using PCR-RFLP analysis of the 5S rDNA region. *Journal of Agricultural and Food Chemistry*, 55(18), 7278–7282.

Foran, J.A., Hites, R.A., Carpenter, D.O. *et al.* (2004). A survey of metals in tissues of farmed Atlantic and wild pacific salmon. *Environmental Toxicology and Chemistry*, 23(9), 2108–2110.

Fox, C.J., Taylor, M.I., Pereyra, R., Villasana, M.I. and Rico, C. (2005). TaqMan DNA technology confirms likely overestimation of cod (*Gadus morhua* L.) egg abundance in the Irish Sea: implications for the assessment of the cod stock and mapping of spawning areas using egg-based methods. *Molecular Ecology*, 14(3), 879–884.

Freire, R., Arias, A., Mendez, J. and Insua, A. (2011). Identification of European commercial cockles (*Cerastoderma edule* and *C. glaucum*) by species-specific PCR amplification of the ribosomal DNA ITS region. *European Food Research and Technology*, 232(1), 83–86.

Gallardo, J.M., Sotelo, C.G., Pineiro, C. and Perez-Martin, R.I. (1995). Use of capillary zone electrophoresis for fish species identification. differentiation of flatfish species. *Journal of Agricultural and Food Chemistry*, 43(5), 1238–1244.

Garcia-Vazquez, E., Alvarez, P., Lopes, P. *et al.* (2006). PCR-SSCP of the 16S rRNA gene, a simple methodology for species identification of fish eggs and larvae. *Scientia Marina*, 70, 13–21.

Hall, G.M. (1997). *Fish Processing Technology*, Springer.

Han, K.P. and Ely, B. (2002). Use of AFLP analyses to assess genetic variation in *Morone* and *Thunnus* species. *Marine Biotechnology*, 4(2), 141–145.

Han J., Lv, F. and Cai H. (2011). Detection of species-specific long VNTRs in mitochondrial control region and their application to identifying sympatric Hong Kong grouper (*Epinephelus akaara*) and yellow grouper (*Epinephelus awoara*). *Molecular Ecology Resources*, 11(1), 215–218.

Harano, Y., Yoshidome, T. and Kinoshita, M. (2008). Molecular mechanism of pressure denaturation of proteins. *The Journal of Chemical Physics*, 129(14), 145103.

Hassanien, H.A. and Gilbey, J. (2005). Genetic diversity and differentiation of Nile tilapia (*Oreochromis niloticus*) revealed by DNA microsatellites. *Aquaculture Research*, 36(14), 1450–1457.

Heindel, U., Scheider, M. and Rubach, K. (1998). Fish species identification by DNA fingerprinting – PCR-RFLP analysis. *Deutsche Lebensmittel-Rundschau*, 94(1), 20–23.

Herrero, B., Lago, F.C., Vieites, J.M. and Espiñeira, M. (2012). Real-time PCR method applied to seafood products for authentication of European sole (*Solea solea*) and differentiation of common substitute species. *Food Additives and Contaminants Part a – Chemistry Analysis Control Exposure and Risk Assessment*, 29(1), 12–18.

Herrero, B., Madriñan, M., Vieites, J.M. and Espiñeira, M. (2010). Authentication of Atlantic cod (*Gadus morhua*) using real time PCR. *Journal of Agricultural and Food Chemistry*, 58(8), 4794–4799.

Hillis, D.M. and Bull, J.J. (1993). An empirical-test of bootstrapping as a method for assessing confidence in phylogenetic analysis. *Systematic Biology*, 42(2), 182–192.

Hird, H.J., Chisholm, J., Kaye, J. *et al.* (2012). Development of real-time PCR assays for the detection of Atlantic cod (*Gadus morhua*), Atlantic salmon (*Salmo salar*) and European plaice (*Pleuronectes platessa*) in complex food samples. *European Food Research and Technology*, 234(1), 127–136.

Hird, H.J., Hold, G.., Chisholm, J. *et al.* (2005). Development of a method for the quantification of haddock (*Melanogrammus aeglefinus*) in commercial products using real-time PCR. *European Food Research and Technology*, 220(5–6), 633–637.

Hisar, O., Erdogan, O., Aksakal, E. and Hisar, SA. (2006). Authentication of fish species using a simple PCR-RFLP method. *Israeli Journal of Aquaculture-Bamidgeh*, 58(1), 62–65.

Hsieh, H.S., Chai, T., Cheng, C.A., Hsieh, Y.W. and Hwang, D.F. (2004). Application of DNA technique for identifying the species of different processed products of swordfish meat. *Journal of Food Science*, 69(1), C1–C6.

Hsieh HS., Chai TJ. and Hwang DF. (2005). Rapid PCR-RFLP method for the identification of 5 billfish species. *Journal of Food Science*, 70(4), C246–C249.

Hsieh, H.S., Chai, T.J. and Hwang, D.F. (2007). Using the PCR-RFLP method to identify the species of different processed products of billfish meats. *Food Control*, 18(4), 369–374.

Hsieh, Y.H.P., Sheu, S.C. and Bridgman, R.C. (1998). Development of a monoclonal antibody specific to cooked mammalian meats. *Journal of Food Protection*, 61(4), 476–481.

Hsu, T.H., Wang, Z.Y., Takata, K. *et al.* (2010). Use of microsatellite DNA and amplified fragment length polymorphism for Cherry salmon (*Oncorhynchus masou*) complex identification. *Aquaculture Research*, 41(9), e316–e325.

Iguchi, J., Takashima, Y., Namikoshi, A. and Yamashita, M. (2012). Species identification method for marine products of *Seriola* and related species. *Fisheries Science*, 78(1), 197–206.

Inoue, K., Waite, J.H., Matsuoka, M., Odo, S. and Harayama, S. (1995). Interspecific variations in adhesive protein sequences of *Mytilus edulis, M-galloprovincialis*, and *M-trossulus*. *Biological Bulletin*, 189(3), 370–375.

Jérôme, M., Lemaire, C., Verrez-Bagnis, V. and Etienne, M. (2003). Direct sequencing method for species identification of canned sardine and sardine-type products. *Journal of Agricultural and Food Chemistry*, 51(25), 7326–7332.

Joensen, H., Steingrund, P., Fjallstein, I. and Grahl-Nielsen, O. (2000). Discrimination between two reared stocks of cod (*Gadus morhua*) from the Faroe Islands by chemometry of the fatty acid composition in the heart tissue. *Marine Biology*, 136(3), 573–580.

Karaiskou, N., Alvarez, P., Lopes, P., Garcia-Vazquez, E. and Triantaphyllidis, C. (2007). Horse mackerel egg identification using DNA methodology. *Marine Ecology – an Evolutionary Perspective*, 28(4), 429–434.

Karaiskou, N., Apostolidis, A.P., Triantafyllidis, A., Kouvatsi, A. and Triantaphyllidis, C. (2003). Genetic identification and phylogeny of three species of the genus *Trachurus* based on mitochondrial DNA analysis. *Marine Biotechnology*, 5(5), 493–504.

Klossa-Kilia, E., Papasotiropoulos, V., Kilias, G. and Alahiotis, S. (2002). Authentication of Messolongi (Greece) fish roe using PCR-RFLP analysis of 16s rRNA mtDNA segment. *Food Control*, 13(3), 169–172.

Knuutinen, J. and Harjula, P. (1998). Identification of fish species by reversed-phase high-performance liquid chromatography with photodiode-array detection. *Journal of Chromatography B*, 705(1), 11–21.

Lago, F.C., Herrero, B., Vieites, J.M. and Espiñeira, M. (2011). FINS methodology to identification of sardines and related species in canned products and detection of mixture by means of SNP analysis systems. *European Food Research and Technology*, 232(6), 1077–1086.

Lago, F.C., Vieites, J.M. and Espiñeira, M. (2012). Authentication of the most important species of freshwater eels by means of FINS. *European Food Research and Technology*, 234(4), 689–694.

Lakra, W.S., Goswami, M., Mohindra, V., Lal, K.K. and Punia, P. (2007). Molecular identification of five Indian sciaenids (Pisces: Perciformes, Sciaenidae) using RAPD markers. *Hydrobiologia*, 583, 359–363.

Larmuseau, M.H.D., Guelinckx, J., Hellemans, B., Van Houdt, J.K.J. and Volckaert, F.A.M. (2008). Fast PCR-RFLP method facilitates identification of *Pomatoschistus* species from the North Atlantic. *Journal of Applied Ichthyology*, 24(3), 342–344.

Larsen, J.B., Frischer, M.E., Rasmussen, L.J. and Hansen, B.W. (2005). Single-step nested multiplex PCR to differentiate between various bivalve larvae. *Marine Biology*, 146(6), 1119–1129.

LeBlanc, E.L., Singh, S. and LeBlanc, R.J. (1994). Capillary zone electrophoresis of fish muscle sarcoplasmic proteins. *Journal of Food Science*, 59(6), 1267–1270.

Lin, W.F. and Hwang, D.F. (2007). Application of PCR-RFLP analysis on species identification of canned tuna. *Food Control*, 18(9), 1050–1057.

Lin, W.F. and Hwang, D.F. (2008). Application of species-specific PCR for the identification of dried bonito product (Katsuobushi). *Food Chemistry*, 106(1), 390–396.

Lockley, A.K. and Bardsley, R.G. (2000). DNA-based methods for food authentication. *Trends in Food Science and Technology*, 11(2), 67–77.

Lopez, I. and Pardo, M.A. (2005). Application of relative quantification TaqMan real-time polymerase chain reaction technology for the identification and quantification of *Thunnus alalunga* and *Thunnus albacares*. *Journal of Agricultural and Food Chemistry*, 53(11), 4554–4560.

Mackie, I., Craig, A., Etienne, M. *et al.* (2000). Species identification of smoked and gravad fish products by sodium dodecylsulphate polyacrylamide gel electrophoresis, urea isoelectric focusing and native isoelectric focusing: a collaborative study. *Food Chemistry*, 71(1), 1–7.

Mackie, I.M. (1968). Species identification of cooked fish by disc electrophoresis. *Analyst*, 93(1108), 458.

Mackie, I.M. (1969). Identification of fish species by a modified polyacrylamide disc electrophoresis technique. *J. Assoc. Public Analysts*, 7, 83–87.

Maldini, M., Marzano, F.N., Fortes, G., Papa, R. and Gandolfi, G. (2006). Fish and seafood traceability based on AFLP markers: Elaboration of a species database. *Aquaculture*, 261(2), 487–494.

Manchado, M., Catanese, G. and Infante, C. (2004). Complete mitochondrial DNA sequence of the Atlantic bluefin tuna *Thunnus thynnus*. *Fisheries Science*, 70(1), 68–73.

Marshall, H.D., Johnstone, K.A. and Carr, S.M. (2007). Species-specific oligonucleotides and multiplex PCR for forensic discrimination of two species of scallops, *Placopecten magellanicus* and *Chlamys islandica*. *Forensic Science International*, 167(1), 1–7.

Martin, I., Garcia, T., Fajardo, V. *et al.* (2007a). Species-specific PCR for the identification of ruminant species in feedstuffs. *Meat Science*, 75(1), 120–127.

Martin, I., Garcia, T., Fajardo, V. *et al.* (2007b). Technical note: detection of cat, dog, and rat or mouse tissues in food and animal feed using species-specific polymerase chain reaction. *Journal of Animal Science*, 85(10), 2734–2739.

Martin, I., Garcia, T., Fajardo, V. *et al.* (2007c). Technical note: detection of chicken, turkey, duck, and goose tissues in feedstuffs using species-specific polymerase chain reaction. *Journal of Animal Science*, 85(2), 452–458.

Martin, R., Azcona, J.I., Casas, C., Hernandez, P.E. and Sanz, B. (1988a). Sandwich ELISA for detection of pig meat in raw beef using antisera to muscle soluble-proteins. *Journal of Food Protection*, 51(10), 790–794.

Martin, R., Azcona, J.I., Garcia, T., Hernandez, P.E. and Sanz, B. (1988b). Sandwich ELISA for detection of horse meat in raw meat mixtures using antisera to muscle soluble-proteins. *Meat Science*, 22(2), 143–153.

Martin, R., Azcona, J.I., Tormo, J., Hernandez, P.E. and Sanz, B. (1988c). Detection of chicken meat in raw meat mixtures by a sandwich enzyme-immunoassay. *International Journal of Food Science and Technology*, 23(3), 303–310.

Matsumoto, J.J. (1980). Chemical deterioration of muscle proteins during frozen storage, in *Chemical Deterioration of Proteins* (eds J.R. Whitaker and M. Fujimaki), American Chemical Society, Washington, DC.

McCormick, R.J., Reeck, G.R. and Kropf, D.H. (1988). Separation and identification of porcine sarcoplasmic proteins by reversed-phase high-performance liquid-chromatography and polyacrylamide-gel electrophoresis. *Journal of Agricultural and Food Chemistry*, 36(6), 1193–1196.

Mendonca, F.F., Hashimoto, D.T., Porto-Foresti, F. *et al.* (2009). Identification of the shark species *Rhizoprionodon lalandii* and *R-porosus* (Elasmobranchii, Carcharhinidae) by multiplex PCR and PCR-RFLP techniques. *Molecular Ecology Resources*, 9(3), 771–773.

Montowska, M. and Pospiech, E. (2007). Species identification of meat by electrophoretic methods. *Acta Scientarum Polonorum, Technologia Alimentaria*, 6(1), 5–16.

Nau, F., Desert, C., Cochet, M.F. *et al.* (2009). Detection of turkey, duck, and guinea fowl egg in hen egg products by species-specific PCR. *Food Analytical Methods*, 2(3), 231–238.

Nebola, M., Borilova, G. and Kasalova, J. (2010). PCR-RFLP analysis of DNA for the differentiation of fish species in seafood samples. *Bulletin of the Veterinary Institute in Pulawy*, 54(1), 49–53.

Ochiai, Y. and Watabe, S. (2003). Identification of fish species in dried fish products by immunostaining using anti-myosin light chain antiserum. *Food Research International*, 36(9–10), 1029–1035.

Osman, M.A., Ashoor, S.H. and Marsh, P.C. (1987). Liquid-chromatographic identification of common fish species. *Journal of the Association of Official Analytical Chemists*, 70(4), 618–625.

Pampoulie, C., Ruzzante, D.E., Chosson, V. *et al.* (2006). The genetic structure of Atlantic cod (*Gadus morhua*) around Iceland: insight from microsatellites, the Pan I locus, and tagging experiments. *Canadian Journal of Fisheries and Aquatic Sciences*, 63(12), 2660–2674.

Perez, M. and Presa, P. (2008). Validation of a tRNA-Glu-cytochrome b key for the molecular identification of 12 hake species (*Merluccius* spp.) and Atlantic cod (*Gadus morhua*) using PCR-RFLPs, FINS, and BLAST. *Journal of Agricultural and Food Chemistry*, 56(22), 10865–10871.

Perez, M., Vieites, J.M. and Presa, P. (2005). ITS1-rDNA-Based methodology to identify worldwide hake species of the genus *Merluccius*. *Journal of Agricultural and Food Chemistry*, 53(13), 5239–5247.

Quinta, R., Gomes, L. and dos Santos, A.T. (2004). A mitochondrial DNA PCR-RFLP marker for population studies of the black scabbardfish (*Aphanopus carbo*). *ICES Journal of Marine Science*, 61(5), 864–867.

Quinteiro, J., Vida, l R., Izquierdo, V. *et al.* (2001). Identification of hake species (*Merluccius* genus) using sequencing and PCR-RFLP analysis of mitochondrial DNA control region sequences. *Journal of Agricultural and Food Chemistry*, 49(11), 5108–5114.

Ram, J.L., Ram, M.L. and Baidoun, F.F. (1996). Authentication of canned tuna and bonito by sequence and restriction site analysis of polymerase chain reaction products of mitochondrial DNA. *Journal of Agricultural and Food Chemistry*, 44(8), 2460–2467.

Rasmussen, R.S., Morrissey, M.T. and Walsh, J. (2010). Application of a PCR-RFLP method to identify salmon species in US commercial products. *Journal of Aquatic Food Product Technology*, 19(1), 3–15.

Rea, S., Storani, G., Mascaro, N., Stocchi, R. and Loschi, A.R. (2009). Species identification in anchovy pastes from the market by PCR-RFLP technique. *Food Control*, 20(5), 515–520.

Recio, I., Ramos, M. and Lopez-Fandino, R. (2001). Capillary electrophoresis for the analysis of food proteins of animal origin. *Electrophoresis*, 22(8), 1489–1502.

Rehbein, H. (2005). Identification of the fish species of raw or cold-smoked salmon and salmon caviar by single-strand conformation polymorphism (SSCP) analysis. *European Food Research and Technology*, 220(5–6), 625–632.

Rehbein, H., Kress, G. and Schmidt, T. (1997). Application of PCR-SSCP to species identification of fishery products. *Journal of the Science of Food and Agriculture*, 74(1), 35–41.

Rezaei, M., Shabani, A., Shabanpour, B. and Kashiri, H. (2011). Microsatellites reveal weak genetic differentiation between *Rutilus frisii kutum* (Kamenskii, 1901) populations south of the Caspian Sea. *Animal Biology*, 61(4), 469–483.

Roca, P. (2003). *Bioqutmica: técnicas y métodos / Pilar Rocá, Jordi Oliver, Ana Mª Rodrtguez*, Hélice, D.L., Madrid.

Rocha-Olivares, A. and Pablo Chavez-Gonzalez, J. (2008). Molecular identification of dolphinfish species (genus Coryphaena) using multiplex haplotype-specific PCR of mitochondrial DNA. *Ichthyological Research*, 55(4), 389–393.

Rodríguez, M.A. (2004) In Facultad de Veterinaria. Departamento de Nutrición, Bromatología y Tecnología de los Alimentos, Vol. Doctor. Umiversidad Complutense de Madrid, Madrid, pp. 319.

Rodríguez Ramos, M.A. (2004) In Departamento de Nutrición, Bromatología y Tecnología de los Alimentos, Universidad Complutense de Madrid, Madrid.

Rojas, M., Gonzalez, I., De la Cruz, S. *et al.* (2011). Application of species-specific polymerase chain reaction assays to verify the labeling of quail (*Coturnix coturnix*), pheasant (*Phasianus colchicus*) and ostrich (*Struthio camelus*) in pet foods. *Animal Feed Science and Technology*, 169(1–2), 128–133.

Royal Decree (1995). Royal Decree 2200/1995 of 28 December 1995 approving the Regulation of Infrastructure for Quality and Industrial Safety, which complements the Royal Decree 2584/1981 of 18 September 1981.

Russell, V.J., Hold, G.L., Pryde, S.E. *et al.* (2000). Use of restriction fragment length polymorphism to distinguish between salmon species. *Journal of Agricultural and Food Chemistry*, 48(6), 2184–2188.

Ruzainah, A., Azizah, M.N.S., Patimah, I. and Amirrudin, A. (2003). RAPID fingerprinting of the eel-loaches *Pangio filinaris* and *Pangio piperata*: preliminary evaluation. *Aquaculture Research*, 34(11), 959–965.

Ryan, A.W., Mattiangeli, V. and Mork, J. (2005). Genetic differentiation of blue whiting (*Micromesistius poutassou* Risso) populations at the extremes of the species range and at the Hebrides-Porcupine Bank spawning grounds. *ICES Journal of Marine Science*, 62(5), 948–955.

Saitou, N. and Nei, M. (1987). The neighbor-joining method – a new method for reconstructing phylogenetic trees. *Molecular Biology and Evolution* 4(4), 406–425.

Sanchez, A., Quinteiro, J., Rey-Mendez, M., Perez-Martin, R.I. and Sotelo, C.G. (2009). Identification of European hake species (*Merluccius merluccius*) using real-time PCR. *Journal of Agricultural and Food Chemistry*, 57(9), 3397–3403.

Sanjuan, A. and Comesana, A.S. (2002). Molecular identification of nine commercial flatfish species by polymerase chain reaction-restriction fragment length polymorphism analysis of a segment of the cytochrome b region. *Journal of Food Protection*, 65(6), 1016–1023.

Sanjuan, A., Raposo-Guillan, J. and Comesana, A.S. (2002). Genetic identification of *Lophius budegassa* and *L-piscatorius* by PCR-RFLP analysis of a mitochondrial tRNA(Glu)/cytochrome b segment. *Journal of Food Science*, 67(7), 2644–2648.

Santaclara, F.J., Cabado, A.G. and Vieites, J.M. (2006). Development of a method for genetic identification of four species of anchovies: *E-encrasicolus, E-anchoita, E-ringens and E-japonicus. European Food Research and Technology*, 223(5), 609–614.

Santaclara, F.J., Espiñeira, M., Cabado, G. *et al.* (2006). Development of a method for the genetic identification of mussel species belonging to *Mytilus, Perna, Aulacomya*, and other genera. *Journal of Agricultural and Food Chemistry*, 54(22), 8461–8470.

Santaclara, F.J., Espiñeira, M. and Vieites, J.M. (2007). Genetic identification of squids (Families Ommastrephidae and Loliginidae) by PCR-RFLP and FINS methodologies. *Journal of Agricultural and Food Chemistry*, 55(24), 9913–9920.

Scobbie, A.E. and Mackie, I.M. (1988). The Use of sodium dodecyl-sulfate polyacrylamide-gel electrophoresis in fish species identification – a procedure suitable for cooked and raw fish. *Journal of the Science of Food and Agriculture*, 44(4), 343–351.

Sebastio, P., Zanelli, P. and Neri, T.M. (2001). Identification of anchovy (*Engraulis encrasicholus* L.) and gilt sardine (*Sardinella aurita*) by polymerase chain reaction, sequence of their mitochondrial cytochrome b gene, and restriction analysis of polymerase chain reaction products in semipreserves. *Journal of Agricultural and Food Chemistry*, 49(3), 1194–1199.

Secor, D.H., Campana, S.E., Zdanowicz, V.S. *et al.* (2002). Inter-laboratory comparison of Atlantic and Mediterranean bluefin tuna otolith microconstituents. *ICES Journal of Marine Science*, 59(6), 1294–1304.

Seibert, J. and Ruzzante, D.E. (2006). Isolation and characterization of eight microsatellite loci for white hake (*Urophycis tenuis*). *Molecular Ecology Notes*, 6(3), 924–926.

Sezaki, K., Itoi, S. and Watabe, S. (2005). A simple method to distinguish two commercially valuable eel species in Japan *Anguilla japonica* and *A-anguilla* using polymerase chain reaction strategy with a species-specific primer. *Fisheries Science*, 71(2), 414–421.

Skarpeid, H.J., Kvaal, K. and Hildrum, K.I. (1998). Identification of animal species in ground meat mixtures by multivariate analysis of isoelectric focusing protein profiles. *Electrophoresis*, 19(18), 3103–3109.

Slattery, W.J. and Sinclair, A.J. (1983). Differentiation of meat according to species by the electrophoretic separation of muscle lactate-dehydrogenase and esterase isoenzymes and isoelectric-focusing of soluble muscle proteins. *Australian Veterinary Journal*, 60(2), 47–51.

Spies, I.B., Gaichas, S., Stevenson, D.E., Orr, J.W. and Canino, M.F. (2006). DNA-based identification of Alaska skates (*Amblyraja, Bathyraja* and *Raja*: Rajidae) using cytochrome c oxidase subunit I (coI) variation. *Journal of Fish Biology*, 69(sb), 283–292.

Sriphairoj, K., Klinbu-nga, S., Kamonrat, W. and Na-Nakorn, U. (2010). Species identification of four economically important Pangasiid catfishes and closely related species using SSCP markers. *Aquaculture*, 308, S47–S50.

Sun, Y.L. and Lin, C.S. (2003). Establishment and application of a fluorescent polymerase chain reaction-restriction fragment length polymorphism (PCR-RFLP) method for identifying porcine, caprine, and bovine meats. *Journal of Agricultural and Food Chemistry*, 51(7), 1771–1776.

Takashima, Y., Morita, T. and Yamashita, M. (2006). Complete mitochondrial DNA sequence of Atlantic horse mackerel *Trachurus trachurus* and molecular identification of two commercially important species *T-trachurus* and *T-japonicus* using PCR-RFLP. *Fisheries Science*, 72(5), 1054–1065.

Taylor, M.I., Fox, C., Rico. I. and Rico, C. (2002). Species-specific TaqMan probes for simultaneous identification of (*Gadus morhua* L.), haddock (*Melanogrammus aeglefinus* L.) and whiting (*Merlangius merlangus* L.). *Molecular Ecology Notes*, 2(4), 599–601.

Teletchea, F. (2009). Molecular identification methods of fish species: reassessment and possible applications. *Reviews in Fish Biology and Fisheries*, 19(3), 265–293.

Toorop, R.M., Murch, S.J. and Ball, R.O. (1997). Development of a rapid and accurate method for separation and quantification of myofibrillar proteins in meat. *Food Research International*, 30(8), 619–627.

Toro, JE. (1998). Molecular identification of four species of mussels from southern Chile by PCR-based nuclear markers: The potential use in studies involving planktonic surveys. *Journal of Shellfish Research*, 17(4), 1203–1205.

Torrejón, R. (1996) In Facultad de Farmacia, Vol. Doctor en farmacia Universidad complutense de Madrid, Madrid, pp. 295.

Trape, S., Blel, H., Panfili, J. and Durand, J.D. (2009). Identification of tropical Eastern Atlantic Mugilidae species by PCR-RFLP analysis of mitochondrial 16S rRNA gene fragments. *Biochemical Systematics and Ecology*, 37(4), 512–518.

Trotta, M., Schonhuth, S., Pepe, T. *et al.* (2005). Multiplex PCR method for use in real-time PCR for identification of fish fillets from grouper (*Epinephelus* and *Mycteroperca* species) and common substitute species. *Journal of Agricultural and Food Chemistry*, 53(6), 2039–2045.

Turan, C., Ozturk, B., Gurlek, M. and Yaglioglu, D. (2009). Genetic differentiation of Mediterranean horse mackerel (*Trachurus mediterraneus*) populations as revealed by mtDNAPCR-RFLP analysis. *Journal of Applied Ichthyology*, 25, 142–147.

Valasek, M.A. and Repa, J.J. (2005). The power of real-time PCR. *Advances in Physiology Education*, 29(3), 151–159.

Valenzuela, M.A., Gamarra, N., Gómez, L. *et al.* (1999). A comparative study of fish species identification by gel isoelectrofocusing two-dimensional gel electrophoresis, and capillary zone electrophoresis. *Journal of Capillary Electrophoresis Microchip Technology*, 6(3–4), 85–91.

Vallejo-Cordoba, B. and Cota-Rivas, M. (1998). Meat species identification by linear discriminant analysis of capillary electrophoresis protein profiles. *Journal of Capillary Electrophoresis*, 5(5–6), 171–175.

Villarreal, B.W., Rosenblum, P.M. and Fries L.T. (1994). Fatty-acid profiles in red drum muscle – comparison between wild and cultured fish. *Transactions of the American Fisheries Society*, 123(2), 194–203.

von der Heyden, S., Lipinski, M.R. and Matthee, C.A. (2007). Species-specific genetic markers for identification of early life-history stages of Cape hakes, *Merluccius capensis* and *Merluccius paradoxus* in the southern Benguela Current. *Journal of Fish Biology*, 70, 262–268.

Wasko, A.P. and Galetti, P.M. (2003). PCR primed with minisatellite core sequences yields species-specific patterns and assessment of population variability in fishes of the genus *Brycon*. *Journal of Applied Ichthyology*, 19(2), 109–113.

Watanabe, S., Minegishi, Y., Yoshinaga, T., Aoyama, J. and Tsukamoto, K. (2004). A quick method for species identification of Japanese eel (*Anguilla japonica*) using real-time PCR: An onboard application for use during sampling surveys. *Marine Biotechnology*, 6(6), 566–574.

Weber, K. and Osborn, M. (1969). Reliability of molecular weight determinations by dodecyl sulfate-polyacrylamide gel electrophoresis. *Journal of Biological Chemistry*, 244(16), 4406–4412.

Weber, K., Pringle, J. and Osborn, M. (1972). Measurement of molecular weights by electrophoresis on SDS-acrylamide gel. *Methods in Enzymology*, 26, 3–27.

Weder, J.K.P., Rehbein, H. and Kaiser, K.P. (2001). On the specificity of tuna-directed primers in PCR-SSCP analysis of fish and meat. *European Food Research and Technology*, 213(2), 139–144.

Wen, J., Hu, Ch., Zhang, L. and Fan, S. (2012). Identification of 11 sea cucumber species by species-specific PCR method. *Food Control*, 27(2), 380–384.

Wen, J., Hu, Ch., Zhang, L. *et al.* (2010). The application of PCR-RFLP and FINS for species identification used in sea cucumbers (Aspidochirotida: Stichopodidae) products from the market. *Food Control*, 21(4), 403–407.

Westermeier, R. (2001). *Electrophoresis in Practice: a Guide to Methods and Applications of DNA and Protein Separations*, John Wiley & Sons, Ltd, Chichester, UK.

Wolf, C., Burgener, M., Hubner, P. and Luthy, J. (2000). PCR-RFLP analysis of mitochondrial DNA: differentiation of fish species. *Lebensmittel-Wissenschaft Und-Technologie-Food Science and Technology*, 33(2), 144–150.

Zhang, J.B. and Cai, Z.P. (2006). Differentiation of the rainbow trout (*Oncorhynchus mykiss*) from Atlantic salmon (*Salmon salar*) by the AFLP-derived SCAR. *European Food Research and Technology*, 223(3), 413–417.

Zhang, J., Huang, H., Cai, Z. and Huang, L. (2007). Species identification in salted products of red snappers by semi-nested PCR-RFLP based on the mitochondria 12S rRNA gene sequence. *Food Control*, 18(11), 1331–1336.

Zhang, J.B., Huang, L.M. and Huo, H.Q. (2004). Larval identification of *Lutjanus Bloch* in Nansha coral reefs by AFLP molecular method. *Journal of Experimental Marine Biology and Ecology*, 298(1), 3–20.

Zhang, J., Wang, H. and Cai, Z. (2007). The application of DGGE and AFLP-derived SCAR for discrimination between Atlantic salmon (*Salmo salar*) and rainbow trout (*Oncorhynchus mykiss*). *Food Control*, 18(6), 672–676.

17

Assuring Safety of Seafood – Risk Assessment

John Sumner,[1] Catherine McLeod[2] and Tom Ross[1]

[1] *University of Tasmania, Tasmanian Institute of Agriculture – School of Agricultural Science, Hobart, Australia*

[2] *South Australian Research and Development Institute, Adelaide, Australia and Seafood Safety Assessment, Tournissan, France*

17.1 Introduction

It is part of food safety history that, in 1959, Howard Bauman of the Pillsbury Company received a call from the Quartermaster Food and Container Institute of the US Armed Forces asking: 'Can Pillsbury produce foods for consumption in the zero gravity of space capsules?' Equally often-told is that the need for the Hazard Analysis Critical Control Point (HACCP) concept was borne out of the cost of testing space foods 'In fact, a large part of the production ... had to be utilized for testing, leaving only a small portion, hence the enormous costs of production ...' (Bauman, 1994).

Though HACCP was in full swing at Pillsbury by 1971, its adoption globally was patchy, often only as a consequence of significant food poisoning incidents, for example the Jack-in-the-Box hamburger outbreak in United Sates caused by *Escherichia coli* O157:H7 (Bell *et al.*, 1994) and an *E. coli* O111 outbreak in Australia following consumption of fermented sausage (Cameron *et al.*, 1995), incidents which prompted mandatory implementation of HACCP in those countries. HACCP became the food safety management system of choice. However, in its earliest versions some of HACCP's seven fundamental principles proved difficult to fulfil, for example Principle 1 requires a hazard analysis, involving an assessment of the severity of the hazard and the likelihood it will occur, in other words a 'risk assessment'. For some HACCP practitioners, assessing risk received only perfunctory attention which, compounded by inconsistent definitions of a Critical Control Point (CCP) and difficulty in setting Critical Limits, resulted in 'paper HACCP plans'. Typically written by a consultant, these were quickly consigned to the QA Manager's bookshelf, to be taken down only on rare occasions, such as a visit from the local regulatory authority.

Seafood Processing: Technology, Quality and Safety, First Edition. Edited by Ioannis S. Boziaris.

17.2 Differentiating risk from hazard

The coupling of HACCP with risk assessment began in 1994 when Notermans and colleagues in Europe published 'The HACCP concept: specification of criteria using quantitative risk assessment' (Notermans *et al.*, 1995). In the United States, Buchanan (1995) similarly linked risk assessment and HACCP and the Council for Agricultural Science (CAST) published an assessment of risks associated with foodborne pathogens (Foegeding and Roberts, 1996). Canadian workers published papers on risk analysis (Todd and Harwig, 1996), and on quantitative risk assessment (QRA) as a tool for emerging pathogens (Lammerding and Paoli, 1997), followed by the first QRA on risks associated with consumption of hamburgers contaminated with *E. coli* O157:H7 (Cassin *et al.*, 1998).

Working parties of the National Advisory Committee on Microbiological Criteria for Foods (NACMCF, 1998) and of the Codex Alimentarius Commission ('Codex', an international body of member nations established within the United Nations for establishment of internationally agreed standards and regulations for foods), prepared documents on principles and guidelines for microbiological risk assessment (CAC, 1999). Meanwhile the International Commission on Microbiological Specifications for Foods (ICMSF) published material on the potential for risk assessment to be applied to microbiological issues related to international trade in foods (ICMSF, 1998). Importantly, the World Trade Organization (WTO), founded by the United Nations to foster economic development of nations through international trade, finalized two fundamentally important agreements in 1995 setting out agreed rules for world trade, namely the Sanitary-PhytoSanitary (SPS) Agreement and Technical Barriers to Trade (TBT) Agreement (WTO, 2007a, 2007b). These mandated that the only basis for restriction of international trade in foods was demonstration of unacceptable risk to human, plant or animal health in an importing nation due to imports from other nations. Science-based risk assessment was nominated by WTO as the tool to evaluate risk. Thus, by the end of its first decade, microbiological risk assessment was a firmly established principle required to underpin international trade in foods and was supported by high-level guidance provided by NACMCF, Codex and ICMSF.

In 1999, the United Nations Food and Agriculture Organization (FAO) undertook the first risk assessment on seafood, assembling an expert consultation on the trade impact of *Listeria* in fish products. The overall findings were published (FAO, 1999) including individual papers in a special edition of the International Journal of Food Microbiology that brought together disparate elements such as predictive microbiology (Ross *et al.*, 2000), exposure assessment (Karunasagar and Karunasagar, 2000) and control options (Huss *et al.*, 2000) to inform a risk assessment (Notermans and Hoornstra, 2000).

The FAO expert consultancy set the scene for a series of risk assessments commissioned by FAO and WHO, known as the Joint FAO/WHO Expert Meetings on Microbiological Risk Assessment (JEMRA), intended to provide advice to Codex. In 2001, JEMRA established a team to work on the risks to human health from vibrios in seafoods which, over the next decade, published assessments on *Vibrio vulnificus* in raw oysters (FAO/WHO, 2005a), *V. cholerae* in warm-water shrimp in international trade (FAO/WHO, 2005b) and *V. parahaemolytius* in seafood (FAO/WHO, 2011).

The first seafood QRA was performed by Lindqvist and Westöö (2000) on *L. monocytogenes* in smoked or gravid salmon and rainbow trout in Sweden, and introduced readers to (for many) the arcane world of dose–response, Monte Carlo simulation, risk estimate and uncertainty.

By contrast, a simple, semi-quantitative tool, Risk Ranger (Ross and Sumner, 2002) was used to profile the Australian seafood industry and to allocate 10 hazard : product pairings to broad risk bands – low, medium and high (Sumner and Ross, 2002).

Following two large food poisoning outbreaks in the United States caused by *V. parahaemolyticus* in oysters in 1997 and 1998, the Food and Drug Administration's Center for Food Safety and Applied Nutrition (CFSAN) undertook a quantitative assessment of the risk to public health from that product : pathogen combination. A protracted period of public/industry ('stakeholder') consultation delayed both publication (FDA, 2005) and implementation of risk-based controls and brings into focus that risk assessment is but one element of risk analysis, along with risk management and risk communication. Concurrently, CFSAN (in cooperation with the US Food and Drug Administration and the Food Safety Inspection Service) undertook a second major risk assessment on *L. monocytogenes* in a range of ready-to-eat (RTE) foods, prompted by a series of listeriosis outbreaks in the United States from RTE meats. In that assessment, smoked seafoods and cooked crustaceans were assessed as high risk on a per serving basis, though they were a moderate risk overall due to relatively low levels of consumption (FDA, 2003).

To elaborate on the theme mentioned here, risk assessment is one aspect of an overall approach to risk management which provides evidence of the magnitude of the risk associated with the hazard and how the risk is influenced by various factors in the harvest-to-consumption pathway. The aim is to support the risk manager to make decisions about the urgency and magnitude of the risk, and potential ways in which it might be reduced to tolerable levels. However, the optimal risk management decision is not always based on the most technically effective solution because risk management can also involve the psychology of those affected by the risk (the stakeholders), for example how tolerant they are of the risk, the extent of 'dread' associated with the hazard, their ability to control the risk themselves, the benefit they receive (if any) by allowing some level of the hazard, etc. Thus, understanding the interests and perceptions of stakeholders is an important dimension to making risk management decisions. This latter aspect involves consulting with stakeholders and is part of 'risk communication', which also includes telling the stakeholders how the risk will be managed and explaining the basis of those decisions.

As JEMRA and CFSAN assessments were being undertaken, a number of problems emerged: in particular, that they were costly and could take several years to complete because of the need to gather relevant data, and the need for interaction with risk managers, risk communicators and peer reviewers. Buchanan *et al.* (2004) recounted what CFSAN had learned from their early risk assessments. While microbial food safety risk assessment (MFSRA) proved challenging for countries such as the United States, for developing countries difficulties were compounded by a lack of human resources with the necessary range of skills. Cahill and Jouve (2004) set out those challenges, for example, the need to comply with an international food trade requirement, and FAO published a text/CD ROM specifically aimed at assisting smaller nations (Sumner *et al.*, 2004). The ability to simplify risk assessment, or at least to use an initial screen to evaluate whether an assessment was even required and to clearly articulate the question, or decision (i.e. the 'food safety problem'), about which the risk manager needed more information, led to the inception of the risk profile, defined by the Codex Committee on Food Hygiene, as 'A description of a food safety problem and its context developed for the purpose of identifying those elements of a hazard or risk which are relevant to risk management decisions'(CAC, 1999). The effectiveness of risk profiling may be seen at the New Zealand Ministry of Primary Industries (www.foodsafety.govt.nz/science-risk/risk-assessment/risk-profiles) where a range of hazard : product profiles are assessed in the context of the New Zealand consumer; where necessary, based on the risk profile, some are progressed into a full QRA.

Now entering its second decade, MFSRA occupies an important position in food safety management: due diligence, import testing and entry to the WTO are all reasons for undertaking risk assessment. Risk assessments are also used in WTO proceedings to challenge import restrictions and to resolve disputes. But how widely are risk estimates used by regulators at local, national and international level? Do regulators all too often equate hazard with risk? Are zero tolerance and precautionary principle merely code for regulators to proclaim zero risk and to set performance standards which accommodate such thinking?

In this chapter we explore the four elements of risk assessment: hazard identification, exposure assessment, hazard characterization and risk characterization and we do so from a viewpoint which questions whether risk managers are enabled to accept, and make decisions based on risk estimates that have 'bad news' conclusions, for example 'Even after risk reduction strategies are implemented, it is likely that consumption of Product A will result in 3 deaths and 25 serious illnesses per annum'.

17.3 Hazards, risks and food safety risk assessment

Risk assessment has been applied in many fields of human activity. It can be done in many different ways from simple, qualitative judgements based on experience, to data-intensive and complex mathematical modelling approaches requiring powerful computers to solve them.

While several nations and organizations had been exploring the application of risk assessment approaches for food safety management (Hathaway and McKenzie, 1991; Buchanan, 1995; Notermans, and Teunis, 1996; Todd and Harwig, 1996; Hathaway and Cook, 1997; NACMCF, 1998; ICMSF, 1998; McNab, 1998), there was no consensus on methods. The decision by WTO to mandate a science-based risk assessment philosophy to harmonize regulations governing international trade in foods became a strong impetus for the development of those method(s), particularly as agreed methods did not exist at that time.

The determination of standards for food additives and chemical contaminants had long been undertaken by international panels (e.g. JECFA) convened by FAO/WHO, using established and agreed methods that were mostly concerned with defining the dose–response relationship and expected levels in typical diets so as to develop tolerable daily intakes, i.e. those levels for which no immediate or long-term harm to the consumer would be expected. From those daily intake levels, maximum tolerable levels in foods could be established. These estimates of tolerable levels are typically of the order of 100- to 1000-fold lower than the concentration of the hazard expected to lead to ill-health in consumers: this safety margin is intended to account for the variability in human susceptibility, the difference in responses in animal models and humans, etc.

While JECFA had agreed methods for setting standards for food-borne chemical hazard's it was considered by some that MFSRA was not feasible due to the dynamic nature of microbial populations in foods, i.e. because microbial hazard levels can change over time and are affected by food composition, storage and processing conditions and the random and heterogeneous distribution of microbial contaminants in, and on, foods. The introduction of predictive microbiology and consequent development of large databases on pathogen responses in food (e.g. ComBase, SymPrevius, Pathogen Modelling Program, Seafood Spoilage and Safety Predictor), together with vastly improved access to computing power, offered a means to overcome at least some of these perceived limitations. Additionally, some argued that the JECFA approach was more hazard-based, than risk-based, and that the ready availability of new computer software for stochastic simulation modelling (which allows variability to be included in calculations) meant that MFSRA could be based on more holistic methods utilizing data in a fully objective manner than the current JECFA approach. The task of establishing those methods was given to Codex.

As noted earlier, Codex convened a series of meetings of international experts to consider methods for MFSRA as part of an overall risk analysis framework. One of the first outputs was the guideline document for the conduct of MFSRA (CAC, 1999). That document did not dictate methods, but rather established principles for risk assessment as a science-based aid to management of microbial food safety. Its application was to include the setting of standards and demonstration of equivalence of food safety management approaches in different nations.

Twelve principles were enunciated, including that:

- Microbiological Risk Assessment should be soundly based upon science.
- Microbiological Risk Assessment should be conducted according to a structured approach that includes:
    - Hazard Identification
    - Hazard Characterization
    - Exposure Assessment
    - Risk Characterization
- The conduct of a Microbiological Risk Assessment should be transparent.
- Any constraints that impact on the Risk Assessment such as cost, resources or time, should be identified and their possible consequences described.
- The risk estimate should contain a description of uncertainty and where the uncertainty arose during the Risk Assessment process.
- A Microbiological Risk Assessment should explicitly consider the dynamics of microbiological growth, survival, and death in foods and the complexity of the interaction (including sequelae) between human and the pathogen or toxin following consumption, as well as the potential for further spread.

Before a risk assessment can begin, two important decisions need to be taken by those proposing the assessment. Firstly, and with great clarity, the purpose of the risk assessment must be articulated. The methods used for risk assessment can vary according to the question requiring an answer, or the decisions for which additional insights are sought. When a risk assessment was proposed after four outbreaks of illness in the United States following consumption of oysters containing *V. parahaemolyticus*, it had two main objectives: (i) to determine the factors that contribute to the risk of becoming ill from the consumption of pathogenic *V. parahaemolyticus* in raw oysters and (ii) to evaluate the likely public health impact of different control measures, including the effectiveness of current and alternative microbiological standards (FDA, 2005).

The second decision involves the composition of the team which will conduct the assessment. There will inevitably be a wide range of skills and disciplines needed. For example, 'the *Vibrio* spp. in seafood' team gathered by FAO/WHO comprised four vibriologists, one predictive microbiologist, one processing expert, one public health expert and one statistician/modeller.

In this chapter we explain and comment on the principles and practice of MFSRA. More comprehensive texts on seafood-borne hazards (Ababouch and Karunasagar, in press), exposure assessment (FAO/WHO, 2008), hazard characterization (FAO/WHO, 2003), risk characterization (FAO/WHO, 2009) and seafood risk assessment modelling (Fazil, 2005) are highly recommended.

17.4 Hazard Identification/Risk Profile

A Hazard Identification is the first stage of a risk assessment in which the evidence that the hazard presents a risk to consumers for the specific situation (country, specific population, product) is documented. As noted earlier, the idea of a Risk Profile was introduced subsequently, and intended as a first screen to evaluate the evidence that the risk is great enough to warrant a full risk assessment, and/or whether there is sufficient information for the risk assessment to be able to provide useful advice for risk managers. The Hazard Identification in earlier risk assessments may have fulfilled the role of a Risk Profile but assessment of risk may now stop at the Risk Profile stage. The Hazard Identification is now regarded as documentation of the evidence that the hazard presents a risk to the population from the food(s) nominated, etc.

One approach to identifying hazards is to search epidemiological data to identify agents causing illness under the storage, marketing and consumption patterns of the country. A typical inter-country comparison is illustrated in Table 17.1 where rates of seafood-borne illness varied greatly between the United States (Smith de Waal *et al.*, 2000) and Australia (Sumner, 2001), most strikingly in the rates of viral illness from bivalve molluscs, ciguatera and shellfish biotoxins. More contemporary Australian public health statistics for the period 2001–2010 (OzFoodNet, 2012) indicate that ciguatera remains the most common cause of outbreaks and, despite the implementation of shellfish quality assurance programmes, there were outbreaks of viral illnesses from locally harvested shellfish; no comparable data are available for the United States for the period 2001–2010. Within the broad hazard categories summarized in Table 17.1 are many individual pathogenic micro-organisms and toxins, but missing from the United States and Australia epidemiological data are two frequent causes of seafood illness. Firstly, parasites are a major cause of seafood-borne illness in many tropical countries. Secondly, seafood allergies are a large, and seemingly increasing, component of seafood illness both in western and Asian countries; in the latter, seafood is a significant sensitizer in up to 40% of children and 33% of adults (Lopata and Lehrer, 2009).

Risk involves both the severity of the hazard and the likelihood of its occurrence. Accordingly, a hazard that produces relatively minor illness in consumers can represent a great risk if all consumers are exposed to it every day. Conversely, a life-threatening hazard, even if present only at very low frequency, can represent a high risk. Thus, another aspect of HI is the relative importance which major trading blocs attribute to specific hazards, whether by virtue of perceived severity of hazard to consumers, or the frequency of exposure, which itself is linked to the amount of the product that is traded

Table 17.1 Annual rates of seafood-related illness in the United States and Australia, 1990–2000

Hazard	USA		Australia	
	Total cases (1990–2000)	Annual rate per 100 000 population	Total cases (1990–2000)	Annual rate per 100 000 population
Ciguatera	328	0.013	616	0.36
Histamine	680	0.027	28	0.017
Viruses	1573	0.063	1737	1.022
Bacterial pathogens	1246	0.05	159	0.094
Biotoxins	125	0.005	102	0.06
Total	3952	0.158	2642	1.552

and consumed. An example is the international trade in warm-water shrimp, comprising 3.5 million tonnes in 2008, around 50% of which was produced by China, Thailand, Vietnam, Indonesia and India (FAO, 2012). In 1974, the United States' Food and Drug Administration (FDA) began a process of blocklisting, described as 'an effective action used against severe or chronic violations or violators'(Reilly *et al.*, 1992) . The practice was aimed at Asian shrimp imports, which were tested for the presence of filth, decomposition and *Salmonella*, detection of which resulted in automatic detention of the shipment and the placement of the supplier on an enhanced testing regime. No other trading bloc has considered it necessary to test imported shrimp for filth.

In 1997, the European Union similarly imposed bans on imports of prawns from Bangladesh because of a perceived public health risk from the presence of *V. cholerae*. This unilateral ban was not followed by other major trading blocs. In 2001 the European Union instituted a ban on seafood, mainly shrimp, containing the broad-spectrum antibiotic, chloramphenicol (CAP). A zero tolerance criterion for CAP was imposed which, with the capacity of modern analytical equipment, yielded positives in the range μg/kg (parts/billion) to ng/kg (parts/trillion). In the absence of information on a maximum allowable level (from the toxicological viewpoint), EU authorities adopted the precautionary principle. Other trading blocs, for example Japan and Australia, also imposed a testing regime for CAP, with a rejection level based on the limit of detection of the laboratory test (ca. 0.1 μg /kg). Other residues, such as nitrofurans, sulphites and cadmium were also added to the suite of hazards perceived by the European Union to be hazardous to consumers.

More importantly, these latter decisions represent risk management actions taken on the basis of the perceived magnitude of a hazard and without the support of a risk assessment to quantify the risk consequent to the presence of the hazards in the products. The latter approaches to risk management, in the absence of risk assessment, introduce two concepts which appear inimical to risk assessment: zero tolerance and the precautionary principle. We return to these approaches later in this chapter, and frame them within a risk-based context.

17.5 Exposure assessment

Exposure assessment (EA) is the qualitative and/or quantitative evaluation of the likely intake (both frequency and amount) of pathogens or their toxins via food. In other words EA aims to answer the question: how often, and how much of, the hazard is ingested in a specific food or group of foods of interest, and by whom?

Coastal communities are most at risk from hazards associated with seafood because of their relatively greater dietary intake of seafoods. In the 1950s, people living around Minimata Bay in Japan consumed shellfish in almost every meal. Unfortunately, factory wastewater led to finfish and shellfish harvested from the area having high methylmercury levels. In all, there were more than 700 cases of poisoning and 46 deaths.

Similarly, Pacific island communities are at increased risk from ciguatera poisoning. A survey of 19 Pacific island countries and territories by Skinner *et al.* (2011) estimated a mean annual incidence of 194 cases/100 000 population during 1998–2008 leading to an estimate that, over the past 35 years, 500 000 Pacific islanders may have suffered from ciguatera.

When risk assessors assemble data for the EA they aim to determine:

(i) how the contamination arose for example at what point in the farm-to-fork pathway;

(ii) how often the food is contaminated and how heavily contaminated it is;
(iii) whether the hazard level changes between the point of harvest or processing and consumption, and by how much;
(iv) how much of the food is eaten, by whom and how often.

Usually, risk assessors approach the first aim by constructing a framework which follows the key processing steps from harvest to consumption – often called a process model. The process model should detail steps in which the level of the hazard (and consequent risk) changes, how that occurs and the magnitude of the change. This framework, following the processes to which the food is subjected between harvest and consumption, and the factors that influence the risk along that supply chain is often termed a 'conceptual model'. To understand the effect of the second aim, it is necessary to gather data on the likely prevalence of contamination of the product and, ideally, the range of concentrations which may be expected. Data on the prevalence of pathogens or toxins in specific products can be found *via* several sources, for example from the literature, shellfish quality assurance programs and websites from countries which carry out import testing, such as the United States (www.accessdata.fda.gov/scripts/importrefusals/), Japan (www.mhlw.go.jp/english/topics/importedfoods/index.html) and the EU Rapid Alert system (http://ec.europa.eu/food/food/rapidalert/rasff_portal_database_en.htm). These databases can be interrogated, for example by sorting the proportion of samples which test positive for the hazard compared with the total number of samples tested, thereby providing an estimate of prevalence, although total test numbers are not always recorded. In addition, some databases also record concentration data for those samples in which the hazard is detected.

At some stages in the process the concentration of the target hazard may change, for example by freezing, drying, mixing, cooking, etc. (point (iii), earlier). To quantify these changes in concentration, data from many sources are combined. Importantly, the data need to be assessed by a risk assessment team which can cover the full spectrum of disciplines. As detailed in FAO/WHO (2005b), the JEMRA *Vibrio* RA team used its combined expertise based on knowledge of the literature and its experience with industrial processes to build a harvest–retail conceptual model for *V. cholerae* in warm-water shrimp in international trade (Figure 17.1), and then populate it with changes to the concentration (Table 17.2).

For consumption estimates (point (iv), earlier), data sources include national nutrition/diet surveys, production figures for various food industries together with import and export statistics, data from supermarkets or market survey companies that estimate the sales of various types of products and foods, etc. National dietary surveys may also include breakdown of consumption into different age and gender classes and geographic regions. Food categories used in those surveys, however, are often based on nutritional considerations, rather than categories that relate to the probable presence of food-borne hazards.

Obtaining relevant and accurate data that are useful for RA is not easy and one method is to request information on the consumption of the specific product by the target group of consumers. That task is made difficult when the target group is large, heterogeneous and covers numerous jurisdictions. The case history in Box 17.1 illustrates the complexity of the task faced by the European Food Safety Authority (EFSA) in providing scientific advice on potentially setting a new standard for the European Union for biotoxins in shellfish: should regulations protect the average consumer, or 80% of consumers, or include those who apparently eat 1500 g of mussel meat at one sitting (around 250 mussels)?

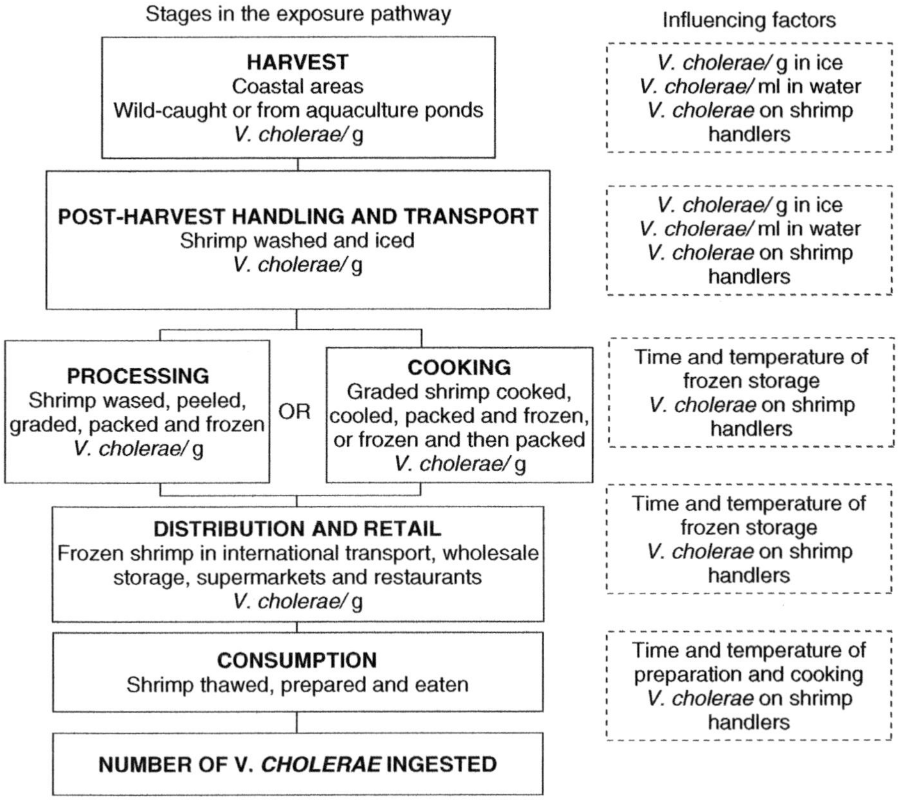

Figure 17.1 Product to consumption pathway for exposure assessment of *V. cholareae* in warm-water shrimp harvested and processed for international trade (*Source*: Food and Agriculture Organization of the United Nations, (2011), Risk assessment of *Vibrio parahaemolyticus* in seafood: Interpretative summary and technical report. Micobiological Risk Assessment Series No 16, 2011, Table 5 on page 28 http://www.fao.org/docrep014j2225e/j2225e00.pdf.)

Box 17.1 Consumption – why quality of the data matter

In 2009 the European Food Safety Authority (EFSA) provided a scientific opinion on marine biotoxins in shellfish, particularly the saxitoxin (STX) group (Anon., 2009). The EFSA panel requested member states to provide consumption data, which was obtained from only 5 of the 27 member states. The panel focused on "high-end" consumers and focused on intake at the 95th percentile (400 g shellfish meat) an intake which was used in the panel's calculation for a new (more stringent) regulatory limit for STX-group toxins in shellfish.

External reviewers noted that, by contrast, an FAO/IOC/WHO panel had cited three consumption levels when generating their 'Derived Guidance Level':

– 100 g (standard portion used in risk assessment, and one which approximated mean consumption)

- 250 g (97.5th percentile for most countries)
- 380 g (97.5th percentile for The Netherlands)

Reviewers also questioned whether 5% of German consumers really ate between 400 g and 1500 g of shellfish meat at one sitting, equating to (using EFSA's estimates of mussel meat mass) 62–80 mussels (400 g) and 231–300 mussels (1500 g).

At the behest of the European Commission, Health and Consumers Directorate-General, EFSA requested further information from member states, of which four responded; other information also became available. While the EFSA Comprehensive European Food Consumption Database revealed a highest 95th percentile for shellfish meat of 251 g, and a 'consumers only' survey in France established a 95th percentile of 300 g the EFSA panel concluded that their earlier estimate of 400 g was appropriate for protecting high risk consumers against acute effects of marine biotoxins (Anon., 2010).

The above illustrates how difficult it is to obtain satisfactory data which may, in turn, prompt a conservative stance when establishing a new regulatory standard.

17.6 Hazard Characterization

FAO/WHO (2003) provides a comprehensive discussion of Hazard Characterization, but here we concentrate on dose–response models used for infection processes. Hazard Characterization describes and, ideally quantifies, the probability of illness or infection associated with microbial hazards that may be present in food. This requires that a dose–response assessment be described if the data are obtainable. In practice, the relationship between pathogens/toxins and illness is better described as a dose-versus-probability-of-infection or probability-of-illness relationship.

For ethical reasons, data to define microbial dose–response relationships are not readily available, although some data sets from volunteer trials do exist. Since these usually involve healthy, young adults they do not reflect the susceptibility of the general population: the very young and the elderly are usually more susceptible to food-borne infections. As well, in many nations increasing numbers of consumers have their immune system weakened by underlying illness or medical treatment. Given this, dose–response relationships for food-borne pathogens tend to be derived from analysis of outbreak data. Animal models have also been used to gain information about dose–infection relationships for some hazards but they do not always reflect infection processes in humans accurately (because of differences in the antigens involved in binding the microbe to the host) or the consequences of toxin ingestion because of differences in physiology and biochemistry between species. There is debate concerning the nature of the infection process and how this can be modelled mathematically. The JEMRA expert consultation on Hazard Characterization (FAO/WHO, 2003) concluded, however, that models based on the single-hit (one viable organism) hypothesis of infection are the most biologically plausible.

Of this class of models, the exponential and Beta-Poisson models are the most widely used, the exponential model being simpler and relying on fewer assumptions. Both models predict that the probability of infection is a function of the dose of cells ingested, up to some limiting dose, above which the probability of illness does not increase further. The exponential model predicts that at some sufficiently high dose, the probability of infection always approaches 1. The Beta-Poisson model can describe a pathogen from which some proportion of the population will never become ill, i.e. the probability of illness may never

Table 17.2 Time and temperature distributions and the effect on levels of choleragenic *V. cholerae* in both wild-caught and aquacultured warm water shrimp (Food and Agriculture Organization of the United Nations, 2011. Reproduced with the permission of the FAO.)

Processing step	Temperature	Time range	Effect on population of *V. cholerae* 01	Source
Harvest				
Handling in period before icing				
Aquaculture shrimp	15–35 °C	0–1 hours	No effect	(a) (b)
Wild-caught shrimp	10–30 °C	0–3 hours	0–1 log increase	
Washing				
Washing and icing of aquaculture shrimp	0–7 °C	1–4 hours		
Washing in sea water of wild-caught shrimp	0–30 °C	1–4 hours	1 log reduction	(c)
Icing				
Icing during transport (including aboard fishing vessel for wild-caught shrimp)to processor	0–7 °C	2–16 hours (aquaculture) 2–48 hours (wild-caught)	2–3 log reduction	(d)
Water Use				
Water use during handling at processing plant	4–10 °C	1–3 hours	No effect	(a) (b)
Temperature				
Temperature during processing before freezing	4–10 °C	2–8 hours	No effect	(a) (b)
Cooking				
Cooking at processing plant	>90 °C	0.5–1.0 minute (This is the holding time at >90 °C)	>6 log reduction	(e) (f)
Freezing				
Freezing of cooked and raw products, storage, and shipment time	−12 °C to −20 °C	15–60 days	2–6 log reduction	(g) (h)

Sources: (a) Industry data for time and temperature from M/S Sterling Seafoods, Mangalore, India, personal communication, 2002. (b) Multiplication from Kolvin and Roberts, (1982). (c) Dinesh (1991). (d) Karunasagar, India, personal communication (2002). (e) Based on industry data on total plate count from M/S Sterling Foods, Mangalore, India, personal communication (2002). (f) In shrimp homogenate, D82.2 = 0.28 (Hinton and Grodner, 1985). (g) INFOFISH, personal communication, for shipment time; Reilly and Hackney (1985). (h) Survival in frozen shrimp from Nascumento *et al.* (1998).

approach 1 no matter what the dose. There are other subtle differences between these two models in the predicted probability of illness when doses approach and exceed the ID_{50} (i.e. the dose at which 50% of an exposed population is predicted to become infected/ill).

The two models have different underlying theories. Whereas the exponential model assumes that all cells have equal virulence, the Beta-Poisson model includes the possibility

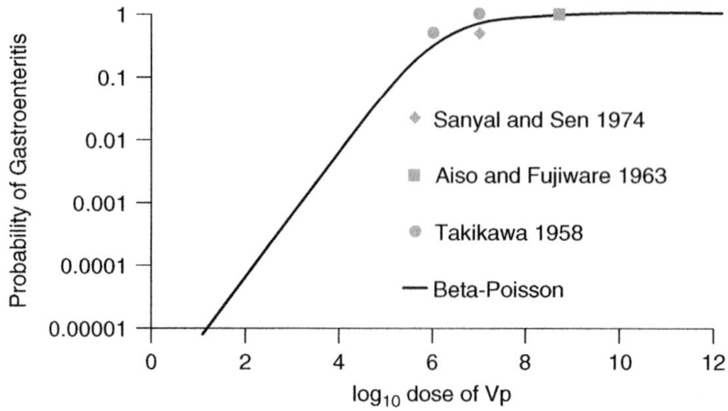

Figure 17.2 Beta-Poisson dose-response curve fitted to data for *Vibrio parahaemolyticus* infection in healthy human volunteers (*Source*: Food and Agriculture Organization of the United Nations, (2011), Risk assessment of Vibrio parahaemolyticus in seafood: Interpretative summary and technical report. Micobiological Risk Assessment Series No 16, 2011 Figure 1 on page 26 http://www.fao.org/docrep014j2225e/j2225e00.pdf.)

that virulence varies between individual cells. That variability is described by a Beta distribution fitted to the data. The Beta distribution can be very asymmetric but, when the distribution is symmetrical, the Beta-Poisson model, in effect, reduces to the exponential model.

The dose–response model chosen for risk assessments carried out on *V. parahaemolyticus* (FAO/WHO, 2011) illustrates the shape of a Beta-Poisson response curve (Figure 17.2), which in this case is quite similar to that of the exponential model (not shown). The data points which estimate the probability of illness given the exposure to the organism are from human volunteer studies and it appears that illness follows ingestion of around one million cells of *V. parahaemolyticus*. However, there are only four data points, all at high dose levels, and this leads to uncertainty around the dose–response estimate.

Some food regulatory authorities are considering the implementation of regulatory limits for viruses in shellfish. Viruses such as hepatitis A (HAV) and norovirus (NoV) in shellfish have been associated with food-borne viral infection throughout the world. For example, in 1991 in Shanghai, almost 300 000 people contracted hepatitis and nine died after consuming cockles contaminated with HAV (Tang *et al.*, 1991). In terms of infection it is known that:

- Virus particles can remain detectable for several months in seawater and food.
- The incubation period for HAV is long, varying from 2–6 weeks (mean duration is 4 weeks).
- Infectious doses are presumed to be low, for example 10–100 virus particles, with the ID_{50} for one strain of NoV estimated at 18 viral genomes (Teunis *et al.*, 2008).

More recently it has become clear that some of the NoV particles present in shellfish are non-infectious, with McLeod *et al.* (2009) demonstrating that ~30% of freshly propagated infectious HAV was rendered non-infectious following uptake by Pacific oysters, a factor which needs to be considered in assessing risk.

The choice of dose–response curve can have a great effect on the final risk estimate, as illustrated by Lindqvist and Westöö (2000) in a quantitative risk assessment of contracting listeriosis from consumption of smoked fish in Sweden. An exponential model predicted 168 cases/annum (range 47–2779 cases) and the Weibull-Gamma model predicted 95 000 cases (range 34 000–1.6 million). The researchers refined their estimates by assuming only 1–10% of *Listeria monocytogenes* strains are virulent, which decreased the annual illness estimates to 9 (exponential) and 5202 (WG). There was an average of 37 listerioses/year reported in Sweden from all food sources, suggesting that the WG model developed is overly conservative and unreliable.

17.7 Risk Characterization

Risk characterization (RC) synthesizes the exposure assessment with the dose–response relationship (Hazard Characterization) to generate an estimate of human health risk. The estimated risk can be expressed in many ways including risk per serving, risk to an individual, risk to an entire population, etc., and may be expressed in absolute terms or relative terms. In the latter case, the relative risk under alternative scenarios can be evaluated to determine, for example, which of a range of alternative risk management options provides the greatest risk reduction, obviating the need to estimate the absolute risk.

17.7.1 Methods for risk characterization

JEMRA have published guidelines for risk characterization and recognize that the most appropriate risk characterization approach depends on the problem being addressed (FAO/WHO, 2009). Thus, qualitative (descriptive) methods may be useful and appropriate in some cases, using terms or rankings like 'higher' or 'lower' to describe the relative risk. Such terms are subjective, however, and in most cases a risk assessment will need to quantify the magnitude of the risk or the magnitude of differences in risk. In essence, risk is a quantitative concept and mathematical approaches to risk assessment are usually needed. In quantitative risk assessments the conceptual model is expressed as a series of mathematical equations.

In a quantitative risk assessment the values of each of the factors that contributes to risk can be defined by a single, representative value. This is termed a deterministic approach. But these factors are rarely at constant levels and the question arises: what is the best representation of the value of each factor? The average value is an obvious choice but can lead the risk assessment to underestimate or ignore risks. For example, if average storage temperature and average storage time were used to estimate the potential growth of *Vibrio* on sushi it might be concluded that there would never be a risk to human health because the pathogen would never grow to unacceptable levels. However, if the product is occasionally kept too warm for too long there may be unacceptable risk and that possibility needs to be included in the risk assessment because it is the scenario of most relevance for identifying risk management needs. An alternative approach is to use the most extreme (worst-case) levels for each factor contributing to the risk, but this will lead to an overestimation because a situation in which each factor is at the most hazardous level is very unusual, and not representative.

In reality, the risk from a particular food and process will vary as it is dependent on many changeable factors. Thus, one approach to characterize the risk is to calculate the risk for a range of scenarios, with different values for each of the risk-affecting factors. From this, an estimate of the range of the risk, and extreme levels of risk, plus some representative measure (e.g. average or mode) can be determined. As more scenarios are evaluated, the more complete and informative the risk estimate will become. While the process of calculation is onerous it can be automated using stochastic simulation software based on Monte Carlo simulation methods. In this approach, each factor in the conceptual model is described by a distribution of possible values, i.e. the full range of values which that variable can take, including the probability that any given value in that range will be observed. Stochastic simulation software is then used to systematically evaluate the model tens or hundreds of thousands of times, each time taking a different combination of values from each of the distributions, and to calculate the expected outcome according to the conceptual model. Each calculation is termed an iteration and represents one possible meal scenario, for example the risk from one unit of food under a particular set of circumstances. As more iterations are performed a distribution of possible levels of risk is generated, representing more and more possible sets of circumstances for that food and its consumers, from which the estimated risk for any particular level of confidence can be calculated.

From the earlier discussion it can be seen that there are three levels at which risk may be estimated: qualitative, semi-quantitative and quantitative. We next describe how the FAO/WHO *Vibrio* RA team used all three approaches to estimate the risk of cholera from consumption of warm-water shrimp in international trade.

17.8 Qualitative Risk Assessment

A risk framework developed by Food Science Australia (2000) was used to generate a qualitative risk estimate. Data collected in the hazard identification, hazard characterization and exposure assessment of a hazard : product pairing informed six questions covering severity of hazard, exposure, processing, consumer preparation and epidemiological links. Hazard severity was categorized as: minor (medical treatment rarely required), mild (sometimes medical treatment is required), moderate (most cases require medical treatment) and severe (life threatening and/or with substantial long-term effects).

As may be seen from Table 17.3 the assessors estimated the risk of cholera, a severe illness, from consumption of shrimp either raw (as sushi) or cooked as "very low". The estimate included the progressive reductions in concentration of the pathogen during processing plus the low level of cholera associated with consumption of shrimp. Although the framework is simple, obtaining the information required to inform each element should not be underestimated. For example, to establish epidemiological linkage, the *Vibrio* RA team gathered public health data for a 6-year period from 11 major countries that import warm-water shrimp. In 2000 for example, there were 88 cases reported (from all food and water sources) of which 75 were imported (contracted while the victim was out of the country). In the same year there were more than 3 billion servings of warm-water shrimp in these eleven countries.

17.9 Semi-quantitative Risk Assessment

To generate a risk rating and an estimate of predicted illness, the *Vibrio* team then used a spreadsheet software tool called Risk Ranger. The tool requires the user to select from qualitative statements and/or to provide quantitative data concerning factors that will affect the food safety risk for a defined population from a hazard : product pairing. The spreadsheet converts the qualitative inputs into numerical values and combines them

Table 17.3 Microbiological hazard risk rating for choleragenic *V. cholerae* in shrimp using the approach of FSA (2000) (Food and Agriculture Organization of the United Nations, 2011. Reproduced with the permission of the FAO.)

Product	Identified hazard	Severity	Occurrence risk	Growth in product required to cause disease	Production/pocess/ handling	Consumer terminal step	Epidemiological link	Risk rating
Raw shrimp	*V. cholerae*	II	Very low	Yes	Inactivation during washing, icing, freezing	No	No	Very low
Shrimp cooked at the plant and eaten without further heat treatment	*V. cholerae*	II	Very low	Yes	Inactivation during washing, icing, cooking(optional), freezing	No	No	Very low
Shrimp cooked immediately before consumption	*V. cholerae*	II	Very low	Yes	Inactivation during washing, icing(optional), freezing, thawing and cooking	Yes	No	Very low

with the quantitative inputs in a series of mathematical and logical steps using standard spreadsheet functions. Risk ratings are estimated on a scale of 0–100 where zero represents no risk and 100 represents every member of the population eating a meal during a single day which contains a lethal dose of the hazard. The scale is logarithmic and an increment of six in the ranking corresponds approximately to a 10-fold increase in risk. Risk Ranger can be downloaded from http://www.foodsafetycentre.com.au/riskranger.php. Ross and Sumner (2002) (who designed the tool), caution that the outputs are only as reliable as the data entered and users should remain aware of the intended uses and limitations of the software; a fuller description of the tool's limitations is presented in FAO/WHO (2011).

The *Vibrio* team estimated risks of contracting cholera from consumption of warm-water shrimp for seven major importing countries (FAO/WHO, 2011). Risk ratings were low, ranging between 25 (Japan and Spain) and 21 (United States, Italy, United Kingdom, France and Germany), the difference reflecting higher consumption *per capita* in the former countries. Annual illnesses were very low with 1–2 cases/decade of cholera caused by consumption of warm-water shrimp predicted for Japan, the United States and Spain and approximately one case every 25 years for the other countries.

17.10 Quantitative Risk Assessment

In the final phase of characterizing risk of cholera from consumption of warm-water shrimp, the *Vibrio* RA team undertook a QRA. Because of limited data in the harvest-processing chain the team modelled the steps from port-of-entry in the importing country to the point of consumption within that country. The main data inputs for the exposure assessment component of this model were the volumes of shrimp imported into selected countries, results of port-of-entry analysis for choleragenic *V. cholerae* in imported warm-water shrimp and the populations of the selected importing countries. The available import testing data for warm-water shrimp included Japan (1995–2000), the United States (2000) and Denmark (1995). An assumption was made that 10% of imported warm-water shrimp would be consumed in the raw state and that, in the case of shrimp consumed after cooking, the hazard was eliminated either during cooking at the processing plant or during meal preparation. It is accepted that, in the former case, recontamination was possible, but the likelihood of this event occurring was considered very low. The Beta-Poisson dose–response model developed using data for the El Tor biotype of *V. cholerae* (Levine *et al.*, 1988) was used and mean exposure and mean risk were evaluated using Monte Carlo simulation.

Using this model, the median predicted risk of becoming ill with cholera following consumption of imported warm-water shrimp varied between countries; for Japan it was 1–5 cases every five years, while for Germany it was 2 cases/century.

17.11 Reality check

When a risk assessment is completed it is wise to undertake a reality check to provide confidence that the estimate is at least of the right order of magnitude. As indicated above, the risk of contracting cholera from consumption of warm-water shrimp was estimated as very low (qualitative), between 1–2 cases/year (Japan) and 1 case/century for other countries (semi-quantitative), and between 1–5 cases every 5 years (Japan) and 2 cases/century (Germany). These estimates accorded with the public health records from eleven major importing countries where, for example, in 2000 there were 88 cases of cholera (from all sources) of which 71 were acquired abroad. The *Vibrio* team concluded that: '... predictions of low risk by each of the approaches taken is supported by absence of epidemiological evidence that imported warm-water shrimp have ever been incriminated in a cholera outbreak in any developed nation in the world'.

On the other hand, as we saw earlier, a mean estimate of annual listerioses from consuming smoked fish in Sweden was 95 000 when total cases averaged 37/year from consumption of all foods (Lindquist and Westöö, 2000). Even allowing for under-reporting, the estimate did not pass the "reality check" and Lindquist and Westöö (2000) concluded that the Weibull-Gamma dose–response model was too conservative, and not useful.

17.12 Uncertainty and variability

Complete information and knowledge needed for a risk assessment is rarely available, so most risk assessments result in an imperfect estimate of risk and of the significance of the various contributing factors. Risk assessments are used to assist decision-making and, accordingly, the level of confidence in the accuracy and reliability of the risk estimate, or the range of the risk assessment, needs to be explained so that risk managers can balance the risk assessment information against other considerations when making risk management decisions.

Uncertainty and variability both affect the confidence in the risk estimate. Uncertainty describes information that is required, but not available, or that the risk assessors have little confidence in the available data (whether because the data are outdated, disputed, or that they have uncertain relevance to the risk being assessed). For example, dose–response information for *Shigella dysenteriae* has often been used as a proxy for the dose–response relationship of enterohaemorrhagic *E. coli* because data for the latter are very scarce. Similarly, temperature data for retail stores in one nation are sometimes used as a surrogate for risk assessments that relate to a different nation. Uncertainty can be reduced by gathering further data.

In any system, slight differences in conditions affect the outcome of the process, for example differences in temperatures and holding times that affect the potential for growth of pathogens. While there may be systematic differences between different countries or regions due to climate or the availability of reliable refrigeration, or consumer choices in the way they manage and maintain their home refrigerator, there will be variability in times and temperatures within each of those regions that cannot be easily predicted from other information. Variability describes variation that cannot readily be explained or further reduced by gathering more data, for example temperature variations over time within an individual refrigerator will depend on the capacity and insulation of the specific refrigerator, the quality of its thermostat and the frequency with which the door is opened. In some cases variability may be decreased by making the model more detailed and/or by discretely recognizing the source of those differences. Doing so, however, makes the conceptual model more complex and may lead to the need for other data, specific to the individual cases, that cannot be obtained, thereby increasing uncertainty. In any risk assessment there will be a compromise between the level of specific detail included that might reduce uncertainty, and the increase in complexity of the model that then demands yet more information. Mathematical methods for systematic treatment and evaluation of variability and uncertainty are considered in JEMRA publications and the published literature. In many MFSRAs, the variability is estimated by the stochastic simulation modelling process, but it is also important to document sources of uncertainty in the risk assessment, for example missing/unavailable data or knowledge, assumptions that were made, and the potential consequences of that uncertainty for the estimate of risk. That evaluation can be addressed by calculating the conceptual model for different assumptions or different values of uncertain factors and comparing the new risk estimates with those under the original set of assumptions. If the risk estimate is little affected by the new set of assumptions, it suggests that the missing, or uncertain, information has little influence on the risk and is not critical to the risk estimate and consequent risk management decision. Further discussion can be found in Fazil (2005).

17.13 Data gaps

As part of any risk assessment, information that is needed, but unavailable ('data gaps'), to improve or complete the risk assessment should be identified. While data gaps are specific for a particular risk assessment, one appears with great regularity: virulence of different strains of a particular pathogen. For example, virulence between strains of *V. vulnificus* was identified as a data gap in FAO/WHO (2005a), while for the *V. parahaemolyticus* risk assessment (FAO/WHO, 2011) the assessors overcame the gap only by assuming that all strains were of equal virulence. Similarly, because of a lack of information, risk assessments for *L. monocytogenes* generally assume all strains are equally virulent. By contrast, Lindqvist and Westöö (2000) provided estimates of listeriosis from consumption of smoked fish in Sweden by making an assumption that only 1–10% of strains are virulent.

If strain virulence does vary, as appears inevitable, questions follow of whether it varies according to season, region or product, cell density (due to quorum sensing phenomena) and whether all members of the at-risk population are equally susceptible.

Identifying data gaps should not be seen as a negative for a risk assessment but rather as an opportunity for re-assessment of estimates if/when more information becomes available.

17.14 Risk management approaches

In the Introduction we posed three questions linking risk assessment and risk management:

1. How widely are risk estimates used by risk managers at local, national and international level?
2. Do risk managers sometimes equate hazard with risk?
3. Are zero tolerance and the precautionary principle merely code for regulators to proclaim zero risk and then set performance standards which accommodate such thinking?

In this concluding section we provide comment on these questions with case studies on how risk managers in various countries used estimates generated in risk assessments to determine risk management strategies.

17.14.1 Case study 1: *Vibrio parahaemolyticus* in oysters consumed raw

In 1997–1998 in the United States there were four outbreaks and 683 cases of vibriosis (Table 17.4) caused by *V. parahaemolyticus* following consumption of raw oysters (FDA, 2005).

In 1999, a risk assessment was undertaken by the CFSAN, the FDA and the US Department of Health and Human Services. A sixteen-person risk assessment team backed by an even larger panel of external experts undertook the work. In 2001, the risk assessment report and model were made available for comment by stakeholders and were modified based on the comments received and on the availability of new data. The revised document and model were again subjected to extensive review.

The assessment estimated that consumption of raw oysters was likely to cause 2826 illnesses each year, mainly in the Gulf of Mexico (2596). It also explored 'what-if' scenarios for risk reduction with interventions focusing on reducing the temperature of oysters as soon after harvest as possible to reduce growth of *V. parahaemolyticus*.

How risk managers mitigate the likelihood of illness from vibrios (i.e., in this case *V. vulnificus* and *V. parahaemolyticus*) in oysters is illustrated by various approaches in different nations. In Canada, the Canadian Food Inspection Agency requires harvesters to establish critical limits in their HACCP plan to '... ensure shellfish are not exposed to sources of contamination or conditions allowing microbiological pathogens to grow to unacceptable levels during harvesting, holding, and transporting from the harvest area to the processing establishment' (CFIA, 2012). In the Canadian province of British Columbia this translated into a requirement to ice oysters within one hour of harvest. In California, by contrast, health authorities banned importation of oysters from the US State of Louisiana during the warmer seasons of Spring and Summer, citing recent deaths associated with consumption of oysters from Louisiana, which supplies 70% of California's oysters (Anon, 2003a). Banning of oyster harvesting in the summer months has also been the risk management option in Japan because of high counts of total *V. parahaemolyticus* in seawater (K. Osaka, *pers. comm.*). Thus, we note two diametrically opposing responses:

Table 17.4 Outbreaks of illness caused by *V. parahaemolyticus* in oysters in United States in 1997–1998 (FDA, 2005)

Location	Number of cases
Pacific Northwest[a]	209[b]
Pacific Northwest[a]	48
Texas	416[c]
Northeast Atlantic	10[b]

[a]The Pacific Northwest includes California, Oregon, Washington State, and British Columbia
[b]Number of cases that were culture-confirmed
[c]Includes 296 cases in Texas and 120 cases in other states traced back to oysters harvested from Texas

one to manage *via* a risk-based HACCP plan, the other to eliminate the risk by not allowing the product to be consumed.

17.14.2 Case study 2: *L. monocytogenes* in cooked crustaceans

In Australia in 2000, the Health Ministers of the Australia New Zealand Food Standards Council (ANZFSC) requested a review of microbiological limits for *L. monocytogenes* in fish and crustacea. As part of the review the Australia New Zealand Food Authority (ANZFA) included consideration of *L. monocytogenes* in cooked crustacea (ANZFA, 2002). However, no epidemiological links were found between this hazard : pathogen pairing and no risk estimate was generated. Nonetheless, a new regulation was promulgated, although the decision may have been influenced by the finding from the United States *L. monocytogenes* risk assessment (Anon, 2003b) that, on a per serving basis, cooked crustaceans (crabs and shrimp) were in the high-risk category.

Following representations from the Australian prawn industry, Food Standards Australia New Zealand (FSANZ) undertook a survey of *L. monocytogenes* in domestically produced ($n = 230$) and imported ($n = 139$) prawns (Marro *et al.*, 2003), in which the prevalence was 2% and 5%, respectively; no positive sample had a count greater than the limit of detection (50 cfu/g). The data, together with information on processing and storage were used for a through-chain QRA (Anon, 2002) with the prediction of <1 case of listeriosis per annum in Australia, reducing to <1 case per millennium if the prawns were consumed on the day of purchase, thereby mitigating the possibility of increase to an infectious dose.

The standards setting body (FSANZ) rescinded the regulation, a case where the risk estimate proved pivotal.

17.14.3 Case study 3: Zero tolerance and the precautionary principle

After two decades of trade, in 2001 the European Union instituted a ban on seafood, mainly shrimp, containing the broad-spectrum antibiotic, chloramphenicol (CAP); some countries destroyed batches of shrimp containing CAP. The ban followed testing using equipment with a limit of detection of around 0.1 μg/kg (0.1 parts per billion). It was stated that CAP was used as an illicit veterinary chemical, an assertion challenged by Hanekamp *et al.* (2003) on the grounds that CAP is a natural chemical, produced by *Streptomyces* spp., a ubiquitous soil bacterium. In Asia, CAP is also a commonly used antibiotic, widely available for over-the-counter purchase. The view of Hanekamp *et al.* (2003), notwithstanding, it became clear that CAP was a commonly used treatment for farmed shrimp, and the regulatory authorities in major exporting countries gradually brought the practice to a close.

Chloramphenicol is active against a range of Gram-positive and Gram-negative bacteria and is prescribed for topical application, especially in eye drops as treatment for bacterial conjunctivitis (Lancaster *et al.*, 1998). However, a side-effect of CAP application treatment is aplastic anaemia, an effect rare, generally fatal and usually occurring weeks or months after treatment has stopped; treatment is also associated with bone marrow suppression.

The perceived severity of the effects of CAP led EU authorities to impose a zero tolerance criterion for CAP and to invoke the precautionary principle.

What, then, is the global burden of aplastic anaemia from CAP therapy? According to Hanekamp *et al.* (2003), JECFA estimated its overall incidence as less than one case per 10 million per year. The same authors estimated exposure during CAP treatment as 25 – 125 μg/kg body weight/day. The Netherlands National Institute for Public Health and Environment (RIVM) undertook an exposure analysis on consumption of imported shrimp with a CAP level of 1–10 μg/kg (1–10 ppb), calculating a level of 0.00000017 mg/kg body weight/day (Janssen *et al.*, 2001). The exposure level difference from therapeutic use and shrimp intake is 150 000 000–735 000 000 leading Janssen *et al.* (2001) to a worst-case risk from shrimp consumption of a 1 : 1 000 000 added cancer risk in the human population.

Zero tolerance (with its implied zero risk) and the precautionary principle are anathema for risk assessment because they consider the hazard so serious that it may not be present at any concentration. The corollary is that, as laboratory detection becomes more sensitive, we can expect shrimp which today are acceptable to become 'unacceptable' for human consumption. The zero tolerance/precautionary principle approach prompts the question: 'To what extent is the public health of European consumers enhanced, and at what social and economic cost to developing nations, the primary producers of farmed shrimp?'

17.15 Final thoughts

Risk assessments are built on a modelled framework and, as the statistician, George Box, famously wrote 'All models are wrong but some are useful' (Box and Draper, 1987). But how useful? In this chapter we described how risk assessment teams go about the task ascribed to them by the statement of purpose. We also examined methods for risk assessment which range from relatively simple qualitative and semi-quantitative, to fully quantitative risk assessments. The latter are very human-, data- and financial-resource intensive and the time required for 'big' risk assessments is also onerous. For example, the US risk assessment on *V. parahaemolyticus* in raw oysters began in January 1999 and finished in July 2005, a large and necessary proportion of which was required for stakeholder comment, expert review and redrafting. By contrast, Lammerding (2006) cites the strategy of using a semi-quantitative approach to complete a risk profile of the Australian meat industry within a 12-month period (Pointon *et al.*, 2005).

Perhaps more importantly, irrespective of how long it takes to complete the assessment, to what extent do risk managers make use of the findings? Some may prefer to maintain expensive testing regimes rather than to respond to estimates of risk (which may have already incorporated the results of these testing regimes). A case in point is the assessment of *V. cholerae* in warm-water shrimp, where cholera cases in importing countries are clearly not linked with shrimp, and estimates of less than one case per decade, or per century, are predicted.

Clearly the *V. cholerae* risk assessment has not yet influenced testing of warm-water shrimp – a major commodity in world seafood trade. Why? Could it be that risk managers consider it 'safer' to maintain testing regimes and, if so, safer for whom – consumers, themselves? Justification for conservative approaches has traditionally been based on the fact that risk managers bear responsibility for public health in their jurisdiction while risk assessors do not. That is no longer necessarily the case. As this chapter was in final edit an Italian court convicted six scientists of manslaughter for underestimating the likely effect of an earthquake which, when it came, registered 6.3 on the Richter scale and killed

309 people in L'Aquila. The case prompts the question: will the six years in jail and the $10 million in costs and damages incurred by the six scientists prove a deterrent to the future of risk assessment in general, and of MRA in particular?

References

Ababouch, L. and Karunasagar, I. (*in press*). Seafood safety and quality: current practices and emerging issues, *FAO Fisheries Technical Paper* No. 574, Food and Agriculture Organization, Rome.

Anonymous. (2002). *Final assessment report Proposal P239 Listeria – risk assessment and risk management strategy*. Food Standards Australia New Zealand. Canberra, Australia.

Anonymous. (2003a). California ban Louisiana oyster industry. *Louisiana Sea Grant News Archive* (May 6, 2003).

Anonymous. (2003b). Listeria monocytogenes *Risk Assessment. Quantitative Assessment of Relative Risk to Public Health from Foodborne Listeria monocytogenes Among Selected Categories of Ready-to-Eat Foods*. FDA/Center for Food Safety and Applied Nutrition USDA/Food Safety and Inspection Service, Washington.

Anonymous. (2009). Marine biotoxins in shellfish – saxitoxin group: Scientific opinion of the panel on contamination in the food chain. *EFSA Journal*, 1019, 1–76.

Anonymous. (2010). Statement on further elaboration of the consumption figure of 400 g shellfish meat on the basis of new consumption data. *EFSA Journal*, 8, 1706.

ANZFA (Australian New Zealand Food Safety Authority). (2002). *Draft microbiological risk assessment. Listeria monocytogenes in cooked crustacea*. Australia New Zealand Food Authority, Canberra, Australia. April, 2002.

Bauman, H. (1994). The origin of the HACCP system and subsequent evolution. *Food Science and Technology Today*, 8, 66–72.

Bell, B., Goldoft, M., Griffin, P., Davis, M., Gordon, D.C., Tarr, P.I., Bartleson, C.A., Lewis, J.H., Barrett, T.J., Wells, J.G., Baron, R. and Kobayashi, J. (1994). A Multistate Outbreak of *Escherichia coli* O157:H7 – Associated Bloody Diarrhea and Hemolytic Uremic Syndrome From Hamburgers: The Washington Experience. *Journal of the American Medical Association*, 272, 1349–1353.

Box, G. and Draper, N. (1987). *Empirical Model-Building and Response Surfaces*. John Wiley & Sons, Ltd, New York.

Buchanan, R. (1995). The role of microbiological criteria and risk assessment in HACCP. *Food Microbiology*, 12, 421–424.

Buchanan, R., Dennis, S. and Miliotis, M. (2004). Initiating and managing risk assessments within a risk analysis framework: FDA/CFSAN's practical approach. *Journal of Food Protection*, 67, 2058–2062.

CAC (Codex Alimentarius Commission) (1999). *Principles and guidelines for the conduct of microbiological risk assessment*. CAC/GL-30 (1999).

Cahill, S. and Jouve, S. (2004). Microbiological risk assessment in developing countries. *Journal of Food Protection*, 67, 2016–2023.

Cameron, S., Walker, C., Beers, M., Rose, N. and Annear, E. (1995). Enterohaemorrhagic *Escherichia coli* outbreak in South Australia associated with the consumption of mettwurst. *Communicable Diseases Intelligence*, 19, 70–71.

Cassin, M., Lammerding, A., Todd, E., Ross, W. and McColl, R. (1998). Quantitative risk assessment for *Escherichia coli* O157:H7 in ground beef hamburgers. *International Journal of Food Microbiology*, 41, 21–44.

CFIA (2012) Canadian shellfish sanitation program – Manual of operations. Canadian Food Inspection Agency. http://www.inspection.gc.ca/food/fish-and-seafood/manuals/canadian-shellfish-sanitation-program/eng/1351609988326/1351610579883 [accessed 7 June 2013].

Dinesh, P. (1991). Effect of iodophor on pathogenic bacteria associated with seafood', MFSc thesis. University of Agricultural Sciences, Bangalore.

FAO (1999). Expert consultation on the trade impact of *Listeria* in fish products. Amherst, 17–20 May 1999. *FAO Fisheries Report No. 604*. Food and Agriculture Organization, Rome, 34 pp.

FAO (2012). Fisheries and Aquaculture Department: Statistics – Introduction. http://www.fao.org/fishery/statistics/en [accessed 18 June 2013].

FAO/WHO (2003). Hazard characterisation for pathogens in food and water. *Microbiological Risk Assessment Series* No. 3, Food and Agriculture Organization, Rome/World Health Organization, 53 pp.

FAO/WHO (2005a). Risk assessment of *Vibrio vulnificus* in raw oysters: Interpretative summary and technical report. *Microbiological Risk Assessment Series* No. 8, Food and Agriculture Organization/World Health Organization, 115 pp.

FAO/WHO (2005b). Risk assessment of *Vibrio cholerae* O1 and O139 in warm-water shrimp in international trade: Interpretative summary and technical report. *Microbiological Risk Assessment Series* No. 9, Food and Agriculture Organization/World Health Organization, 91 pp.

FAO/WHO (2008). Exposure assessment of microbiological hazards in food. *Microbiological Risk Assessment Series* No. 7, Food and Agriculture Organization/World Health Organization, 89 pp.

FAO/WHO (2009). Risk Characterization of Microbiological Hazards in Foods, Guidelines. *Microbiological Risk Assessment Series* No. 17, Food and Agriculture Organization/World Health Organization, 119 pp.

FAO/WHO (2011). Risk assessment of *Vibrio parahaemolyticus* in seafood: Interpretative summary and technical report. *Microbiological Risk Assessment Series* No. 16, Food and Agriculture Organization/World Health Organization, 193 pp.

Fazil, A. (2005). A primer on risk assessment modelling: focus on seafood products. *FAO Fisheries Technical Paper* No. 462, Food and Agriculture Organization, Rome, 56 pp.

FDA (2003). *Quantitative assessment of relative risk to public health from foodborne* Listeria monocytogenes *among selected categories of ready-to-eat foods*. United States Food and Drug Administration, Center for Food Safety and Applied Nutrition, USDA/Food Safety and Inspection Service. 272 pp.

FDA (2005). *Quantitative risk assessment on the public health impact of pathogenic Vibrio parahaemolyticus in raw oysters*. United States Food and Drug Administration, US Department of Health and Human Services, 141 pp.

Foegeding, P. and Roberts, T. (1996). Assessment of the risks associated with foodborne pathogens: an overview of a Council for Agricultural Science and Technology report. *Journal of Food Protection*, 1996 supplement, 19–23.

FSA (2000). *Final Report – Scoping study on the risk of plant products (SafeFood NSW)*, Food Science Australia, North Ryde, NSW, Australia.

Hanekamp, J., Frapporti, G. and Olieman, K. (2003). Chloramphenicol, food safety and precautionary thinking in Europe. *Environmental Liability*, 11, 209–221.

Hathaway, S. and Mckenzie, A. (1991). Meat Inspection in New Zealand – prospects for change. *New Zealand Veterinary Journal*, 39, 1–7.

Hathaway, S. and Cook, R. (1997). A regulatory perspective on the potential uses of microbial risk assessment in international trade. *International Journal of Food Microbiology*, 36, 127–133.

Hinton, A. and Grodner, R. (1985). Determination of the thermal death time of *V. cholerae* in shrimp (*Penaeus setiferus*). Proceedings 9th Annual Tropical and Subtropical Fisheries Conference of the Americas, 109–119. Texas A & M University.

Huss, H., Jorgensen, L. and Vokel, B. (2000). Control options for *Listeria moncytogenes* in seafoods. *International Journal of Food Microbiology*, 62, 267–274.

ICMSF (International Commission on Microbiological Specifications for Foods) (1998). Potential application of risk assessment techniques to microbiological issues related to international trade in food and food products. *Journal of Food Protection*, 61, 1075–1086.

Janssen, P., Baars, A. and Pieters, M. (2001). Advies met betrekking tot chloramphenicol in garnalen (Advice regarding chloramphenicol in shrimp). RIVM/CSR, Bilthoven, The Netherlands.

Karunasagar, I. and Karunasagar, I. (2000). *Listeria* in tropical fish and fishery products. *International Journal of Food Microbiology*, 62, 177–181.

Kolvin, J. and Roberts, D. (1982). Studies on the growth of *Vibrio cholerae* biotype El Tor and biotype Classical in foods. *Journal of Hygiene Cambridge*, 89, 243–252.

Lammerding, A. (2006). Modeling and risk assessment for *Salmonella* in meat and poultry. *Journal of AOAC International*, 89, 543–552.

Lammerding, A. and Paoli, G. (1997). Quantitative risk assessment: an emerging tool for emerging foodborne pathogens. *Emerging Infectious Diseases*, 3, 483–487.

Lancaster, T., Stewart, A. and Jick, H. (1998). Risk of serious haematological toxicity with use of chloramphenicol eye drops in a British general practice database. *British Medical Journal*, 316, 667.

Levine, M., Kaper, J., Herrington, E. *et al.* (1998). Volunteer studies of deletion mutants of *Vibrio cholerae* O1 prepared by recombinant techniques. *Infection and Immunity*, 56, 161–167.

Lindqvist, R. and Westöö, A. (2000). Quantitative risk assessment for *Listeria monocytogenes* in smoked or gravid salmon and rainbow trout in Sweden. *International Journal of Food Microbiology*, 58, 181–196.

Lopata, A. and Lehrer, S. (2009). New insights into seafood allergy. *Current Opinion in Allergy and Clinical Immunology*, 9, 270–277.

McLeod, C., Hay, B., Grant, C., Greening, G. and Day, D. (2009). Inactivation and elimination of human enteric viruses by Pacific oysters. *Journal of Applied Microbiology*, 107, 1809–1818.

McNab, W. (1998). A general framework illustrating an approach to quantitative microbial food safety risk assessment. *Journal of Food Protection*, 161, 1216–28.

Marro, M., Hasell, S., Boorman, J. and Crerar, S. (2003). Survey of *Listeria monocytogenes* in cooked prawns. Annual Conference, Australian Institute of Food Science and Technology, March, 2003.

NACMCF (National Advisory Committee on Microbiological Criteria for Foods) (1998). Principles of risk assessment for illness caused by foodborne biological agents. *Journal of Food Protection*, 61, 1071–1074.

Nascumento, D., Viera, R., Almeida, H., Patel, T. and Laria, S. (1998). Survival of *Vibrio cholerae* O1 strains in shrimp subjected to freezing and boiling, *Journal of Food Protection*, 61, 1317–1320.

Notermans, S., Gallhoff, G., Zwietering, M. and Mead, G. (1995). The HACCP concept: specification of criteria using quantitative risk assessment. *Food Microbiology*, 12, 81–90.

Notermans, S. and Teunis, P. (1996). Quantitative risk analysis and the production of microbiologically safe food: An introduction. *International Journal of Food Microbiology*, 30, 3–7.

Notermans, S. and Hoornstra, E. (2000). Risk assessment of *Listeria monocytogenes* in fish products: some general principles, mechanism of infection and the use of performance standards to control human exposure. *International Journal of Food Microbiology*, 62, 223–229.

OzFoodNet (2012). *Foodborne disease in Australia Annual reports of the OzFoodNet network.* http://www.health.gov.au/internet/main/publishing.nsf/Content/cda-pubs-annlrpt-ozfnetar.htm [accessed on 18 June 2013].

Pointon, A., Jenson, I., Jordan, D. *et al.* (2005). A risk profile of the Australian red meat industry: approach and management. *Food Control*, 17, 712–718.

Reilly, A. and Hackney, C.R. (1985). Survival of *Vibrio cholerae* during storage in artificially contaminated seafoods. *Journal of Food Science*, 50, 838–839.

Reilly, A., Twiddy, D. and Fuchs, R. (1992) Review of the occurrence of Salmonella in cultured tropical shrimp. *FAO Fisheries Circular* No. 851, Food and Agriculture Organization, Rome.

Ross, T., Dalgaard, P. and Tienungoon, S. (2000). Predictive modelling of the growth and survival of *Listeria* in fishery products. *International Journal of Food Microbiology*, 62(SI), 231–245.

Ross, T. and Sumner, J. (2002). A simple, spreadsheet-based, food safety risk assessment tool. *International Journal of Food Microbiology*, 77, 39–53.

Skinner, M., Brewer, T., Johnstone R., Fleming, L. and Lewis, R. (2011). Ciguatera fish poisoning in the Pacific islands (1998 to 2008). *PLoS Neglected Tropical Diseases*, 5, e1416. doi:10.1371/journal.pntd.0001416.

Smith de Waal, C., Alderton, L. and Jacobsen, M. (2000). *Closing the Gaps in our Federal Food-Safety Net*, Center for Science in the Public Interest, Washington, DC.

Sumner, J. (2001). *Australian Seafood Safety Risk Assessment*, Seafood Services Australia, Brisbane, Australia.

Sumner, J. and Ross, T. (2002). A semi-quantitative seafood safety risk assessment. *International Journal of Food Microbiology*, 77, 55–59.

Sumner, J, Ross, T. and Ababouch, L. (2004). Application of risk assessment in the fish industry. *FAO Fisheries Technical Paper* No. 442, Food and Agriculture Organization, Rome.

Takikawa, I. (1958). Studies on pathogenic halophilic bacteria. *Yokohama Medical Bulletin*, 9, 313–322.

Tang, Y., Wang, J., Xu, Z.. *et al.* (1991). A serologically confirmed, case-control study, of a large outbreak of hepatitis A in China, associated with consumption of clams. *Epidemiology and Infection*, 107, 651–657.

Teunis, P., Moe, L., Liu, P. *et al.* (2008). Norwalk virus: How infectious is it?. *Journal of Medical Virology*, 80, 1468–1476.

Todd, E. and Harwig, J. (1996). Microbial risk analysis of food in Canada. *Journal of Food Protection*, 1996 supplement, pp. 10–18.

WTO (2007a). *The WTO agreement on technical barriers to trade*. World Trade Organization. http://www.wto.org/english/docs_e/legal_e/17-tbt_e.htm [accessed 18 June 2013].

WTO (2007b). *The WTO Agreement on the application of sanitary and phytosanitary measures*. World Trade Organization. http://www.wto.org/english/tratop_e/sps_e/spsagr_e.htm [accessed 18 June 2013].

Index

Seafood Processing: Technology, Quality and Safety, First Edition. Edited by Ioannis S. Boziaris.

Printed and bound by CPI Group (UK) Ltd, Croydon, CR0 4YY
16/09/2026
14963476-0001